Introductory Statistics

Roger E. Kirk
Baylor University

Brooks/Cole Publishing Company
Monterey, California
A Division of Wadsworth Publishing Company, Inc.

© 1978 by Wadsworth Publishing Company, Inc., Belmont, California 94002. All rights reserved. No part of this book may be reproduced, stored in a retrieval system, or transcribed, in any form or by any means—electronic, mechanical, photocopying, recording, or otherwise—without the prior written permission of the publisher: Brooks/Cole Publishing Company, Monterey, California 93940, a division of Wadsworth Publishing Company, Inc.

Printed in the United States of America

10 9 8 7 6 5 4 3

Library of Congress Cataloging in Publication Data

Kirk, Roger E.
    Introductory statistics.

    Bibliography: p. 428
    Includes index.
    1. Statistics.   I. Title.
HA29.K548   519.5   77-6683
ISBN 0-8185-0226-6

Manuscript Editor: *Phyllis Niklas*
Production Editor: *Joan Marsh*
Interior Design: *John Edeen*
Cover Design: *Jamie S. Brooks*

Introductory Statistics

Roger E. Kirk, Consulting Editor

Exploring Statistics: An Introduction for Psychology and Education
Sarah M. Dinham, The University of Arizona

Methods in the Study of Human Behavior
Vernon Ellingstad, The University of South Dakota
Norman W. Heimstra, The University of South Dakota

An Introduction to Statistical Methods in the Behavioral Sciences
Freeman F. Elzey, San Francisco State University

Experimental Design: Procedures for the Behavioral Sciences
Roger E. Kirk, Baylor University

Introductory Statistics
Roger E. Kirk, Baylor University

Statistical Issues: A Reader for the Behavioral Sciences
Roger E. Kirk, Baylor University

The Practical Statistician: Simplified Handbook of Statistics
Marigold Linton, The University of Utah
Philip S. Gallo, Jr., San Diego State University

Nonparametric and Distribution-Free Methods for the Social Sciences
Leonard A. Marascuilo, University of California, Berkeley
Maryellen McSweeney, Michigan State University

Basic Statistics: Tales of Distributions
Chris Spatz, Hendrix College
James O. Johnston, University of Arkansas at Monticello

Multivariate Analysis with Applications in Education and Psychology
Neil H. Timm, University of Pittsburgh

Preface

This book was written for an introductory course in statistics. Its aim is to provide a sound noncalculus presentation for students in the behavioral sciences and education. A number of good once-over-lightly books have been written for the same audience. Their aims are to introduce the student to the most frequently used statistical procedures and develop an appreciation of the role of statistics in the scientific enterprise. Such books typically emphasize rule of thumb procedures, gloss over underlying assumptions, omit algebraic proofs, and provide only a superficial introduction to probability. For many students, developing an appreciation of statistics along with a modest grasp of the mechanics is a legitimate goal, but this book was written with different students in mind—those who plan to take additional statistics courses and perhaps eventually to enter graduate school and become productive researchers. Such students usually have not gone beyond college algebra, but they are willing and sometimes eager to follow intuitive arguments and elementary algebraic demonstrations in grappling with complex concepts. This kind of motivation along with a working knowledge of high school algebra is sufficient to follow the presentation in this book. For those whose mathematical skills are weak, Appendix A briefly reviews elementary mathematics, including rules for solving algebraic equations and inequalities. Appendix A also contains a diagnostic mathematics skills test.

Learning statistics is rather like learning a foreign language—both require mastering a new vocabulary. In this book the definitions of new terms are set apart. Other learning aids include an extensive glossary of statistical symbols (Appendix B), chapter summaries, and 503 review exercises following chapter sections. The exercises indicate which concepts and computational procedures are most important, present interesting real-life examples of the way statistics are used, and provide practice in applying what has been learned. Answers are given for 206 of the exercises to provide the student with feedback.

A list of recommended supplemental readings from my book *Statistical Issues* (Brooks/Cole, 1972) is given in Appendix E. These articles and the editorial commentaries that introduce them will help broaden students' understanding of important concepts and issues in statistics.

I see the underlying logic of statistics as both beautiful and simple, and students can understand this logic from a verbal presentation. With this in mind I have striven throughout for clarity and teachability. But I have also tried to present topics in depth and to show interrelationships among concepts. This is reflected, for example, in the extensive coverage given to probability (including Bayes' theorem), the logic of hypothesis testing, factors affecting the Pearson product-moment correlation coefficient, and procedures for rationally determining sample size. Also, the underpinnings of statistics—its supporting structure of derivations and assumptions—are examined when this can be done using high school mathematics. Algebraic derivations that might disrupt the intuitive development of topics are placed in technical notes at the ends of many chapters. The presentation is complete without these notes, but the interested student may perhaps find them and the optional starred sections useful.

It is a pleasure to acknowledge the collaboration of Juliet Dickason Swieczkowski during the writing of this book. Her contributions to its style and technical exposition appear on every page. Also, it is a pleasure to express my appreciation to James N. Bowen of the University of Texas at Arlington, James K. Brewer of Florida State University, and Henry M. Halff of the University of Illinois for reading the manuscript and for their thoughtful comments. Four of my former graduate assistants, Margaret Covington, Paul Huie, Vickie Maynard, and Cathy Raevsky, deserve special mention for working the problems in the review exercises. The fine production staff at Brooks/Cole was always there when I needed help. Phyllis Niklas, who edited the manuscript, and John Edeen, the designer, deserve special recognition.

I am grateful to the Literary Executor of the late Sir Ronald A. Fisher, F.R.S., to Frank Yates, F.R.S., and to Longman Group Ltd., London, for permission to reprint Tables D.1, D.2, D.3, D.4, D.6, and D.8 from their book *Statistical Tables for Biological, Agricultural and Medical Research* (6th edition, 1974).

I am also grateful to E. S. Pearson and H. O. Hartley, editors of *Biometrika Tables for Statisticians*, Vol. 1, and to the *Biometrika* trustees for permission to reprint Table D.5; and to the editor of *Annals of Mathematical Statistics* for permission to reprint Table D.7.

Portions of this book were written while on sabbatical from Baylor University and while teaching as a visiting professor of psychology and statistics at Seinan Gakuin University, Fukuoka, Japan. I am grateful to the administrations of both universities and particularly to Abner V. McCall and Herbert H. Reynolds of Baylor and to Eiichi Funakoshi and Sanshiro Shirakashi of Seinan for providing environments that were so conducive to writing.

And finally, I want to express my appreciation to my statistics classes for what I trust has been a mutually rewarding learning experience.

Roger E. Kirk

Contents

1 Introduction to Statistics 1

1.1 Introduction 2
1.2 Studying Statistics 3
1.3 Basic Concepts 5
1.4 Describing Characteristics by Numerals 8
1.5 Historical Development of Statistics 18
1.6 Summary 20

2 Frequency Distributions and Graphs 22

2.1 Introduction 23
2.2 Frequency Distributions 23
2.3 Introduction to Graphs 33
2.4 Graphs for Qualitative Variables 33
2.5 Graphs for Quantitative Variables 36
2.6 Shapes of Distributions 40
2.7 Misleading Graphs 42
2.8 Summary 44

3 Measures of Central Tendency 45

3.1 Introduction 46
3.2 The Mode 46
3.3 The Mean 47
3.4 The Median 51
3.5 Relative Merits of the Mean, Median, and Mode 54
3.6 Location of $\bar{X}$, Mdn, and Mo in a distribution 59
3.7 Mean of Combined Subgroups 60

3.8 Summary 61
3.9 Technical Notes 61

4 Measures of Dispersion, Skewness, and Kurtosis 65

4.1 Introduction to Measures of Dispersion 66
4.2 Four Measures of Dispersion 67
4.3 Relative Merits of the Measures of Dispersion 78
4.4 Dispersion and the Normal Distribution 82
4.5 Tchebycheff's Theorem 83
4.6 Skewness and Kurtosis 84
4.7 Summary 87
4.8 Technical Notes 88

5 Correlation 94

5.1 Introduction 95
5.2 A Numerical Index of Correlation 98
5.3 Pearson Product-Moment Correlation Coefficient 100
5.4 Interpretation of r: Explained and Unexplained Variation 105
5.5 Some Common Errors in Interpreting r 107
5.6 Factors that Affect the Size of r 109
5.7 Spearman Rank Correlation 116
5.8 Other Kinds of Correlation Coefficients 120
5.9 Summary 120
5.10 Technical Notes 121

6 Regression 127

6.1 Introduction 128
6.2 Criterion for the Line of Best Fit 129
6.3 Another Measure of Ability to Predict: The Standard Error of Estimate 136
6.4 Alternative Formula for $S_{Y \cdot X}$ 138
6.5 Assumptions Associated with Estimate of Prediction Error 140
6.6 Summary 141
6.7 Technical Notes 142

7 Probability 146

7.1 Introduction to Probability 147
7.2 Basic Concepts 149
7.3 Probability of Combined Events 152
7.4 Summary 160

8 More about Probability 162

8.1 Counting Simple Events 163
8.2 Revising Probabilities Using Bayes' Theorem 167
8.3 Summary 170

9 Random Variables and Probability Distributions 172

9.1 Random Sampling 173
9.2 Random Variables and Probability Distributions 175
9.3 Binomial Distribution 182
9.4 Summary 187

10 Normal Distribution and Sampling Distributions 189

10.1 Normal Distribution 190
10.2 Interpreting Scores in Terms of z Scores and Percentile Ranks 197
10.3 Sampling Distributions 201
10.4 Summary 208
10.5 Technical Note 209

11 Statistical Inference: One Sample 212

11.1 Introduction to Hypothesis Testing 213
11.2 Hypothesis Testing 218
11.3 One-Sample z Test for μ When σ^2 Is Known 222
11.4 More about Hypothesis Testing 224
11.5 Summary 234
11.6 Technical Note 235

12 Statistical Inference: Other One-Sample Test Statistics 237

12.1 Introduction to Other One-Sample Test Statistics 238
12.2 One-Sample t Test for μ When σ^2 Is Unknown 238
12.3 One-Sample Chi-Square Test for a Population Variance 244
12.4 One-Sample z Test for a Population Proportion 247
12.5 One-Sample t and z Tests for a Population Correlation 250
12.6 Summary 253

13 Statistical Inference: Two Samples 255

13.1 Introduction to Hypothesis Testing for Two Samples 256
13.2 Two-Sample z Test for $\mu_1 - \mu_2$ 256
13.3 Two Randomization Strategies: Random Sampling and Random Assignment 257
13.4 Two-Sample t Test for $\mu_1 - \mu_2$ 261
13.5 The z and t Tests for $\mu_1 - \mu_2$ Using Dependent Samples 265
13.6 Summary 271

14 Statistical Inference: Other Two-Sample Test Statistics 274

14.1 Two-Sample Tests for Equality of Variances 275
14.2 Two-Sample Tests for $p_1 - p_2$ 278
14.3 Summary 283

15 Interval Estimation 284

15.1 Introduction to Interval Estimation 285
15.2 Confidence Intervals for μ and $\mu_1 - \mu_2$ When σ^2 Is Unknown 285
15.3 Confidence Intervals for σ^2 and σ_1^2/σ_2^2 292
15.4 Confidence Intervals for p and $p_1 - p_2$ 295
15.5 Confidence Intervals for ρ and $\rho_1 - \rho_2$ 296
15.6 Summary 300

16 Introduction to the Analysis of Variance 301

16.1 Purposes of Analysis of Variance 302
16.2 Basic Concepts of ANOVA 303
16.3 Completely Randomized Design 313
16.4 Assumptions Associated with a Type CR-k Design 318
16.5 Introduction to Multiple Comparisons 322
16.6 Introduction to Factorial Designs 325
16.7 Summary 328

17 Statistical Inference for Frequency Data 330

17.1 Applications of Pearson's Chi-Square Statistic 331
17.2 Testing Goodness of Fit 332
17.3 Testing Independence 336
17.4 Testing Equality of $c \geq 2$ Proportions 342
17.5 Summary 346
17.6 Technical Notes 348

18 Statistical Inference for Rank Data 352

18.1 Introduction to Assumption-Freer Tests 353
18.2 Mann–Whitney U Test for Two Independent Samples 354
18.3 Wilcoxon T Test for Dependent Samples 360
18.4 Comparison of Parametric Tests and Assumption-Freer Tests for Rank Data 365
18.5 Summary 366

Appendix A/Review of Basic Mathematics 368
Appendix B/Glossary of Symbols 379
Appendix C/Answers to Starred Exercises 385
Apendix D/Tables 403
Appendix E/Recommended Supplemental Readings 426
References 428
Index 432

1
Introduction to Statistics

1.1 Introduction
   Some Misconceptions
   What Is Statistics?
   Why Study Statistics?
   Kinds of Statisticians
1.2 Studying Statistics
   Read More Slowly
   Don't Worry If You Weren't an Ace in Math
   Resolve to Review Often
   Master Foundation Concepts before Going on to New Material
   Strive for Understanding
1.3 Basic Concepts
   Population and Sample Defined
   Descriptive and Inferential Statistics
   Random Sampling
   Review Exercises for Section 1.3
1.4 Describing Characteristics by Numerals
   Variables and Constants
   Perspectives on Numbers
   Classification of Variables in Mathematics
   Measuring Operations in the Behavioral Sciences and Education
   Nominal Measurement
   Ordinal Measurement
   Interval Measurement
   Ratio Measurement
   Implications of the Two Ways of Thinking about Numerals
   Review Exercises for Section 1.4
1.5 Historical Development of Statistics
   National Statistics
   Probability Theory
   Experimental Statistics
   Review Exercises for Section 1.5
1.6 Summary

1.1 Introduction

Some Misconceptions

It is widely believed that statistics can be used to prove anything—which implies, of course, that it can prove nothing. Furthermore, the word "statistics" conjures up visions of numbers piled upon numbers, uninterpretable charts, and computers cranking out gloomy predictions. To the ordinary person, besieged from all sides by advertising claims, statistics is hocus-pocus with numbers. It was Benjamin Disraeli who said "there are three kinds of lies—lies, damned lies, and statistics." In primitive cultures, exaggeration was common. One writer, with tongue in cheek, reasoned that, since primitive people did not have a science of statistics, they were forced to rely on exaggeration, which is a less effective form of deception. Another writer remarked that "if all the statisticians in the world were laid end to end—it would be a good thing." Regardless of its public image, statistics endures as a required course, and my students continue to refer to it, affectionately no doubt, as Sadistics 2402.

What Is Statistics?

In spite of frequent misuse, statistics can be an elegant and powerful tool for making decisions in the face of uncertainty. The word "statistics" comes from the Latin *status*, which is also the root for our modern term, "state," or political unit. Statistics was a necessary tool of the state, since to levy a tax or wage war a ruler had to know the number of subjects in the state and the amount of their wealth. Gradually the meaning of the term expanded to include any type of data. *Today the word* **statistics** *has four distinct meanings. Depending on the context, it can mean: (1) data; (2) functions of data, such as the mean and range; (3) techniques for the collection, analysis, and interpretation of data for subsequent decision-making; and (4) the science of creating and applying such techniques.*

Why Study Statistics?

A knowledge of statistics yields more than the obvious benefits. For example, it generates new ways of thinking about questions and provides effective tools for answering them. It takes only a cursory examination of the professional literature in your own field to see the inroads made by statistical techniques and ways of thinking. Statistics is undoubtedly an indispensable tool for experimenters, but its usefulness is not limited to research. In many fields, it is virtually impossible to keep up with new developments without an understanding of elementary statistics. Also, statistics is an interesting subject—some people even find it fascinating.

In all likelihood you are reading this book because it was assigned in your required statistics course. You have been told that the study of statistics is necessary, and there is a strong implication that it will be good for you. At this point you may be skeptical. Just what can you expect to learn by studying statistics? A quick scanning of this book will give you an idea. You will acquire a new vocabulary, since in many ways learning statistics is like learning a foreign language, and you will learn to manipulate numbers according to symbolic instructions. But more important, you will learn when and how to apply statistics to research problems in the behavioral sciences and in education. Your study of statistics should enable you to read the literature in your professional field with greater understanding, and it will prepare you to learn more complex procedures in the design and analysis of experiments. You will probably become more critical of statistical presentations in your field and in the mass media. And you should gain a greater appreciation of the probabilistic nature of scientific knowledge. Statistics involves a special way of thinking that can be used not only in research but in daily life; hopefully, you will add this mode of thinking to your conceptual ability.

Kinds of Statisticians

Users of statistics fall into four categories: (1) those who must be able to understand statistical presentations of findings in their field; (2) those who select, apply, and interpret statistical procedures in their work; (3) professional statisticians; and (4) mathematical statisticians. This book is addressed to those in the first two categories, including psychologists, educators, sociologists, engineers, speech therapists, biologists, counselors, business people, physicians, medical researchers, political scientists, geologists, and city, state, and federal government officials, to mention only a few. In each case the person's primary interest is in his or her own field, be it sociology or city planning; he or she is interested in statistics because it is a useful tool for answering questions in that field. Such persons are both consumers and users of statistics. Their knowledge of statistics may range from meager to expert.

The professional statistician helps professionals in substantive areas to use statistics effectively.[1] He or she may work for industry or a government agency, engage in a private consulting practice, or teach in a university. Unlike individuals in the first two categories, a professional statistician usually has advanced degrees in mathematics and/or statistics.

The mathematical statistician is not interested in applied statistics but, instead, in pure (mathematical) statistics and probability theory. Most likely this statistician teaches in a university and makes contributions to the theoretical foundations of statistics that may ultimately be used by those with applied interests.

1.2 Studying Statistics

Read More Slowly

Statistics cannot be read like assignments in history, English, or political science. Statistics uses a specialized vocabulary that must be learned, and ideas and computational procedures are presented in a highly symbolic form. Consequently, a 30 page

---

[1] A pamphlet, *Careers in Statistics*, which describes the work of statisticians, is available from the American Statistical Association, 806 15th St. N.W., Washington, D.C. 20005.

assignment may take three or four times as long to read as a comparable assignment in history. You will understand many sections of the book on a first reading; others will require two or more readings, lots of concentration, and perhaps some time between readings for the ideas to soak in.

Don't Worry If You Weren't an Ace in Math

If you're concerned about the level of mathematics required to understand statistics, stop worrying. Most statistical procedures in this book involve nothing more complicated than addition, subtraction, multiplication, and division. Although some use is made of high school algebra, the level is very elementary. The essential arithmetic and algebra are reviewed in Appendix A.

Resolve to Review Often

Frequent reviews of the material are a must or it will slip away. The exercises at the end of each section should be a helpful review because they provide (1) feedback about what you know and what you don't, (2) an indication of which concepts and computational procedures are the most important, (3) numerous examples of how statistics is used, and (4) practice in applying what you are learning. Answers to the starred exercises are given in Appendix C. The chapter summaries should also be useful for reviewing, since they briefly present the major points and place the topics in perspective.

Master Foundation Concepts before Going on to New Material

In statistics, as in mathematics or a foreign language, the material presented first is the foundation for what follows. Each chapter should be mastered before going on to the next. Cramming is effective in some subjects, at least as far as tests are concerned. But in statistics it inevitably results in superficial understanding of basic concepts and subsequent learning problems. Periodic reviews require considerable discipline, but they pay off. Also, different teaching approaches are effective with different students. If you are having difficulty with a topic, ask your professor to recommend a book that gives an alternative explanation.

Strive for Understanding

This book contains hundreds of formulas. I have not memorized all of them and neither should you. Some, such as the one for the arithmetic mean, $\bar{X} = \sum X/n$, appear so often that you really can't help but learn them; the others aren't worth the effort. I decided a long time ago, when faced with my own inability to remember telephone numbers, addresses, and lock combinations, that books are better repositories for such things than my head. Instead of memorizing formulas, strive to understand concepts and think about their applications. In what situations is a statistic useful? How is the statistic interpreted? What assumptions must be fulfilled to interpret the statistic? When you read about an experiment in your field, consider how you would have designed it and analyzed the data. And talk about your research ideas with your professor.

1.3 Basic Concepts

Population and Sample Defined

Many statistical terms are a legacy from the time when statistics was concerned only with the condition of the state. "Population," for example, originally meant, and still does, the total number of inhabitants of the state. Its meaning in statistics is broader. *A **population** is the collection of all objects or observations having one or more specified characteristics.* The population is identified when we specify the common characteristics. All the people listed in a telephone directory constitute a population, as do the number of heads and tails obtained in tossing a coin for eternity. *A single object or observation is called an **element** of the population.* The population of telephone book listees contains a *finite* number of elements; the population resulting from tossing the coin contains an *infinite* number. A population is either concrete or conceptual. The population of telephone book listees is *concrete*—given sufficient time we could contact each person, since the number is finite and well defined. The population of heads and tails is *conceptual*—no matter how long we work at it, we can't record all the possible results of tossing a coin.

The elements of a population can be people, objects, events, or, as is usually the case, observations based on measurable characteristics of people, objects, or events. For example, a population could consist of all the students in a university, their cars, their pep rallies, or their IQ scores. In identifying a population for a particular experiment, an investigator considers the nature of the research problem as well as such practical matters as the availability of population elements.

*A **sample** is a proper subset of a population.* That is, a sample can contain a single element or all but one of the population elements. For practical reasons, such as limited resources and time, or because the population is infinite in size, most research is carried out with samples rather than with populations. It is assumed that the study of a sample will reveal something about the population. This leap of faith often appears to be justified, as when a laboratory technician analyzes a sample of a patient's blood or when an automobile manufacturer crash tests a sample of bumpers. Occasionally, however, samples lead us astray. Later we'll see how and why.

Descriptive and Inferential Statistics

It is convenient to divide statistical techniques into two categories—descriptive and inferential. ***Descriptive statistics** are tools for depicting or summarizing data so they can be more readily comprehended.* When we say a player's lifetime batting average is

0.420 or when we graph the gas consumption of American cars, we are using descriptive statistics. A computer printout listing the IQs of all college students in California would boggle our minds—not so, a statement that the mean IQ is 116. Large masses of data are difficult to comprehend. Descriptive statistics reduce data to some form, usually a number, that is easily comprehended. We will go into the details of descriptive statistics in the first half of this book.

We saw that it is usually impossible for experimenters to observe all the elements in a population. Instead they observe a sample of elements and generalize from the sample to all the elements, a process called *induction*. They are aided in this process by **inferential statistics**, *which are tools for inferring the properties of one or more populations from an inspection of samples drawn from them.* Inferential statistics were developed to improve decision-making in cases where successive observations exhibit some degree of variation although they are taken under conditions that appear to be identical. The variation may be due to inherent variability in the phenomenon or in the subjects, errors of measurement, undetected changes in conditions, or a combination of these. In the behavioral sciences and education, the variability among subjects is the major stumbling block to induction. Suppose that a physiologist wants to know if a new drug will arrest the development of cancer in humans. It is impossible to administer the drug to the population of all cancer victims, but it is possible to administer it to a sample. The physiologist would probably attempt to control attitudinal and other extraneous factors by administering an inert druglike substance, a *placebo*, to half the sample and the new drug to the other half. Suppose that remission of cancer occurred in 100% of the sample receiving the new drug and in only 8% of those receiving the placebo. The difference, 100% versus 8%, between the drug and placebo samples is dramatic. The physiologist would probably conclude that if the drug had been administered to the population of all cancer victims, the remission rate would have been higher than if the population received the placebo. But what if the remission rate were 12% for the new drug and 8% for the placebo? We know that chance factors can produce a difference between two samples even though the samples are taken from the same population and receive identical treatments. Is the difference, 12% versus 8%, greater than would be expected by chance? Stated another way, if the experiment were repeated many, many times, could the physiologist predict with confidence that the difference would consistently favor the sample receiving the drug? This is the kind of question that can be answered using inferential statistics. We will deal with procedures for answering such questions in the latter half of this book.

Random Sampling

Some samples provide a sound basis for drawing conclusions about populations, while others do not. The difference lies in the method by which the samples are selected. *The method of drawing samples from a population such that every possible sample of a particular size has an equal chance of being selected is called* **random sampling**, *and the resulting samples are* **random samples**. Sampling methods based on haphazard or purposeless choices, such as soliciting volunteers, students enrolled in introductory psychology, or every tenth person in an alphabetical listing of names, produce *nonrandom* *samples*. Such samples, unlike random samples, do not provide a sound basis for deducing the properties of populations. Hence, whenever sampling is mentioned in this book, it will refer to random sampling. A detailed discussion of random sampling must await the development of other basic concepts. At this point, we will simply illustrate several characteristics of random samples.

Consider a box containing 300 balls, each identified by a number stamped on its surface. Two hundred of them are red (R), and 100 are black (B). If you didn't know the proportion of red to black balls, you could estimate it from a random sample of balls. You close your eyes, shake the box, reach in, withdraw a ball, note its color and number, and replace it. You do this six times and obtain the following sample: R_{102}, R_{75}, B_{39}, R_{62}, B_{37}, R_{50}; the subscripts, 102, 75, and so on, denote the numbers stamped on the balls. From this sample you would infer that the box contains more red than black balls; in fact, twice as many red balls. Suppose you drew four more samples, each time replacing the balls drawn, and obtained R_{154}, B_{62}, R_{35}, R_{143}, R_4, R_{29} (sample 2); R_{104}, B_{41}, B_{21}, R_{50}, R_{192}, R_{67} (sample 3); B_{28}, B_{41}, R_{150}, B_{61}, R_{88}, R_{148} (sample 4); and R_{152}, R_{120}, B_{88}, R_{33}, R_{36}, B_5 (sample 5). The results of the five random samples are summarized in Table 1.3-1. This simple experiment illustrates several points about random samples. First, the elements obtained (and the proportion of R to B balls) differ from sample to sample. This is referred to as *sampling fluctuation* or *chance variability*. Second, the characteristics of a sample do not necessarily correspond to those in the population. It turns out, however, that the larger a random sample, the more likely it is to resemble closely the population. Hence, experimenters prefer to work with large samples if it is economically feasible. Although there is no guarantee that large random samples will resemble the population, in the long run they are more likely to do so than small ones.

Table 1.3-1. Outcomes of Drawing Five Random Samples

Number of balls	Sample				
	1	2	3	4	5
Red	4	5	4	3	4
Black	2	1	2	3	2
Proportion of red to black	2:1	5:1	2:1	1:1	2:1

Review Exercises for Section 1.3[2]

1. How does the original meaning of the term "population" differ from today's statistical definition?
2. (a) In your local newspaper find two examples of populations and two examples of samples.
   (b) Classify the populations according to whether they are finite or infinite and whether they are concrete or conceptual.

[2] Problems or portions thereof for which answers are given in Appendix C are denoted, respectively, by * or **.

**3. For each of the following statements, indicate (a) the population, (b) the element, and (c) the datum to be recorded.
   *a. At least 50% of White female students in this university are ambivalent about having a career.
   b. Tequila Tech students are involved in more automobile accidents than other drivers in their age group.
   c. At least 80% of the homes in Chickasha, Oklahoma, have color televisions.
   d. Students at Ginebra University who hold outside jobs have higher grade point averages than those who don't hold outside jobs.
*4. What are the lower and upper limits on the size of a sample?
5. (a) Why is most research conducted on samples rather than populations? (b) How is sample size related to the resemblance between a random sample and the population?
6. Distinguish between descriptive statistics and inferential statistics.
*7. Indicate whether each of the following procedures would produce a random sample (R) or a nonrandom sample (NR) of students in an introductory sociology class.
   a. Write each student's name on a slip of paper, place the slips in a hat, shake the hat, and draw out ten names.
   b. Place the blindfolded instructor in the middle of a circle made up of all the class members. Have the instructor point to ten people around the circle. The student nearest the position pointed to becomes an element of the sample.
   c. For each student, flip a coin. If the coin lands heads, the student is in the sample.
   d. Line the students up from the tallest to the shortest. The third, fifth, seventh,..., twenty-first students become members of the sample.
8. Find two newspaper stories citing survey or poll results. Do the articles mention the method of sampling? Do they reveal where the financial support for the surveys came from? Why is this information—sampling method and source of financial support—important?
9. Terms to remember:
   a. Population
   b. Element
   c. Sample
   d. Random sample
   e. Sampling fluctuation

1.4 Describing Characteristics by Numerals

People, objects, and events have many distinguishable characteristics. Early in the design of an experiment, two key decisions must be made: What characteristics should be observed? And how should the characteristics be measured? The answer to the first question is determined by the experimenter's research problem and interests. Suppose an experimenter is interested in the IQs of male and female college students. College students differ in sex, age, IQ, major, hair color, family income, and so forth, but only two characteristics are of interest in this example—sex and IQ. These are the characteristics that are measured. The others are ignored. The second question, which involves the actual measurement of the selected characteristic, is outside the scope of this book. We are interested here in the more basic issues surrounding the assignment of numerals to characteristics. In the process of examining these issues we will discuss variables and constants and see how mathematicians classify variables.

Variables and Constants

A ***variable*** *is a symbol that is used to stand for an unspecified element of a set.* The set of elements for which the variable stands is called the *range* of the variable, and each element of the range is called a *value*. When we replace a variable by one of the elements in its range, we say that the variable "takes" this value. For example, the variable of sex can take one of two values, male or female. *A* ***constant*** *is a symbol with a range that consists of a single element.* The ratio of the circumference of a circle to its diameter, denoted by π, is a constant because its range consists of the single value 3.1415926536....

Perspectives on Numbers

We observed that selection of the characteristics to measure is relatively straightforward once the research problem has been identified. The second key decision, deciding how the characteristics should be measured, isn't as simple. For example, the variable of student IQ could be measured by (1) ranking or ordering IQ scores from highest to lowest and assigning each student the number of his or her rank; (2) assigning each student a label such as average, high average, or superior based on IQ test performance; or (3) assigning each student her or his actual IQ score. Depending on the measuring scheme adopted, Jonathan Whiz would be designated 3, superior, or 132. The number 3 would represent his rank among students in the sample and 132 would be his IQ score. Sex can be classified by assigning a unique symbol such as M, 2, or ♂ to males and F, 1, or ♀ to females.

Assignment of numerals to characteristics of people, objects, or events and the accuracy of the representation are central concerns of researchers but not of mathematicians and mathematical statisticians. Only during its formative years was mathematics tied to the real world. Then it seemed perfectly natural to prove mathematical theorems by recourse to counting and measuring. But in recent times mathematics has shed its real-world ties—mathematicians are now free to manipulate symbols that are totally void of empirical meaning. They are interested in the formal properties of the systems they create; applications in the real world are left to other specialists. The theoretical work of mathematicians and mathematical statisticians laid the foundation for statistical tools, but the scientists who use them must decide whether a particular tool is appropriate for his or her research applications and whether the numbers assigned to variables accurately represent the characteristics of interest. This division of interests between the developers and users of statistics has led to considerable confusion about the correct uses of statistical tools and also to two ways of thinking about numerals.

Classification of Variables in Mathematics

Mathematicians classify variables as qualitative or quantitative. *A* ***qualitative variable*** *is one with a range that consists of attributes or nonquantitative characteristics of people, objects, or events,* for example, sex (male, female), race (Caucasian, Negro,

Oriental, other), and grade in a course (A, B, C, D, F). The categories of a qualitative variable are nonoverlapping and exhaustive and may or may not suggest an order or rank. Grades in a course, A, B, C, D, F, clearly order academic achievement from highest to lowest, but no order is suggested by the categories for sex, race, religious preference, or blood type. Course grade is an example of an *ordered* qualitative variable; sex, race, and so on, are *unordered* qualitative variables.

A **quantitative variable** *is one with a range that consists of a count or a numerical measurement of a characteristic.* Quantitative variables can be discrete or continuous. A variable is *discrete* if its range can assume only a finite number of values or an infinite number of values that is countable. (That is, the infinite number of values can be placed in a one-to-one correspondence with the counting or natural numbers.) Family size is an example of a variable with a finite range. It can assume values 1, 2, 3, 4, and so on, but not 200, 8000, or any noninteger value such as 0.5 and 4.3. The rational numbers—numbers that can be expressed as the ratio of two integers, for example, $\frac{2}{2}$, $-\frac{2}{3}$, $\frac{7}{4}$—illustrate countably infinite numbers. There is no largest number, no smallest number, and between, say, 1 and 2 an infinite number of rationals can be inserted, for example, $\frac{3}{2}, \frac{4}{3}, \frac{5}{4}, \ldots$. Other examples of discrete quantitative variables are the number of parking tickets received, the number of trials required to learn a list of nonsense syllables, and one's score on a standardized achievement test. In each of these, the value assigned to the variable is obtained by counting, and the counting units—family members, parking tickets, learning trials, or achievement test items—are equivalent in arriving at the total count.

By contrast, a variable is *continuous* if its range is uncountably infinite. Such a range can be likened to points on a line that have no interruptions or intervening spaces between them. Examples of continuous variables are temperature in Bangor during January, length of fish caught off the Florida Keys, and speed of cars on the New Jersey Turnpike. Although a variable is continuous, measurement of it is by necessity discrete because of the calibration of measuring instruments. For example, the thermometer is

Table 1.4-1. Mathematical Classification of Variables

Type of Variable	Characteristics
Qualitative variable	Range consists of nonoverlapping and exhaustive categories that represent attributes or nonquantitative characteristics.
Unordered	Categories do not suggest an order or rank.
Ordered	Categories suggest an order or rank.
Quantitative variable	Range consists of a count or a numerical measurement of a characteristic.
Discrete	Range consists of only a finite number of values or an infinite number of values that is countable.
Continuous	Range consists of an uncountably infinite number of values.

usually calibrated in 1° steps, the ruler in $\frac{1}{16}$ in., and the speedometer in 1 mph. Consequently, our measurement of continuous variables is always approximate. Discrete variables, on the other hand, can be measured exactly. A husband and wife with two children are a family of exactly four, but a temperature of 20°C refers to a range of temperatures—20°C give or take $\frac{1}{2}°$.

The classification scheme for variables is summarized in Table 1.4-1. It is useful to mathematicians and statisticians because different mathematical tools and assumptions are required in derivations and proofs depending on the classification of the variables. The classification scheme is a convenience; it was not devised to mirror characteristics in the real world. When we use statistical methods to answer real-world questions, we must remember that they were developed to analyze numbers as numbers. If the numbers analyzed bear no relation to the characteristics we are interested in, the resulting answers will be meaningless.

Measuring Operations in the Behavioral Sciences and Education

Numbers are used for a variety of purposes, three of which are of particular interest to behavioral scientists and educators: (1) to serve as labels, (2) to indicate rank in a series, and (3) to represent quantity. For example, a football player is identified by the number 10 on his uniform, a team is ranked number two in the UPI poll, and the winning touchdown play covered 20 yd. Without thinking, we treat these numbers differently. It doesn't take a football fan to know that player 30 is not three times player 10 and that the number two team is not necessarily twice as good as the number four team, but that a 20 yd touchdown play did indeed move the ball twice as far down the field as a 10 yd play. We intuitively treat the numbers differently because they involve different levels of measurement. **Measurement** *is the process of assigning numerals to people, objects, or events according to a set of rules.* We shall see that the rules used to assign the numerals determine the level of measurement. S. S. Stevens (1946), a behavioral scientist, identified four levels of measurement: nominal, ordinal, interval, and ratio.

Nominal Measurement

Nominal measurement *is the simplest form. It consists of assigning observations to mutually exclusive and exhaustive* **equivalence classes** *so that those in the same class are considered to be qualitatively the same and those in different classes qualitatively different. The classes are then denoted by unique labels.* The set of labels constitutes a *nominal scale.* Assignment of women to one equivalence class called "women" and men to the other called "men" is nominal measurement. Numbers may be used instead of words to identify the two classes, for example, "1" for women and "2" for men. Numbers used in this way are simply alternative names for the classes. We could just as well have assigned the numbers "9" and "6." A transformation of the scale in which 9 is substituted for 1 and 6 for 2 would be as useful for distinguishing between the classes as any other one-to-one transformation.[3] The numbers in a nominal scale could be added, subtracted, averaged, and so on, but the numbers obtained would tell us nothing about the equivalence classes represented by the numbers. For example, $1 + 2 = 3$ and $9 + 6 = 15$, but neither 3 nor 15 corresponds to any characteristic of men and women. This follows since we didn't utilize the properties of size and order of numbers when we assigned them to the classes. The only property of numbers that we did utilize is that 1 is distinct from 2, 3, ... There are many examples of nominal scales in psychology and education, for example, Eysenck's four personality types (stable–extrovert, stable–introvert, unstable–extrovert, unstable–introvert), the primary taste qualities (sweet, sour, salty, bitter), and categories of psychoses (organic, functional). We can see a similarity between nominal measurement and one of the mathematician's types of variables. The nominal scale corresponds to the range of an unordered qualitative variable.

Ordinal Measurement

*The next higher level is **ordinal measurement**. It consists of assigning observations to equivalence classes that are ranked or ordered with respect to each other. The classes can be identified by numbers or any other ordered symbols that reflect the rank of the classes, such as letters.* Girls in a beauty contest, for example, can be ordered with respect to attractiveness.[4] If Diane is judged to be the most attractive, followed by Nancy, Fannie, and then Alma, we could assign Diane the number 1; Nancy, 2; Fannie, 3; and Alma, 4. We have no reason to believe that Nancy, ranked second, is half as attractive as Diane or that the difference in attractiveness between Diane and Nancy, 1 versus 2, is the same as the difference between Nancy and Fannie, 2 versus 3. The numbers indicate rank order but not magnitude or difference in magnitude between classes. The numbers assigned to the equivalence classes can be subjected to any transformation that preserves the order-isomorphism of the numbers and classes.[5] For example, the set of ordered numbers 2, 16, 39, 40 would serve just as well to rank the girls since only the order and not the distance between any two numbers is important. Alternatively, we could assign the letter A to Diane, B to Nancy, C to Fannie, and D to Alma.

Some characteristics, such as people's heights, can be measured in several ways, like ranking from tallest to shortest or recording actual feet and inches. The latter procedure assigns numbers that represent the magnitudes of the equivalence classes and therefore has several advantages over ordinal measurement, as we shall see later. For the moment we simply note that ordinal measurement is most often used when it is difficult or impossible to apply a more refined procedure.

Numerous examples of ordinal scales can be found in the behavioral sciences and education. Mentally subnormal children are sometimes classified as borderline, edu-

---

[3] A one-to-one transformation associates with each element in one set, one and only one element in a second set, and vice versa. For example, if the two sets are $a_1, a_2, a_3, \ldots$, and $b_1, b_2, b_3, \ldots$, each a is paired with one and only one b, and each b is paired with one and only one a.

[4] When my editor asked whether this example might not be a little sexist, I immediately sought the opinions of ten male colleagues and one female beauty-contest winner, who all assured me that it is definitely *not* sexist.

[5] Such a transformation is called a *monotonic transformation*.

cable, trainable, or profoundly retarded. Other examples are rank in high school class and a supervisor's ranking of employees. Such ordinal scales correspond, in the language of the mathematician, to the range of an ordered qualitative variable.

Interval Measurement

Interval measurement is much more sophisticated than nominal and ordinal measurement. *In **interval measurement**, the numbers assigned to equivalence classes have the properties of distinctness and order, and in addition, equal differences between numbers reflect equal magnitude differences between the corresponding classes. The measurement procedure consists of defining a unit of measurement, such as a calendar year or 1°F, and determining how many units are required to represent the difference between equivalence classes.* In our measurement of calendar time, the same amount of time elapsed between 1970 and 1971 as between 1971 and 1972, and similarly the temperature difference between 70 and 75°F is the same as that between 80 and 85°F. A given numerical interval, say 1 yr or 5°F, represents the same difference in the characteristic measured irrespective of the location of that interval along the measurement scale. In other words, numerically equal distances along the measurement continuum represent empirically equal differences among the corresponding equivalence classes—that is, the measured characteristic.

Since the units along interval scales are empirically equal, we can meaningfully perform most arithmetic operations on the numbers. For example, we can say that the difference between 80 and 60°F is twice as great as that between 60 and 50°F. That is, the ratio of intervals $(80 - 60°F)/(60 - 50°F) = 2$ has meaning with respect to temperature. However, not all operations are permissible because the starting point or origin of an interval scale is always arbitrarily defined and does not necessarily correspond to an absence of the measured characteristic. In the case of the Fahrenheit scale, zero corresponds to the temperature produced by mixing equal quantities by weight of snow and salt. It does not indicate an absence of molecular action and hence an absence of heat. Therefore although $80°F/40°F = 2$, we cannot say that 80°F is twice as hot as 40°F. The ratio of temperatures is meaningless with respect to temperature. The situation is the same with calendar time, which is calculated from the birth of Christ, and altitude, which is calculated from sea level.

The numbers in an interval scale can be subjected to any positive linear transformation.[6] Degrees Fahrenheit, F, can be transformed into degrees Celsius, C, by means of the linear transformation

$$C = \frac{5}{9}(-32) + \frac{5}{9}F.$$

Although the variable represented by an interval scale may be continuous, our measurement of it is always discrete, since measuring instruments are calibrated in discrete steps. Thus, in practice an interval scale corresponds to the range of a discrete quantitative variable.

---

[6] A positive linear transformation of a variable X consists of multiplying X by a positive constant b and adding a constant a to the product. The equation for a transformed value thus is $X' = a + bX$.

Ratio Measurement

Ratio scales *constitute the highest level of measurement. The numbers assigned to equivalence classes have the properties of distinctness, order, equivalence of intervals, and, in addition, the origin of the scale represents the absence of the measured characteristic.* Thus, ratio scales have all the properties of interval scales plus an absolute zero. Most scales in the physical sciences are ratio scales—height in inches, weight in pounds, temperature on the Kelvin scale, and elapsed time such as age in years, years of experience, and reaction time in seconds.

Table 1.4-2. Overview of Levels of Measurement

Level of Measurement	*Characteristics*
Nominal	Symbols serve as labels for mutually exclusive and exhaustive equivalence classes. The symbols have the property of distinctness. Appropriate transformation: any one-to-one substitution. Examples: sex, eye color, racial origin, personality types, and primary taste qualities.
Ordinal	Ordered symbols, usually numbers, indicate rank order of equivalence classes. The symbols have the properties of distinctness and order. Differences between ordered symbols provide no information about differences between equivalence classes. Appropriate transformation: monotonic. Examples: military ranks, classification of mentally retarded children, rank in high school, and a supervisor's ranking of employees.
Interval[a]	Equal differences among numbers reflect equal magnitude differences among equivalence classes, but the origin or starting point of the scale is arbitrarily determined. Numbers have the properties of distinctness, order, and equivalence of intervals. Appropriate transformation: positive linear. Examples: Fahrenheit and Celsius temperature scales, calendar time, and altitude.
Ratio[a]	All the properties of interval scales apply, and, in addition, the origin of the scale reflects the absence of the measured characteristic. Appropriate transformation: multiplication by a positive constant. Examples: height, weight, Kelvin temperature scale, and measures of elapsed time.

[a] These two levels are sometimes referred to collectively as *metric measurement* or *numerical measurement*.

Not only is the difference between 5 and 6 in. the same distance as between 10 and 11 in., but also 10 in. is twice as long as 5 in. Ratio scales permit us to make meaningful statements about the ratio of two numbers, for example, 10 in./5 in. = 2; hence 10 in. is twice as long as 5 in. The properties of a ratio scale permit us to perform all arithmetic operations on the numbers. However, the only transformation of a ratio scale that preserves these properties is multiplication by a positive constant. For example, we can transform inches into centimeters by multiplying inches by the constant 2.54: 10 in. = 25.4 cm and 5 in. = 12.7 cm. Ten inches is twice as long as 5 in. and, similarly, 25.4 cm is twice as long as 12.7 cm. As we move from the simplest measurement (nominal scales) to the most sophisticated (ratio scales), more and more constraints are placed on the transformations that can be meaningfully applied. This is because the numbers in higher-measurement scales contain more information that can be altered or destroyed. In practice, a ratio scale, like the interval scale, corresponds to the range of a discrete quantitative variable. The major characteristics of the four scales are summarized in Table 1.4-2.

Implications of the Two Ways of Thinking about Numerals

Two ways of thinking about numerals have been described—one that reflects the concerns of mathematicians and statisticians and the other, the concerns of behavioral scientists and educators. Statistical methods were developed for analyzing numbers as numbers, whether or not they are true measures of some characteristic. If the assumptions associated with the statistical methods are fulfilled, they will produce answers that are formally correct as numbers. And this is true regardless of the degree of isomorphism between the numbers and the characteristic they represent. The problem comes in translating statistical results into statements about the real world. If numbers representing a nominal scale are manipulated arithmetically, the result will be numbers that are numerically correct but uninterpretable. If nonsense is put into the equation, nonsense will indeed come out.

Experimenters concerned with problems of people and nature are very sensitive to the difficulties of interpreting numbers produced by statistical procedures—and rightfully so. Some authors have even gone so far as to prescribe the statistical procedures that can be used with each level of measurement.[7] Except in the physical sciences, few scales have equal intervals, so the number of statistical techniques on the approved list is relatively small. This position, it seems, fails to recognize that measurement of many variables in the behavioral sciences and in education lies somewhere between the ordinal and interval levels. The IQ scale is a good example. The 10 point difference between IQs of 90 and 100 is probably not the same as that between 130 and 140, but the scores contain some magnitude information, since most test administrators would agree that the intellectual difference between children with IQs of 80 and 90 is less than that between children with IQs of 100 and 120. Even though the measurement units along the scale are not equal, those familiar with IQ tests believe that the magnitude of the difference between scores does provide useful information about differences in

[7] Examples can be found in Senders (1958), Siegel (1956), and Stevens (1946, 1951).

intellectual aptitude. Should we avoid performing arithmetic operations on IQ scores because the measurement is below interval level? Or will mathematical manipulations yield interpretable answers and enable us to make better predictions concerning intelligence? Such questions have been heatedly debated.[8] We cannot look to mathematicians and statisticians for answers since these questions are outside their province. The answers must come from users of statistics who are acquainted with the problems of translating numerical answers into statements about the real world. An examination of the professional literature reveals that most experts in the behavioral sciences and education do apply arithmetic operations to numbers even though the measurement is somewhere between the ordinal and interval levels; further, they interpret the results as if differences between the numbers reflect something about the differences in the measured characteristics. Apparently, experts prefer to utilize whatever magnitude information the numbers contain even though differences among numbers only grossly approximate true magnitude differences. If an experimenter feels that any order-preserving transformation of a set of numbers adequately represents the equivalence classes, he or she should treat the scale as ordinal. If the experimenter's intuition is correct, the numbers contain no magnitude information, and they should not be treated as though they do. It is the experimenter, the person most familiar with the data, who must decide how much information the numbers contain.

If all our measurements were nominal or ratio, interpreting the outcomes of research would be fairly straightforward. Unfortunately, ratio measurement and even interval measurement are rare in the behavioral sciences and education. At best our measurement falls somewhere between the ordinal and interval levels. We can avoid some obvious and not so obvious interpretation errors by being sensitive to the degree of isomorphism between a set of numbers and the characteristic they represent. Suppose that on a standardized arithmetic-achievement test, Mortimer received a score of 0, Dude a score of 30, and Reginald a score of 60. Can we conclude that Mortimer knows nothing about arithmetic? Obviously not; a score of 0 means that he couldn't answer any questions on the test, but easier questions may exist that he could answer. Achievement tests, as well as many other tests, have arbitrary rather than absolute zero points, and therefore fall short of ratio measurement. It follows that although Reginald's score of 60 is twice as high as Dude's 30, Reginald's arithmetic achievement isn't necessarily twice Dude's.

The interpretation problem that results from lack of equal intervals is more subtle. Suppose we compare the effectiveness of two methods of teaching arithmetic. Students in a class using method A gained on the average 10 points; those in a class using method B, 7 points. The results seem straightforward, but suppose that originally the two groups were not equal in arithmetic achievement. Let the average score for group A be 40 and the score for group B be 10. Is it possible that a 7 point change from 10 to 17 represents more improvement in arithmetic achievement than a 10 point change from 40 to 50? Unless we know that, say, a 10 point change anywhere on the measurement scale represents the same empirical change, the interpretation of the experiment

[8] The major issues in this debate have been presented by Anderson (1961), Boneau (1961), Gaito (1960), and Stevens (1968). These articles are reproduced in Kirk (1972a, chap. 2) along with suggestions for further reading on the issue. Also of interest are articles by Gardner (1975) and Wainer (1976).

is equivocal. The greater the differences between the groups' initial performances, the more the problem is accentuated.

Because numbers don't always mean what they appear to mean, they should be carefully scrutinized. Consider a test that doesn't have enough ceiling to differentiate adequately among high-scoring subjects. Suppose two individuals make the top score of 60. For one subject, this may represent maximum capability, but the other person may be capable of a much higher performance. The measuring instrument is simply incapable of showing it.

It is evident that the experimenter must be guided by two sets of rules. When the tools of statistics are used, the mathematician's and statistician's rules must be followed. When the numbers are interpreted as statements about the real world, measurement rules must be used as a guide. Of course, this calls for considerable informed judgment, since the application of measurement rules is the subject of unresolved controversy.

Review Exercises for Section 1.4

10. Mathematicians and behavioral scientists have somewhat different interests in numbers; discuss these differences.
**11. Ignoring for the moment limitations of measuring instruments, classify measures of the following according to the mathematician's scheme (unordered qualitative, U; ordered qualitative, O; discrete quantitative, D; continuous quantitative, C).
   *a. Size of family
   *b. Race
   *c. Marital compatibility
   *d. Seeding of tennis players
    e. Employee production on an assembly line
    f. Creativity
    g. Political party affiliation
    h. Final standing of football teams in the Southwest conference
    i. Weight loss after jogging for 3 mi
    j. Number of reported suicides in 1975
    k. Major in college
    l. Religious preference
    m. Grading scale in school (A, B, C, D, F)
    n. Amount of rainfall
12. Due to limitations of measuring instruments, measurement of some variables is of necessity approximate. Classify the variables in Exercise 11 according to whether our measurement is exact (E) or approximate (A).
**13. Reclassify the variables in Exercise 11 according to the mathematician's scheme, taking into account limitations in our ability to measure some of the variables.
14. (a) In what three ways do behavioral scientists use numbers in measurement? (b) Give two examples of each use.
**15. Classify the variables in Exercise 11 with respect to level of measurement, taking into account limitations in our ability to measure some of the variables.
16. For each level of measurement, list the properties that characterize the numbers assigned to the equivalence classes.
*17. For each level of measurement, indicate the appropriate transformation that can be performed on the numbers.

*18. What level of measurement is most often achieved (a) in the physical sciences, and (b) in the behavioral sciences and education?
19. Who is in the best position to determine the degree of isomorphism between a set of numbers and the corresponding equivalence classes and hence to determine the arithmetic operations that can meaningfully be applied?
20. What does a score of 0 on an aptitude test mean?
*21. Suppose that achievement test scores for a control group increased from 62 to 65 while those for the experimental group increased from 68 to 74. What must be true to conclude unequivocally that the experimental group improved twice as much as the control group?
22. Terms to remember:
   a. Variable
   b. Constant
   c. Qualitative variable, ordered and unordered
   d. Quantitative variable, discrete and continuous
   e. Measurement
   f. One-to-one transformation
   g. Monotonic transformation
   h. Positive linear transformation

†1.5 *Historical Development of Statistics*[9]

National Statistics

The science of statistics grew out of an attempt to solve practical problems associated with raising taxes, producing insurance tables, and determining the odds in games of chance. Its subject matter was shaped by three lines of development—national statistics, probability theory, and experimental statistics. The oldest of these is *national statistics*, which was enumerative and descriptive in character; it can be traced to the beginning of recorded history. David numbered his people, and the Egyptians and Romans kept detailed records of taxes and other state resources. Caesar Augustus simplified the enumerative process by ordering all citizens to report to the nearest statistician, better known as the tax collector. The descriptive use of statistics came of age in the work of an English Army captain, John Graunt (1620–1674), who published in 1662 a small book of birth and death statistics for London from 1604 to 1661. Unlike earlier works, such as William the Conqueror's *Domesday Book*, which simply contained data compiled for purposes of taxation and military service, Graunt's book summarized and interpreted the data. His was the first work to shed light on the regularity of social phenomena. It marked the beginning of a theory of annuities and led to the founding of insurance societies.

Probability Theory

A second and independent line of development in statistics is *probability theory*. The earliest traces of probability, found in the Orient around 200 B.C., concerned whether an expected child would be male or female. However, the real impetus for development of probability came not from prospective parents but from gamblers who

† This and other sections so marked can be omitted without loss of continuity.
[9] The development of this section was strongly influenced by Dudycha and Dudycha (1972). Their article provides a more complete introduction to the history of statistics with special emphasis on the behavioral sciences.

wanted to know the odds of winning at various games of chance. Leading mathematicians and scientists of the day—Pierre de Fermat (1601–1665), Blaise Pascal (1623–1662), Christianus Huygens (1629–1695), and James Bernoulli (1654–1705)—responded to the problem. Gradually they chiseled out the foundation of a theory of probability. A milestone in this development was the discovery of the normal curve of errors by Abraham de Moivre (1667–1754), a mathematics tutor who supplemented a meager income by calculating odds for gamblers at the coffeehouses he frequented. Apparently, de Moivre did not appreciate the significance of his discovery; it was published in 1733 only obscurely as a supplement written in Latin to a limited reprinting of a book he had published 3 years earlier. Therefore, it remained for others to demonstrate the pervasiveness of the *normal distribution*. For more than a century it was attributed to a later discoverer, Carl Friedrich Gauss (1777–1885), one of the greatest mathematicians of all time. It was also discovered independently by Pierre-Simon de Laplace (1749–1827), who forsook a cleric's robe for his lifework in celestial mechanics and probability. Both Laplace and Gauss used the normal distribution in investigating errors of observation in astronomy. Lambert Adolphe Jacques Quetelet (1796–1874), who is considered the father of social science, saw that the normal distribution and probability theory could be applied to all observational sciences—astronomy, anthropology, physics, the census, and the statistics of mental and moral traits. He used the normal curve, for example, to predict the number and type of crimes committed. His work integrated national statistics and probability theory and paved the way for the third line of development—experimental statistics.

Experimental Statistics

The emerging interest in the sciences in the early 1800s created a need for new statistical procedures and principles to guide the design of experiments. The result was *experimental statistics*. Its development was dominated by such intellectual giants as Sir Francis Galton (1822–1911). Terman, the developer of the Stanford–Binet intelligence test, estimated Galton's IQ at about 200. Galton, more than anyone before, used statistics in investigating problems of people and nature. His major statistical contributions were regression and correlation procedures (see Chapters 5 and 6), which he used to unravel mysteries of heredity. Karl Pearson (1857–1936) refined the mathematical theory of regression and made an astonishing number of other contributions to statistical theory and practice. Perhaps his greatest was the development in 1900 of the chi-square test for goodness of fit (see Chapter 17), which is used in testing the significance of differences between observed data and those expected on the basis of some hypothesis.

The modern era in experimental statistics was ushered in by William Sealey Gosset (1876–1937), who derived the t distribution (see Chapter 12) in 1908. Thus began the development of exact inductive procedures appropriate for both large and small samples. Heretofore experimenters had relied on large-sample statistical procedures. Gosset, who published under the pseudonym of *Student*, was a brewer for Messrs. Guinness. His discovery, like others in statistics, resulted from a practical need—in this case the need for inductive procedures appropriate for small samples. He was involved in brewing research, where variable materials and susceptibility to temperature changes precluded the use of large samples.

The modern era matured in the work of Sir Ronald A. Fisher (1890–1962), whose contributions to statistics are legion. He is best remembered for his derivation of the F distribution, contributions to the design and analysis of experiments, and heated exchanges about statistical theory with Jerzy Neyman (1894–   ) and Egon Pearson (1895–   ). Fisher's work was a unique blend of the rigor of the mathematician with a common-sense approach; the latter was undoubtedly due to his applied work in agriculture, biology, and genetics.

Neyman and Pearson carefully consolidated the work of Fisher and others while developing their own theory of statistical inference. The bulk of the statistical arsenal of today's researcher can be traced to Fisher, Neyman, and Pearson. But in response to changing research needs, there have been many new developments. The electronic computer has made possible the solution of problems that were heretofore intractable and has sparked new lines of inquiry. It seems unlikely, however, that a new era could be dominated to the extent that Fisher, Neyman, and Pearson dominated the one from 1920 to the present.

Review Exercises for Section 1.5

**23. *(a) What three lines of development shaped the subject matter of contemporary statistics?
(b) Briefly summarize the major characteristics of each line.
24. For each of the following men, give at least one major contribution to statistics.
   a. Abraham de Moivre (1667–1754)
   b. Lambert Adolphe Jacques Quetelet (1796–1874)
   c. Francis Galton (1822–1911)
   d. Karl Pearson (1857–1936)
   e. William Sealey Gosset (1876–1937)
   f. Ronald A. Fisher (1890–1962)
   g. Jerzy Neyman (1894–   )
   h. Egon Pearson (1895–   )
*25. What distinguished the modern era in experimental statistics from the previous period?

1.6 Summary

This book is addressed to consumers and users of statistics and, more specifically, to students in the behavioral sciences and education. One course doesn't make a statistician, but it can develop basic understanding and fluency in the technical jargon of statistics. This is useful, since statistics is the universal language for communicating research findings.

The word "statistics" has several meanings; usually it refers either to a collection of techniques for making decisions based on data or to functions describing data, such as the mean and range. Your study of statistics should help you to (1) read the professional literature in your field, (2) design and analyze simple experiments, and (3) detect statistical fallacies in the mass media and technical reports. In addition, you may learn new, more critical and analytical ways of thinking.

Research questions usually concern characteristics of populations. For example, what do people of voting age think about an issue? Is one instructional technology for fifth graders more effective than another? Do 21-year-old women prefer smaller families than men of the same age? The populations are, respectively, the attitudes of voters on the issue, the achievement scores of fifth graders, and preferred family sizes of 21-year-old women and men. The term "population" originally referred to all the inhabitants of the state. In statistics it refers to the collection of all objects or observations having one or more specified characteristics. For practical reasons or because the population is infinite in size, it is rarely possible to observe all the elements of a population. Instead, we conduct research on a sample of elements. A sample can contain a single element or all but one of the population elements. If every sample of a particular size has an equal chance of being selected from the population, the sampling process is said to be random.

Statistics can be applied to data from samples or from populations to obtain a clearer understanding of their characteristics. If we obtained a random sample of 21-year-old women and men, we might find that on the average they prefer, respectively, 2.2 and 2.4 children. In addition, we might learn that the range for women was 0–14 and for men 0–8.

The numbers, 2.2 and 2.4, and the ranges, 0–14 and 0–8, are descriptive statistics; they summarize properties of the two samples. Description is one important application of statistics. A second is the inferring of characteristics of a population by observing a sample. The sample statistics for preferred family size, for example, provide our best guess about the corresponding population values. These two uses of statistics, descriptive and inferential, are discussed in the first and second halves of this book.

Once an experimenter has identified the population of interest and the characteristic to be observed, he or she must decide how the characteristic should be measured. Mathematicians and statisticians have historically classified variables as qualitative (ordered or unordered) or quantitative (discrete or continuous). This scheme evolved because different mathematical tools are used in derivations and proofs for different kinds of variables.

Behavioral scientists, on the other hand, developed a classification scheme that reflected their concern with the degree to which numerals are isomorphic with the characteristics they represent. A four-level classification of measurement resulted: nominal, ordinal, interval, and ratio. Today we recognize that there are more than four levels of measurement and that the measurement of many variables in the behavioral sciences and education lies somewhere between the ordinal and interval levels. Much of the controversy surrounding the use of arithmetic operations with data between these levels can be resolved by adhering to two sets of rules. When the tools of statistics are used, the mathematician's and statistician's rules must be followed; when the numbers are interpreted as statements about the real world, the behavioral scientist's measurement rules must be followed.

Modern statistics is the culmination of three historical lines of development—national statistics, probability theory, and experimental statistics. Its origins are in antiquity, yet most of the material in this book is the product of the twentieth century. The changing research practices and requirements of experimenters make continuing demands for the development of new statistical tools.

1.6 Summary

2
Frequency Distributions and Graphs

2.1 Introduction
2.2 Frequency Distributions
    Ungrouped Frequency Distribution for Quantitative Variables
    Grouped Frequency Distribution for Quantitative Variables
    Determining the Number and Size of Class Intervals for a Quantitative Variable
    The Pros and Cons of Grouping Data
    Relative Frequency Distributions
    Cumulative Frequency Distribution
    Frequency Distributions for Qualitative Variables
    Review Exercises for Section 2.2
2.3 Introduction to Graphs
2.4 Graphs for Qualitative Variables
    Bar Graphs
    Pie Charts
    Review Exercises for Section 2.4
2.5 Graphs for Quantitative Variables
    Histograms
    Frequency Polygons
    Cumulative Polygons
    Review Exercises for Section 2.5
2.6 Shapes of Distributions
    Bell-Shaped Distributions
    Skewed Distributions
    Bimodal Distributions
    J, U, and Rectangular Distributions
    Review Exercises for Section 2.6
2.7 Misleading Graphs
    Review Exercises for Section 2.7
2.8 Summary

2.1 Introduction

Variation in the behavior of subjects under the same apparent conditions is inevitable. The problem is more troublesome in the behavioral sciences and education than in the physical sciences. A chemist can be confident that different samples of H_2O will react with another substance the same way under controlled tests. But this kind of uniformity in behavioral research is rare. The variation problem is usually handled by observing many subjects and collecting volumes of data. This in turn calls for procedures for depicting and summarizing data so they can be more readily comprehended. Two kinds of descriptive tools serve the purpose—graphical and numerical methods. This chapter is devoted to graphical methods; numerical methods will be described in Chapters 3–6.

2.2 Frequency Distributions

The first step in summarizing data is to order the data in some logical fashion. This involves defining equivalence classes and counting the number of observations in each class. *An equivalence class can be a single score value, a collection of score values, or a collection of elements in a qualitative category. A table showing the classes and the frequency of occurrence of their elements is called a* **frequency distribution**. The equivalence classes of a frequency distribution are often called *class intervals*. If each of the class intervals is a single score value, the distribution is said to be *ungrouped*.

Ungrouped Frequency Distribution for Quantitative Variables

Suppose we administered a test of leadership aptitude to all high school football coaches in Punt County, Iowa. Their test scores are shown in Table 2.2-1. If we examine the table carefully, we can see that the smallest score is 30 and the largest is 68, and that most of the scores are in the high 40s or low 50s. The same information can be

Table 2.2-1. Leadership Aptitude Scores

Coach	Score	Coach	Score	Coach	Score
John Walker	55	William Kleis	39	Mark Anderson	45
Harold Wilson	46	David Hays	47	Dave Baker	33
Jack Thornton	50	Thomas Brown	52	Bill Hicks	50
Heyward Helm	51	Bill Adams	54	Paul Busby	51
John Hansen	48	Alan Ellis	48	Bob Dean	54
Terry Hendrix	50	Norman Jones	46	Ed Black	59
Chuck Carter	30	Douglas Kahn	68	Jeff Nance	49
Stephen Shanon	53	Bill Fluet	44	Don Campbell	42
Jack Patterson	57	Robert Vandorn	49	Glen Sheriff	56
Juliet Bussey	62	Joseph Schmidt	52	Mac Barker	53

Table 2.2-2. Ungrouped Frequency Distribution for Leadership Scores from Table 2.2-1

Score, X	Frequency, f	Score, X	Frequency, f	Score, X	Frequency, f	Score, X	Frequency, f
68	I	58	0	48	II	38	0
67	0	57	I	47	I	37	0
66	0	56	I	46	II	36	0
65	0	55	I	45	I	35	0
64	0	54	II	44	I	34	0
63	0	53	II	43	0	33	I
62	I	52	II	42	I	32	0
61	0	51	II	41	0	31	0
60	0	50	III	40	0	30	I
59	I	49	II	39	I		

extracted much more easily from the ungrouped frequency distribution in Table 2.2-2, which associates with each score value, X, the frequency of its occurrence, f. In constructing the frequency distribution we followed the convention of putting the largest score at the upper left of the table. In addition, each number between the largest and the smallest scores is listed in the distribution so that every possible score can be tallied and the gaps between scores can be easily detected.

The frequency distribution is an effective organizing device, but some information is lost. We cannot tell from Table 2.2-2 which coach made the highest score, which coach made the lowest score, or that one of the coaches is a woman. We must refer to the original data for this information.

Grouped Frequency Distribution for Quantitative Variables

If the spread of scores for a quantitative variable is large, it is often helpful to construct a *grouped frequency distribution* in which the class intervals span two or more score values, as shown in Table 2.2-3. A class interval for a quantitative variable has a *nominal lower limit* and a *nominal upper limit*; for the class interval 66–68 they are, respectively, 66 and 68. However, the interval 66–68 actually includes any number *equal to or greater than* 65.5 and *less than* 68.5. The numbers 65.5 and 68.5 are called the *real limits* of the interval.[1] They extend 0.5 below the nominal lower limit and, in actuality, 0.4999... above the nominal upper limit. The nominal limits are used for convenience in the tables; the real limits indicate the underlying continuity of the class intervals and are used to compute class interval size. The size of a class interval, denoted by i, is equal to the real upper limit minus the real lower limit; for example, $i = 68.5 - 65.5 = 3$. The score values in an ungrouped frequency distribution like Table 2.2-2

---

[1] If our measurement were accurate to the nearest tenth, so that we had scores such as 6.7, the class interval nominal limits would be 6.6 and 6.8 and the real limits would be 6.55 and 6.85. These values are obtained by adding and subtracting 0.05 instead of 0.5 from the nominal limits. Similarly, if our measurement were accurate to the nearest hundredth, the nominal limits would be 0.66 and 0.68 and the real limits would be 0.655 and 0.685, which differ from the nominal limits by ±0.005.

Table 2.2-3. Grouped Frequency Distribution for Leadership Aptitude Scores from Table 2.2-1

Class Interval	Frequency, f
66–68	1
63–65	0
60–62	1
57–59	2
54–56	4
51–53	6
48–50	7
45–47	4
42–44	2
39–41	1
36–38	0
33–35	1
30–32	1
	$n^a = 30$

[a] n denotes the total number of scores in the frequency distribution.

can be thought of as class intervals with upper and lower real limits. For the class interval 68, they are 67.5 and 68.5.[2] The class interval size is $68.5 - 67.5 = 1$.

Several conventions are followed in constructing a frequency distribution. They are not inviolate rules, but rather are guidelines for constructing easily interpreted tables.

1. For quantitative variables, the class interval containing the largest score should be at the top of the table. For qualitative variables, the order of class intervals should be determined by convenience and/or logic.
2. The class intervals should be mutually exclusive; that is, class intervals should be chosen so that a score belongs to one and only one interval.
3. For quantitative variables, there should be no gaps between the class intervals. In Table 2.2-3, even though no scores fall in the class intervals 63–65 and 36–38, the intervals are included for completeness.
4. All quantitative class intervals should have the same width or size.[3]
5. There should be 10–20 class intervals unless the number of scores is very small, in which case it may be desirable to use fewer class intervals. For qualitative variables, the number of class intervals is usually dictated by the nature of the variable. For example, if the variable is sex, there may be three class intervals—male, female, and unknown.

[2] Some variables do not follow this convention. A common example is age. If a person is 21, this means that the twenty-first birthday has passed, but the twenty-second has not. The real limits for the age 21 are 21.0 and 21.999....

[3] Sometimes this is not possible or desirable. Suppose one subject was unable to learn a list of nonsense syllables in the usual number of trials, 6–10, required by most subjects. After the twentieth trial the subject was still unable to meet the learning criterion and gave up. This subject can't be given an exact score; he or she falls into the top class interval "20 or more." This interval is *open*, since one real limit can be specified but not the other. Or suppose the class intervals represent family income. It might be desirable to make the bottom and top class intervals open to include the few families with extremely small or extremely large incomes.

6. For quantitative variables, one of the preferred class interval sizes should be used: 1, 2, 3, 5, 10, 15, 20,....
7. The nominal lower limit of each quantitative class interval should be an integer multiple of the class interval size. For example, the nominal lower limit of the class interval 30–32 is equal to $3 \times 10 = 30$, where $i = 3$ and 10 is the integer multiplier.

Determining the Number and Size of Class Intervals for a Quantitative Variable

The conventions for constructing a grouped frequency distribution provide general guidelines for the number and size of class intervals. We know there should be 10–20 class intervals (unless there are only a few scores) and the preferred class interval sizes are 1, 2, 3, 5, 10, 15, 20,.... With this in mind, we can estimate the number and size of class intervals in a trial and error fashion using the formula

$$\frac{\text{Range}}{\text{Preferred } i} = \text{Number of class intervals}.$$

The range is equal to the real upper limit of the largest score minus the real lower limit of the smallest score. A preferred class interval size ($i = 1$ or 2 or 3 or ...) is selected so that the formula yields between 10 and 20 class intervals. To illustrate, the largest and smallest scores in Table 2.2-1 are 68 and 30. The range is $68.5 - 29.5 = 39$. If a class interval size of 2 is tried in the formula, there would be $\frac{39}{2} \simeq 20$ class intervals. Since there are only 30 scores, a smaller number of class intervals would be preferable. If a class interval size of 3 is selected, the formula yields $\frac{39}{3} = 13$ class intervals, the number used in Table 2.2-3. A class interval size of 5 should not be used because it would give only $\frac{39}{5} \simeq 8$ class intervals. For most sets of data there will be no more than two class interval sizes that give the desired 10–20 class intervals. As a general rule when the number of scores is small, use less than 15 class intervals; when the number is large, use 15–20.

Suppose we have administered a test of reading readiness to 26 children enrolled in the first grade. The highest and lowest scores on the test are 132 and 73; the range is $132.5 - 72.5 = 60$. How many class intervals of what size should the frequency distribution have? By trial and error and the formula, Range/Preferred i = Number of class intervals, we see that two grouping schemes are possible: $\frac{60}{3} = 20$ and $\frac{60}{5} = 12$. The one in which $i = 5$ is preferred because there are only 26 scores. The lowest class interval, following convention 7, given above, would be 70–74 since 70 is an integer multiple of $i = 5$—that is, $70 = 5 \times 14$. The highest class interval would be 130–134. Thus, this grouping scheme actually results in 13 instead of 12 class intervals because the smallest and largest scores don't fall at the nominal lower and upper limits of their class intervals. If the extreme scores had been 70 and 129, the distribution would have had 12 class intervals.

Suppose we had tested 221 children instead of 26. In this case a class interval size of 3 should be used. The lowest and highest class intervals would be 72–74 and 132–134, since 72 and 132 are integer multiples of 3. Even though the use of $i = 3$

results in 21 class intervals, it is preferred to $i = 5$ because of the large number of scores. The purpose of graphical methods is to make data easier to comprehend, and sometimes this can best be done by departing from the conventions.

The Pros and Cons of Grouping Data

Grouping scores into class intervals has its disadvantages. First, some information inevitably is lost. For example, we know from Table 2.2-3 that four scores occur in the class interval 54–56, but we don't know their individual values.

A second disadvantage is that grouping produces nonunique frequency distributions. For any set of data, only one ungrouped frequency distribution can be constructed, but for grouped frequency distributions, there is often a choice between two or more class interval sizes. As we will see later, the grouping scheme that is adopted affects the values of descriptive statistics such as the range and mode.

The disadvantages of loss of information and lack of uniqueness must be weighed against the simplicity achieved in grouping. If the spread of scores is large, a grouped frequency distribution is much more easily interpreted. Also, grouping may be ultimately more convenient if graphs are to be used, since graphs are usually constructed from grouped frequency distributions.

Relative Frequency Distributions

It is often helpful to express either the proportion or the percentage of the total number of scores contained in each class interval. The formulas for proportionate frequency and percentage frequency are

$$\text{Prop } f = \frac{f}{n} \quad \text{and} \quad \%f = \frac{f}{n} \times 100,$$

where f is the frequency in the class interval and n is the total number of scores. *A distribution giving **Prop f** or **%f** is called a **relative frequency distribution.*** Usually the frequency associated with each class interval is given along with Prop f or $\%f$. Both kinds of relative frequencies are shown in Table 2.2-4.

The transformations Prop f and $\%f$ convert the total number of scores, n, into standard numbers—either 1.0 or 100. We see that the sum of Prop f is 1.0 and the sum of $\%f$ is 100. Relative frequencies indicate whether a frequency is "relatively large" rather than whether it is "absolutely large." For example, the class interval 48–50 in Table 2.2-4 contains only seven scores, but this is a relatively large proportion (almost one-fourth) of the total number of scores. Also, relative frequencies can be used to compare two frequency distributions with different n's. Consider the history achievement scores shown in Table 2.2-5 for high school students taught by two methods. Because of the great difference in n's, a comparison of $\%f$'s is more meaningful than a comparison of f's.

Table 2.2-4. Relative Frequency Distributions for Leadership Aptitude Scores

Class Interval	f	Prop f	%f
66–68	1	0.03	3
63–65	0	0	0
60–62	1	0.03	3
57–59	2	0.07	7
54–56	4	0.13	13
51–53	6	0.20	20
48–50	7	0.23	23
45–47	4	0.13	13
42–44	2	0.07	7
39–41	1	0.03	3
36–38	0	0	0
33–35	1	0.03	3
30–32	1	0.03	3
	n = 30	Sum = 0.98[a]	Sum = 98[a]

[a] The sums do not equal 1.00 and 100 due to errors introduced by rounding numbers.

Table 2.2-5. History Achievement Scores for Classes Taught by Different Methods

Achievement Scores	Method A		Method B	
	f	%f	f	%f
150–154	1	1	1	3
145–149	0	0	2	6
140–144	2	3	2	6
135–139	4	5	4	12
130–134	6	8	6	19
125–129	8	11	8	25
120–124	9	12	5	16
115–119	10	14	2	6
110–114	8	11	1	3
105–109	8	11	0	0
100–104	6	8	1	3
95–99	5	7	0	0
90–94	3	4	0	0
85–89	2	3	0	0
80–84	1	1	0	0
	n = 73	Sum = 99[a]	n = 32	Sum = 99[a]

[a] Sum is not equal to 100 due to errors introduced by rounding numbers.

Cumulative Frequency Distribution

*A **cumulative frequency distribution** shows the number, proportion, or percentage of scores that occur below the real upper limit of each class interval.* Such a distribution is helpful in answering certain kinds of questions. Susan's score is 62—how many students did better and how many did worse? What scores divide the top 10% or the bottom 25% of students from the rest of the class?

To construct a cumulative frequency distribution, we begin with a frequency distribution like the one in the first two columns of Table 2.2-6. A given cumulative frequency is obtained by adding the frequency in column 2 for the class interval to the cumulative frequency recorded in column 3 for the class interval below it. For example, in the class interval 30–32, $f = 1$ and there are no scores below, so the Cum f for that class interval is $1 + 0 = 1$. For the class interval 33–35, $f = 1$, which added to the Cum f below yields a Cum f of $1 + 1 = 2$. The cumulative frequency recorded for the top class interval should equal n.

Cumulative frequencies can be transformed into Cum prop f and Cum %f by the formulas Cum prop $f = $ Cum f/n and Cum %$f = $ Cum $f/n \times 100$. These relative frequencies are shown in columns 4 and 5 of Table 2.2-6.

Table 2.2-6. Cumulative Frequency Distributions for Leadership Aptitude Scores

(1) Class Interval	(2) f	(3) Cum f	(4) Cum prop f	(5) Cum %f
66–68	1	30	1.00	100
63–65	0	29	0.97	97
60–62	1	29	0.97	97
57–59	2	28	0.93	93
54–56	4	26	0.87	87
51–53	6	22	0.73	73
48–50	7	16	0.53	53
45–47	4	9	0.30	30
42–44	2	5	0.17	17
39–41	1	3	0.10	10
36–38	0	2	0.07	7
33–35	1	2	0.07	7
30–32	1	1	0.03	3
	$n = 30$			

Frequency Distributions for Qualitative Variables

The construction of frequency distributions for qualitative variables is relatively simple since no decisions about size and number of class intervals have to be made—the equivalence classes of the variable become the class intervals. Consider the unordered qualitative variable of political party affiliation—Democrat, Independent, Republican, and Unspecified or other. If we obtained a random sample of college students at Ohio

State University and determined their political affiliation, we could construct a frequency distribution like the one in columns 1 and 2 of Table 2.2-7. The equivalence classes are ordered alphabetically for lack of a more logical sequence. For ordered qualitative variables, class intervals should preserve the order inherent in the original equivalence classes.

The frequencies in column 2 of Table 2.2-7 are converted to Prop f in column 3 and $\%f$ in column 4. Cumulative frequencies are not shown; they are not meaningful since the order of the class intervals was arbitrarily determined.

Table 2.2-7. Political Affiliation of Students at Ohio State University

(1) Political Affiliation	(2) f	(3) Prop f	(4) $\%f$
Democrat	92	0.42	42
Independent	33	0.15	15
Republican	85	0.38	38
Unspecified or other	11	0.05	5
	$n = 221$	Sum = 1.00	Sum = 100

Review Exercises for Section 2.2

*1. A marriage counselor asked his clients to keep a record of the number of arguments they had during the week. The following data for 23 couples were obtained; construct an ungrouped frequency distribution for these data.

```
2  5   4  9   6
4  3   3  5  10
5  0  13  4   2
1  7   6  3
4  5   4  4
```

2. Construct an ungrouped frequency distribution for study-abroad candidates' ages at last birthday. The data are as follows.

```
18  20  19  20
20  19  19  19
23  18  20  21
17  20  18  20
```

*3. Assembly-line workers were asked to complete a job-satisfaction questionnaire. Construct an ungrouped frequency distribution for the following scores.

```
 7   8   4  25   9   8   4  15  11   9
 6   9   7   7  10  17   5  10   5   8
 3   7  11   8  13  22   7   8   7   6
10   6   7   9   4   8   6   6   8  11
 7   5   6   6   7   7  12   5   5   6
15  21   5  11   6   9   5  12  10   8
```

*4. List the guidelines for constructing an ungrouped frequency distribution.
**5. For the following nominal class intervals, give the real limits and the class interval size.
   *a. 50–54   *b. 74   c. 16
   *d. 18.0–19.9   e. 60–69   f. 18.00–19.49
   g. 12.0–14.9   h. 0–0.4   i. 1.50–1.74
**6. For each of the following give (a) the number of class intervals, (b) the size of the class interval, and (c) the nominal limits of the lowest class interval.

	Largest Score	Smallest Score	Number of Scores
*a.	68	22	53
*b.	260	106	21
*c.	254	92	91
d.	37	8	106
e.	62	23	273
f.	164	126	29
g.	52	0	22

*7. A test of mechanical aptitude was given to seniors at Middlecenter High School. Construct a grouped frequency distribution for the following data.

```
80  73  51  81  46  85  84
75  44  84  77  95  48  88
50  35  52  93  43  59  63
47  66  55  58  62  51  75
86  82  89  51  77  73  59
```

8. First-line supervisors were asked to complete the job-satisfaction questionnaire mentioned in Exercise 3. Construct a grouped frequency distribution for the following data.

```
25  23  18  24  14
21  17  12  19
15   6  22  16
20  20  21  18
```

*9. In a traffic safety research project the reaction time in milliseconds of 27 subjects to the onset of a light was measured. For the following data, (a) construct two grouped frequency distributions having different i's, and (b) discuss the relative merits of the two grouping schemes.

```
186  187  211  185  196  193
184  185  191  188  192  190
188  190  202  199  189
193  186  180  205  187
189  195  184  198  202
```

10. What are the advantages and disadvantages of grouped and ungrouped frequency distributions?
*11. For the data in Exercise 7, construct a relative frequency distribution using Prop f.
12. For the data in Exercise 8, construct a relative frequency distribution using $\% f$.
13. Construct a relative frequency distribution for comparing the job satisfaction of assembly-line workers in Exercise 3 with that of first-line supervisors in Exercise 8.
14. Under what conditions is a relative frequency distribution more informative than an ordinary frequency distribution?
*15. Thirty-two subjects participated in a paired-associates learning experiment in which they were shown 12 nouns written in hiragana (a Japanese writing system) and asked to learn the corresponding English words. The number of trials required to learn to a criterion of two consecutive

correct anticipations is shown below. Construct a cumulative frequency distribution for the data.

```
10   9  11  12   6  14  10  12
11  10  12  10   9  11  16   8
 9   7   8  11  10   8  12  12
13  10  10   9  11  13   7  11
```

*16. For the data in Exercise 1, construct a cumulative proportionate frequency distribution.

17. For the data in Exercise 7, construct a cumulative frequency distribution.

18. For the data in Exercise 8, construct a cumulative percentage frequency distribution.

*19. A random sample of 29 students from each of the classifications: freshman, sophomore, junior, senior, and graduate student were asked whether they believed in ESP. The classifications of students who believed in ESP are listed below. Construct a frequency distribution for these data.

Junior	Senior	Junior	Sophomore	Junior
Freshman	Junior	Freshman	Junior	Sophomore
Sophomore	Senior	Senior	Senior	Junior
Graduate	Sophomore	Junior	Freshman	Senior
Freshman	Junior	Junior	Senior	Sophomore
Junior	Senior	Sophomore	Senior	

20. Students enrolling in Introductory Sociology were randomly assigned to one of three classes: traditional lecture (TL), guided reading (GR), or audiotutorial (AT). Following are the class assignments of the top 30 students on the final examination; construct a frequency distribution for these data.

```
AT  GR  AT  TL  GR  AT
AT  TL  GR  AT  AT  GR
TL  TL  TL  AT  AT  AT
GR  AT  AT  AT  TL  TL
AT  AT  TL  GR  AT  AT
```

21. Twenty-five physicians were asked what they felt was the main health threat to male executives. The most common responses were occupational stress (OS), obesity (OB), smoking (S), lack of exercise (LE), and other (O). Construct a frequency distribution for these data.

```
OB  OS  S   OB  LE
S   OB  OB  OS  O
LE  S   OS  S   OB
O   LE  O   LE  LE
OB  LE  O   O   OB
```

22. Toss a die 50 times and construct a frequency distribution showing the number of times each die face occurred.

23. Contrast the procedures for constructing frequency distributions for qualitative variables with those for quantitative variables.

24. Under what condition would it be meaningless to construct a cumulative frequency distribution for a qualitative variable?

25. Terms to remember:
   a. Frequency distribution
   b. Nominal limits
   c. Real limits
   d. Relative frequency distribution

e. Proportionate frequency
f. Percentage frequency
g. Cumulative frequency distribution

2.3 Introduction to Graphs

Frequency distributions present the main features of data succinctly, but they are still abstract numerical representations and require effort to interpret. Graphs can impart the same information and speak to us more directly. Their ease of interpretation makes them particularly useful when we want to present data to the general public.

There are many ways to graph data. In fact, whole books have been devoted to the subject.[4] Our presentation is limited to the five most common graphs: bar graphs, pie charts, histograms, frequency polygons, and cumulative polygons. Qualitative variables are usually represented by bar graphs and pie charts, and quantitative variables by histograms, frequency polygons, and cumulative polygons.

2.4 Graphs for Qualitative Variables

Bar Graphs

Once a frequency distribution has been made, most of the work of constructing a bar graph has been done. The only step remaining is to cast the data into two dimensions. This is illustrated in Figure 2.4-1 for the data in Table 2.2-7. Class intervals are represented along the horizontal axis (*abscissa*, or X axis) and frequency along the vertical axis (*ordinate*, or Y axis). A vertical bar is erected over each class interval such that its height corresponds to the number of scores in the interval. The bars can be any width, but they should not touch, since their noncontiguity points up the discrete, qualitative character of the class intervals. By convention, the height of the tallest bar should be 66–75% of the length of the horizontal axis. This results in a rectangular figure with proportions that are the most esthetically pleasing (according to the ancient Greeks). The X and Y axes of the graph should be labeled and a figure caption provided to help the reader interpret the graph.

The Y axis can be used to represent proportionate frequency or percentage frequency, depending on the questions of interest to the experimenter. We have seen that these transformations are useful in determining whether a frequency is relatively large and in comparing graphs with different total numbers of scores.

Pie Charts

Perhaps the most easily interpreted graph is a **pie chart**, which is merely a circle divided into sectors representing proportionate frequency or percentage frequency of the class intervals. A pie chart is illustrated in Figure 2.4-2 for the data in Table 2.2-7. To

---

[4] One example is Arken and Colton (1938).

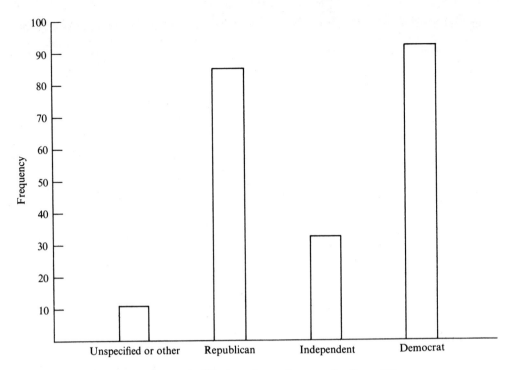

Figure 2.4-1. Political affiliation of a random sample of $n = 221$ students at Ohio State University.

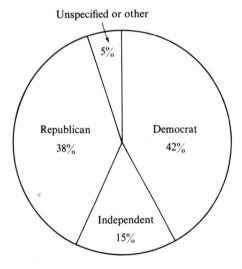

Figure 2.4-2. Political affiliation in percentage frequency of a random sample of $n = 221$ students at Ohio State University.

construct a pie chart it is convenient to think of it as a circle having 60 min like a clock. The sections corresponding to proportionate frequency or percentage frequency are marked off in minutes according to the formulas Prop $f \times 60$ and $\% f/100 \times 60$. For Figure 2.4-2, the minutes corresponding to the four percentage frequencies are

$$42\%/100 \times 60 = 25.2$$
$$15\%/100 \times 60 = 9.0$$
$$38\%/100 \times 60 = 22.8$$
$$5\%/100 \times 60 = 3.0.$$

Thus, 42% corresponds to 25.2 min after 12 o'clock; the next 15% corresponds to $25.2 + 9.0 = 34.2$ min after 12 o'clock; the next 38% to $25.2 + 9.0 + 22.8 = 57$ min; and so on. The last steps in constructing a pie chart are to label the sections of the pie and provide an appropriate figure caption.

Review Exercises for Section 2.4

*26. College students were asked to give their favorite leisure-time activity. The five most commonly mentioned activities were rapping with friends (RF), reading (R), watching television (TV), participating in a sport (PS), and drinking (D). Construct a bar graph for the following data.

RF	PS	D	RF	R	TV	RF	D	PS
RF	RF	R	TV	RF	D	TV	RF	TV
D	TV	RF	RF	D	RF	R	R	RF
R	R	TV	D	TV	D	D	RF	TV
TV	RF	PS	TV	RF	TV	TV	D	
D	D	TV	RF	PS	RF	RF	D	

27. Information from a biographical inventory was used to compute a socioeconomic index for students in the university marching band. Scores above 72 were classified as very high (VH); scores from 61 to 72 as high (H); from 43 to 60 as middle (M); and below 43 as low (L). Construct a bar graph for the following data.

H	H	H	H	M	VH	VH	H	M
M	L	H	M	VH	H	H	H	VH
H	M	M	H	H	VH	H	M	
VH	H	H	M	M	VH	M	L	
H	VH	VH	H	H	M	VH	M	
M	M	VH	L	M	H	H	VH	

28. Construct a bar graph for the data in Exercise 20 (page 32).
29. Construct a bar graph for the data in Exercise 21 (page 32); plot percentage frequency on the Y axis.
30. Describe the procedure for constructing a bar graph from a frequency distribution.
*31. Construct a pie chart for the data in Exercise 26.
32. Construct a pie chart for the data in Exercise 27.
33. Construct a pie chart for the data in Exercise 20 (page 32).
34. Describe the procedure for constructing a pie chart from a frequency distribution.

35. Terms to remember:
   a. Bar graph
   b. Abscissa
   c. X axis
   d. Ordinate
   e. Y axis
   f. Pie chart

2.5 Graphs for Quantitative Variables

Histograms

A ***histogram*** is similar in appearance and construction to a bar graph, but it is used for quantitative variables rather than qualitative variables. It is constructed by erecting vertical bars over the ***real limits*** of each class interval, with the height of each bar corresponding to the number of scores in the interval. The bars are contiguous rather than separated; this emphasizes the continuity of the class intervals. Otherwise histograms and bar graphs are constructed in the same manner: (1) the class intervals are represented along the horizontal axis and frequency is represented along the vertical axis, (2) the height of the tallest bar is 66–75% of the length of the horizontal axis, and (3) the two axes are appropriately labeled and a figure caption is given to help the reader interpret the graph.

A histogram for the data in Table 2.2-3 is shown in Figure 2.5-1. Note that the edges of the bars coincide with the real limits of the class intervals rather than with the nominal limits, for example, 29.5–32.5 and not 30–32. Either frequency or relative frequency can be represented along the vertical axis. (Transformation of frequencies to relative frequencies is discussed in Section 2.2.)

Frequency Polygons

To construct a frequency polygon from a frequency distribution we begin as though we were making a histogram. The horizontal axis is marked off into class intervals and the vertical axis into numbers representing frequencies. However, frequency is represented not by vertical bars, but by dots placed at the proper height over the midpoints of the class intervals. Finally, adjacent dots are joined by straight lines. At the extremes of the graph, two additional class intervals containing no scores are identified and lines are dropped to their midpoints so as to anchor the graph to the horizontal axis. The midpoint of a class interval is given by

$$\text{Midpoint} = \frac{(\text{Upper limit of class interval} + \text{Lower limit of class interval})}{2}.$$

For the data in Table 2.2-3, the midpoint of the class interval 30–32 is $(32 + 30)/2 = 31$. A frequency polygon for these data is shown in Figure 2.5-2. Frequency polygons and histograms impart the same information; the choice between them is largely a matter

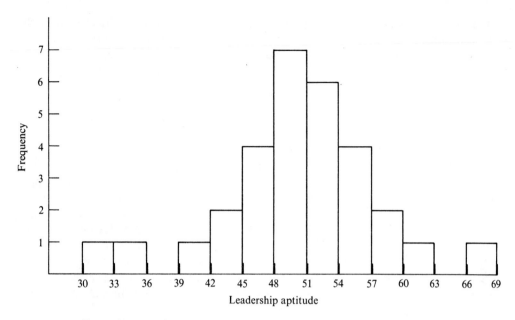

Figure 2.5-1. Histogram for leadership aptitude scores for $n = 30$ football coaches.

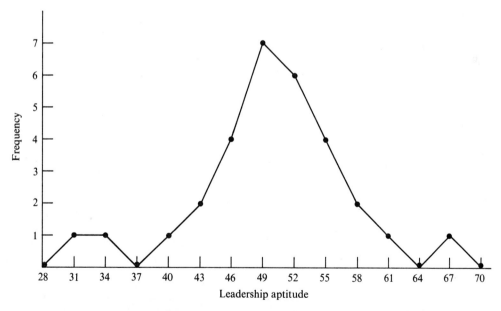

Figure 2.5-2. Frequency polygon for leadership aptitude scores for $n = 30$ football coaches.

2.5 Graphs for Quantitative Variables

of personal preference. The histogram is probably a little easier for the general public to interpret, but the stepwise bars tend to obscure the shape of the distribution. The frequency polygon is preferred when two or more sets of data are represented in the same graph, since superimposed histograms, even markedly different ones, are difficult to interpret.

Cumulative Polygons

We saw in Section 2.2 that a cumulative frequency distribution can be used to show the number, proportion, or percentage of scores that lie below the real upper limit of each class interval. This information can be represented graphically by a cumulative polygon. Instead of placing dots over the midpoints of class intervals, they are placed over the real upper limits. The vertical axis can represent Cum f, Cum prop f, or Cum %f. A cumulative percentage polygon for the data in Table 2.2-6 is shown in Figure 2.5-3. As is usually the case in the behavioral sciences and education, the cumulative polygon has the characteristic S shape. The S shape occurs whenever there are more scores in the middle of the corresponding frequency distribution than at the extremes. Graphs that are S shaped are called *ogives*.

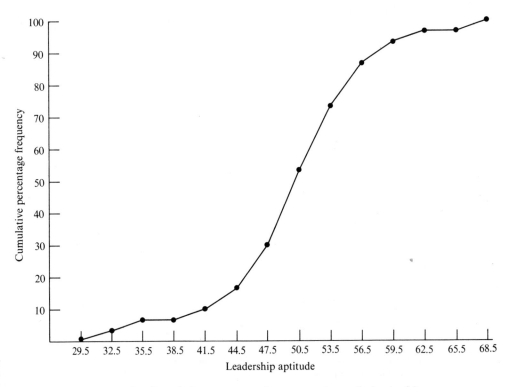

Figure 2.5-3. Cumulative percentage frequency polygon for leadership aptitude scores for $n = 30$ football coaches.

Review Exercises for Section 2.5

*36. The following data represent the number of cigarettes smoked per day by mothers whose first babies were stillborn. Construct a histogram for these data.

```
27  25  31  22   3  16  15
21  32  29  30  12  14  26
 9  27  25  27  30  28  31
30  18   0  23  20  21
28  16  10  19  13  19
```

37. Rats were shown three illuminated symbols; their task was to press the lever below the symbol that differed from the other two. The dependent measure was the number of trials required to learn to the criterion of eight consecutive correct responses. Construct a histogram for these data.

```
52  34  57  47  54  56  46
60  63  42  20  50  81  41
43  51  36  73  56  77  59
50  42  58  65  42  58  63
66  55  53  63  53  54  61
```

38. Construct a histogram for the data in Exercise 8 (page 31). Plot percentage frequency on the ordinate.
39. Construct a histogram for the data in Exercise 9 (page 31). Plot proportionate frequency on the ordinate.
40. How does the construction of histograms and bar graphs differ?
**41. Determine the midpoints of the following class intervals.
   *a. 20–24   b. 8–11
   *c. 132–133   d. 1.50–1.74
   *e. 15–29   f. 100–104
    g. 0–2   h. 60–69
*42. Construct a frequency polygon for the data in Exercise 36.
43. Construct a frequency polygon for the data in Exercise 37.
44. Construct a frequency polygon for the data in Exercise 8 (page 31). Plot percentage frequency on the ordinate.
45. What are the relative merits of histograms and frequency polygons?
*46. (a) Construct a cumulative polygon for the data in Exercise 36; plot Cum %f on the ordinate.
   (b) Estimate the score above which 50% of the cases fall.
47. (a) Construct a cumulative polygon for the data in Exercise 37; plot Cum prop f on the ordinate.
   (b) Estimate the score below which 50% of the cases fall and the score below which 20% of the cases fall.
48. Construct a cumulative polygon for the data in Exercise 9 (page 31).
*49. How can we tell from a frequency distribution whether a cumulative polygon for the data would have an S shape?
50. Terms to remember:
   a. Histogram
   b. Frequency polygon
   c. Class interval midpoint
   d. Cumulative polygon
   e. Ogive

2.6 Shapes of Distributions

Graphs come in many different shapes. Some shapes occur with enough regularity that they have been given special names. These are shown in Figure 2.6-1.

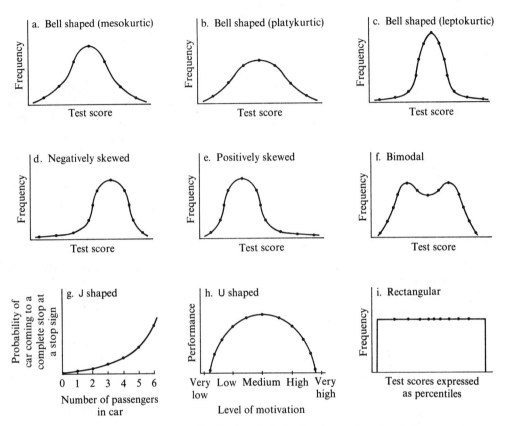

Figure 2.6-1. Common distributions in behavioral and educational research.

Bell-Shaped Distributions

The frequency polygon in Figure 2.6-1(a) approximates the shape of the normal distribution. The normal curve is symmetrical—the right half is the mirror image of the left half—and it has a particular degree of peakedness. The height of the curve above the X axis is described mathematically by an equation.[5] *The property of being peaked,*

---

[5] This equation associates a Y value with an X value according to the function rule

$$Y = \frac{1}{\sigma\sqrt{2\pi}} e^{-(X-\mu)^2/2\sigma^2}.$$

The normal distribution is further discussed in Chapter 10.

flat, or in between is referred to as **kurtosis**. The normal curve is *mesokurtic*; "meso-" means intermediate. Distributions that are flatter than the normal curve are called *platykurtic*; "platy-" means flat or broad. Those that are more peaked are designated "lepto-," meaning slender or narrow. Examples of these curves are shown in Figures 2.6-1(b) and (c).

Skewed Distributions

Distributions are either symmetrical or asymmetrical. *If the longer tail of an asymmetrical distribution extends toward the X and Y origin, as in Figure 2.6-1(d), the distribution is **negatively skewed**. If the longer tail extends away from the intercept, as in Figure 2.6-1(e), the distribution is **positively skewed**.* A negatively skewed distribution results, for example, if a very easy test is given to subjects. Since most of the subjects score high and only a few score low, the longer tail trails off toward the X and Y origin. A positively skewed distribution results if the test is very hard.

Bimodal Distributions

*A distribution is **bimodal** if it has two humps each having the same maximum frequency.* Bimodal distributions often result when two distinct populations are represented on a single graph. For example, a graph like Figure 2.6-1(f) would result if we plotted the masculinity scores of 50 men and 50 women.

*A graph with three or more humps each having the same maximum frequency is **multimodal**.* Technically, a distribution is bimodal or multimodal only if its humps have the same frequency. Nevertheless, distributions with pronounced but unequal humps are commonly described as bimodal.

J, U, and Rectangular Distributions

J and U distributions are so named because their shapes resemble those letters. A J-shaped curve like the one in Figure 2.6-1(g) is obtained, for example, if the probability of coming to a complete stop at a stop sign is plotted as a function of the number of passengers in the car. A reversed J curve is obtained if the number of people arriving for church is plotted as a function of the number of minutes they are late. Similar results are obtained in most studies of conforming social behavior; most people conform to social conventions and laws, so fewer and fewer people exhibit larger degrees of nonconformity.

A U curve like the one in Figure 2.6-1(h) is obtained, for example, if performance on a difficult task is plotted as a function of the level of motivation of the subjects.

*A **rectangular** or **uniform** distribution is one in which each class interval has the same frequency.*

Review Exercises for Section 2.6

**51. Indicate whether the following statements are true or false.
  *a. A distribution that is symmetrical and mesokurtic is called a normal distribution.
  *b. If the upper half of a distribution is not the mirror image of the lower half, the distribution is asymmetrical.
  c. A distribution that is flatter than the normal distribution is called mesokurtic.
  d. Lepto in leptokurtic means slender or narrow.
  e. The tail of a negatively skewed distribution extends away from the X and Y intercept.
  f. A distribution with three maximum humps each having the same frequency is bimodal.

**52. Draw the shape of a frequency polygon that would occur in each of the following experiments. Identify each distribution.
  *a. Beauty contestants take a masculinity test.
  *b. An intelligence test is given to a large sample of sixth-grade children.
  c. Students at Juilliard School of Music take a test of musical aptitude.
  d. Students are surprised with a pop quiz immediately after the Easter vacation.
  e. Subjects attempt to solve 20 complex puzzles under three levels of motivation: low, medium, and high.
  f. Crime statistics are collected for five cities; it turns out that the cities have the same crime rate.
  g. Engineering majors' and business majors' scores on a test of mechanical aptitude are plotted.
  h. Strength of grip is measured for males; the sample contains all ages but a preponderance of young boys, men in their early 20s, and men over 65.
  i. Arrival time is recorded for people who are late for a concert.
  j. The number of persons contracting polio in the United States from 1940 to 1970 is determined from hospital records.

53. Terms to remember:
  a. Normal distribution
  b. Symmetrical distribution
  c. Kurtosis
  d. Mesokurtic
  e. Leptokurtic
  f. Platykurtic
  g. Skewness (negative and positive)
  h. Bimodal
  i. Multimodal
  j. J distribution
  k. U distribution
  l. Rectangular distribution

2.7 Misleading Graphs

Graphs should be constructed so they accurately portray the essential characteristics of data. Not all graphs do this—some even defy correct interpretation. Graphs of the same data can convey entirely different impressions as shown in Figures 2.7-1(a) and (b), which report crime statistics for three similar neighborhoods. In neighborhood A, cruising patrol cars were eliminated during a 3 month trial period; B had 5 cruising cars during the period; while C was flooded with 15 cars. Your conclusions about the

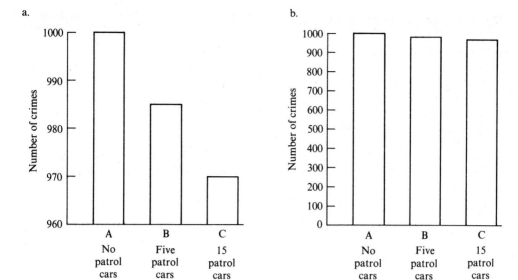

Figure 2.7-1. Number of reported crimes in three similar neighborhoods during a 3 month test period. Note how graph "a" falsely gives the impression of a great difference in crime rate across the three conditions.

effects of patrol cars would probably depend on which graph you saw. Figure 2.7-1(a) gives the impression that the presence or absence of patrol cars is associated with a dramatic difference in crime rate. Note, however, that the largest difference—1000 versus 970—is only 3%. Such a small difference could just as well be attributed to chance factors or to differences in crime reporting procedures. The graph is misleading because it violates the height–width rule and because the Y axis begins with a frequency of 960 crimes instead of 0 crimes.[6] This is inimical to the aim of statistics, which is to help the user make sense out of data.

Review Exercises for Section 2.7

*54. Prepare two bar graphs for the following data—one deliberately designed to suggest that government spending has been stable, the other to suggest a dramatic increase in government spending.

Month	Spending	Month	Spending
June	$29,400,000	October	$29,500,000
July	29,200,000	November	29,600,000
August	29,300,000	December	29,800,000
September	29,600,000	January	30,200,000

[6] Huff (1954) illustrates many more misleading techniques.

55. The following data are sales figures for vacuum cleaner salespeople. Prepare a graph which suggests that (a) all the salespeople are producing at a uniformly high level, (b) Chapman should be fired, and (c) they should all be fired.

Chapman	$66,000	Mayes	$68,200
Hays	$67,300	Reynolds	$71,000
Toland	$69,900	McCall	$71,100

56. a. Collect examples from newspapers and news magazines of graphs that appear to be misleading.
    b. Why is each one misleading?
57. In the news media, how do graphs reporting financial statistics differ from those used in advertising? Note whether the X and Y axes of the graphs are clearly labeled and whether they have figure captions.

2.8 Summary

This chapter presented two descriptive devices which make data easily comprehensible: frequency distributions and graphs. As a first and sometimes final step in summarizing data an experimenter may report the various equivalence classes and the number of observations that fall into each. This results in a table called a "frequency distribution." If each equivalence class is a single score value, the distribution is ungrouped; if the classes contain two or more score values, the distribution is grouped. Grouping simplifies the interpretation of data by collapsing them into 10–20 class intervals.

A graph is a pictorial representation of a frequency distribution and hence is much easier to interpret. The most common graphs for qualitative variables are bar diagrams and pie charts. Histograms, frequency polygons, and cumulative polygons are commonly used to represent quantitative variables.

A graph should present data accurately and unambiguously in such a way that its main characteristics can be seen at a glance. To achieve this, certain conventions are followed: (1) plotting frequency on the Y axis and class intervals on the X axis; (2) placing the zero point (or origin) at the X and Y intercept; (3) making the height of the graph 66–75% of its width; (4) labeling the X and Y axes; (5) and providing a figure caption.

3

Measures of Central Tendency

3.1 Introduction
3.2 The Mode
    Review Exercises for Section 3.2
3.3 The Mean
    Computational Formula Using Summation Notation
    Computing the Mean from a Grouped Frequency Distribution
    Review Exercises for Section 3.3
3.4 The Median
    Computation of the Median in a Frequency Distribution
    Review Exercises for Section 3.4
3.5 Relative Merits of the Mean, Median, and Mode
    Mean
    Median
    Mode
    Summary of the Properties of the Mean, Median, and Mode
    Review Exercises for Section 3.5
3.6 Location of $\bar{X}$, Mdn, and Mo in a Distribution
    Review Exercises for Section 3.6
3.7 Mean of Combined Subgroups
    Review Exercises for Section 3.7
3.8 Summary
3.9 Technical Notes
    Summation Rules
    Proof that the Mean Is a Balance Point
    Effect on the Mean of Adding a Constant to Each Score
    Effect on the Mean of Multiplying Each Score by a Constant
    Proof that $\sum (X_i - \bar{X})^2 = $ Minimum

3.1 Introduction

A frequency distribution or graph summarizes data, but sometimes it is desirable to summarize further by describing certain properties of the data numerically. The most interesting property is usually *central tendency,* the score value around which a distribution tends to center. This value is popularly called the *average,* and it connotes what is typical, usual, representative, normal, or expected. Because of these different connotations, statisticians prefer more precise terms. The *mode, mean,* and *median* are three conceptions of central tendency described in this chapter.

Close behind central tendency in importance is *dispersion*—the extent to which scores differ from one another—that is, their scatter or heterogeneity. Several ways of describing dispersion are discussed in Chapter 4. Chapter 4 also covers two other properties of data—*skewness* and *kurtosis.* Measures of skewness tell us whether a distribution is symmetrical or asymmetrical; measures of kurtosis tell us whether it is peaked or flat. These four properties of data—central tendency, dispersion, skewness, and kurtosis—can be measured to yield a relatively complete summary of the information presented by frequency distributions and graphs. In many cases, a knowledge of only two of these—central tendency and dispersion—is sufficient for our purposes.

3.2 The Mode

The simplest of the three conceptions of central tendency is the mode, denoted by *Mo.* The **mode** *is the score or qualitative category that occurs with greatest frequency.* Consider the following scores, which represent the number of times in September that 11 students called their parents long-distance: 0, 3, 1, 1, 2, 0, 0, 9, 1, 2, 1. We note that 0 occurs three times; 1 four times; 2 twice; and 9 once. The mode is 1, since it occurs with the greatest frequency. If data are tabulated in an ungrouped frequency distribution, we can determine the mode at a glance. This can be seen for the distribution of family size of college professors shown in Table 3.2-1. The largest frequency, 10, is associated with a family size of four, hence the mode is 4. This tells us that the most

Table 3.2-1. Frequency Distribution of Family Size of College Professors

X	f
11	1
10	0
9	0
8	1
7	1
6	2
5	4
4	10
3	8
2	8
1	5
	$n = 40$

typical family size for this sample is four, an easy-to-understand concept. As these examples show, the mode is determined by inspection rather than by computation. The mode can be used to describe the central tendency of both qualitative and quantitative variables, but it is most often used for qualitative variables. We will see why this is true later, when we compare the three measures of central tendency.

The mode should be computed from an ungrouped frequency distribution if possible. If this can't be done, the midpoint of the class interval with the greatest frequency is designated as the mode. This is, of course, imprecise since the midpoint would be different for different grouping schemes.

As a measure of central tendency, the mode has one particularly serious limitation. As we saw in Section 2.6, a distribution can have two nonadjacent scores or two non-adjacent class intervals with the same maximum frequency. Such distributions are called bimodal and cannot be described by a single mode. It is customary in such cases to cite two modes, the scores associated with the two maximum frequencies, but then the concept of the mode as the most typical score no longer applies.

Review Exercises for Section 3.2

*1. The behavior of members of the university wine-tasting club was rated following their biweekly learn-by-doing meeting. The following scale was used: N = no change in behavior, S = slight change in verbal and/or emotional expressions, M = marked change in verbal and/or emotional expressions, C = clumsiness in locomotion, and G = gross intoxication. (a) Determine the mode for the following data: N, S, S, G, M, N, S, M, M, C, G, N, S, M, C, S, S, M, S, S. (b) What type of variable do the data represent?
2. In a paired-associates learning experiment, data representing the number of trials to reach the criterion of three consecutive errorless trials were 10, 6, 11, 10, 9, 8, 10, 11, 14, 12, 10, 9, 11, 10, 12, 9, 8, 9. (a) Determine the mode. (b) What type of variable do the data represent?
3. The electoral systems of 11 emerging nations were classified as N = noncompetitive, P = partially competitive, and C = competitive. (a) Determine the mode for the following data: N, P, N, C, N, P, P, N, N, C, N. (b) What type of variable do the data represent?
*4. The ruling structures of 11 emerging nations were classified as 1 = elitist, 2 = moderately elitist, and 3 = nonelitist. (a) Determine the mode for the following data: 1, 3, 1, 1, 2, 3, 1, 3, 3, 1, 3. (b) What type of variable do the data represent?
5. Why should the mode be computed from ungrouped rather than grouped data whenever possible?

3.3 The Mean

The most widely used and familiar measure of central tendency is the **arithmetic mean**—*the sum of scores divided by the number of scores.* The mean[1] is commonly known as the average. The usual symbol for a sample mean, $\bar{X}$, is read "X bar."[2] The letter X identifies the variable that has been measured, and the bar above it denotes the mean. Other letters toward the end of the English alphabet, for example, Y and Z, are also used as symbols for variables, and the corresponding means are denoted by $\bar{Y}$ and $\bar{Z}$. In later chapters on inferential statistics, it is important to distinguish between the

---
[1] There are several kinds of means, but only the arithmetic mean is discussed in this book.
[2] Some books in the behavioral sciences and education denote the mean by M; $\bar{X}$ or $\bar{x}$ is preferred by statisticians (Halperin, Hartley, & Hoel, 1965).

mean computed for a sample and the population mean. It is customary to use lowercase Greek letters to represent population characteristics. Following this convention, the mean of a population is denoted by μ, the Greek letter mu, and is read "mew." Different population means can be identified by number or letter subscripts, for example, μ_1 and μ_2 or μ_X and μ_Y. The distinction between samples and populations appears in another way—a descriptive measure for a sample is called a *statistic*; a descriptive measure for a population is called a *parameter*.

Computational Formula Using Summation Notation

The mean of a sample is obtained by dividing the sum of the scores by the number of scores. At this point, we will describe a useful notation for the sum of scores. Suppose we are interested in frequency of movie attendance of college students. We can denote this variable by the capital letter X and individual values of the variable by X and a subscript: $X_1, X_2, \ldots, X_i, \ldots, X_n$. According to this notation, X_1 is the frequency of movie attendance for person 1, X_2 is the frequency for person 2, and X_n denotes the frequency for the nth, and last, person in the sample. We will let i be a general subscript that designates an unspecified one of the $i = 1, \ldots, n$ persons. The i in X_i can be replaced by any integer between 1 and n inclusive.[3] Suppose the following values of X_i were obtained: $X_1 = 3, X_2 = 1, X_3 = 4, X_4 = 2$. The mean of these $n = 4$ scores is given by

$$\bar{X} = \frac{X_1 + X_2 + X_3 + X_4}{n} = \frac{3 + 1 + 4 + 2}{4} = \frac{10}{4} = 2.5.$$

This formula is tedious to write, so it is customary to abbreviate it by using the *summation* symbol $\sum$, the Greek capital sigma. The symbol $\sum$, like $+$, indicates addition. However, $+$ indicates the addition of only two numbers, while $\sum_{i=1}^{n}$ means to perform addition until all $i = 1, \ldots, n$ numbers have been added. The equation

$$\sum_{i=1}^{n} X_i = X_1 + X_2 + \cdots + X_n$$

says to let the first value of X_i be X_1; add to this the second value X_2; and so forth until the X_nth value has been added. In the notation $\sum_{i=1}^{n}$, i is called the *index of summation*, $i = 1$ is the initial value of i, and n is its terminal value. Using summation notation, the formula for the mean movie attendance of four students is written $\bar{X} = \sum_{i=1}^{4} X_i/4$, which is equivalent to $\bar{X} = (X_1 + X_2 + X_3 + X_4)/4$. The general formula for $\bar{X}$ is written $\bar{X} = \sum_{i=1}^{n} X_i/n$. When the initial and terminal values for the summation are clearly understood, the formula may be simplified to $\bar{X} = \sum X_i/n$ or $\sum X/n$.

Computing the Mean from a Grouped Frequency Distribution

The formula $\bar{X} = \sum_{i=1}^{n} X_i/n$ is appropriate for data in their original unordered state. *If the data have been ordered by a frequency distribution, the mean can be computed from* $\bar{X} = \sum_{j=1}^{k} f_j X_j/n$, *where X_j is the midpoint of the jth class interval, f_j is the fre-*

---

[3] The letter i is also used to denote the size of a class interval; this use is discussed in Section 2.2. (Since there are only 26 letters, it is not surprising that they have multiple meanings.)

quency in the jth class interval, $\sum_{j=1}^{k}$ is the sum over the $j = 1, \ldots, k$ class intervals, and n is the number of scores. Because virtually all researchers have access to electronic calculators and computers, means are rarely computed from grouped data, but most teachers feel compelled to teach it. I have on occasion used it—when I forgot my calculator and could tolerate a small error in my answer. Consider the computation illustrated in Table 3.3-1. The midpoint of a class interval, $X_j =$ (Upper limit of X + Lower limit of X)/2, is used to represent all the scores in the class interval. This inevitably introduces some inaccuracy when $i \geq 2$, and as a result the mean for a grouped frequency distribution rarely equals that for unordered data or for an ungrouped frequency distribution. The more symmetrical the distribution and the larger the number of class intervals in the grouped frequency distribution, the smaller is the discrepancy, or *grouping error*.[4]

Table 3.3-1. IQ Data for a Sample of Sixth Graders

(i) Data:

X	X_j	f_j	$f_j X_j$
160–169	164.5	1	164.5
150–159	154.5	0	0
140–149	144.5	2	289.0
130–139	134.5	3	403.5
120–129	124.5	6	747.0
110–119	114.5	8	916.0
100–109	104.5	5	522.5
90–99	94.5	5	472.5
80–89	84.5	1	84.5
70–79	74.5	1	74.5
		$n = 32$	$\sum_{j=1}^{k} f_j X_j = 3674.0$

(ii) Computation of $\bar{X}$ from grouped frequency distribution:

$$\bar{X} = \frac{\sum_{j=1}^{k} f_j X_j}{n} = \frac{3674}{32} = 114.8$$

Review Exercises for Section 3.3

**6. Identify:
  *a. X_1 b. $\bar{Y}$ *c. μ_1
  d. Z_3 e. $\bar{X}$ f. μ_Z
  g. Y_2 *h. X_i i. Y_j
  j. Z_k k. Y_n l. n

**7. Write out the following, listing individual values of the variable.
  *a. $\sum_{i=1}^{n} X_i$ b. $\sum_{i=1}^{5} Y_i/n$

[4] For an excellent discussion of why the grouping error occurs, see Blommers and Lindquist (1960, pp. 105–108).

*c. $\sum_{\substack{i=1 \\ i \neq 3}}^{4} Z_i/n$ *d. $\sum_{j=1}^{k} f_j X_j/n$

e. $\sum_{j=1}^{6} f_j Y_j/n$ f. $\sum_{\substack{j=1 \\ j \neq 2}}^{4} f_j Z_j$

g. $\sum_{i=1}^{n} n_i \bar{X}_i/n_i$

*8. The socioeconomic level of white families in a predominantly black neighborhood was rated on the basis of their income, educational attainment, physical condition of dwelling, and number of home appliances. Compute the mean using $\sum_{i=1}^{n} X_i/n$ for the following socioeconomic scores.

5	4	9	5	3	4
4	6	7	5	3	2
6	2	5	1	7	

9. In the same neighborhood described in Exercise 8, socioeconomic ratings were obtained on black families. Compute the mean using $\sum_{i=1}^{n} X_i/n$.

5	6	4	5	10	6	3	5	7	6
3	4	5	8	5	4	7	1	6	7

10. The following data represent the number of suicides in predominantly rural prefectures in Japan. Compute the mean using $\sum_{i=1}^{n} X_i/n$.

22	10	12	2	10	9	16	11	8
14	11	8	13	10	9	12	0	10
12	8	11	5	7	10	7	9	9
9	8	7	8	5	14	3	10	11

11. Suicide data for predominantly urban prefectures in Japan are as follows. Compute the mean using $\sum_{i=1}^{n} X_i/n$.

23	24	21	19	23	24	25	22	21	27
24	23	23	22	20	23	26	25	24	22
20	17	26	23	21	25	14	21	23	
26	24	23	22	25	23	25	28	24	

**12. Compute the midpoint for each of the following class intervals.
  *a. 10–14 b. 12–13 c. 33–35 d. 20–29
  *e. 5 f. 15–17 g. 9

*13. For the data in Exercise 10, compute the mean using $\sum_{j=1}^{k} f_j X_j/n$. Let i, the size of the class interval, equal 2.

14. For the data in Exercise 11, compute the mean using $\sum_{j=1}^{k} f_j X_j/n$. Let i, the size of the class interval, equal 2.

15. (a) What is a grouping error and why does it occur? (b) Under what conditions is the grouping error minimal?

16. Terms to remember:
  a. Mu
  b. Statistic
  c. Parameter
  d. Index of summation
  e. Class interval midpoint
  f. Grouping error

3.4 The Median

*The **median** is the point in a distribution below which 50% of the scores fall.* It is denoted by Mdn. As its name suggests, it is the middle score when data have been rank ordered; it divides the data into two groups having equal frequency. The procedure for determining the median is slightly different depending on whether n, the number of scores, is odd or even and whether a frequency distribution has been constructed for the data. If the number of scores is small, the median can be determined by inspection. Consider the case in which n is odd, and the scores are 5, 3, 8, 9, 11, 12, 2. When the scores are rank ordered from smallest to largest along the number line, as in Figure 3.4-1, it is immediately apparent that 8 is the middle score; half the scores fall above 8 and half below. Therefore, 8 is the median. Figure 3.4-1, which represents scores as the distance between real limits on a number line, illustrates that the median does not divide the interval from 2 to 12 in half, rather it divides the number of scores in half. When, as in Figure 3.4-1, the number of scores is odd, the median is the $(n + 1)/2$th score from either end of the number line. For example, $(n + 1)/2 = (7 + 1)/2 = 4$; hence the median is the fourth score counting from either end.

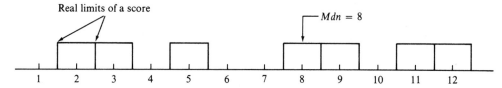

Figure 3.4-1. Determination of the median when n is odd.

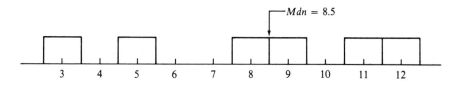

Figure 3.4-2. Determination of the median when n is even.

Figure 3.4-2 illustrates the location of the median along the line when n is even and the scores are 3, 5, 8, 9, 11, 12. Any point along the number line larger than 8 and less than 9 would qualify as the median, since there are two scores below 8 and two scores above 9. By convention, the median is taken as the midway point between the $n/2$th score and the $(n/2) + 1$th score. For example, $n/2 = 3$ and $(n/2) + 1 = 4$, and the midway point is $(8 + 9)/2 = 8.5$, which is the median.

Frequencies greater than 1 at the middle score value may present special problems. The median for Figure 3.4-3(a) is obviously 8, but what about Figure 3.4-3(b)? According to our definition, the median should be the $(n + 1)/2 = (7 + 1)/2 = $ 4th score from either end. This score is 8, but below 8 there are three scores and above only two scores. The problem is resolved by arbitrarily assigning half the interval 7.5–8.5 to each score.

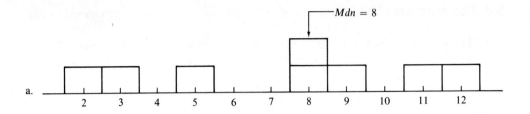

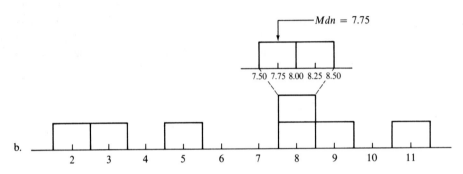

Figure 3.4-3. Determination of the median when the frequency of the middle score value is greater than 1.

This results in two smaller intervals, 7.5–8 and 8–8.5, as shown in the upper part of Figure 3.4-3(b). Going four scores from the lower end, we reach the score defined by 7.5–8, which has a midpoint at $(7.5 + 8)/2 = 7.75$; similarly, four scores from the upper end is also the score defined by 7.5–8. Thus, the median is 7.75, the midpoint of the score defined by the interval 7.5–8. Now consider the scores in Figure 3.4-4. Each of the three scores in the interval 7.5–8.5 are assumed to occupy one-third of the interval, resulting in the smaller intervals 7.500–7.833, 7.833–8.167, 8.167–8.500, as shown in the upper part of the figure. Since n is even, the median is the score value that is midway

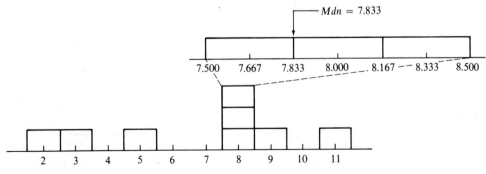

Figure 3.4-4. Determination of the median when the frequency of the middle score value is greater than 1.

between the $n/2 = $ 4th and the $(n/2) + 1 = $ 5th scores. These scores are defined by the intervals 7.500–7.833 and 7.833–8.167, respectively. The median is the midway point $(7.667 + 8.000)/2 = 7.833$.

Computation of the Median in a Frequency Distribution

We determined the median in Figures 3.4-3 and 3.4-4 by interpolating—finding the value within an interval such that the definition was satisfied. For data that have been ordered in a frequency distribution, this can be done automatically by a formula. This computation is described in Table 3.4-1 for the data in Figure 3.4-4. The computational procedure works fine as long as the median falls in a class interval with frequency greater than zero. The formula will not yield an answer if f_i, the number of scores in

Table 3.4-1. Procedure for Computing the Median from a Frequency Distribution

(i) Data and computational formula:

X	f	Cum f[a]	
11	1	8	$Mdn = X_{ll} + i\left(\dfrac{n/2 - \sum f_b}{f_i}\right)$
10	0	7	
9	1	7	
8	3	6	$= 7.5 + 1\left(\dfrac{8/2 - 3}{3}\right)$
7	0	3	
6	0	3	
5	1	3	$= 7.5 + 1\left(\dfrac{4 - 3}{3}\right)$
4	0	2	
3	1	2	
2	1	1	$= 7.5 + 0.33 = 7.83$
	$n = 8$		

(ii) Definition of terms:
X_{ll} = real lower limit of class interval containing the median
i = class interval size
n = number of scores
$\sum f_b$ = number of scores below X_{ll}
f_i = number of scores in the class interval containing the median

(iii) Computational sequence:
1. Compute $n/2 = 8/2 = 4$.
2. Locate the class interval containing the $n/2 = $ 4th score in the Cum f column. The median will fall somewhere in this class interval. The fourth score occurs in the class interval 8. This class interval contains the fourth, fifth, and sixth scores; X_{ll} for this class interval is 7.5.
3. Compute $i: i = $ (Real upper limit of class interval − Real lower limit of class interval), for example, $i = 8.5 - 7.5 = 1$.
4. Determine $\sum f_b$.
5. Determine f_i.

[a] Cumulative frequency is discussed in Section 2.2.

the class interval containing the median, is equal to zero because then the fraction $i(n/2 - \sum f_b)/f_i$ is undefined. (See Appendix A, Rule 9d.) In this case, the median is taken to be the midpoint of the class interval in which it falls.

Review Exercises for Section 3.4

17. For a small number of scores, how is the median determined (a) when n is odd and (b) when n is even?
**18. Determine the median for the following scores.
   *a. 9, 3, 16, 5, 21
   b. 2, 8, 11, 19, 3, 26, 28
   *c. 16, 19, 17, 31
   d. 3, 1, 3, 4
   *e. 3, 1, 3, 4, 5
   *f. 3, 4, 4, 2, 8
   g. 3, 5, 5, 4, 8
   h. 3, 5, 5, 4, 8, 5
*19. For the data in Exercise 10 (page 50), compute the median using $X_{ll} + i\left(\dfrac{n/2 - \sum f_b}{f_i}\right)$; let $i = 2$.
20. For the data in Exercise 11 (page 50), compute the median using $X_{ll} + i\left(\dfrac{n/2 - \sum f_b}{f_i}\right)$; let $i = 2$.
*21. The computational procedure for the median illustrated in Table 3.4-1 calculates the median from below—that is, by coming halfway through the scores starting from the lowest class interval. Alternatively, the median can be computed by coming down halfway from above—from the highest class interval. The computational formula is $X_{ul} - i\left(\dfrac{n/2 - \sum f_a}{f_i}\right)$. By analogy with the definitions in Table 3.4-1, define each of the symbols in the alternative formula.
22. For the data in Table 3.4-1, compute the median using $X_{ul} - i\left(\dfrac{n/2 - \sum f_a}{f_i}\right)$.
23. For the data in Exercise 11 (page 50), compute the median using $X_{ul} - i\left(\dfrac{n/2 - \sum f_a}{f_i}\right)$.

3.5 Relative Merits of the Mean, Median, and Mode

Computation is fairly simple for all three measures of central tendency. Which one should an experimenter use for a given problem? The choice should be based on (1) the shape of the distribution, (2) the intended uses of the statistic, (3) the nature of the variable, and (4) the mathematical properties and merits of the three measures.

Although they are all measures of central tendency, the mean, median, and mode impart somewhat different information. Consider the scores in Figure 3.5-1. By inspection, we see that the mode is 3. The median is the $(n + 1)/2 = $ 3rd score from either end of the number line. This score falls in the interval with real limits 2.5–3.5. When the

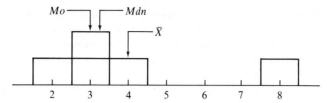

Figure 3.5-1. Comparison of $\bar{X}$, Mdn, and Mo. The number line can be thought of as a teeter-totter whose balance point is the mean.

interval is divided in half, the real limits of the third score are 3–3.5 and its midpoint is 3.25; hence, the median is 3.25. The mean is $\bar{X} = (2 + 3 + \cdots + 8)/5 = 4$. The three values obtained represent different conceptions of the point around which scores cluster. For a unimodal set of data plotted as a histogram,

1. the *mode* is the score value with the highest frequency—the most typical score;
2. the *median* is the score point that divides the number of scores in half; and
3. the *mean* is the score point at which the distribution balances—its center of gravity.

If a distribution is asymmetrical, as in Figure 3.5-1, the mean and the median are unequal; the value of the mode may or may not differ from those for the mean and the median. If a distribution is symmetrical, the mean and median are equal; if, in addition, the distribution is unimodal, all three measures are equal.

Mean

The mean has a number of mathematical properties that make it the preferred measure of central tendency for relatively symmetrical distributions and for quantitative variables. One of these is its sampling stability.[5] Suppose that from an extremely large population we repeatedly drew random samples of size n. If we computed the means for the samples, we would expect them to be similar but not identical. Suppose we also computed the median and the mode for each sample. The variability from sample to sample would be greatest for the mode and least for the mean. The better *sampling stability* of the mean is an important advantage, especially when one uses inferential statistics to draw conclusions about the central tendency of a population by observing a single sample.

Another advantage of the mean is that it is amenable to arithmetic and algebraic manipulations in ways that the other measures are not. Therefore, if further statistical computations are to be performed, the mean is usually the measure of choice. This property accounts for the appearance of the mean in the formulas for many important statistics.

The mean is the only one of the three measures that reflects every score and its value. Obviously, since $\bar{X} = \sum X/n$, the mean is a function of the value of every score

[5] See Technical Notes, Section 3.9, for more about these properties.

in addition to the number of scores. On the other hand, the median is independent of the values of scores (other than the median value itself) so long as the number of scores above and below the median is not altered. If, for example, the score of 8 in Figure 3.5-1 is changed to 5, the values of the median and the mode are unchanged; the mean, however, is changed from 4 to 3.4.

It is no accident that the balance point of the scores in Figure 3.5-1 coincides with the mean. This fulcrum property of the mean follows from the mathematical statement, $\sum_{i=1}^{n} (X_i - \bar{X}) = 0$, the sum of the deviations of scores from the mean is zero. In Figure 3.5-1, for example, $\sum_{i=1}^{n} (X_i - \bar{X}) = (2 - 4) + (3 - 4) + \cdots + (8 - 4) = 0$, and this will be true for any distribution. It is also true that the sum of squared deviations of scores from the mean is less than the sum of squared deviations from any other point. This can be stated mathematically: the mean is the point for which $\sum_{i=1}^{n} (X_i - \bar{X})^2 =$ Minimum. If we think of a deviation as a distance, the mean is the point from which (1) the sum of the distances to all the scores is zero and (2) the sum of the squared distances is a minimum. For proofs of these properties see Sections 3.9-2 and 3.9-5.

There are two situations in which the mean is not the preferred measure of central tendency—when the distribution is very skewed and when the data are qualitative in character. Suppose that the following data were obtained for the number of minutes required to solve math problems: 10.1, 10.3, 10.5, 10.6, 10.7, 10.9, 56.9. The mean is $120/7 = 17.1$; the median is 10.6. Which number best represents the seven scores? Most would agree that it is 10.6, the median. The mean is unduly affected by the lone extreme score of 56.9. Any time a distribution is extremely asymmetrical the mean is strongly affected by the extreme scores and, as a result, falls further away from what would be considered the distribution's central area.

The mean is not the preferred measure of central tendency when the data are qualitative in character. Suppose that the dependent variable is eye color and we collect the following data: blue, brown, brown, gray, blue, brown. There is no meaningful way to represent these data by a mean; we could, however, compute the mode and say the most typical eye color is brown.

Median

Although the mean is usually the preferred measure of central tendency, there are several situations in which the median is preferred. As mentioned earlier, the median is not sensitive to the values of the scores above and below it—only to the number of such scores. Unlike the mean, it is not affected by extreme scores and thus is a more representative measure of central tendency for very skewed distributions. Also, it can be computed when the values of the extreme scores are unknown. Suppose, for example, that we recorded the number of trials required to learn a list of paired adjectives and Japanese kana (writing) symbols. The data are as follows: 12, 17, 17, 18, 21, 24, >41. After the forty-first trial, the poorest learner was still unable to learn the list and gave up; his score is some number greater than 41. The distribution is *open-ended*, since one of the extreme values is unknown. The median for these data is 18, but the mean cannot be computed because the value of the extreme score is unknown.

The median has the added advantage of being easy to compute; when the number of scores is small, it can be determined by inspection.

We saw earlier that the mean is the point for which the sum of the squared deviations is a minimum. We expressed this idea mathematically by saying that $\sum_{i=1}^{n}(X_i - \bar{X})^2 =$ Minimum. We can think of the median as the point closest to the aggregate of all scores, disregarding the sign of the differences. We can state this mathematically: the median is the point for which $\sum_{i=1}^{n}|X_i - Mdn| =$ Minimum; that is, the sum of the deviations of scores from the median, disregarding their signs, is as small or smaller than from any other point. Note the similarity between this and the mathematical statement describing the minimum squared deviation property of the mean. For the data in Figure 3.5-1, $\sum_{i=1}^{n}|X_i - Mdn| = |2 - 3.25| + |3 - 3.25| + \cdots + |8 - 3.25| = 1.25 + 0.25 + \cdots + 4.75 = 7.25$. If we take the deviations from, say, the mean, and disregard their signs, the sum is 8, which is larger than the sum for the median (or any other point in the interval that defines the median).

The principal disadvantages of the median relative to the mean are (1) its poorer sampling stability and (2) its poorer mathematical tractability. For these and other reasons it is not widely used in advanced descriptive and inferential statistical procedures.

Mode

The mode is the only measure of central tendency that can be used with qualitative variables such as eye color, blood type, race, and political party affiliation. For quantitative variables that are inherently discrete, such as family size, it is sometimes a more meaningful measure of central tendency than the mean or the median. Who ever heard of a family with 4.2 members? It makes more sense to say that the most typical family size is 4, the mode. Other than these two applications, the mode has little to recommend it except its ease of estimation.

Let us consider why the mode is called the most typical score. We have seen that the mean is the value for which $\sum_{i=1}^{n}(X_i - \bar{X})^2 =$ Minimum and the median is the value for which $\sum_{i=1}^{n}|X_i - Mdn| =$ Minimum. The mode is the score value that occurs most often. We can state this mathematically: the mode is the value for which the number of $(X_i \neq Mo) =$ Minimum; that is, the number of scores not equal to the mode is a minimum. In Figure 3.5-1, for example, there are three scores that differ from the mode, but four that differ from the mean and five that differ from the median; hence the mode is the most typical score.

The mode has a number of limitations. Its sampling stability is much poorer than that of the mean and the median, and it is also less mathematically tractable. Therefore, it is rarely used in advanced descriptive and inferential statistics. However, the mode, like the median, can be computed for an open-ended distribution if the distribution is known to be unimodal and if the unknown scores don't have the greatest frequency. However, because of the median's superior mathematical properties, it is preferred for this application.

A mode may not exist for a set of data, as when the distribution is bi- or multimodal. In such cases, it may be useful to report two or more modes. Since most variables in the behavioral sciences are approximately normally distributed, the existence of two modes suggests the presence of two underlying distributions. This would occur if we administered a test of masculinity to a sample containing an equal number of males and females. To report a mean or median for such data would be misleading without also reporting that the distribution is bimodal and the values of the modes.

3.5
Relative Merits of the Mean, Median, and Mode

Summary of the Properties of the Mean, Median, and Mode

The mean is

1. the balance point of a distribution—the point for which $\sum_{i=1}^{n}(X_i - \bar{X}) = 0$ and the point for which $\sum_{i=1}^{n}(X_i - \bar{X})^2 = $ Minimum;
2. the preferred measure for relatively symmetrical distributions and quantitative variables;
3. the measure with the best sampling stability;
4. widely used in advanced statistical procedures;
5. mathematically tractable;
6. the only measure with a value that is dependent on the value of every score in the distribution;
7. more sensitive to extreme scores than the median and mode and hence is not recommended for markedly skewed distributions; and
8. not appropriate for qualitative data.

The median is

1. the point that is closest to the aggregate of all scores, disregarding their signs—the point for which $\sum_{i=1}^{n}|X_i - Mdn| = $ Minimum;
2. second to the mean in usefulness;
3. widely used for markedly skewed distributions, since it is sensitive only to the number, rather than the values of scores above and below it;
4. the most stable measure that can be used with open-ended distributions;
5. more subject to sampling fluctuation than the mean;
6. less mathematically tractable than the mean; and
7. less often used in advanced statistical procedures.

The mode is

1. the score value that occurs most often and therefore is most typical—the value for which the number of $(X_i \neq Mo) = $ Minimum;
2. the only measure appropriate for qualitative variables;
3. more appropriate than the mean or median for quantitative variables that are inherently discrete;
4. the easiest measure to compute;
5. much more subject to sampling fluctuation than the mean and the median;
6. less mathematically tractable than the mean and the median;
7. not necessarily existing, as when a distribution has two or more points with the same maximum frequency; and
8. rarely used in advanced statistical procedures.

Review Exercises for Section 3.5

**24. For the following sets of data, what measures of central tendency would you compute? Justify your choices.
  *a. 9, 6, 5, 7, 1, 6, 7, 8, 10, 6, 5, 4, 3, 6, 9, 7, 4, 5, 6, 8, 3, 2
   b. 4, 3, 7, 5, 4, 2, 12, 6, 5, 4, 3, 3, 2, 7, 1, 6, 4, 5, 3, 5
  *c. 6, 5, 9, 6, 7, 5, 6, 8, 3, 4, 5, 7, 5, 4, 8, 5
   d. Eye color: blue, brown, brown, blue, green, brown, gray, brown, blue

e. 7, 8, 6, 7, 8, 9, 1, 6, 5, 3, 7, 8, 7, 6, 7, 8, 5, 7
f. Family size: 4, 3, 5, 4, 1, 2, 4, 6, 5

**25. Rank the three measures of central tendency with respect to the following characteristics; let 1 = most or hardest and 3 = least or easiest.
  *a. Sampling stability
   b. Suitability for advanced applications
   c. Mathematical tractability
   d. Sensitiveness to value of each score
  *e. Appropriateness for qualitative variables
   f. Ease of computation
   g. Likelihood that it can be computed for a set of quantitative data

26. State in your own words the meaning or significance of the following.
   a. The mean is the point for which $\sum (X_i - \bar{X}) = 0$.
   b. The median is the point for which $\sum |X_i - Mdn| =$ Minimum.
   c. The mode is the score for which the number of $(X_i \neq Mo) =$ Minimum.

27. Term to remember:
    Sampling stability

3.6 Location of $\bar{X}$, Mdn, and Mo in a Distribution

If a distribution is unimodal and symmetrical, the mean, median, and mode have the same value. If the distribution is unimodal but skewed, the three measures will be arranged in a predictable order. This is illustrated in Figure 3.6-1. In both examples, the mean is on the side of the distribution having the longest tail and the median falls about $\frac{1}{3}$ of the distance from the mean to the mode. To remember the order—mean, median, mode—note that it is alphabetical, starting from the longer tail. This order occurs because the mean is affected by the value of extreme scores; the median is affected by the presence of extreme scores but not their value; and the mode is not affected by extreme scores unless they happen to have the greatest frequency of occurrence.

The relative position of the measures provides a rough guide as to whether a distribution is positively or negatively skewed. For negatively skewed distributions, it

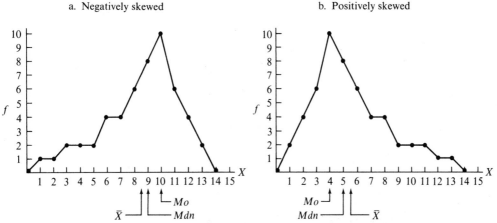

Figure 3.6-1. Location of the $\bar{X}$, Mdn, and Mo for skewed distributions.

is virtually always true that $Mdn > \bar{X}$; for positively skewed distributions, $\bar{X} > Mdn$. If, for example, we know that the median is 25 and the mean is 20, we would strongly suspect that the distribution is negatively skewed. The greater the discrepancy between the two values, the greater the departure from symmetry.[6]

Review Exercises for Section 3.6

28. For each of the following distributions, indicate on the X axis the approximate location of $\bar{X}$, Mdn, and Mo.

a.
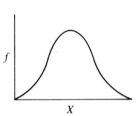

b.
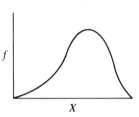

c.
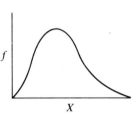

d.
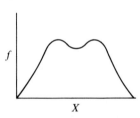

**29. Determine the shape of each distribution from the following measures of central tendency. Assume a unimodal distribution.
  *a. $\bar{X} = 16$, $Mdn = 10$
  b. $\bar{X} = 21$, $Mdn = 21$, $Mo = 21$
  c. $Mdn = 109$, $Mo = 116$
  d. $\bar{X} = 73$, $Mdn = 84$
  *e. $\bar{X} > Mdn$
  f. $Mo > Mdn$
  g. $Mdn > \bar{X}$
  h. $\bar{X} = Mdn = Mo$

3.7 Mean of Combined Subgroups

Suppose that two introductory sociology classes obtained the following mean scores on a departmental examination: 80 and 90. What is the mean of the two means? If each class had the same number of students, we could compute the mean of the means by $\bar{X} = (\bar{X}_1 + \bar{X}_2)/2 = (80 + 90)/2 = 85$. If, as is more likely, the classes contain different numbers of students, we must weight the means proportionally to their respective sample sizes. Assume that $\bar{X}_1 = 80$ and $n_1 = 20$, and that $\bar{X}_2 = 90$ and

[6] A more sophisticated measure of skewness is described in Section 4.6.

$n_2 = 40$. The weighted mean, $\bar{X}_w$, is given by

$$\bar{X}_w = \frac{n_1\bar{X}_1 + n_2\bar{X}_2 + \cdots + n_n\bar{X}_n}{n_1 + n_2 + \cdots + n_n} = \frac{20(80) + 40(90)}{20 + 40} = \frac{5200}{60} = 86.7.$$

The weighted mean is closer to 90 than to 80; this reflects the larger n_2 associated with $\bar{X}_2 = 90$.

Review Exercises for Section 3.7

**30. For the following data, compute weighted means.
  *a. $\bar{X}_1 = 30, n_1 = 10; \bar{X}_2 = 50, n_2 = 20$
  b. $\bar{X}_1 = 50, n_1 = 20; \bar{X}_2 = 100, n_2 = 30$
  c. $\bar{X}_1 = 8, n_1 = 10; \bar{X}_2 = 12, n_2 = 30; \bar{X}_3 = 18, n_3 = 20$
  d. $\bar{X}_1 = 100, n_1 = 20; \bar{X}_2 = 200, n_2 = 20$

3.8 Summary

Three measures of central tendency are described in this chapter—the mean, median, and mode. The different indexes result from different ways of conceptualizing the point around which scores cluster. The mean is the point around which the distribution balances—its center of gravity; the median is the point that divides the number of scores in half; and the mode is the score value with the greatest frequency—the most typical score.

The mean is the most widely used of the indexes, partly because of its superior sampling stability and partly because many advanced statistical procedures are based on it. The median and mode, by contrast, are *terminal statistics*; their usefulness in advanced descriptive and inferential procedures is very limited.

There are two situations in which the mean is not the preferred measure of central tendency: when the distribution is markedly skewed and when the variable is qualitative in character. For markedly skewed distributions, the median is preferred because it is not as sensitive as the mean to the presence of extreme scores. For qualitative variables, the only measure that can be computed is the mode. In addition, the mode may be more meaningful for inherently discrete variables such as family size.

3.9 Technical Notes

The following notes are designed to provide additional insights about selected topics in this chapter. Except for the first note on summation rules, the book is complete without them. In keeping with the rest of the book, no knowledge of mathematics beyond high school math is assumed, and the presentation is appropriately simplified.

3.9-1 Summation Rules

Four simple summation rules are widely used in statistical proofs and derivations. An understanding of these rules will go far toward taking derivations out of the realm of magic.

Rule 3.9-1-1 The sum of a constant. Let c be a constant; the sum over $i = 1, \ldots, n$ of the constant can be written as the product of the upper limit of the summation, n, and c. That is,

$$\sum_{i=1}^{n} c = \underbrace{c + c + \cdots + c}_{n \text{ terms}} = nc.$$

For example, let $c = 2$ and $i = 1, \ldots, 3$; then

$$\sum_{i=1}^{3} 2 = 2 + 2 + 2 = 3(2) = 6.$$

Rule 3.9-1-2[7] The sum of a variable. Let V be a variable with values $V_1, V_2, \ldots, V_n$; the sum over $i = 1, \ldots, n$ of the variable is

$$\sum_{i=1}^{n} V_i = V_1 + V_2 + \cdots + V_n.$$

For example, let $V_1 = 2$, $V_2 = 3$, and $V_3 = 4$; then

$$\sum_{i=1}^{3} V_i = 2 + 3 + 4 = 9.$$

Rule 3.9-1-3 The sum of the product of a constant and a variable. The sum $\sum_{i=1}^{n} cV_i$ can be written as the product of the constant and the sum of the variable—that is,

$$\sum_{i=1}^{n} cV_i = c \sum_{i=1}^{n} V_i.$$

For example, let $c = 2$ and $V_1 = 2$, $V_2 = 3$, and $V_3 = 4$; then

$$\sum_{i=1}^{3} cV_i = 2(2) + 2(3) + 2(4)$$

$$= c \sum_{i=1}^{n} V_i$$

$$= 2(2 + 3 + 4) = 18.$$

Rule 3.9-1-4 Distribution of summation. If the only operation to be performed before summation is addition or subtraction, the summation sign can be distributed among the separate terms of the sum. Let V and W be two variables; then

$$\sum_{i=1}^{n} (V_i + W_i) = \sum_{i=1}^{n} V_i + \sum_{i=1}^{n} W_i.$$

For example, let $V_1 = 2$, $V_2 = 3$, $V_3 = 4$, $W_1 = 5$, $W_2 = 6$, and $W_3 = 7$; then

$$\sum_{i=1}^{3} (V_i + W_i) = (2 + 5) + (3 + 6) + (4 + 7)$$

$$= \sum_{i=1}^{3} V_i + \sum_{i=1}^{3} W_i$$

$$= (2 + 3 + 4) + (5 + 6 + 7) = 27.$$

---

[7] This rule was introduced in Section 3.3.

This rule applies to any number of terms. For example, let V, W, and X be variables and a, b, and c be constants; then according to Rules 3.9-1-1, 3.9-1-2, and 3.9-1-4,

$$\sum_{i=1}^{n} (V_i + W_i + X_i + a + b + c) = \sum_{i=1}^{n} V_i + \sum_{i=1}^{n} W_i + \sum_{i=1}^{n} X_i + na + nb + nc.$$

3.9-2 Proof that the Mean Is a Balance Point

In Section 3.5 we said that the mean is the point such that $\sum_{i=1}^{n} (X_i - \bar{X}) = 0$. We can construct a simple proof of this using Rules 3.9-1-1, 3.9-1-2, and 3.9-1-4. In the term $\sum_{i=1}^{n} (X_i - \bar{X})$, X_i is a variable; but for any set of scores, $\bar{X}$ is a constant. Hence,

$$\sum_{i=1}^{n} (X_i - \bar{X}) = \sum_{i=1}^{n} X_i - n\bar{X} \qquad \text{By definition}$$

$$\bar{X} = \sum_{i=1}^{n} X_i/n, \text{ so } n\bar{X} = \sum_{i=1}^{n} X_i.$$

$$= \sum_{i=1}^{n} X_i - \sum_{i=1}^{n} X_i$$

$$= 0.$$

3.9-3 Effect on the Mean of Adding a Constant to Each Score

Let $\bar{X}_{X+c}$ be the mean of a distribution that has been altered by adding a constant c to each score—that is, $X_1 + c, X_2 + c, \ldots, X_n + c$. Then

$$\bar{X}_{X+c} = \frac{\sum_{i=1}^{n} (X_i + c)}{n} = \frac{\sum_{i=1}^{n} X_i + nc}{n} = \frac{\sum_{i=1}^{n} X_i}{n} + c = \bar{X} + c.$$

The effect of adding a constant c to each score is to change $\bar{X}$, the mean of the original scores, to $\bar{X} + c$. Similarly, it can be shown that if a constant is subtracted for each score, the effect is to change $\bar{X}$ to $\bar{X} - c$.

3.9-4 Effect on the Mean of Multiplying Each Score by a Constant

Let $\bar{X}_{cX}$ be the mean of a distribution that has been altered by multiplying each score by a constant c—that is, $cX_1, cX_2, \ldots, cX_n$. Then

$$\bar{X}_{cX} = \frac{\sum_{i=1}^{n} (cX_i)}{n} = \frac{c \sum_{i=1}^{n} X_i}{n} = c\bar{X}.$$

The effect of multiplying each score by a constant is to change $\bar{X}$, the mean of the original scores, to $c\bar{X}$. Similarly, it can be shown that the effect of dividing each score by a constant is to change $\bar{X}$ to $\bar{X}/c$.

3.9-5 Proof that $\sum (X_i - \bar{X})^2 = $ Minimum

Because the mean is the balance point of a distribution, the sum of the squared deviations from the mean, $\sum_{i=1}^{n} (X_i - \bar{X})^2$, is smaller than the sum about any other point. Let c be a constant not equal to zero that is added to the mean; then

$$\sum_{i=1}^{n} [X_i - (\bar{X} + c)]^2 = \sum_{i=1}^{n} [(X_i - \bar{X}) - c]^2 \qquad \text{According to the associative law.}$$

$$= \sum_{i=1}^{n} [(X_i - \bar{X})^2 - 2c(X_i - \bar{X}) + c^2]$$

$$= \sum_{i=1}^{n} (X_i - \bar{X})^2 - 2c \sum_{i=1}^{n} (X_i - \bar{X}) + nc^2. \qquad \text{Rules 3.9-1-1--3.9-1-4.}$$

But we saw in Section 3.9-2 that $\sum_{i=1}^{n} (X - \bar{X}) = 0$. Therefore,

$$\sum_{i=1}^{n} [X_i - (\bar{X} + c)]^2 = \sum_{i=1}^{n} (X_i - \bar{X})^2 + nc^2.$$

Thus, for $c \neq 0$, $nc^2 > 0$ and

$$\sum_{i=1}^{n} (X_i - \bar{X})^2 + nc^2 > \sum_{i=1}^{n} (X_i - \bar{X})^2.$$

The sum of squared deviations about the mean appears in many advanced statistics. We will see in Chapter 4 that $S = \sqrt{\sum_{i=1}^{n} (X_i - \bar{X})^2 / n}$ is used to describe the dispersion of scores in a distribution; S is called the standard deviation.

4

Measures of Dispersion, Skewness, and Kurtosis

4.1 Introduction to Measures of Dispersion
4.2 Four Measures of Dispersion
    Range
    Semi-Interquartile Range
    Percentile
    Standard Deviation
    Index of Dispersion
    Review Exercises for Section 4.2
4.3 Relative Merits of the Measures of Dispersion
    Standard Deviation
    Semi-Interquartile Range
    Range
    Index of Dispersion
    Summary of the Properties of the Four Measures of Dispersion
    Review Exercises for Section 4.3
4.4 Dispersion and the Normal Distribution
    Review Exercises for Section 4.4
4.5 Tchebycheff's Theorem
    Review Exercises for Section 4.5
4.6 Skewness and Kurtosis
    Skewness
    Kurtosis
    Review Exercises for Section 4.6
4.7 Summary
4.8 Technical Notes
    Effect on the Standard Deviation of Adding a Constant to Each Score
    Effect on the Standard Deviation of Multiplying Each Score by a Constant
    Proof of the Equivalence of Raw-Score and Deviation Formulas for S
    Minimum and Maximum Values of S
    Derivation of Formula for the Index of Dispersion, D

4.1 Introduction to Measures of Dispersion

Mr. Jacques and Ms. Shapley are taking a well-deserved break in the teachers' lounge. The conversation turns to Ms. Shapley's third-grade class. "I've got a bunch of little monsters this year. I can't seem to keep their interest for more than 10 minutes. I had to discipline Emerson twice this morning for flying paper airplanes during arithmetic, and Waldo is still picking fights. I just can't understand it; this class has the same average IQ as my class last year, and you remember how good those kids were." As Ms. Shapley contemplates her options—should I face the class for 7 more months, drop out and become a topless dancer, or go back to school in molecular physics—we wonder what makes one class a joy and the other a disaster. The frequency polygon in Figure 4.1-1 illustrates the reason. Although the two classes have almost identical mean IQs, this year's class is much more heterogeneous in learning aptitude. Last year, for example, there were no children with IQs below 90, this year there are two. That's Waldo in the class interval 75–79—moderately retarded. At the other end of the distribution in the 140–144 class interval is our paper plane thrower—a potential genius. It is small wonder that this year's class, with its wide range of aptitude, is giving Ms. Shapley fits.

Information about central tendency is important, but this tells only part of the story; the heterogeneity or dispersion of scores is often just as informative. Four numerical indexes of dispersion are presented in this chapter: range, semi-interquartile range, standard deviation, and index of dispersion.

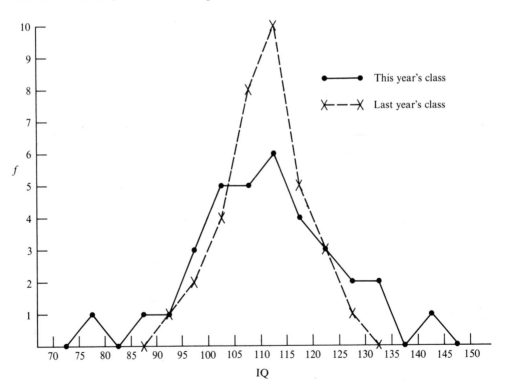

Figure 4.1-1. Frequency polygons for two third-grade classes with the same central tendency but different dispersions.

The measures of central tendency in Chapter 3 represent points around which a distribution tends to center; dispersion measures, with the exception of the index of dispersion, represent distances and thereby summarize the extent to which the scores differ from each other. Not surprisingly, there are as many measures of dispersion as there are conceptions.

4.2 Four Measures of Dispersion

Range

Intuitively, the simplest measure of dispersion is the **range**—the distance between the largest and smallest scores, inclusive. The range is denoted by R and is computed from the formula

$$R = X_{ul\text{(largest score)}} - X_{ll\text{(smallest score)}},$$

where X_{ul} is the real upper limit of the largest score and X_{ll} is the real lower limit of the smallest score. (The midpoints of these scores are sometimes used instead of their real limits.) If, for example, Emerson's 144 is the highest IQ and Waldo's 76 is the lowest, the range is $144.5 - 75.5 = 69$. The range of 69 IQ points is a distance that includes 100% of the scores.

Whenever possible the range should be computed from an ungrouped frequency distribution. If it must be computed from a grouped frequency distribution, the range is taken to be the real upper limit of the highest class interval minus the real lower limit of the lowest class interval. A range computed from a grouped frequency distribution usually overestimates the correct value, and the size of the error is influenced by the scheme used in grouping the data.

In spite of its simplicity, the range is not widely used. For one thing, its value is determined by the two most extreme scores, and so its sampling stability is quite poor. It also is mathematically intractable and is not meaningful for qualitative data. These and other disadvantages discussed in Section 4.3 limit its usefulness as a measure of dispersion.

As we will see, each measure of dispersion is typically reported with a particular measure of central tendency. For quantitative data, the range can be reported with the mode, thereby giving a more complete picture of data.[1] However, since the mode is often used with qualitative data, a different measure of dispersion is needed. The index of dispersion described later fills this need.

Semi-Interquartile Range

The sampling stability of R is poor because it is computed from the two most extreme scores in a distribution. A second measure of dispersion, the semi-interquartile range, is based on two scores closer to the center of the distribution. Hence, it is considerably more stable than R. The **semi-interquartile range**, denoted by Q, is defined as one-half of the distance between the 1st quartile point, Q_1, and the 3rd quartile point, Q_3.

---

[1] The range can also be reported with a rarely used measure of central tendency called the *midrange*. For a set of scores, $X_1, X_2, \ldots, X_n$, arranged in order of magnitude, the midrange is defined as $(X_1 + X_n)/2$.

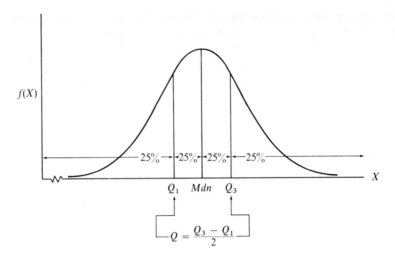

Figure 4.2-1. As seen in the figure, Q_1 is a point below which 25% of the scores fall and above which 75% fall; Q_3 is a point below which 75% fall and above which 25% fall. The median is sometimes referred to as Q_2, since it is a point that divides the distribution of scores in half. The semi-interquartile range, Q, is one-half the distance from Q_3 to Q_1.

These points and the median are shown in Figure 4.2-1. The formula for Q is

$$Q = \frac{Q_3 - Q_1}{2}.$$

The computation of Q_1 and Q_3 is similar to that for the median, and is illustrated in Table 4.2-1. The data are IQ scores from Ms. Shapley's class. The semi-interquartile range for these data is 8.4. The larger the value of Q, the greater the distance between Q_3 and Q_1. This suggests greater heterogeneity among scores.

The semi-interquartile range is often reported along with the median to give a more complete description of data. For a symmetrical distribution, the median plus and minus the semi-interquartile range, $Mdn \pm Q$, is a distance that contains 50% of scores. This is evident from Figure 4.2-1. For the data in Table 4.2-1, $Mdn \pm Q$ is 110.3 ± 8.4 (or 101.9–118.7). This interval, however, does not contain exactly 50% of the scores, since the distribution is asymmetrical.

The semi-interquartile range, like the median, is a terminal statistic; its usefulness in advanced descriptive and inferential procedures is very limited. It shares both the advantages and disadvantages of the median, since it is computed from "medianlike" descriptive statistics, Q_1 and Q_3. We will now digress for a moment to describe another medianlike statistic—the percentile.

Percentile

A **percentile point**, also called a **percentile** or **centile**, is a score point in a distribution below which a specified percentage of scores fall. The term **percentile rank** denotes the percentage of scores that fall below the percentile point. Procedures for computing

Table 4.2-1. Computational Procedures for Q_1, Q_3, and Q

(i) Data and computational formulas:

X	f	Cum f
140–144	1	
135–139	0	
130–134	2	
125–129	2	
120–124	3	
115–119	4	26
110–114	6	22
105–109	5	16
100–104	5	11
95–99	3	6
90–94	1	3
85–89	1	2
80–84	0	1
75–79	1	1

$n = 34$

$$Q_1 = X_{ll} + i\left(\frac{n/4 - \sum f_b}{f_i}\right)$$

$$= 99.5 + 5\left(\frac{8.5 - 6}{5}\right)$$

$$= 99.5 + 2.5 = 102.0$$

$$Q_3 = X_{ll} + i\left(\frac{3n/4 - \sum f_b}{f_i}\right)$$

$$= 114.5 + 5\left(\frac{25.5 - 22}{4}\right)$$

$$= 114.5 + 4.375 = 118.9$$

$$Q = \frac{Q_3 - Q_1}{2}$$

$$= \frac{118.9 - 102.0}{2} = 8.4$$

(ii) Definition of terms:
X_{ll} = real lower limit of class interval containing Q_1 or Q_3
i = class interval size
n = number of scores
$\sum f_b$ = number of scores below X_{ll}
f_i = number of scores in the class interval containing Q_1 or Q_3

(iii) Computational sequence illustrated for Q_1:
1. Compute $n/4 = \frac{34}{4} = 8.5$.
2. Locate the class interval containing the $n/4 = 8.5$th score in the Cum f column; the 8.5th score occurs in the class interval 100–104. For this class interval, X_{ll} is 99.5.
3. Compute i: i = real upper limit of class interval − real lower limit of class interval = 104.5 − 99.5 = 5.
4. Determine $\sum f_b$.
5. Determine f_i.

percentile points corresponding to the 25th, 50th, and 75th percentile ranks have already been described, since these points correspond to Q_1, Mdn, and Q_3, respectively. Percentiles corresponding to other percentile ranks can be computed from the formula

$$P_{\%} = X_{ll} + i\left(\frac{n\%/100 - \sum f_b}{f_i}\right),$$

where the subscript in $P_{\%}$ identifies the rank of the percentile point and the % that appears in the fraction denotes the percentile rank of interest. The remaining symbols are defined in Table 4.2-1; $P_{\%}$ should be substituted for Q_1 where it appears.

To determine P_{60} for the data in Table 4.2-1, first compute $n(60)/100 = 34(60)/100 = 20.4$. By following the computational sequence illustrated for Q_1, we obtain

$$P_{60} = 109.5 + 5\left(\frac{20.4 - 16}{6}\right) = 109.5 + 3.7 = 113.2.$$

This tells us that 60% of the scores fall below an IQ of 113.2. Percentiles are widely used in reporting the performance of individuals on psychological tests. We will return to percentiles in Chapter 10.

Standard Deviation

The standard deviation, denoted by S for a sample and by σ for a population, is the most important and most widely used measure of dispersion. The formulas for S and σ are, respectively,

$$S = \sqrt{\frac{\sum_{i=1}^{n}(X_i - \bar{X})^2}{n}} \quad \text{and} \quad \sigma = \sqrt{\frac{\sum_{i=1}^{n}(X_i - \mu)^2}{n}}.$$

We can develop an intuitive understanding of the standard deviation by examining the formula for S. First note that, unlike R and Q, S is computed from every score in a distribution. Each score is expressed as a deviation from the mean, $(X_i - \bar{X})$, squared, and then summed. What would happen if we didn't square the deviations? We know from Chapter 3 that for any distribution,[3]

$$\sum_{i=1}^{n}(X_i - \bar{X}) = 0,$$

so squaring or some other operation on the deviations is necessary for the sum to equal a value other than zero.[4] The sum of the squared deviations is divided by n; this gives us the mean squared distance by which the scores deviate from the mean. To convert $\sum(X_i - \bar{X})^2/n$ back into deviations expressed in the original unit of measurement we take its square root.[5]

---

[2] When the population standard deviation, σ, is estimated from sample data, a better estimator is

$$\hat{\sigma} = \sqrt{\frac{\sum_{i=1}^{n}(X_i - \bar{X})^2}{n - 1}},$$

denoted by $\hat{\sigma}$; this is used in Chapters 12–16 on inferential statistics.

[3] For a proof see Section 3.9-2.

[4] An obvious alternative would be to take the absolute value of each deviation, that is, to ignore the sign of $(X_i - \bar{X})$. Years ago, a statistic called the *mean deviation* (*MD*), or *average deviation* (*AD*), was developed that used this procedure. Its formula is $MD = \sum |X_i - \bar{X}|/n$, where $|X_i - \bar{X}|$ means to treat all differences as positive. The mean deviation has fallen into disuse, because it lacks certain advantages possessed by the standard deviation.

[5] The statistics $\sum(X_i - \bar{X})^2/n$ and $\sum(X_i - \bar{X})^2/(n - 1)$ are called *variance* and are denoted, respectively, by S^2 and $\hat{\sigma}^2$; $\hat{\sigma}^2$ is said to be an unbiased estimator of σ^2, since the mean of an indefinite but very large number of $\hat{\sigma}^2$'s computed from random samples is equal to σ^2. This is discussed in Section 10.3. The estimator $\hat{\sigma}^2$ is widely used in inferential statistics. We will return to $\hat{\sigma}^2$ in Chapters 12 and 14, and in Chapter 16 when we discuss analysis of variance.

To summarize, the standard deviation is a number that (1) is based on every score in a distribution and (2) represents the square root of the mean squared distance of scores from the mean. The larger the value of S, the larger the mean squared deviation of scores from the mean. But more important, since the standard deviation is based on every score in the distribution, its sampling stability is much better than that of other measures of dispersion. For this reason and because it is mathematically tractable, S is widely used in advanced descriptive and inferential statistics.

As we saw earlier, each measure of dispersion is typically reported with a particular measure of central tendency—R with Mo and Q with Mdn. The standard deviation is reported with the mean. One special type of distribution, called the *normal distribution*, is often approximated by behavioral science data (see Section 2.6 and Chapter 10). For this distribution, the mean plus and minus the standard deviation ($\bar{X} \pm S$) is an interval that contains 68.27% of scores, as Figure 4.2-2 illustrates. The other two dispersion measures are shown in the figure for comparison.

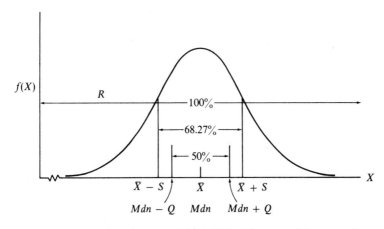

Figure 4.2-2. A region that contains 68.27% of the area of the normal distribution is marked off by $\bar{X} \pm S$; $Mdn \pm Q$ contains 50% of the area; and the R contains 100% of the area.

Computation of the standard deviation using the formula

$$S = \sqrt{\frac{\sum_{i=1}^{n}(X_i - \bar{X})^2}{n}}$$

is illustrated in Table 4.2-2 (see columns 2 and 3). The data are socioeconomic level ratings of white families in a predominantly black neighborhood. For these data, $\bar{X} = 5$ and $S = 2.3$. If we take $\bar{X} \pm S$, we obtain an interval of 2.7–7.3. This distance contains 7.6 of the 12 scores, or 63.3%, which is reasonably close to the 68.27% we would find for the normal distribution even though the distribution in the example deviates appreciably from the normal form.

The formula for S just illustrated is called the *deviation formula*, since it requires the subtraction of $\bar{X}$ from each X_i. It is usually necessary to round off the mean. Since

Table 4.2-2. Computation of the Standard Deviation

(i) Data:

(1) X_i	(2) $X_i - \bar{X}$	(3) $(X_i - \bar{X})^2$	(4) X_i^2
5	0	0	25
9	4	16	81
2	−3	9	4
8	3	9	64
6	1	1	36
5	0	0	25
4	−1	1	16
7	2	4	49
4	−1	1	16
3	−2	4	9
1	−4	16	1
6	1	1	36
$\sum_{i=1}^{n} X_i = 60$		$\sum_{i=1}^{n} (X_i - \bar{X})^2 = 62$	$\sum_{i=1}^{n} X_i^2 = 362$

$\bar{X} = 5$

(ii) Computational formulas:

Deviation formula

$$S = \sqrt{\frac{\sum_{i=1}^{n}(X_i - \bar{X})^2}{n}}$$

$$= \sqrt{\frac{62}{12}}$$

$$= 2.3$$

Raw-score formula

$$S = \sqrt{\frac{n \sum_{i=1}^{n} X_i^2 - \left(\sum_{i=1}^{n} X_i\right)^2}{n^2}}$$

$$= \sqrt{\frac{12(362) - (60)^2}{(12)^2}}$$

$$= \sqrt{\frac{4344 - 3600}{144}} \; ^a$$

$$= 2.3$$

[a] The numerator of this ratio is always positive; if a negative value is obtained, it indicates a computational error. One common error is to use the value of $\sum X_i^2$ for $(\sum X_i)^2$ and vice versa.

the rounding error occurs in every deviation, it is magnified as the deviations are summed and squared in $\sum (X_i - \bar{X})^2$. An alternative formula that circumvents this problem and also is more convenient to use is

$$S = \sqrt{\frac{n \sum_{i=1}^{n} X_i^2 - \left(\sum_{i=1}^{n} X_i\right)^2}{n^2}}.$$

This is called the *raw-score formula*; it is algebraically equivalent to the deviation formula.[6] Use of the raw-score formula is illustrated in columns 1 and 4 of Table 4.2-2. It looks more complicated than the deviation formula, but is actually simpler. Many pocket calculators automatically give $\sum X_i^2$, one of the two required sums, when $\sum X_i$ is computed.[7]

It is sometimes necessary to compute S from a frequency distribution. In this case, the two formulas are modified as follows:

Deviation formula

$$S = \sqrt{\frac{\sum_{j=1}^{k} f_j(X_j - \bar{X})^2}{n}}$$

and

Raw-score formula

$$S = \sqrt{\frac{n \sum_{j=1}^{k} f_j X_j^2 - \left(\sum_{j=1}^{k} f_j X_j\right)^2}{n^2}},$$

where X_j is the midpoint of the jth class interval, f_j is the frequency of scores in the jth class interval, and summation is performed over the $j = 1, \ldots, k$ class intervals. For frequency distributions with $i > 2$, the midpoint of a class interval is used to represent all the scores in the class interval. This inevitably introduces some inaccuracy, and the resulting standard deviation is too large. *An adjustment for the grouping error can be applied to make the computed value of S closer to the correct value. It is called* **Sheppard's correction** *and is given by*

$$S_c = \sqrt{S^2 - \frac{i^2}{12}},$$

where S_c denotes the corrected standard deviation, S the standard deviation computed from a grouped frequency distribution, and i the class interval size. This correction works best when the distribution is approximately normal.[8]

Index of Dispersion

The three measures of dispersion that have been described are distance measures, which of course are appropriate for quantitative data. If data do not contain distance information, as in the case of qualitative variables, how can we describe dispersion? One approach is to think of it as the distinguishability of the observations—that is, as the number of pairs of observations actually distinguishable relative to the maximum possible number.[9] Consider $c = 2$ qualitative categories, A and B, with observations denoted by a_i and b_j, as shown in Figure 4.2-3(a). The two observations in A are indistinguishable; the four in B are indistinguishable. However, among the $n = 6$ observations there are eight distinguishable pairs (*DP*): $a_1b_1, a_1b_2, a_1b_3, a_1b_4, a_2b_1, a_2b_2, a_2b_3, a_2b_4$. The lower limit of *DP*, which represents minimum dispersion, is zero; this occurs when all the observations are in one category and hence are indistinguishable. The

---

[6] The equivalence is shown in Section 4.8-3.
[7] Many scientific and statistical pocket calculators will automatically compute the standard deviation after all scores have been entered in the keyboard. However, they compute $\hat{\sigma}$ instead of S. To convert $\hat{\sigma}$ to S, multiply $\hat{\sigma}$ by $\sqrt{(n-1)/n}$.
[8] The normal distribution was introduced in Section 2.6; it is discussed in Chapter 10.
[9] Such an index has been suggested by several people, including Hammond, Householder, and Castellan (1970, pp. 111–117) and Mueller and Schuessler (1961, pp. 177–179).

a. 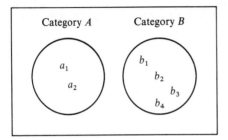 b.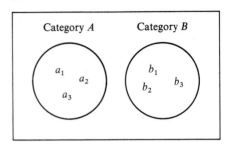

Figure 4.2-3. Observations are assigned to different qualitative categories such that those within a category are indistinguishable with respect to some characteristic.

upper limit of DP, denoted by DP', occurs when the observations are equally divided among the c categories, as in Figure 4.2-3(b). For this example, representing maximum dispersion, there are nine distinguishable pairs (DP'): a_1b_1, a_1b_2, a_1b_3, a_2b_1, a_2b_2, a_2b_3, a_3b_1, a_3b_2, a_3b_3. The ratio DP/DP'—the number of distinguishable pairs to the maximum possible number of distinguishable pairs for c categories—is called the **index of dispersion** and is denoted by D. For the data in Figure 4.2-3(a), $D = DP/DP' = \frac{8}{9} = 0.89$.

The minimum value of $D = DP/DP'$ is 0 and occurs when $DP = 0$, which indicates that all the observations are in one category. The maximum value of D is 1 and occurs when $DP = DP'$, which indicates that the observations are equally divided among the c categories. Thus, D ranges over values 0–1; the larger D, the larger the observed number of distinguishable pairs of observations relative to the maximum number for c categories and, hence, the greater the dispersion.

When the number of observations is large, it is tedious to enumerate DP and DP'. A simple alternative formula for D is

$$D = \frac{c\left(n^2 - \sum_{j=1}^{c} n_j^2\right)}{n^2(c-1)},$$

where c is the number of categories, n is the number of observations, and n_j is the number of observations in each of the $j = 1,\ldots, c$ categories.[10] For the data in Figure 4.2-3(a),

$$D = \frac{2[(6)^2 - (2)^2 - (4)^2]}{(6)^2(2-1)} = 0.89,$$

the same value obtained previously.

The value $D \times 100$ can be interpreted as a percentage, since it is the ratio of the number of distinguishable pairs of observations relative to the maximum possible number of pairs for the c categories. For example, if $D = 0.25$, we know that the number of observed distinguishable pairs is $0.25 \times 100 = 25\%$ of the maximum possible number. The index is particularly useful in comparing the dispersions of several distri-

[10] See Section 4.8-5 for the derivation of this formula.

Table 4.2-3. Marital Happiness Ratings of Women with Either a High School or College Education

(i) Data:

Rating	High School Graduates	College Graduates
Very happy	15	12
Moderately happy	28	36
Neutral	16	30
Unhappy	13	12
Very unhappy	8	6
	$n_1 = 80$	$n_2 = 96$
	Mo_1 = Moderately happy	Mo_2 = Moderately happy
	$D_1 = 0.96$	$D_2 = 0.91$

(ii) Computation of D:

$$D = \frac{c\left(n^2 - \sum_{j=1}^{c} n_j^2\right)}{n^2(c-1)}$$

$$D_1 = \frac{5[(80)^2 - (15)^2 - (28)^2 - (16)^2 - (13)^2 - (8)^2]}{(80)^2(5-1)} = \frac{24{,}510}{25{,}600} = 0.96$$

$$D_2 = \frac{5[(96)^2 - (12)^2 - (36)^2 - (30)^2 - (12)^2 - (6)^2]}{(96)^2(5-1)} = \frac{33{,}480}{36{,}864} = 0.91$$

butions based on the same set of c categories. Suppose we have asked married women with either a high school or a college education to rate their marital happiness. The results of the survey along with the mode and index of dispersion are shown in Table 4.2-3. We see that the modes are identical; however, the dispersion of the college graduates' distribution is smaller than that for the high school graduates.

When data are qualitative in character, the only appropriate measure of central tendency is the mode; the appropriate measure of dispersion is D. The index of dispersion has two disadvantages: (1) it is a terminal statistic (its usefulness in advanced descriptive and inferential statistics is limited) and (2) it is less familiar than R, Q, and S, which are based on the concept of distance rather than on the number of distinguishable pairs of observations.

Review Exercises for Section 4.2

**1. Compute the range for the following sets of numbers.
   *a. 11, 6, 5, 2, 9, 14, 17, 4   *b. 7, 1, 6, 6, 6, 7, 7, 16
   c. 3, 9, 5, 6, 5, 4, 5   d. 26, 18, 30, 24, 23, 24
2. The ranges in Exercises 1a and 1b are identical, although the first set of numbers appears to be more heterogeneous than the second. Why doesn't the range reflect this difference?
3. What are the disadvantages of computing the range from a grouped frequency distribution?
4. For what kind of data can we compute the mode but not the range?

**5. Data representing the length of time required to notice the onset of a warning light during the performance of a simulated driving test are listed in the table. *(a) Compute the median and the semi-interquartile range for these data. *(b) Compute P_{10} and P_{90}. (c) Construct a histogram.

Time (Seconds)	f	Time (Seconds)	f
32	1	26	3
31	1	25	2
30	2	24	1
29	3	23	0
28	4	22	0
27	6	21	1

*6. (a) For the data in Exercise 5, what percentage of the scores falls in the interval $Mdn \pm Q$? (*Hint:* See formula for % in Exercise 8.) (b) Why doesn't this interval contain exactly 50% of the scores?

7. The emotional stability of a random sample of encounter group participants at Nelase Institute was measured. (a) Compute the median and semi-interquartile range for the emotional stability scores listed in the table. (b) What percentage of the scores falls in the interval $Mdn \pm Q$? (*Hint:* See formula for % in Exercise 8.) (c) Compute P_{20}. (d) Construct a histogram.

X_i	f	X_i	f
30–31	1	18–19	7
28–29	0	16–17	6
26–27	1	14–15	4
24–25	2	12–13	4
22–23	4	10–11	2
20–21	5	8–9	1

**8. If we are given a score, its percentile rank (denoted by %) can be computed from

$$\% = \frac{100}{n}\left[\sum f_b + \frac{f_i(P_\% - X_{ll})}{i}\right].$$

a. Derive this formula from

$$P_\% = X_{ll} + i\left(\frac{n\%/100 - \sum f_b}{f_i}\right).$$

*b. For the data in Exercise 5, compute the percentile rank for $X = 30$ using the formula. For these data, note that $i = 1$ and the real limits of a score, say 27, are 26.5 and 27.5.

9. Describe the nature of the distance represented by the standard deviation.

*10. Preschool children, particularly those who are very intelligent, often create imaginary companions. Compute the mean and standard deviation using $\sqrt{\sum(X_i - \bar{X})^2/n}$ for the following data, which represent the number of companions per child.

```
4 2 5 3 1
3 2 1 2 3
2 4 3 2 0
```

*11. Compute S for the data in Exercise 10 using $\sqrt{[n\sum X_i^2 - (\sum X_i)^2]/n^2}$.

12. Infants are not as passive and undiscriminating about stimulation as we once thought; they show distinct preferences when given an opportunity to control stimuli presented to them. The following data are the number of trials required for infants to learn to control visual stimuli by varying their sucking responses. (a) Compute the mean and standard deviation for these data. Compute S using both the deviation and the raw-score formulas. (b) Construct a frequency polygon.

$$
\begin{array}{cccccc}
81 & 73 & 75 & 72 & 76 & 74 \\
77 & 72 & 71 & 74 & 72 & 73 \\
73 & 70 & 78 & 73 & 71 & 69 \\
75 & 74 & 68 & 70 & 69 & 73 \\
72 & 70 & 66 & 71 & 75 & 72 \\
76 & 74 & 73 & 77 & & \\
\end{array}
$$

13. In a concept learning experiment, chimpanzees were taught to recognize a triangle in different orientations. Compute the mean and standard deviation using $\sqrt{[n \sum f_j X_j^2 - (\sum f_j X_j)^2]/n^2}$ for the data in the table, which represent the number of trials to reach the learning criterion.

Number of Trials	f	Number of Trials	f
56–57	1	42–43	5
54–55	1	40–41	5
52–53	2	38–39	3
50–51	3	36–37	2
48–49	3	34–35	2
46–47	4	32–33	0
44–45	6	30–31	1

*14. Apply Sheppard's correction to the standard deviation in Exercise 13.

**15. The attitudes of a random sample of white female college students toward having a career were surveyed. *(a) For the data in the table, compute the mode and index of dispersion. (b) Construct a bar graph.

Category	f
Strongly desire career	16
Moderately desire career	23
Undecided about career	19
Don't want career	10

16. The following data for black female high school students were obtained in the survey described in Exercise 15. Compute the mode and index of dispersion.

Category	f
Strongly desire career	38
Moderately desire career	19
Undecided about career	5
Don't want career	17

4.2 Four Measures of Dispersion

**17. Interpret the following. *(a) $\bar{X} = 100$, $S = 15$, and the distribution is approximately normal, (b) $Mdn = 50$, $Q = 8$, (c) $Mo = 30$, $R = 5$, (d) $Mo =$ Category of Ford cars, $D = 0.20$.

18. Terms to remember:
    a. Deviation formula for S
    b. Raw-score formula for S
    c. Sheppard's correction

4.3 Relative Merits of the Measures of Dispersion

Standard Deviation

The standard deviation, which is typically reported with the mean, is the most important and most widely used measure of dispersion for quantitative variables with distributions that are relatively symmetrical. Its popularity is due largely to its superior sampling stability and its mathematical tractability. There are two situations, however, in which the standard deviation is not the preferred measure of dispersion: when a distribution is very skewed and when the data are qualitative. Consider the case of a skewed distribution. The value of the standard deviation is determined by squaring the deviation of each score from the mean. The squaring operation gives undue weight to extreme scores in the longer tail and results in a much larger standard deviation than would have been obtained in the absence of extreme scores. This is a disadvantage. Suppose we wished to compare the dispersion of two distributions that are similar except that one contains several more extreme scores in the longer tail. In spite of their similarity, the standard deviations of the distributions would be quite different, and the comparison would be misleading. A few extreme scores exert an influence that is disproportionate to their number.

The standard deviation shouldn't be used with qualitative variables, since it is a distance measure, actually the root-mean-squared distance by which scores deviate from the mean. If data don't contain distance information, as in the case of qualitative variables, the standard deviation is meaningless.

Semi-Interquartile Range

The semi-interquartile range is reported with the median and shares many of its advantages and disadvantages. It is limited to descriptive applications with quantitative variables and is relatively intractable mathematically. Nevertheless, it is preferred over the standard deviation in two situations. We saw in Section 3.5 that the median can be computed for open-ended distributions. So can the semi-interquartile range if the unknown scores lie above Q_3 or below Q_1. The standard deviation can also be computed when there are unknown scores, but none of the procedures for doing so is entirely satisfactory. The semi-interquartile range is also preferred for skewed distributions—it is sensitive to the number but not the value of scores lying above Q_3 and below Q_1, so it is less influenced by the extreme scores in the longer tail of a distribution than the standard deviation. However, it is only when distributions are markedly skewed or open-ended that the semi-interquartile range is preferred over the standard deviation.

Range

The range is used for quantitative variables and may be reported with the mode. The great advantage of the range is its simplicity—it is easy to understand and compute. As a result, it is widely used as a preliminary measure of dispersion. It is also used in deciding how to group data in a frequency distribution, an application that was described in Section 2.2.

The major deficiency of the range is its poor sampling stability. Its value is determined by only two scores (the largest and the smallest), which means it is not sensitive to most of the score values.

Another deficiency is its dependency on sample size. If scores are randomly sampled from a population, the range will tend to be larger for larger samples since large samples are more likely to include extreme scores. These deficiencies, plus its poor mathematical tractability, limit the range to descriptive applications.

Index of Dispersion

The index of dispersion is the only measure that is appropriate for qualitative variables. Unlike other dispersion measures, it represents not distance but the number of distinguishable pairs of observations relative to the maximum possible number in c categories. It is usually reported with the mode, which is the appropriate measure of central tendency for qualitative variables. However, the index of dispersion is less familiar than the other measures of dispersion and is rarely used in advanced statistical procedures.

Summary of the Properties of the Four Measures of Dispersion

The standard deviation is

1. a distance measure—the root-mean-squared distance by which scores deviate from the mean;
2. the preferred measure for quantitative variables with distributions that are relatively symmetrical;
3. often reported with the mean—for a normal distribution, $\bar{X} \pm S$ is an interval that contains 68.27% of scores;
4. the measure with the best sampling stability;
5. widely used, implicitly or explicitly, in advanced statistics;
6. mathematically tractable;
7. the only measure with a value that is affected by the value of every score in the distribution;
8. fairly sensitive to extreme scores, so it is not recommended for markedly skewed distributions; and
9. not appropriate for qualitative variables.

The semi-interquartile range is

1. a distance measure—one-half the distance between the 1st and 3rd quartiles;
2. often reported with the median for quantitative variables;
3. closely related to the median, since both are defined in terms of quartile points;

4. sensitive only to the number and not the value of scores above Q_3 and below Q_1, hence it is widely used for markedly skewed distributions;
5. the only relatively stable measure that is appropriate for open-ended distributions;
6. more subject to sampling fluctuation than the standard deviation;
7. less mathematically tractable than the standard deviation; and
8. rarely used in advanced statistical procedures.

The range is

1. a distance measure—the distance between the largest and smallest scores, inclusive;
2. often reported with the mode for quantitative variables;
3. the simplest measure of dispersion to compute and interpret;
4. used in deciding how to group data in a frequency distribution;
5. much more subject to sampling fluctuation than the other measures of dispersion;
6. dependent on sample size—the larger the sample size, the larger, on the average, the range;
7. less mathematically tractable than the standard deviation; and
8. rarely used in advanced statistical procedures.

The index of dispersion is

1. a measure of the distinguishability of observations—that is, the number of distinguishable pairs of observations relative to the number possible. The index is zero when all observations are in one qualitative category (minimum dispersion) and has its maximum value of one when the observations are evenly distributed over the categories (maximum dispersion);
2. the only measure appropriate for qualitative variables;
3. reported with the mode;
4. rarely used in advanced statistical procedures; and
5. less familiar than R, Q, and S, which are based on the concept of distance.

Review Exercises for Section 4.3

**19. What measure of central tendency and dispersion would you compute for the following data? Defend your choice.

*a.

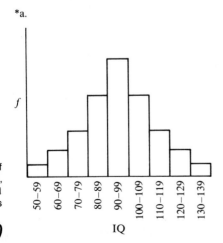

b.

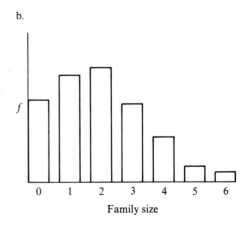

*c.

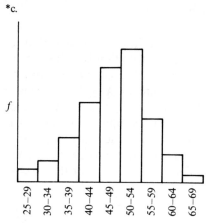

Creativity scores of doctoral candidates in English

*d.

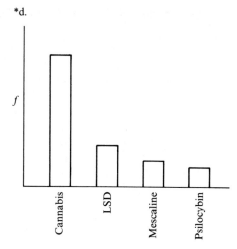

Frequency of drug use among teenagers

e.

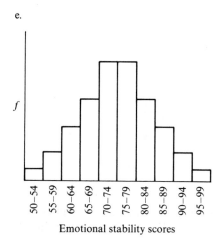

Emotional stability scores

f.

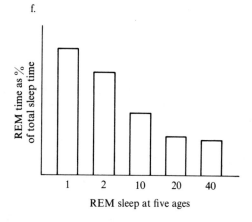

REM sleep at five ages

g.

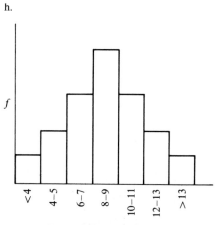

Assembly line productivity during a week

h.

Income (thousand dollars)

4.3 Relative Merits of the Measures of Dispersion

4.4 Dispersion and the Normal Distribution

The distribution of many variables in the behavioral sciences and education resembles the bell-shaped normal distribution. Because this distribution is so important, its properties have been extensively studied by mathematicians. We saw earlier that 50% of scores fall above and 50% below the mean and that the interval $\bar{X} \pm S$ is a distance that includes 68.27% of scores. Suppose that we are interested in the interval $\bar{X} \pm 2S$ or $\bar{X} \pm 3S$. The percent of scores included within these intervals is shown in Figure 4.4-1. It can be seen that an interval of six standard deviations includes almost all of the scores, 99.73%. Also, $\bar{X} \pm S$ gives the scores that mark the inflection points of the normal distribution—that is, where the curve changes from convex to concave or the reverse.

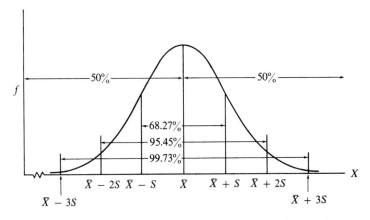

Figure 4.4-1. Percentage of scores contained in selected intervals around the mean for a normal distribution.

For a normal distribution, a knowledge of one dispersion measure enables us to determine the other two. For example, $Q = 0.6745S$ and $S = 1.483Q$. Computing the range from Q or from S is not as simple, since the range is a function of sample size as the following approximations illustrate.[11]

$$S \cong \frac{R}{3} \quad \text{and} \quad R \cong 3S \quad \text{when} \quad n = 10$$

$$S \cong \frac{R}{4} \quad \text{and} \quad R \cong 4S \quad \text{when} \quad n = 30$$

$$S \cong \frac{R}{5} \quad \text{and} \quad R \cong 5S \quad \text{when} \quad n = 100$$

$$S \cong \frac{R}{6} \quad \text{and} \quad R \cong 6S \quad \text{when} \quad n = 500$$

[11] Adapted from Tippett (1925).

For any distribution, the minimum value of S is $R/\sqrt{2n}$ and the maximum is $R/2$.[12] This information can be used to check quickly the value of S for computational errors. For example, if $n = 30$, $R = 60$, and the distribution is approximately normal, the standard deviation should be close to $R/4 = \frac{60}{4} = 15$. If your computed S is 25, you should be suspicious; if it is greater than $R/2 = \frac{60}{2} = 30$, you know you have made a computational error.

Review Exercises for Section 4.4

**20. For a normal distribution, what percentage of the scores fall *(a) below $\bar{X} + S$? (b) above $\bar{X} - 3S$? (c) above $\bar{X} + 2S$?

*21. Assume a normal distribution, $S = 5$, and $n = 30$. Estimate (a) Q and (b) R.

22. Assume a normal distribution, $S = 10$, and $n = 100$. Estimate (a) Q and (b) R.

*23. If $R = 30$ and $n = 50$, determine the minimum and maximum values for S.

24. If $R = 40$ and $n = 32$, determine the minimum and maximum values for S.

**25. Which of the following values of S are incorrect?
   *(a) $S = 10$, $R = 42$, $n = 30$
   (b) $S = 15$, $R = 25$, $n = 10$
   (c) $S = 18$, $R = 210$, $n = 50$

26. Term to remember:
   Inflection point

†4.5 Tchebycheff's Theorem

The percentage of scores contained in an interval bounded by $\bar{X} \pm S$, $\bar{X} \pm 2S$, and $\bar{X} \pm 3S$ is shown in Figure 4.4-1 for a normal distribution. *For nonnormal distributions, what is the minimum percentage of scores contained in these intervals?* This question can be answered by **Tchebycheff's theorem**, which states that no matter how skewed the distribution, $\bar{X} \pm k$ standard deviation units, where $k \geq 1$, will include at least $(1 - 1/k^2) \times 100\%$ of the scores. For example, if $k = 2$, $\bar{X} \pm 2S$ will contain no less than $(1 - 1/2^2) \times 100 = 75\%$ of the scores; the corresponding percentage for the normal distribution is 95.45. Table 4.5-1 gives a comparison for selected values of k.

Table 4.5-1. Illustrative Values of $1 - 1/k^2$ Giving the Minimum Percentage of Scores Bounded by $\bar{X} \pm kS$

k	Normal Distribution (%)	Nonnormal Distribution (%)
1.0	68.27	0
1.5	86.64	55.56
2.0	95.45	75.00
2.5	98.76	84.00
3.0	99.73	88.89

[12] See Section 4.8-4 for a proof.
† This section can be omitted without loss of continuity.

Since Tchebycheff's theorem applies to any distribution, it is very conservative. It and the normal distribution can be used to bracket the percentage of scores that would be expected to fall in the interval $\bar{X} + kS$. For example, if $k = 2$, at least 75% will fall within $\bar{X} \pm 2S$; as the distribution approaches the normal form, the percentage approaches 95.45.

Review Exercises for Section 4.5

**27. Compute the minimum percentage of scores that can be expected to lie within k standard deviations of the mean when the distribution is not normal and k is *(a) 1.75 (b) 2.25 (c) 4 (d) 4.50.

**28. Suppose a distribution is believed to be approximately normal. Give the minimum and probable percentage of scores that can be expected to lie within k standard deviations of the mean when k is equal to *(a) 1 *(b) 2 (c) 2.5 (d) 3.

4.6 Skewness and Kurtosis

To complete our description of a distribution we need two more statistics: indexes of skewness and kurtosis. We saw in Section 2.6 that skewness refers to the asymmetry of a distribution and kurtosis to its peakedness or flatness.

Skewness

A number of indexes of skewness have been developed; the most widely used one is[13]

$$Sk = \frac{\dfrac{\sum_{i=1}^{n}(X_i - \bar{X})^3}{n}}{\left[\dfrac{\sum_{i=1}^{n}(X_i - \bar{X})^2}{n}\right]\sqrt{\dfrac{\sum_{i=1}^{n}(X_i - \bar{X})^2}{n}}}.$$

If a distribution is symmetrical, $Sk = 0$; if it is positively skewed, $Sk > 0$; and if it is negatively skewed, $Sk < 0$. Computation of the index is illustrated in Table 4.6-1. For these data, $Sk = -0.7$, which indicates that the distribution is negatively skewed, as Figure 4.6-1 shows.

The value of Sk can be used to compare the type and degree of skewness of two distributions independent of any differences in central tendency and dispersion. However, in practice Sk is rarely computed, since asymmetry can easily be detected by looking at a frequency distribution or graph.

[13] This index, developed by Karl Pearson, is sometimes denoted by $\sqrt{\beta_1}$ and sometimes, by g_1.

Table 4.6-1. Example Illustrating Computation of Measures of Skewness and Kurtosis

(i) Data:

X_i	$(X_i - \bar{X})$	$(X_i - \bar{X})^2$	$(X_i - \bar{X})^3$	$(X_i - \bar{X})^4$
6	2	4	8	16
5	1	1	1	1
5	1	1	1	1
5	1	1	1	1
4	0	0	0	0
3	−1	1	−1	1
2	−2	4	−8	16
1	−3	9	−27	81

$\sum_{i=1}^{n} X_i = 36 \quad \sum_{i=1}^{n}(X_i - \bar{X}) = 0 \quad \sum_{i=1}^{n}(X_i - \bar{X})^2 = 22 \quad \sum_{i=1}^{n}(X_i - \bar{X})^3 = -24 \quad \sum_{i=1}^{n}(X_i - \bar{X})^4 = 118$

$\bar{X} = 4$

(ii) Computation of Sk:

$$Sk = \frac{\dfrac{\sum_{i=1}^{n}(X_i - \bar{X})^3}{n}}{\left[\dfrac{\sum_{i=1}^{n}(X_i - \bar{X})^2}{n}\right]\sqrt{\dfrac{\sum_{i=1}^{n}(X_i - \bar{X})^2}{n}}} = \frac{\dfrac{-24}{9}}{\left(\dfrac{22}{9}\right)\sqrt{\dfrac{22}{9}}} = \frac{-2.67}{3.82} = -0.7$$

(iii) Computation of Kur:

$$Kur = \frac{\dfrac{\sum_{i=1}^{n}(X_i - \bar{X})^4}{n}}{\left[\dfrac{\sum_{i=1}^{n}(X_i - \bar{X})^2}{n}\right]^2} = \frac{\dfrac{118}{9}}{\left(\dfrac{22}{9}\right)^2} = \frac{13.11}{5.98} = 2.2$$

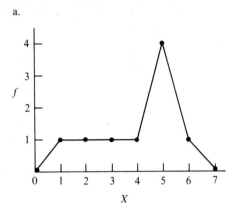 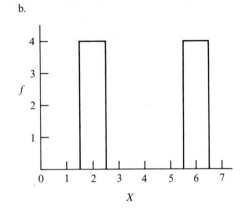

Figure 4.6-1. (a) Frequency polygon for data in Table 4.6-1; $Sk = -0.7$ and $Kur = 2.2$. (b) Histogram for perfectly symmetrical bimodal distribution; $Sk = 0$ and $Kur = 1$.

Kurtosis

The most common index of kurtosis is[14]

$$Kur = \frac{\dfrac{\sum_{i=1}^{n}(X_i - \bar{X})^4}{n}}{\left[\dfrac{\sum_{i=1}^{n}(X_i - \bar{X})^2}{n}\right]^2}.$$

If a distribution is flatter than the normal distribution, it is called *platykurtic* and $Kur < 3$. If its peakedness is the same as that of the normal distribution, it is *mesokurtic* and $Kur = 3$. If it is more peaked than the normal distribution, it is *leptokurtic* and $Kur > 3$. Computation of the index is illustrated in Table 4.6-1.

A graph for these data and one for a perfectly symmetrical bimodal distribution are given in Figure 4.6-1. In Figure 4.6-1(a) $Kur = 2.2$, and in Figure 4.6-1(b) $Kur = 1$. It should be apparent that the interpretation of Kur is not as straightforward as that for Sk. It turns out that the value of Kur is dependent not only on the central peak of a distribution, but on the shape of its tails as well. Therefore, for distributions that deviate appreciably from the normal form, like those in Figure 4.6-1, the interpretation of Kur is ambiguous.[15] It is doubtful whether any single statistic can adequately measure the quality of peakedness.

[14] This index, also developed by Karl Pearson, is sometimes denoted by β_2. Some writers use the index $Kur - 3$ for kurtosis and denote it by g_2.
[15] This point is discussed in detail by Chissom (1970) and Darlington (1970).

Review Exercises for Section 4.6

*29. Age at onset of Parkinson's disease, a degenerative brain disorder, was determined for a sample of adults between 60 and 70 years old. (a) Determine the type and degree of skewness for these data. (b) Construct a histogram. (c) Does it support your decision based on Sk?

67	68	60	64	68	63
68	70	63	70	68	69
70	69	69	69	69	68
62	70	70	64	66	66
66	69	67	67	70	67

30. The reading readiness of preschool children in two neighborhoods was measured. (a) Determine the type and degree of skewness for these data. (b) Which set of data has the greatest skewness? (c) Construct histograms. (d) Do the histograms support your decision based on Sk?

Neighborhood A						Neighborhood B				
30	33	32	31	35	33	29	32	28	29	29
32	29	33	30	32	28	30	31	26	30	28
31	31	29	31	26	30	28	29	29	34	30
32	30	33	32	27	32	29	27	30	31	35

*31. One theory predicts that the distribution of reaction times in a paired-associates learning task will be leptokurtic. Do the following learning data support the prediction?

28	28	27	29	29
29	31	32	28	30
27	28	30	31	25
24	27	28	25	27
28	24	32	27	28
29	28	27	28	29

*32. Determine the type and degree of kurtosis for the data in Exercise 29.
33. Determine which set of data in Exercise 30 deviates most from the normal distribution in terms of kurtosis.
34. Why is Kur not an entirely satisfactory measure of peakedness?
35. Terms to remember:
   a. Positively and negatively skewed
   b. Platykurtic
   c. Mesokurtic
   d. Leptokurtic

4.7 Summary

Measures of dispersion summarize the extent to which scores differ from each other, either quantitatively in terms of the spread of scores or qualitatively in terms of their distinguishability.

Of the four measures of dispersion discussed in this chapter, three are based on the concept of distance and are appropriate for quantitative variables. They are the range, the semi-interquartile range, and the standard deviation. The most important and widely used of the three is the standard deviation, which is typically reported with the mean.

The index of dispersion, which is reported with the mode, describes the distinguishability of observations. Specifically, it indicates the number of distinguishable pairs of observations relative to the maximum possible number of distinguishable pairs in c categories. The lower bound of the index, 0, occurs when all observations are in one category; its upper bound, 1, occurs when the observations are evenly distributed over the categories.

Dispersion and central tendency are generally the most important characteristics of a distribution, and they completely describe a normal distribution, which is by definition symmetrical and mesokurtic. For nonnormal distributions, Sk and Kur provide interesting but somewhat less important information about, respectively, skewness (symmetry) and kurtosis (peakedness).

†4.8 Technical Notes

4.8-1 Effect on the Standard Deviation of Adding a Constant to Each Score

Let S_{X+c} be the standard deviation of a distribution that has been altered by adding a constant c to each score—that is, $X_1 + c, X_2 + c, \ldots, X_n + c$. To determine the effect of adding a constant, we replace X_i by $(X_i + c)$ and $\bar{X}$ by $\sum_{i=1}^{n}(X_i + c)/n$ in the formula for S as follows.

$$S_{X+c} = \sqrt{\frac{\sum_{i=1}^{n}\left[(X_i + c) - \sum_{i=1}^{n}(X_i + c)/n\right]^2}{n}}$$

$$= \sqrt{\frac{\sum_{i=1}^{n}\left(X_i + c - \sum_{i=1}^{n} X_i/n - nc/n\right)^2}{n}} \quad \text{Rules 3.9-1-1, 3.9-1-2, 3.9-1-4}$$

$$= \sqrt{\frac{\sum_{i=1}^{n}(X_i + c - \bar{X} - c)^2}{n}} \quad \text{By definition } \bar{X} = \sum_{i=1}^{n} X_i/n.$$

$$= \sqrt{\frac{\sum_{i=1}^{n}(X_i - \bar{X})^2}{n}}$$

$$= S$$

Thus, adding a constant to each score doesn't affect the standard deviation. Similarly, it can be shown that subtracting a constant doesn't affect the standard deviation.

† These notes, which are not essential to the text, can be omitted without loss of continuity.

4.8-2 Effect on the Standard Deviation of Multiplying Each Score by a Constant

Let S_{cX} be the standard deviation of a distribution that has been altered by multiplying each score by a positive constant—that is, $cX_1, cX_2, \ldots, cX_n$. The effect of this alteration can be shown in the same way as in Section 4.8-1.

$$S_{cX} = \sqrt{\frac{\sum_{i=1}^{n}\left(cX_i - \sum_{i=1}^{n} cX_i/n\right)^2}{n}}$$

$$= \sqrt{\frac{\sum_{i=1}^{n}\left(cX_i - c\sum_{i=1}^{n} X_i/n\right)^2}{n}} \quad \text{Rule 3.9-1-3}$$

$$= \sqrt{\frac{\sum_{i=1}^{n}(cX_i - c\bar{X})^2}{n}} \quad \text{By definition } \bar{X} = \sum_{i=1}^{n} X_i/n.$$

$$= \sqrt{\frac{\sum_{i=1}^{n} c^2(X_i - \bar{X})^2}{n}}$$

$$= c\sqrt{\frac{\sum_{i=1}^{n}(X_i - \bar{X})^2}{n}}$$

$$= cS$$

Thus, the effect of multiplying each score by a positive constant is to change S, the standard deviation of the original scores, to cS. Similarly, it can be shown that the effect of dividing each score by a positive constant is to change S to S/c.

If c is a negative constant, $S_{cX} = |c|S$. The use of $|c|$ ensures that $|c|S$ is positive and is consistent with the definition of the standard deviation as the positive square root of $\sum_{i=1}^{n}(X_i - \bar{X})^2/n$.

4.8-3 Proof of the Equivalence of Raw-Score and Deviation Formulas for S

The algebraic equivalence of

$$\sqrt{\frac{\sum_{i=1}^{n}(X_i - \bar{X})^2}{n}} \quad \text{and} \quad \sqrt{\frac{n\sum_{i=1}^{n} X_i^2 - \left(\sum_{i=1}^{n} X_i\right)^2}{n^2}}$$

can be shown by deriving the formula on the right from the formula on the left as follows.

$$S = \sqrt{\frac{\sum_{i=1}^{n}(X_i - \bar{X})^2}{n}}$$

$$= \sqrt{\frac{\sum_{i=1}^{n}(X_i^2 - 2X_i\bar{X} + \bar{X}^2)}{n}}$$

$$= \sqrt{\frac{\sum_{i=1}^{n} X_i^2 - 2\bar{X}\sum_{i=1}^{n} X_i + n\bar{X}^2}{n}} \qquad \text{Rules 3.9-1-1–3.9-1-4}$$

$$= \sqrt{\frac{\sum_{i=1}^{n} X_i^2 - \dfrac{2\sum_{i=1}^{n} X_i}{n}\sum_{i=1}^{n} X_i + \dfrac{n\left(\sum_{i=1}^{n} X_i\right)^2}{n^2}}{n}} \qquad \begin{array}{l}\text{By definition}\\[4pt]\bar{X} = \sum_{i=1}^{n} X_i/n.\end{array}$$

$$= \sqrt{\frac{\sum_{i=1}^{n} X_i^2 - \dfrac{2\left(\sum_{i=1}^{n} X_i\right)^2}{n} + \dfrac{\left(\sum_{i=1}^{n} X_i\right)^2}{n}}{n}}$$

$$= \sqrt{\frac{\sum_{i=1}^{n} X_i^2 - \dfrac{\left(\sum_{i=1}^{n} X_i\right)^2}{n}}{n}}$$

$$= \sqrt{\frac{n\sum_{i=1}^{n} X_i^2 - \left(\sum_{i=1}^{n} X_i\right)^2}{n^2}}$$

4.8-4 Minimum and Maximum Values of S

The minimum and maximum possible values of the standard deviation are determined by the range. For a nonzero range, the minimum occurs when all but two scores are equal to the mean; this results in two nonzero deviations from the mean, as shown in Figure 4.8-1. We can substitute $-R/2$ and $R/2$ for two of the $(X_i - \bar{X})$'s and zeros for the rest in the standard deviation formula $\sqrt{\sum(X_i - \bar{X})^2/n}$ as follows.

$$S_{min} = \sqrt{\frac{1}{n}\left(\frac{-R}{2}\right)^2 + \frac{1}{n}(0)^2 + \cdots + \frac{1}{n}(0)^2 + \frac{1}{n}\left(\frac{R}{2}\right)^2}$$

$$= \sqrt{\frac{1}{n}\frac{R^2}{4} + \frac{1}{n}\frac{R^2}{4}}$$

$$= \sqrt{\frac{1}{n}\frac{2R^2}{4}}$$

$$= \frac{R}{\sqrt{2n}}$$

Thus, the minimum value of S is $R/\sqrt{2n}$.

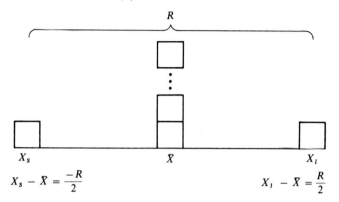

Figure 4.8-1. The midpoint of the smallest score is denoted by X_s; the midpoint of the largest score is denoted by X_l. For purposes of the deviation, the range is defined as $X_l - X_s$.

The maximum value of S occurs when half the scores are equal to the smallest score and the other half to the largest score. For this case, all the deviations from the mean are equal to $-R/2$ or $R/2$. Since the deviations in the formula $\sqrt{\sum (X_i - \bar{X})^2/n}$ are squared, we can substitute $R/2$ for $(X_i - \bar{X})$ as follows.

$$S_{max} = \sqrt{\frac{1}{n} \sum_{i=1}^{n} (R/2)^2}$$

$$= \sqrt{\frac{1}{n} \sum_{i=1}^{n} (R^2/4)}$$

$$= \sqrt{\frac{1}{n} n(R^2/4)} \qquad \text{Rule 3.9-1-1}$$

$$= \frac{R}{2}$$

Thus, the maximum value of S is $R/2$.

4.8.5 Derivation of Formula for the Index of Dispersion, D

Two formulas for D were presented in Section 4.2, $D = DP/DP'$ and

$$D = \frac{c\left(n^2 - \sum_{j=1}^{c} n_j^2\right)}{n^2(c-1)}.$$

We saw that the first formula reflects the observed number of distinguishable pairs of observations relative to the number that could occur if the c categories contained the same number, n_j, of observations. We will now show that the second formula reflects the same information.

If we have n observations, how many different pairs can be formed? Let $a_1, a_2, a_3, \ldots, a_n$ be n observations; a_1 can be paired with any of the other $n - 1$ observations to produce $n - 1$ pairs. Similarly, a_2 can be paired with the remaining $n - 2$ observations to produce $n - 2$ pairs, and finally a_{n-1} can be paired with a_n to produce 1 pair. Thus, among n observations there are $(n - 1) + (n - 2) + (n - 3) + \cdots + 1$ pairs. This sum is known to equal $n(n - 1)/2$. Thus, if the n observations are distinguishable, there are $n(n - 1)/2$ distinguishable pairs that can be formed.

Suppose now that the n observations are classified in two categories, 1 and 2, the first containing n_1 indistinguishable observations and the second n_2 indistinguishable observations, and $n_1 + n_2 = n$. We want to know the number of distinguishable pairs among the n observations, taking into account that those in category 1 are indistinguishable and those in category 2 are indistinguishable. Using the formula $n(n - 1)/2$, we can arrive at this sum as follows.

$$\begin{pmatrix}\text{Number of} \\ \text{distinguishable pairs}\end{pmatrix} = \begin{pmatrix}\text{Total number} \\ \text{of pairs}\end{pmatrix} - \begin{pmatrix}\text{Number of} \\ \text{indistinguishable} \\ \text{pairs in 1}\end{pmatrix} - \begin{pmatrix}\text{Number of} \\ \text{indistinguishable} \\ \text{pairs in 2}\end{pmatrix}$$

$$= \frac{n(n-1)}{2} - \frac{n_1(n_1-1)}{2} - \frac{n_2(n-1)}{2}.$$

For $j = 1, \ldots, c$ categories, the number of distinguishable pairs, DP, is

$$DP = \frac{n(n-1)}{2} - \sum_{j=1}^{c} \frac{n_j(n_j-1)}{2} = \left(\frac{n^2}{2} - \frac{n}{2}\right) - \frac{1}{2}\sum_{j=1}^{c} n_j^2 + \frac{1}{2}\sum_{j=1}^{c} n_j.$$

But $\sum_{j=1}^{c} n_j = n$; hence,

$$DP = \left(\frac{n^2}{2} - \frac{n}{2}\right) - \frac{1}{2}\sum_{j=1}^{c} n_j^2 + \frac{n}{2} = \frac{n^2}{2} - \frac{1}{2}\sum_{j=1}^{c} n_j^2$$

$$= \frac{1}{2}\left(n^2 - \sum_{j=1}^{c} n_j^2\right).$$

The lower limit of DP is 0 and occurs when all n observations are in one category, say 1, in which case $n = n_1$ and $DP = 1/2(n^2 - n_1^2) = 0$. The upper limit of the ratio occurs when the categories contain an equal number of observations—that is, when $n_1 = n_2 = \ldots = n_c$ and $n_j = n/c$. Let us denote the theoretical maximum of DP by DP'. We can compute the theoretical maximum by substituting n/c, a constant, for n_j in the formula for DP:

$$DP' = \frac{1}{2}\left[n^2 - \sum_{j=1}^{c}\left(\frac{n}{c}\right)^2\right] = \frac{1}{2}\left[n^2 - c\left(\frac{n}{c}\right)^2\right]$$

$$= \frac{1}{2}\left(n^2 - \frac{n^2}{c}\right).$$

Since D is equal to DP/DP', substituting for DP and DP', we obtain the index of dispersion as follows.

$$D = \frac{DP}{DP'}$$

$$= \frac{\frac{1}{2}\left(n^2 - \sum_{j=1}^{c} n_j^2\right)}{\frac{1}{2}\left(n^2 - \frac{n^2}{c}\right)} = \frac{c\left(n^2 - \sum_{j=1}^{c} n_j^2\right)}{cn^2 - c\frac{n^2}{c}}$$

$$= \frac{c\left(n^2 - \sum_{j=1}^{c} n_j^2\right)}{n^2(c-1)}$$

5

Correlation

5.1 Introduction
   Correlation and Regression Distinguished
   A Bit of History
   Review Exercises for Section 5.1
5.2 A Numerical Index of Correlation
   Review Exercises for Section 5.2
5.3 Pearson Product-Moment Correlation Coefficient
   Information Contained in the Cross Product
   Review Exercises for Section 5.3
5.4 Interpretation of r: Explained and Unexplained Variation
   Review Exercises for Section 5.4
5.5 Some Common Errors in Interpreting r
   Error: Interpreting r in Direct Proportion to Its Size
   Error: Interpreting r in Terms of Arbitrary Descriptive Labels
   Error: Inferring Causation from Correlation
   Review Exercises for Section 5.5
5.6 Factors that Affect the Size of r
   Nature of the Relationship between X and Y
   Truncated Range
   Spurious Effects Due to Subgroups with Different Means or Standard Deviations
   Normality and Heterogeneity of Array Variances
   Review Exercises for Section 5.6
5.7 Spearman Rank Correlation
   The Problem of Tied Ranks
   Review Exercises for Section 5.7
5.8 Other Kinds of Correlation Coefficients
5.9 Summary
5.10 Technical Notes
   Proof of Algebraic Equivalence of the Definitional Formula and Computational Formula for r
   Equivalence of r and r_s Formulas when Two Sets of Consecutive Untied Ranks $1,\ldots,n$ Are Substituted for X_i and Y_i in the Formula for r
   Effect on r of Adding a Constant to Each X Score
   Effect on r of Multiplying Each X Score by a Constant

5.1 Introduction

Correlation and Regression Distinguished

Correlation and regression, which are described in this chapter and the next, are procedures for describing the relationship between two sets of paired scores. Because the procedures have much in common, perhaps the simplest way to distinguish between them is by means of examples. Suppose we perform an experiment in which different dosages of amphetamine, the independent variable, denoted by X, are administered to children suffering from hyperkinesis, a behavioral disorder characterized by restlessness, inattention, and disruptive behavior. Children are assigned randomly to the dosage conditions. Following administration of the drug, changes in frequency of hyperkinetic behavior, the dependent variable, denoted by Y, are recorded. For each child the experimenter has paired X and Y scores representing, respectively, dosage and behavior change. The experimenter is interested in knowing whether the two variables are related and if so, in predicting Y from a knowledge of X. This experiment illustrates the key features of a problem in regression analysis. First, there is a clearly defined independent variable—amount of amphetamine—with dosage levels determined in advance by the experimenter. Second, the value of the dependent variable for a given dosage is free to take on different values. This is in contrast to the independent variable with values that were fixed in advance. Finally, the experimenter is interested in predicting Y from a knowledge of X.

Contrast this experiment with one in which tests of reading readiness and intelligence are administered to a sample of children yielding paired X and Y scores, respectively, for each child. The experimenter is interested in knowing whether variables X and Y are related, and if so, the strength of their association. In addition we might want to predict either variable from a knowledge of the other. This experiment illustrates a problem in correlation analysis. How does it differ from the regression problem? First, there is no obvious independent variable. Second, since the experimenter did not preselect the values of either X or Y, both X and Y are free to take on different values. Finally, the experimenter is interested in assessing the strength of association between X and Y and possibly in predicting either variable from a knowledge of the other.

To summarize, both correlation and regression procedures are concerned with assessing the relationship between sets of paired data. They differ in the way in which the samples are drawn, the status of one variable as the independent or predictor variable, and to some extent the kinds of conclusions drawn.

A Bit of History

The concepts of correlation and regression were developed by Sir Francis Galton during his investigations of the genetic transmission of natural characteristics. He was intrigued by the question "How is it possible for a whole population to remain alike in its features during many successive generations, if the average produce of each couple resembles the parents?" Data from one of his studies on the inheritance of stature are reproduced in Table 5.1-1. Parents' height is plotted on the horizontal or

Table 5.1-1. Scatter Diagram of Midparent Height and Height of Adult Offspring[a] (Female Heights Multiplied by 1.08)

Height of Adult Offspring	Midparent Height (Inches)[b]										
	<64	64.5	65.5	66.5	67.5	68.5	69.5	70.5	71.5	72.5	≥73
≥73.7							5	3	2	4	
73.2						3	4	3	2	2	③
72.2			1		4	11	4	9		⑦	1
71.2				2		18	20	7	4	2	
70.2				5	4	19	21	25	14	⑩	1
69.2	1	2		7	13	38	48	㉝	⑱	5	2
68.2	1		7	14	28	㉞	20	12	3	1	
67.2	2	5	⑪	⑰	㊳	31	27	3	4		
66.2	2	⑤	11	17	36	25	17	1	3		
65.2	①	1	7	2	15	16	4	1	1		
64.2	4	4	5	5	14	11	16				
63.2	2	4	9	3	5	7	1	1			
62.2		1		3	3						
<61.7	1	1	1			1		1			

[a] Galton (1889, p. 208). I am grateful to Edward W. Minium for bringing these data to my attention.
[b] Circles denote the class intervals containing the median of each column.

X axis and offspring's height on the vertical or Y axis. This representation of the joint frequency of two variables is called a *bivariate frequency distribution* or *scatter diagram* (*scattergram*). Consider the entry in the cell at the intersection of column 68.5 and row 69.2; the frequency is 48. This means that for parents whose height was 68–69, there were 48 offspring whose height was 68.7–69.7 in. The circles in Table 5.1-1 denote the class intervals containing the median of each column. We see, as did Galton, that the relationship between height of offspring and height of parents is approximately *linear*; that is, the set of circles approximates a straight line. Galton developed a procedure for finding the "straight line of best fit," thereby laying the foundation for correlation and regression.

A straight line provides a reasonably good fit for many relationships found in behavioral and educational research. Even those that are nonlinear are often approximately linear over some portion of their range. But let us return to the question that sparked Galton's interest. How is it that a population remains alike? The answer is in the trend represented by the circles in Table 5.1-1. We note that on the average, short parents have slightly taller offspring, while tall parents have slightly shorter offspring. This is shown more clearly in Figure 5.1-1. Galton referred to this tendency as *regression* or *reversion* toward the mean and called the best-fitting straight line the *regression* or *reversion line*.

In the discussion that follows we'll focus our attention on variables like those in Table 5.1-1 that appear to be linearly related.[1] We find that many relationships of interest in the behavioral sciences and education fall into this category.

Correlation

[1] The nonlinear case is treated in a number of excellent advanced texts such as Glass and Stanley (1970, pp. 150–152) and Hays (1973, pp. 683–684).

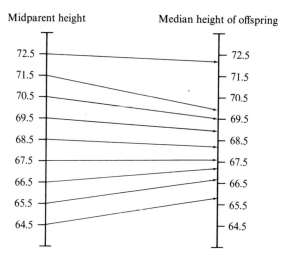

Figure 5.1-1. Arrows relate height (in inches) of parents to median height of offspring. Tall parents tend to have slightly shorter offspring, and short parents slightly taller offspring. Galton referred to this as reversion toward the mean.

Review Exercises for Section 5.1

*1. A speech therapist interested in the relationship between two tests of articulation disorders administered the tests to 26 children. (a) Construct a scatter diagram like Table 5.1-1 for the data in the following table. (b) Does the relationship appear to be linear or nonlinear?

Subject	Test A	Test B	Subject	Test A	Test B
1	26	36	14	33	39
2	28	35	15	22	33
3	25	34	16	24	32
4	21	32	17	27	35
5	25	33	18	29	36
6	26	32	19	32	39
7	26	34	20	28	36
8	31	37	21	25	36
9	27	34	22	24	34
10	20	30	23	25	34
11	23	32	24	27	36
12	30	38	25	26	34
13	29	37	26	26	35

2. A job-satisfaction questionnaire was administered to a random sample of 36 males between the ages of 29 and 34. The investigator was interested in the relationship between years of formal education completed and job satisfaction. (a) Construct a scatter diagram for the data in the following table. (b) Does the relationship appear to be linear or nonlinear?

Subject	Years of Education	Job Satisfaction	Subject	Years of Education	Job Satisfaction
1	14	36	19	12	43
2	11	38	20	11	46
3	10	36	21	18	53
4	15	51	22	8	30
5	7	30	23	9	35
6	8	37	24	12	40
7	12	40	25	13	40
8	13	43	26	13	41
9	16	47	27	10	32
10	12	44	28	14	50
11	12	37	29	12	33
12	11	40	30	14	47
13	9	32	31	10	38
14	12	42	32	11	37
15	13	45	33	12	40
16	11	38	34	14	50
17	12	42	35	13	42
18	11	37	36	13	45

*3. Discuss the meaning of the term "regression toward the mean."
 4. Terms to remember:
    a. Bivariate frequency distribution or scatter diagram
    b. Linear relationship
    c. Regression line

5.2 A Numerical Index of Correlation

The degree of association or strength of relationship between two variables is represented by a number called a **correlation coefficient**. A correlation coefficient for linearly related sample data is denoted by r_{XY}, or simply r; for a population, it is denoted by the Greek letter ρ (rho).[2] If the relationship is perfect, the value of r is either $+1$ or -1. A value of $+1$ denotes a perfect positive relationship; this is depicted in the scatter diagram in Figure 5.2-1(a). For this case, all the data points fall on a straight line such that high scores on one variable are paired with high scores on the other, while low scores are paired with low scores. A coefficient of -1 denotes a perfect negative or inverse relationship. For this case, the data points also fall on a straight line, but high scores on one variable are paired with low scores on the other and vice versa, resulting in a line that slopes down instead of up. This is shown in Figure 5.2-1(b). If there is no association between variables, $r = 0$. In this case, the data points tend to fall in a circle as shown in Figure 5.2-1(c). Intermediate degrees of association are represented by

Correlation

[2] Galton originally used the letter r from the word *reversion* to denote the slope of the straight line of best fit.

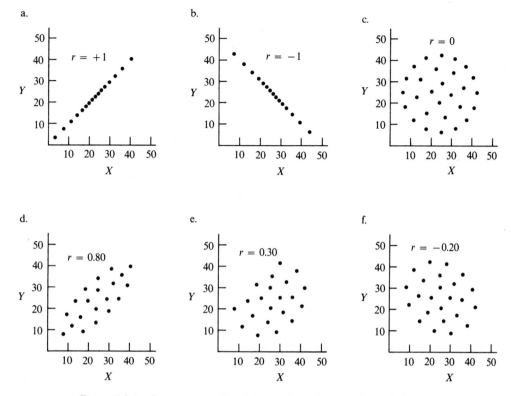

Figure 5.2-1. Scattergrams illustrating various degrees of correlation.

coefficients between -1 and 0 ($-1 < r < 0$) or by coefficients between 0 and 1 ($0 < r < 1$). Some examples of intermediate degrees of association for normally distributed X and Y variables are depicted in Figures 5.2-1(d)–(f). As shown in the figures, the data points for intermediate values of r tend to form an ellipse; the lower the degree of association, the more the ellipse resembles a circle.

Review Exercises for Section 5.2

5. Distinguish between r and ρ.
**6. Match the r values 1, -1, 0, 0.4, and -0.9 with the scatter diagrams shown here.

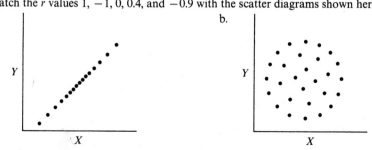

5.2
A Numerical Index
of Correlation

c.

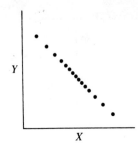

d.

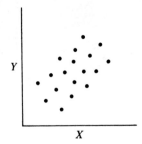

*e.

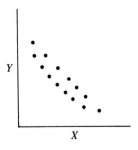

**7. Would you expect the correlation between the following to be positive, negative, or essentially zero?
   *a. Masculinity of fathers and sons
   *b. Reaction time and number of choice alternatives
   c. Mechanical aptitude and birth order
   d. Verbal intelligence and number of trials to learn a list of nonsense syllables
   e. Grades in college and annual income 10 years after graduation
   f. Number of letters in last name and musical aptitude
8. Terms to remember:
   a. Correlation coefficient
   b. Positive relationship
   c. Negative relationship

5.3 Pearson Product-Moment Correlation Coefficient

The most widely used index of correlation is called the **Pearson product-moment correlation coefficient** after Karl Pearson (1859–1936), who contributed so much to its development. The coefficient is appropriate for describing the relationship between two quantitative variables that are linearly related. The definitional formula for Pearson's r is

$$r = \frac{\sum_{i=1}^{n}(X_i - \bar{X})(Y_i - \bar{Y})}{\sqrt{\left[\frac{\sum_{i=1}^{n}(X_i - \bar{X})^2}{n}\right]\left[\frac{\sum_{i=1}^{n}(Y_i - \bar{Y})^2}{n}\right]}}.$$

A more convenient computational formula, called the raw-score formula,[3] is

$$r = \frac{n \sum_{i=1}^{n} X_i Y_i - \sum_{i=1}^{n} X_i \sum_{i=1}^{n} Y_i}{\sqrt{\left[n \sum_{i=1}^{n} X_i^2 - \left(\sum_{i=1}^{n} X_i\right)^2\right]\left[n \sum_{i=1}^{n} Y_i^2 - \left(\sum_{i=1}^{n} Y_i\right)^2\right]}}.$$

Calculation of r using the raw-score formula is illustrated in Table 5.3-1. The data are 20 paired father–son scores on a test of authoritarianism, which measures rigidity, dependency, and ethnocentrism. The coefficient is equal to 0.85. This tells us two things

Table 5.3-1. Computation of r for Father and Son Authoritarianism Scores

(i) Data:

Family	Father's Score, X	Son's Score, Y	XY	X^2	Y^2
1	25	28	700	625	784
2	32	31	992	1,024	961
3	40	41	1,640	1,600	1,681
4	29	33	957	841	1,089
5	31	25	775	961	625
6	16	18	288	256	324
7	28	26	728	784	676
8	36	38	1,368	1,296	1,444
9	33	34	1,122	1,089	1,156
10	29	36	1,044	841	1,296
11	23	20	460	529	400
12	27	28	756	729	784
13	37	30	1,110	1,369	900
14	30	26	780	900	676
15	27	22	594	729	484
16	20	23	460	400	529
17	28	29	812	784	841
18	38	36	1,368	1,444	1,296
19	35	32	1,120	1,225	1,024
20	19	19	361	361	361
	$\sum X_i = 583$	$\sum Y_i = 575$	$\sum X_i Y_i = 17{,}435$	$\sum X_i^2 = 17{,}787$	$\sum Y_i^2 = 17{,}331$

(ii) Computational procedure:

$$r = \frac{n \sum X_i Y_i - \sum X_i \sum Y_i}{\sqrt{[n \sum X_i^2 - (\sum X_i)^2][n \sum Y_i^2 - (\sum Y_i)^2]}} = \frac{20(17{,}435) - (583)(575)}{\sqrt{[20(17{,}787) - (583)^2][20(17{,}331) - (575)^2]}}$$

$$= \frac{13{,}475}{15{,}922.837} = 0.85$$

---

[3] The algebraic equivalence of this formula and the definitional formula is shown in Section 5.10-1.

about the relationship: its strength, represented by the extent to which the value of r differs from zero, and the direction of the relationship, represented by the sign of r. In the following discussion, we will see why r reflects this information. The interpretation of r is discussed in Section 5.4.

Information Contained in the Cross Product

The standing of a person with respect to the mean of variables X and Y can be expressed in terms of the size and algebraic sign of $(X_i - \bar{X})(Y_i - \bar{Y})$, which is the product of the deviation of the X_i and Y_i scores from their respective means. If a person is above the mean on both variables, the algebraic sign of $(X_i - \bar{X})(Y_i - \bar{Y})$, which is called a *cross product*, is positive and the associated data point falls in quadrant 2 of Figure 5.3-1. If a person is below the mean on both variables, $(X_i - \bar{X})(Y_i - \bar{Y})$ is also positive since it is the product of two negative numbers, but the corresponding data point falls in quadrant 4 of Figure 5.3-1. If a person is above the mean on one variable but below the mean on the other, the sign of $(X_i - \bar{X})(Y_i - \bar{Y})$ is negative and the data point falls in either quadrant 1 or 3.

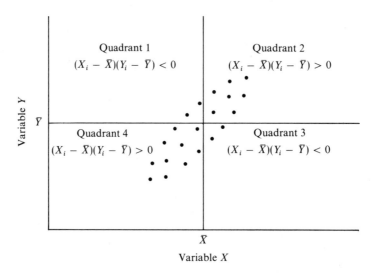

Figure 5.3-1. All the cross products in quadrants 1 and 3 are negative; those in quadrants 2 and 4 are positive.

In Figure 5.3-1 most of the data points are in quadrants 2 and 4; hence the algebraic sign of the sum $\sum_{i=1}^{n} (X_i - \bar{X})(Y_i - \bar{Y})$ is positive. When this sum is positive, the two variables are said to be positively related; that is, an increase in one variable is accompanied by an increase in the other. If an inverse relationship exists between X and Y, most of the data points fall in quadrants 1 and 3 and the sign of $\sum_{i=1}^{n} (X_i - \bar{X})(Y_i - \bar{Y})$ is negative. From the foregoing, it is apparent that the algebraic sign of $\sum_{i=1}^{n}(X_i - \bar{X})(Y_i - \bar{Y})$ in the numerator of

$$r = \frac{\dfrac{\sum_{i=1}^{n}(X_i - \bar{X})(Y_i - \bar{Y})}{n}}{\sqrt{\left[\dfrac{\sum_{i=1}^{n}(X_i - \bar{X})^2}{n}\right]\left[\dfrac{\sum_{i=1}^{n}(Y_i - \bar{Y})^2}{n}\right]}}$$

indicates whether X and Y are positively or inversely related.

The greater the strength of the relationship between X and Y, the larger the absolute value of $\sum_{i=1}^{n}(X_i - \bar{X})(Y_i - \bar{Y})$. A large positive value of $\sum_{i=1}^{n}(X_i - \bar{X})(Y_i - \bar{Y})$ occurs when the largest $(X_i - \bar{X})$ is paired with the largest $(Y_i - \bar{Y})$, the second largest $(X_i - \bar{X})$ with the second largest $(Y_i - \bar{Y})$, and so on. A large negative value of $\sum_{i=1}^{n}(X_i - \bar{X})(Y_i - \bar{Y})$ occurs when the largest $(X_i - \bar{X})$ is paired with the smallest $(Y_i - \bar{Y})$, and so on. If the correlation is positive but less than one, the cross product sum may be composed of some large positive values of $(X_i - \bar{X})$ multiplied by some moderate-sized $(Y_i - \bar{Y})$'s or even some negative $(Y_i - \bar{Y})$'s. The resulting value of $\sum_{i=1}^{n}(X_i - \bar{X})(Y_i - \bar{Y})$ is smaller than if the largest $(X_i - \bar{X})$ were paired with the largest $(Y_i - \bar{Y})$, and so on. If $r = 0$, positive $(X_i - \bar{X})$'s are as likely to be paired with negative $(Y_i - \bar{Y})$'s as with positive $(Y_i - \bar{Y})$'s, resulting in a sum of $\sum_{i=1}^{n}(X_i - \bar{X})(Y_i - \bar{Y})$ that is equal to zero.

It should be apparent after some reflection that the absolute value of the sum of the cross product reflects the strength of the association between X and Y, but it also reflects, or is affected by, the number of X and Y scores and the size of the standard deviations of X and Y. To obtain a measure of the strength of the association that is independent of the number of pairs of scores, we compute the mean of the cross product sum:

$$S_{XY} = \frac{\sum_{i=1}^{n}(X_i - \bar{X})(Y_i - \bar{Y})}{n},$$

where n is the number of paired X and Y scores. This mean is called the *covariance* of X and Y and is denoted by S_{XY}. If we divide the covariance by the standard deviations of X and of Y, S_X and S_Y, we obtain a measure of strength of association that is independent of the dispersions of the two variables as well. The resulting statistic,

$$r = \frac{\dfrac{\sum_{i=1}^{n}(X_i - \bar{X})(Y_i - \bar{Y})}{n}}{\sqrt{\left[\dfrac{\sum_{i=1}^{n}(X_i - \bar{X})^2}{n}\right]\left[\dfrac{\sum_{i=1}^{n}(Y_i - \bar{Y})^2}{n}\right]}} = \frac{S_{XY}}{S_X S_Y},$$

was defined earlier as the definitional formula for the Pearson product-moment correlation coefficient.

5.3 Pearson Product-Moment Correlation Coefficient

Review Exercises for Section 5.3

*9. A reading readiness test and an intelligence test were administered to a random sample of first-grade children. Compute r for the data in the table.

Child	Reading Readiness Score	IQ Score	Child	Reading Readiness Score	IQ Score
1	45	102	11	43	104
2	40	100	12	50	108
3	48	106	13	42	96
4	45	101	14	40	99
5	38	98	15	41	96
6	43	100	16	48	102
7	36	92	17	47	104
8	41	102	18	37	94
9	42	102	19	42	98
10	50	110	20	45	100

*10. Calculate r for the data in Exercise 1 (page 97).

11. Calculate r for the data in Exercise 2 (page 97–98).

**12. Calculate $\sum (X_i - \bar{X})(Y_i - \bar{Y})$ for the following data points. In which quadrants of Figure 5.3-1 would the data points fall? Are the variables related, and if so, is the relationship positive or negative?

*a. X Y   *b. X Y   c. X Y   d. X Y

 9 14   9 14   9 17   9 17
11 17   11 14   11 14   11 17
13 17   11 16   13 10   13 13
 7 12   9 16   7 19   7 13

13. What does the covariance S_{XY} tell us about the relationship between X and Y? In computing r, why is S_{XY} divided by $S_X S_Y$?

14. For a set of data with $S_X = 4$ and $S_Y = 5$, what is the largest possible value that S_{XY} can be? (*Hint*: The maximum value of $r = +1$ and $r = S_{XY}/S_X S_Y$.)

*15. If $n = 2$, what are the possible values of r? Make a scatter diagram that supports your answer.

*16. We are sometimes interested in knowing the variance of the sum of two paired variables $X + Y$. We might want to know this if, say, we added students' English and history grades together. The mean $\overline{X + Y}$ of the sum of paired scores, $\sum (X_i + Y_i)$, is equal to $(\bar{X} + \bar{Y})$. The variance of the sum is denoted by S^2_{X+Y}. Show that $S^2_{X+Y} = S_X^2 + S_Y^2 + 2rS_X S_Y$. (*Hint*: $S^2_{X+Y} = \sum [(X_i + Y_i) - (\bar{X} + \bar{Y})]^2/n = \sum [(X_i - \bar{X}) + (Y_i - \bar{Y})]^2/n$.)

17. Show that the variance of the difference between two variables, S^2_{X-Y}, is $S_X^2 + S_Y^2 - 2rS_X S_Y$. (*Hint*: $S^2_{X-Y} = \sum [(X_i - Y_i) - (\bar{X} - \bar{Y})]^2/n$.)

18. Terms to remember:
   a. Cross product
   b. Covariance

Correlation

5.4 Interpretation of r: Explained and Unexplained Variation

As we have seen, a correlation coefficient reflects the nature and strength of the linear association between two variables. However, two other statistics, both functions of r, are more useful for getting an intuitive notion for the strength of association represented by r. These statistics are the *coefficient of determination*, r^2, which is equal to the square of the correlation coefficient, and the *coefficient of nondetermination*, k^2, which is equal to $1 - r^2$.

If we examine the authoritarianism scores in Table 5.3-1, we see that there are differences among the X scores and among the Y scores. How can we account for intersubject variability? One reason why subjects differ on Y is that they differ on X. Since X and Y are correlated, $r = 0.85$, a person who is high on X is also likely to be high on Y. Because of the linear relationship between X and Y, some of the variation among the Y scores can be accounted for or explained by variation among the X scores. However, not all the variation can be explained in this way, since some subjects who have the same X score have different Y scores. For a given linear relationship between X and Y we would like to know how much of the Y score variance is accounted for by the X score variance and how much is not accounted for. We will see that this information is given by, respectively, r^2 and k^2. Let us denote the total variance of X and of Y by S_X^2 and S_Y^2, respectively. (Recall from the discussion of the standard deviation, page 70, that variance is a measure of dispersion: the mean squared distance by which scores deviate from the mean.) If we divide S_X^2 by itself and S_Y^2 by itself, we change both variances into proportions with values of one, each of which can be partitioned into two components as follows.[4]

$$\frac{S_X^2}{S_X^2} = r^2 + k^2$$

$$\begin{pmatrix}\text{Total } X \text{ variance} \\ \text{expressed as a} \\ \text{proportion}\end{pmatrix} = \begin{pmatrix}\text{Proportion of} \\ X \text{ variance explained} \\ \text{by } Y \text{ variance}\end{pmatrix} + \begin{pmatrix}\text{Proportion of} \\ X \text{ variance not explained} \\ \text{by } Y \text{ variance}\end{pmatrix}$$

$$\frac{S_Y^2}{S_Y^2} = r^2 + k^2$$

$$\begin{pmatrix}\text{Total } Y \text{ variance} \\ \text{expressed as a} \\ \text{proportion}\end{pmatrix} = \begin{pmatrix}\text{Proportion of} \\ Y \text{ variance explained} \\ \text{by } X \text{ variance}\end{pmatrix} + \begin{pmatrix}\text{Proportion of} \\ Y \text{ variance not explained} \\ \text{by } X \text{ variance}\end{pmatrix}$$

Thus, the total variance expressed as a proportion is equal to the coefficient of determination, r^2, plus the coefficient of nondetermination, k^2. To compute r^2 we square the correlation coefficient; k^2 is computed from $k^2 = 1 - r^2$.

For the authoritarianism data in Table 5.3-1, $r^2 = (0.85)^2 = 0.72$ and $k^2 = 1 - 0.72 = 0.28$. This means that 0.72 of the variance in the Y scores, for example, can

---

[4] The formal derivation is deferred to Section 6.7-3, since it uses concepts developed in Chapter 6.

be explained by the linear relationship between this variable and the corresponding X scores, as estimated by the regression line for X and Y, but 0.28 of the Y variance is not explained by the X variance. To put it more simply, the linear relationship between the fathers' and sons' scores enables us to account for much of the variance (0.72) in the sons' authoritarianism scores; however, 0.28 of the variance cannot be accounted for. In all likelihood we could find other variables such as the sons' levels of education that would enable us to reduce the proportion of unaccounted-for variance. The index k^2 is a measure of how much of the variance is unaccounted for.

A visual representation of the proportion of explained and unexplained variance is presented in Figure 5.4-1(a), where the proportions $S_Y^2/S_Y^2 = 1$ and $S_X^2/S_X^2 = 1$ are represented by the areas of circles. The area where the circles overlap corresponds to r^2; the nonoverlap areas correspond to k^2.

If $r = +1$ or -1, the circles completely overlap and all the variance in one variable is explained by that in the other variable. If $r = 0$, the circles do not overlap and none of the variance in either variable is explained by that in the other variable.

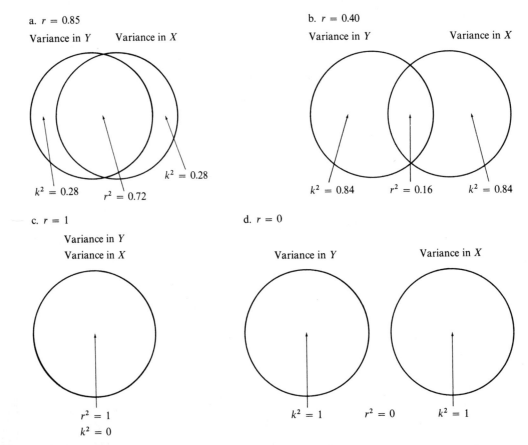

Figure 5.4-1. Visual representation of r^2, the proportion of variance in one variable that is explained by the variance in the other variable, and k^2, the proportion that is not explained.

Most variables of interest to behavioral scientists and educators are affected by a multiplicity of factors. School performance, for example, is affected by academic aptitude, scholastic motivation, health, and parental support for achievement, to name only a few. The correlation between performance and academic aptitude is 0.30, which tells us we have accounted for $(0.30)^2 = 0.09$ of performance variance and we have to look to other variables to account for the remaining $1 - 0.09 = 0.91$ of the variance. Note that since $r^2 \leq |r|$, values of $|r|$ close to 1 are required to account for an appreciable proportion of variance. Not until $r = 0.71$ does $r^2 = 0.50$.

Review Exercises for Section 5.4

19. What do r^2 and k^2 tell us about the relationship between X and Y?
**20. For the following experiments, compute r^2 and k^2 and interpret them verbally and by means of diagrams like those in Figure 5.4-1.
   *a. The correlation between freshman English grades and grades in a physical education bowling class was 0.22.
   b. The correlation between the number of hours rats had been deprived of food and time to traverse a maze with sunflower seeds in the goal box was 0.80.
   c. The correlation between the last two digits of students' social security numbers and number of trials to learn nonsense syllables was 0.02.
21. Terms to remember:
   a. Coefficient of determination, r^2
   b. Coefficient of nondetermination, k^2

5.5 Some Common Errors in Interpreting r

Error: Interpreting r in Direct Proportion to Its Size

Correlation coefficients are often incorrectly interpreted. A common error is to interpret the size of r as directly reflecting the percentage of association. For example, it is incorrect to say there is a 60% association between two variables when $r = 0.60$. This statement is meaningless. Does it mean that 60% of the elements are associated? An r does not indicate the percentage of association but rather is a measure of strength of association on a scale of -1 to $+1$. A related misconception is saying, for example, that $r = 0.80$ represents twice the relationship indicated by $r = 0.40$ or that an increase in correlation from 0.10 to 0.20 is equivalent to an increase in correlation from 0.60 to 0.70.

Error: Interpreting r in Terms of Arbitrary Descriptive Labels

Various schemes have been suggested to help students interpret correlation coefficients. A common but misleading scheme is the classification of certain r values as "very high" (for example, $r \geq 0.90$), "high" ($r = 0.70-0.89$), "medium" ($r = 0.30-0.69$), or "low" ($r < 0.30$). The problem with such classifications is that what constitutes a high or low correlation depends on what is being correlated with what, and what use is to be made of r once it has been computed. This will be illustrated for the concepts of *reliability* and *validity*—two desirable characteristics of psychological tests. One type of reliability called "test–retest reliability" is determined by administering a test to a group of subjects, waiting, and then readministering the test to the same subjects. The test's reliability, or consistency of measurement, is the correlation between the two sets of scores. Reliability coefficients of 0.90 or higher are common for tests of intellectual aptitude. A test–retest reliability coefficient below 0.80 would raise serious questions about the reliability of an intelligence test; however, the system described above would classify $r = 0.80$ as high. Equally misleading designations result when these schemes are used to interpret validity coefficients. The validity of a test is the degree to which it measures what it is supposed to measure. To assess the validity of, say, a college aptitude test, students' aptitude scores can be correlated with their grade point averages. The best aptitude tests rarely have validity coefficients above 0.60. It is misleading to label a validity coefficient of 0.60 as medium when higher coefficients are seldom ever obtained. An $r = 0.60$ is an extremely high validity coefficient, but a very, very low reliability coefficient. It should be apparent from these examples that no single classification scheme for interpreting r is applicable to all situations.

Error: Inferring Causation from Correlation

Another common error in interpreting a correlation coefficient is to infer that because two variables are correlated, one causes the other. *A nonzero correlation coefficient simply means there is a* **concomitant relationship** *between X and Y; that is, variation in one variable is associated in some way with variation in the other.* It is true that if X causes Y, there must be a correlation between the variables. However, the converse of this statement is not true. A concomitant relationship is necessary but not sufficient for inferring causality. A concomitant relationship often exists because both variables are caused by a third variable. For example, it does not necessarily follow from the positive correlation between Sunday school attendance and honesty that attending Sunday school causes honesty. In all likelihood both variables are caused by a third variable—certain early childhood training practices in the home.

It is easy to fall into the trap of inferring causality from correlation, especially when one variable occurs before the other. Consider the well-publicized positive correlation between years of formal education and income. Does this mean that going to college causes one to earn more money? Before giving an affirmative answer we would have to know how much college graduates would have earned if they hadn't gone to college. A causal relationship may in fact exist, but this can't be ascertained from the correlation. Some or all of the correlation between education and income might be explained in terms of other causal variables. For example, colleges attract two kinds of students—the bright and the rich. We know that bright individuals tend to rise to better-paying jobs whether or not they have gone to college and few children of rich parents end up poor.

Review Exercises for Section 5.5

**22. Which of the following are incorrect interpretations of a correlation coefficient and why?
  *a. The strength of association between test forms L and M is 0.96.
  b. The correlation between height and weight at age 6 is 0.40; this is twice as high as that at age 16 when $r = 0.20$.

*c. There is a medium correlation, $r = 0.67$, between the age at which babies can roll over and the age at which they can sit up alone.

d. Since the correlation between reaction time and number of automobile accidents is 0.20, 96% of the variance in frequency of accidents is unaccounted for.

e. We can conclude from the high correlation between level of motivation and number of elective offices sought that office-seeking behavior is caused at least in part by motivation.

23. What is wrong with interpreting r
    a. in direct proportion to its size?
    b. in terms of arbitrary descriptive labels?
    c. as indicating causality?

*24. In an attempt to help children with low IQs improve their school performance, a special perceptual awareness program was instituted. Suppose the program was completely ineffective. The group's mean IQ before the program was 72. Would you expect it to change after the special program, and if so in what direction? (If you don't see the issue, reread Section 5.1.)

25. Employees with the worst accident rates were required to complete a safety course. Following the course, the employees had fewer accidents. Can we conclude that the course was effective? What controls could be used in the experiment to make the outcome easier to interpret?

26. Terms to remember:
    a. Reliability
    b. Validity
    c. Concomitant relationship

5.6 Factors that Affect the Size of r

Nature of the Relationship between X and Y

There are many ways in which two variables can be related. It is sufficient for our purposes to classify them as linear or nonlinear. Three examples showing the straight or curved lines of best fit for bivariate data are presented in Figure 5.6-1. In general, the more closely data points hug the line of best fit, whether it is a straight or curved line, the higher the correlation. We saw in Section 5.2 that when $r = +1$ or -1 the data points fall on a straight line. If X and Y are normally distributed, as the absolute value of r decreases, the points form fatter and fatter ellipses until finally, when $r = 0$, they tend to form a circle. The Pearson product-moment correlation always fits data

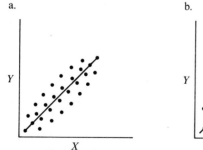
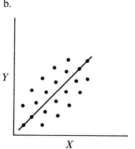
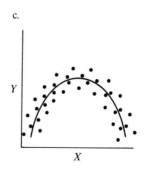

Figure 5.6-1. Parts a and b illustrate linear relationships and part c illustrates a nonlinear relationship. The higher the correlation, the better the data points hug the line of best fit.

points by a straight line. This works fine if the relationship is linear but not so well if the relationship is nonlinear, as in Figure 5.6-1(c). If a nonlinear relationship is fitted by a straight line, the data points will not hug the line as closely as they would an appropriate curved line; consequently, r underestimates the strength of association. In fact, an $r = 0$ can be obtained even though X and Y are highly correlated.

A different correlation measure called the **correlation ratio** *or* **eta squared (η^2)** *has been developed for determining the strength of association between nonlinearly related variables.* Eta squared fits data points by whatever line is appropriate. If the relationship is linear, a straight line is used and $\eta^2 = r^2$. For nonlinear relationships in which the correlation is not equal to zero, η^2 fits the points by a curved line, and its value is always larger than that for r^2. A discussion of the correlation ratio can be found in more advanced texts.[5]

How can we determine whether the relationship between X and Y is linear or nonlinear and hence whether to use r or η^2? There are statistical tests that can be used;[6] however, the simplest method is to examine the scatter diagram for evidence of non-linearity—the so-called "eyeball" test. Usually, visual inspection is adequate to detect cases in which r would underestimate strength of association.

In summary, r is a measure of the linear relationship between two quantitative variables. If the relationship is not linear, r underestimates the strength of association.

Truncated Range

The size of the Pearson product-moment correlation coefficient is affected by the range of the X and Y variables. If the range of either variable is restricted, the size of r will be reduced. Suppose we have administered an aptitude test to assembly line job applicants at a new factory. Because of the large number of jobs to be filled, all the applicants were hired regardless of their scores. Six months later we construct a scatter diagram like that in Figure 5.6-2, compute the correlation between aptitude scores and employee productivity, and find that $r = 0.55$. This is a very respectable validity coefficient. In the future if we had a surplus of applicants, we could improve productivity by only hiring those with high aptitude scores. Suppose that instead of hiring all the applicants when the plant opened, we had artificially restricted the range of aptitude scores by hiring only applicants with scores of 60 or above. For this case, the correlation between aptitude and productivity would have been 0.06 instead of 0.55. In this case, we would have incorrectly concluded that the test is of little value in selecting employees. The reason why the restriction or *truncation* of the range of the X variable results in a misleadingly low correlation coefficient can be seen from Figure 5.6-2. The effect would have been the same had the range of the Y variable been truncated.

The truncated range problem is common in behavioral and educational research, since much of it is conducted with college students who have been carefully screened for intelligence and related variables and consequently constitute a relatively homogeneous population. It's not surprising that college aptitude scores do not correlate very highly with grades, since admission offices truncate the range by admitting only students with high aptitude scores.

[5] See Glass and Stanley (1970, pp. 150–152) and Hays (1973, pp. 683–686).
[6] See, for example, Hays (1973, pp. 684–686).

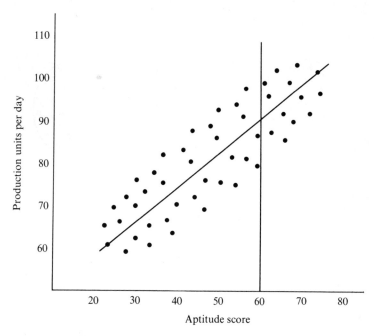

Figure 5.6-2. Scatter diagram illustrating the effect on *r* of restricting the range of *X* to scores of 60 or above. The *r* for the unrestricted range is 0.55; that for the restricted range is 0.06.

Spurious Effects Due to Subgroups with Different Means or Standard Deviations

A substantial correlation between X and Y may occur because the sample of subjects contains two or more subgroups with means that differ for both variables.[7] Suppose we are interested in the correlation between school achievement (Y) and anxiety level (X) as measured by the *Taylor manifest anxiety scale* and we obtain random samples of students from lower- and middle-class families. The correlation coefficient computed for the combined samples will be much higher than that for either sample taken alone. This occurs because the means for the two subgroups differ with respect to both X and Y. Those from middle-class families tend to perform better in school and to be somewhat more anxious than children from lower-class families. When the subgroups are combined, the correlation between achievement and anxiety is misleadingly high because of the differing means. The reason for this can be seen in Figure 5.6-3(a), where the letters L and M denote data points, respectively, for children from lower- and middle-class families.

A spurious correlation may occur when the standard deviations of the subgroups but not their means differ for both variables. This situation is depicted in Figure 5.6-3(b), where the letters A and B are used to denote the subgroups. Figure 5.6-3(c) and (d) depict other ways in which subgroups can produce spurious correlations.

[7] This problem has been examined in detail by Sockloff (1975).

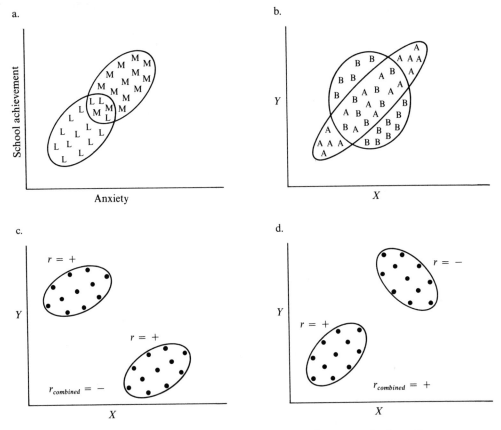

Figure 5.6-3. Scatter diagrams illustrating the effects on *r* of subsamples with means that differ on both variables (parts a, c, and d) or standard deviations that differ on both variables (part b). In part a the correlation is spuriously high; in part b it is too low for one subsample and too high for the other. Parts c and d show that the sign of the coefficient for the combined samples may differ from that for one or both of the subsamples.

From the foregoing, it is apparent that the inclusion of subgroups with different means or standard deviations on X and Y can affect the size and sign of r. Unfortunately, we are not always aware that our sample contains distinct subgroups. Our first clue may come when we construct a scatter diagram and note in retrospect that the scores clustered together tend to come from subjects who have some common distinguishing attribute.

Sometimes an experimenter intentionally conducts research with *extreme groups*—groups at opposite ends of a continuum. The use of introverts and extroverts, high and low achievers, or normals and neurotics enhances the likelihood of detecting other variables on which the groups differ. This is a useful research strategy, but it may lead to spuriously high correlation coefficients. Frequently, the means of the groups differ on both X and Y, and the data points have the shape illustrated in Figure 5.6-4. The data were selected from Table 5.3-1 so as to contain two extreme groups: the four

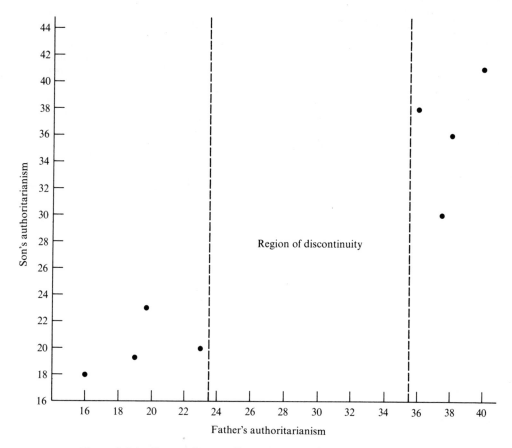

Figure 5.6-4. Scatter diagram illustrating the effects on *r* of using extreme groups. The data are taken from Table 5.3-1, with the eight data points representing the four highest and four lowest authoritarianism scores based on the fathers' data.

fathers with the highest authoritarianism scores and the four with the lowest scores. The correlation for all 20 father–son pairs in Table 5.3-1 is 0.85; the correlation based on the two extreme groups is 0.94.

Extreme groups constitute one type of *discontinuous distribution*. A discontinuous distribution results any time we restrict our samples to extreme groups or to a relatively small number of points along a continuum. It should be apparent from this discussion that correlation coefficients involving discontinuous distributions should be viewed with suspicion.

Normality and Heterogeneity of Array Variances

If the distributions of X and Y are markedly skewed, the value of r will be less than if the variables are approximately normally distributed. The reason for this can be seen in Figure 5.6-5, which shows various combinations of skewed X and Y distributions.

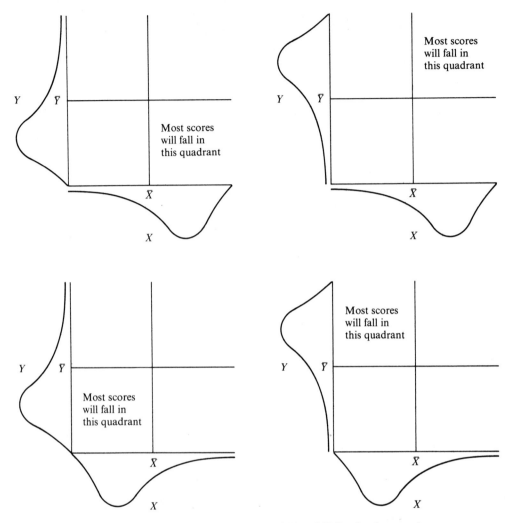

Figure 5.6-5. Effects of markedly skewed X and Y distributions on the distribution of data points in a scatter diagram.

The presence of skewed X and Y distributions is often accompanied by unequal dispersion of the Y scores for different values of X and similarly unequal dispersion of the X scores for different values of Y. This condition is called **heterogeneity of array (row and column) variances** or **heteroscedasticity**. This is illustrated in Figure 5.6-6. We saw earlier that r reflects the average degree to which scores hug the line of best fit. If the dispersion around the line differs for different values of, say, X and Y, the correlation coefficient will not have the same meaning as for homogeneous array variances. For example, in Figure 5.6-6(a) and (b) the correlation coefficient will underestimate the magnitude of association for low X scores and overestimate it for high X scores.

The use of r as a descriptive measure of association doesn't require any assumptions regarding the shape of the X and Y distributions. As we have seen, however, if X and Y are markedly skewed, the value of r will be closer to zero than if the distributions

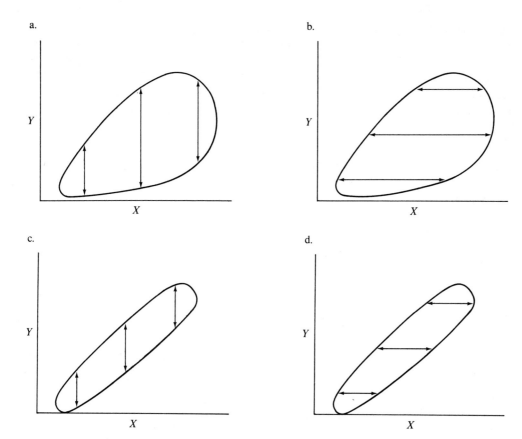

Figure 5.6-6. Parts a and b illustrate heterogeneity of column and row dispersion, respectively; parts c and d illustrate homogeneity of dispersion.

are approximately normal. Furthermore, under these conditions the interpretation of r is altered, since it no longer describes the average degree to which the data points hug the line of best fit. Finally, the presence of skewed X and Y distributions is often accompanied by a nonlinear relationship between the variables. This condition calls for the computation of η^2 instead of r.

Table 5.6-1. Factors that Affect the Size of r

r Underestimates Magnitude of Relationship when:	r Overestimates Magnitude of Relationship when:
1. The relationship between X and Y is nonlinear 2. The range of either X or Y is truncated 3. The distributions of X and Y are skewed	1. The sample contains subgroups with means that differ for both variables 2. The sample is composed of extreme groups

It is apparent from this discussion that the interpretation of r as a descriptive measure is simplified if X and Y are approximately normally distributed. It should be emphasized, however, that normality is not required for purely descriptive purposes, since regardless of the shapes of the X and Y distributions, r is a measure of how closely data points hug a straight line of best fit. We will see in Chapter 12 that normality is required to draw inferences about the population correlation from the sample coefficient.

The factors that affect the size of r are summarized in Table 5.6-1.

Review Exercises for Section 5.6

**27. What effects do the following factors have on r as a measure of strength of association? Draw figures like Figures 5.6-1–5.6-5 to represent the data.
  *a. Relationship between X and Y looks like an inverted U.
  b. Range of X is reduced by deleting subjects with scores below $\bar{X}$.
  c. Sample contains subgroups, denoted by a and b, with means $\bar{X}_a = 16$, $\bar{X}_b = 22$, $\bar{Y}_a = 31$, and $\bar{Y}_b = 42$.
  *d. Sample contains subgroups a and b with means $\bar{X}_a = 16$, $\bar{X}_b = 22$, $\bar{Y}_a = 31$, and $\bar{Y}_b = 37$.
  e. Sample contains subgroups a and b with standard deviations $S_{X_a} = 13$, $S_{X_b} = 22$, $S_{Y_a} = 22$, and $S_{Y_b} = 13$.
  f. Sample contains subgroups a and b with standard deviations $S_{X_a} = 18$, $S_{X_b} = 18$, $S_{Y_a} = 26$, and $S_{Y_b} = 34$.
  g. The distribution of the X variable is negatively skewed; that for the Y variable is positively skewed.

28. How can we detect cases in which η^2 should be used instead of r?
29. What are the potential advantages and disadvantages of using extreme groups in research?
30. The correlation between IQ and grade point average (GPA) for high school seniors was 0.63. For seniors who went on to college the correlation between IQ and college GPA was 0.51. Explain why this correlation is lower.
31. The correlation between IQ and ratings of the creativity of 50 highly creative individuals was 0.18. Can we conclude that IQ is a relatively unimportant factor in creativity? Discuss.
32. Terms to remember:
  a. Correlation ratio
  b. Truncated range
  c. Extreme groups
  d. Discontinuous distribution
  e. Heterogeneity of array variance (heteroscedasticity)

5.7 Spearman Rank Correlation

The **Spearman rank correlation coefficient**, denoted by r_s, is used to describe the degree of agreement between paired data that are in the form of ranks.[8] Such data may occur as a result of ranking scores, as when grade point averages are converted to rank in graduating class, or because rank data are obtained in the original instance, as when freshman English themes are ranked on the basis of creativity. Ranking can often be done when it is difficult or impossible to apply more refined measuring procedures.

[8] This coefficient was first used by Sir Francis Galton, but was named for the British psychologist Charles Spearman, who made more extensive use of it.

The formula for r_s is

$$r_s = 1 - \frac{6 \sum_{i=1}^{n} (R_{X_i} - R_{Y_i})^2}{n(n^2 - 1)},$$

where $R_{X_i} - R_{Y_i}$ is the difference between the ith person's rank on X and rank on Y, and n is the number of pairs of ranks. The computation of r_s is illustrated in Table 5.7-1, where 14 graduate school applicants have been ranked by tenured faculty (R_{X_i}) and nontenured faculty (R_{Y_i}).

The index r_s is a measure of the agreement between two sets of ranks and is interpreted in much the same way as the Pearson product-moment coefficient. The range of r_s is from -1 to $+1$. Values of r_s greater than 0 indicate that large R_X's tend to be paired with large R_Y's; values less than 0 that large R_X's are paired with small R_Y's, and so on; $r_s = 1$ if and only if each person's X and Y ranks are equal. It can be shown[9] that the formula for r_s is equivalent to that for r when two sets of consecutive untied ranks $1, \ldots, n$ are substituted for X_i and Y_i in the Pearson formula. However, the use of ranks instead of scores alters the meaning of the correlation coefficient.

We saw earlier that r is a measure of the linear relationship between two quantitative variables; r_s is a measure of the *monotonic relationship* between two sets of ranks.

Table 5.7-1. Computation of r_s for Ranks Assigned to Applicants by Tenured Faculty (R_{X_i}) and Nontenured Faculty (R_{Y_i})

(i) Data:

Applicant	Rank, R_{X_i}	Rank, R_{Y_i}	$R_{X_i} - R_{Y_i}$	$(R_{X_i} - R_{Y_i})^2$
1	6	8	-2	4
2	3	2	1	1
3	4	5	-1	1
4	12	11	1	1
5	10	9	1	1
6	1	1	0	0
7	5	4	1	1
8	7	7	0	0
9	14	14	0	0
10	2	3	-1	1
11	8	10	-2	4
12	11	12	-1	1
13	9	6	3	9
14	13	13	0	0

$\sum (R_{X_i} - R_{Y_i})^2 = 24$

(ii) Computational procedure:

$$r_s = 1 - \frac{6 \sum_{1}^{n} (R_{X_i} - R_{Y_i})^2}{n(n^2 - 1)} = 1 - \frac{6(24)}{14[(14)^2 - 1]} = 1 - \frac{144}{2730} = 0.95$$

[9] See Section 5.10-2.

A function $Y = f(X)$ is said to be **strictly monotonic increasing** if an increase in the value of X is always accompanied by an increase in Y.[10] A **strictly monotonic decreasing** function is one in which a decrease in X is accompanied by a decrease in Y. Monotonic functions include linear functions ($Y = a + bX$) as well as a number of other functions that are nonlinear ($Y = X^3$; $Y = \log X$). Thus, Spearman's rank correlation coefficient does not necessarily reflect the linear relationship between the variables represented by ranks. It does reflect the strength of the monotonic relationship—a more general relationship. If $r_s = 0$, either the variables represented by ranks are not related or the form of the relationship is nonmonotonic.

The Problem of Tied Ranks

Occasionally, two or more objects or individuals are tied with respect to the variable being ranked. The usual practice is to give them the mean of the ranks they would have received collectively if they had been distinguishable. For example, if Jack, Terry, and Bill are considered equally gregarious, each is given the mean of the ranks they would have occupied, say, $(8 + 9 + 10)/3 = 9$. Unfortunately, the presence of tied ranks violates the assumptions underlying the derivation of the computational formula for r_s. A correction for ties can be incorporated in the formula, but the computation is tedious. The most desirable solution is to force those making ratings to discern differences among the objects or individuals, thereby eliminating tied ranks. If this is done, the uncorrected formula can be used. If raters persist in assigning tied ranks, the next best solution is to treat the sets of ranks as though they were scores and compute a Pearson product-moment correlation coefficient. The result can be regarded as a Spearman rank correlation coefficient that has been corrected for ties.

Review Exercises for Section 5.7

33. List the similarities and differences between r and r_s.
*34. A random sample of freshman psychology majors rank ordered various fields of psychology according to vocational attractiveness. The students again ranked the fields when they were seniors. Compute the correlation between their freshman and senior rankings.

Field	Freshman Rank	Senior Rank
Social	5	2
Experimental	7	6
Human factors	6	7
Clinical	1	1
Statistics and measurement	8	8
Industrial	3	3
Educational	4	4
Counseling	2	5

[10] If for all $X_n < X_{n+1}$, $f(X_n) \leq f(X_{n+1})$, the function $f(X)$ is said to be monotonic increasing. If for all $X_n < X_{n+1}$, $f(X_n) < f(X_{n+1})$, the function is strictly monotonic increasing.

35. A psychiatric social worker and an occupational therapist ranked 11 Veterans' Administration (VA) patients with respect to extent of recovery following 3 months of therapy. Compute the correlation between the two sets of ranking.

Patient	Social Worker	Occupational Therapist
1	7	7
2	2	1
3	1	2
4	3	5
5	8	9
6	10	10
7	4	3
8	9	8
9	11	11
10	6	6
11	5	4

36. Subjects rated the attractiveness of one set of geometric shapes before smoking marijuana and a similar set after smoking the marijuana. One shape in the two sets was the same. The following data are the ratings for that shape. A rating of 1 means very attractive; a rating of 20 very unattractive. Transform the ratings to ranks and compute the correlation between the two sets of ranks.

Subject	Before Smoking	After Smoking
1	6	3
2	8	7
3	14	16
4	7	2
5	10	12
6	9	15
7	5	1
8	15	20
9	12	17

37. Suppose that for the data in Exercise 36, subject 6 assigned a rating of 12 instead of 15 to the geometric shape after smoking marijuana. This results in tied ranks. How would this affect the computational procedure for the correlation coefficient?

**38. Which of the following are strictly monotonic functions?
   *a. $Y = 1 + 2X$ *b. $Y = X^2$
   c. $Y = X^3$ d. $Y = 1/X$

39. Terms to remember:
   Strictly monotonic increasing and decreasing functions

†5.8 Other Kinds of Correlation Coefficients

Three correlation coefficients have been mentioned thus far: r, η^2, and r_s. A fourth coefficient, Cramer's $\hat{\phi}'$, appropriate for unordered qualitative variables is discussed in Section 17.3. Other coefficients are also available, but they are beyond the scope of this book. A summary of the more widely used coefficients is given in Table 5.8-1.

Table 5.8-1. Summary of the Major Correlation Coefficients

Coefficient	Symbol	Characteristics	Reference[a] and Pages
1. Pearson product moment	r	X and Y quantitative, linear relationship	(a) 97–103 (b) 109–116 (c) 631–635
2. Eta squared (correlation ratio)	η^2	X and Y quantitative, curvilinear relationship	(a) 136–140 (b) 150–152 (c) 683–684
3. Spearman rank coefficient	r_s	X and Y ranked, monotonic relationship	(a) 133–136 (b) 172–176 (c) 788–792
4. Kendall's tau	τ	X and Y ranked, monotonic relationship	(b) 176–179 (c) 792–800
5. Point biserial	r_{pb}	One variable quantitative, the other dichotomous[b]	(a) 123–126 (b) 163–164
6. Biserial	r_b	X and Y quantitative, but one variable forced into a dichotomy	(a) 128–130 (b) 168–172
7. Tetrachoric	r_t	X and Y quantitative, but both forced into dichotomies	(a) 131–132 (b) 165–167
8. Phi or fourfold	ϕ	X and Y both dichotomous	(a) 126–128 (b) 158–162

[a] (a) Edwards (1967), (b) Glass and Stanley (1970), (c) Hays (1973).
[b] Dichotomous classifications assign elements to one of two categories, for example, pass–fail, male–female, married–not married, and IQ > 100–IQ ≤ 100.

5.9 Summary

The term "correlation" denotes the concomitance between two or more quantitative or qualitative variables. A correlation coefficient is a measure of the degree of concomitance. The two most widely used coefficients in the behavioral sciences and education are the Pearson product-moment correlation coefficient, r, and the Spearman rank correlation coefficient, r_s. Pearson's r reflects the strength and nature of the linear relationship between two quantitative variables. It is a number that varies between -1 and 1, with 0 indicating the absence of a linear relationship. Negative values indicate

† This section can be omitted without loss of continuity.

an inverse relationship between the variables; positive values a positive or direct relationship. Spearman's r_s measures the strength and nature of the monotonic relationship between two ordered qualitative variables—that is, ranked data. It, like r, varies between -1 and 1, with 0 indicating the absence of a monotonic relationship.

Two statistics, both functions of r, are useful in interpreting a particular r value—the coefficient of determination, r^2, and the coefficient of nondetermination, $k^2 = 1 - r^2$. For a given linear relationship between X and Y, r^2 reflects the proportion of the X score variance that can be explained by the Y score variance and vice versa; k^2 reflects the proportion that can't be explained. If, for example, $r = 0.50$, we know that based on the linear relationship between the variables 25% of the variance in one variable can be explained by the variance in the other variable, while 75% remains to be accounted for.

The Pearson product-moment correlation coefficient is appropriate for linearly related quantitative variables. For descriptive purposes, no other assumptions regarding the variables are required. However, in interpreting r, one should keep in mind that the size of r can be affected by such factors as the shape of the X and Y distributions, the presence of a truncated X or Y range, and the presence of subgroups with standard deviations or means that differ on both variables.

†5.10 Technical Notes

5.10-1 Proof of Algebraic Equivalence of the Definitional Formula and Computational Formula for r

The definitional formula for r given in Section 5.3 is

$$r = \frac{\dfrac{\sum_{i=1}^{n}(X_i - \bar{X})(Y_i - \bar{Y})}{n}}{\sqrt{\left[\dfrac{\sum_{i=1}^{n}(X_i - \bar{X})^2}{n}\right]\left[\dfrac{\sum_{i=1}^{n}(Y_i - \bar{Y})^2}{n}\right]}}$$

$$= \frac{\dfrac{1}{n}\left[\sum_{i=1}^{n}(X_i Y_i - \bar{Y}X_i - \bar{X}Y_i + \bar{X}\bar{Y})\right]}{\dfrac{1}{n}\sqrt{\left[\sum_{i=1}^{n}(X_i^2 - 2\bar{X}X_i + \bar{X}^2)\right]\left[\sum_{i=1}^{n}(Y_i^2 - 2\bar{Y}Y_i + \bar{Y}^2)\right]}}$$

$$= \frac{\sum_{i=1}^{n}X_i Y_i - \bar{Y}\sum_{i=1}^{n}X_i - \bar{X}\sum_{i=1}^{n}Y_i + n\bar{X}\bar{Y}}{\sqrt{\left(\sum_{i=1}^{n}X_i^2 - 2\bar{X}\sum_{i=1}^{n}X_i + n\bar{X}^2\right)\left(\sum_{i=1}^{n}Y_i^2 - 2\bar{Y}\sum_{i=1}^{n}Y_i + n\bar{Y}^2\right)}},$$

† These notes, which are not essential to the text, can be omitted without loss of continuity.

but $\bar{X} = \sum X/n$, $\bar{X}^2 = (\sum X_i)^2/n^2$, $\bar{Y} = \sum Y_i/n$, and $\bar{Y}^2 = (\sum Y_i)^2/n^2$. Substituting,

$$r = \frac{\sum X_i Y_i - \frac{\sum Y_i}{n}\sum X_i - \frac{\sum X_i}{n}\sum Y_i + n\frac{\sum X_i}{n}\frac{\sum Y_i}{n}}{\sqrt{\left[\sum X_i^2 - \frac{2\sum X_i}{n}\sum X_i + \frac{n(\sum X_i)^2}{n^2}\right]\left[\sum Y_i^2 - \frac{2\sum Y_i}{n}\sum Y_i + \frac{n(\sum Y_i)^2}{n^2}\right]}}$$

$$= \frac{\sum X_i Y_i - \frac{2\sum X_i \sum Y_i}{n} + \frac{\sum X_i \sum Y_i}{n}}{\sqrt{\left[\sum X_i^2 - \frac{2(\sum X_i)^2}{n} + \frac{(\sum X_i)^2}{n}\right]\left[\sum Y_i^2 - \frac{2(\sum Y_i)^2}{n} + \frac{(\sum Y_i)^2}{n}\right]}}.$$

Multiplying the numerator and denominator by n gives the computational formula for r:

$$r = \frac{n\sum X_i Y_i - \sum X_i \sum Y_i}{\sqrt{[n\sum X_i^2 - (\sum X_i)^2][n\sum Y_i^2 - (\sum Y_i)^2]}}.$$

5.10-2 Equivalence of r and r_s Formulas when Two Sets of Consecutive Untied Ranks $1,\ldots,n$ Are Substituted for X_i and Y_i in the Formula for r

We begin the derivation with the quantity

(5.10-2-1) $\quad \sum [(X_i - \bar{X}) - (Y_i - \bar{Y})]^2,$

which we will show later is equal to $\sum (R_{X_i} - R_{Y_i})^2$ when X_i and Y_i are ranks.

(5.10-2-2) $\sum [(X_i - \bar{X}) - (Y_i - \bar{Y})]^2$
$= \sum [(X_i - \bar{X})^2 + (Y_i - \bar{Y})^2 - 2(X_i - \bar{X})(Y_i - \bar{Y})]$
$= \sum (X_i - \bar{X})^2 + \sum (Y_i - \bar{Y})^2 - 2\sum (X_i - \bar{X})(Y_i - \bar{Y}).$

A formula for r equivalent to the one given in Section 5.3 is

$$r = \frac{\sum (X_i - \bar{X})(Y_i - \bar{Y})}{\sqrt{[\sum (X_i - \bar{X})^2][\sum (Y_i - \bar{Y})^2]}}.$$

By rearranging terms and multiplying by 2, we obtain

$$2\sum (X_i - \bar{X})(Y_i - \bar{Y}) = 2r\sqrt{[\sum (X_i - \bar{X})^2][\sum (Y_i - \bar{Y})^2]}.$$

Substituting in (5.10-2-2) gives

(5.10-2-3) $\sum [(X_i - \bar{X}) - (Y_i - \bar{Y})]^2$
$= \sum (X_i - \bar{X})^2 + \sum (Y_i - \bar{Y})^2 - 2r\sqrt{[\sum (X_i - \bar{X})^2][\sum (Y_i - \bar{Y})^2]}.$

Rearranging terms, we obtain

(5.10-2-4) $\quad r = \frac{\sum (X_i - \bar{X})^2 + \sum (Y_i - \bar{Y})^2 - \sum [(X_i - \bar{X}) - (Y_i - \bar{Y})]^2}{2\sqrt{[\sum (X_i - \bar{X})^2][\sum (Y_i - \bar{Y})^2]}}.$

If X_i and Y_i are ranks instead of scores, you may remember from your first algebra course that the sum of the first n consecutive integers starting with 1 is

(5.10-2-5) $$\sum X_i = 1 + 2 + \cdots + n = \frac{n(n+1)}{2}$$

and that their mean is

(5.10-2-6) $$\bar{X} = \frac{n+1}{2}.$$

The sum of squares of the first n consecutive integers is

(5.10-2-7) $$\sum X_i^2 = (1)^2 + (2)^2 + \cdots + n^2 = \frac{(n+1)(2n+1)}{6}.$$

We now show that $\sum (X_i - \bar{X})^2 = n(n^2 - 1)/12$. We know from Section 4.8-3 that $\sum (X_i - \bar{X})^2 = \sum X_i^2 - (\sum X_i)^2/n$. Substituting (5.10-2-5) and (5.10-2-7) in the formula for $\sum X_i^2 - (\sum X_i)^2/n$ gives

$$\sum X_i^2 - \frac{(\sum X_i)^2}{n} = \frac{n(n+1)(2n+1)}{6} - \frac{[n(n+1)/2]^2}{n}.$$

Squaring the last term on the right gives

$$\sum X_i^2 - \frac{(\sum X_i)^2}{n} = \frac{n(n+1)(2n+1)}{6} - \frac{[n^2(n+1)^2/4]}{n},$$

and obtaining a common denominator leads to

$$\sum X_i^2 - \frac{(\sum X_i)^2}{n} = \frac{2n(n+1)(2n+1)}{12} - \frac{3n(n+1)(n+1)}{12}$$

$$= \frac{n(n+1)(4n+2) - n(n+1)(3n+3)}{12}$$

$$= \frac{n(n+1)(4n+2-3n-3)}{12}$$

$$= \frac{n(n+1)(n-1)}{12}$$

$$= \frac{n(n^2-1)}{12}.$$

By the same process we could show that

$$\sum (Y_i - \bar{Y})^2 = \sum Y_i^2 - \frac{(\sum Y_i)^2}{n} = \frac{n(n^2-1)}{12}.$$

We stated earlier that if X_i and Y_i represent ranks, then $\sum [(X_i - \bar{X}) - (Y_i - \bar{Y})]^2 = \sum (R_{X_i} - R_{Y_i})^2$. We can show this as follows. According to (5.10-2-6), $\bar{X}$ and $\bar{Y} = (n+1)/2$; thus,

$$\sum [(X_i - \bar{X}) - (Y_i - \bar{Y})]^2 = \sum [(X_i - Y_i - \bar{X} + \bar{Y})]^2$$

$$= \sum \left[\left(X_i - Y_i - \frac{n+1}{2} + \frac{n+1}{2}\right)\right]^2$$

$$= \sum (X_i - Y_i)^2.$$

The variables X_i and Y_i are ranks rather than scores; to make this obvious, we can use the symbol R_{X_i} for X_i and R_{Y_i} for Y_i. Thus,

$$\sum [(X_i - \bar{X}) - (Y_i - \bar{Y})]^2 = \sum (R_{X_i} - R_{Y_i})^2.$$

If we have ranks instead of scores, we can substitute the above terms in (5.10-2-4). The correlation coefficient is then called r_s instead of r.

$$r = \frac{\sum (X_i - \bar{X})^2 + \sum (Y_i - \bar{Y})^2 - \sum [(X_i - \bar{X}) - (Y_i - \bar{Y})]^2}{2\sqrt{[\sum (X_i - \bar{X})^2][\sum (Y_i - \bar{Y})^2]}}$$

$$r_s = \frac{\frac{n(n^2 - 1)}{12} + \frac{n(n^2 - 1)}{12} - \sum (R_{X_i} - R_{Y_i})^2}{2\sqrt{\left[\frac{n(n^2 - 1)}{12}\right]\left[\frac{n(n^2 - 1)}{12}\right]}}$$

$$= \frac{\frac{2n(n^2 - 1)}{12} - \sum (R_{X_i} - R_{Y_i})^2}{\frac{2n(n^2 - 1)}{12}}$$

$$= 1 - \frac{\sum (R_{X_i} - R_{Y_i})^2}{\frac{n(n^2 - 1)}{6}}$$

(5.10-2-8) $$= 1 - \frac{6\sum (R_{X_i} - R_{Y_i})^2}{n(n^2 - 1)}$$

Equation (5.10-2-8) is the Spearman rank correlation coefficient. We have just shown that it is algebraically equivalent to Pearson's r when X and Y scores are replaced by the first n consecutive integers starting with 1.

5.10-3 Effect on r of Adding a Constant to Each X Score

Let $r_{(X+c)Y}$ be the correlation between variables X and Y, where X has been altered by adding a constant c to each score $(X_i + c)$. The mean of the distribution of $(X_i + c)$ is denoted by $\bar{X}_{X+c}$ and the standard deviation by S_{X+c}.

(5.10-3-1) $$r_{(X+c)Y} = \frac{\frac{\sum [(X_i + c) - \bar{X}_{X+c}](Y_i - \bar{Y})}{n}}{\sqrt{\left\{\frac{\sum [(X_i + c) - \bar{X}_{X+c}]^2}{n}\right\}\left[\frac{\sum (Y_i - \bar{Y})^2}{n}\right]}}$$

$$= \frac{\frac{\sum [(X_i + c) - \bar{X}_{X+c}](Y_i - \bar{Y})}{n}}{S_{X+c}S_Y}$$

We show in Section 3.9-3 that $\bar{X}_{X+c} = \bar{X} + c$. In Section 4.8-1 we show that $S_{X+c} = S$; here we will denote S by S_X to distinguish it from the standard deviation of Y. Substituting in (5.10-3-1) we obtain

$$r_{(X+c)Y} = \frac{\frac{\sum[(X_i + c) - \bar{X} - c](Y_i - \bar{Y})}{n}}{S_X S_Y}$$

$$= \frac{\frac{\sum[(X_i - \bar{X} + c - c)](Y_i - \bar{Y})}{n}}{S_X S_Y}$$

$$= \frac{\frac{\sum(X_i - \bar{X})(Y_i - \bar{Y})}{n}}{S_X S_Y}$$

$$= \frac{S_{XY}}{S_X S_Y} = r.$$

Thus, adding a constant to each X score does not affect the value of the coefficient. By the same line of reasoning, it can be shown that the value of r isn't affected by subtracting a constant from each X and Y score.

5.10-4 Effect on r of Multiplying Each X Score by a Constant

Let $r_{(cX)Y}$ be the correlation between variables X and Y, where X has been altered by multiplying each X score by a positive constant c. The mean of the distribution of cX_i is denoted by $\bar{X}_{cX}$ and the standard deviation by S_{cX}.

$$r_{(cX)Y} = \frac{\frac{\sum(cX_i - \bar{X}_{cX})(Y_i - \bar{Y})}{n}}{\sqrt{\left[\frac{\sum(cX_i - \bar{X}_{cX})^2}{n}\right]\left[\frac{\sum(Y_i - \bar{Y})^2}{n}\right]}}$$

(5.10-4-1)
$$= \frac{\frac{\sum(cX_i - \bar{X}_{cX})(Y_i - \bar{Y})}{n}}{S_{cX} S_Y}$$

We show in Section 3.9-4 that $\bar{X}_{cX} = c\bar{X}$. In Section 4.8-2 we show that $S_{cX} = cS$; here we will denote cS by cS_X to distinguish it from the standard deviation of Y. Substituting in (5.10-4-1) we obtain

$$r_{(cX)Y} = \frac{\frac{\sum(cX_i - c\bar{X})(Y_i - \bar{Y})}{n}}{cS_X S_Y}$$

$$= \frac{\frac{c\sum(X_i - \bar{X})(Y_i - \bar{Y})}{n}}{cS_X S_Y}$$

$$= \frac{S_{XY}}{S_X S_Y} = r.$$

Thus, multiplying each X score by a positive constant doesn't affect the value of the coefficient. By the same line of reasoning, it can be shown that the value of r is not affected by dividing each X and Y score by a positive constant. Combining Sections 5.10-3 and 5.10-4 tells us that variables can be subjected to any positive linear transformation (multiplication by a positive constant and/or addition of a constant) without affecting the value of r.

If c is a negative constant, $r_{(cX)Y} = -r$.

6

Regression

6.1 Introduction
   An Overview of Prediction
6.2 Criterion for the Line of Best Fit
   Predicting Y from X
   Predicting X from Y
   Relationship between r and the Slopes of the Regression Lines
   Review Exercises for Sections 6.1 and 6.2
6.3 Another Measure of Ability to Predict: The Standard Error of Estimate
   Interpretation of $S_{Y \cdot X}$
6.4 Alternative Formula for $S_{Y \cdot X}$
   Coefficient of Alienation, k
   Descriptive Application of $S_{Y \cdot X}$
6.5 Assumptions Associated with Estimate of Prediction Error
   Review Exercises for Sections 6.3, 6.4, and 6.5
6.6 Summary
6.7 Technical Notes
   Standard Error of Estimate as the Standard Deviation of Prediction Errors
   Derivation of Alternative Formula for $S_{Y \cdot X}$
   Partition of S_Y^2/S_Y^2 into r^2 and k^2

6.1 Introduction

Jean's score on the Law School Aptitude Test (LSAT) is 69. What grade point average can she expect to make in law school? Bertha is on a 750 calorie diet; how many pounds should she be able to lose in a month? Since the variables are correlated, we can predict GPA from an LSAT score and weight loss from calorie intake with better than chance accuracy. The higher the correlation, the more accurate the prediction. For $r = \pm 1$, the dependent variable, denoted by Y, can be predicted from the independent variable, X, with perfect accuracy; but if $r = 0$, a knowledge of X is useless in predicting Y. Although a correlation coefficient is indicative of our ability to predict, the actual prediction is made using regression analysis, the subject of this chapter. Strictly speaking, *regression analysis* applies to paired data (X_i, Y_i), where X is the independent variable with values X_i that are selected in advance and Y is the dependent variable with values Y_i that are free to vary. However, regression procedures are also applicable when both X and Y are free to vary, as they are in correlation.

An Overview of Prediction

George, who is taking statistics, copies down the grades from last semester's class and constructs the scatter diagram shown in Figure 6.1-1. He finds that the correlation between the midterm and the final exam was 0.80. His midterm grade was 82, and he

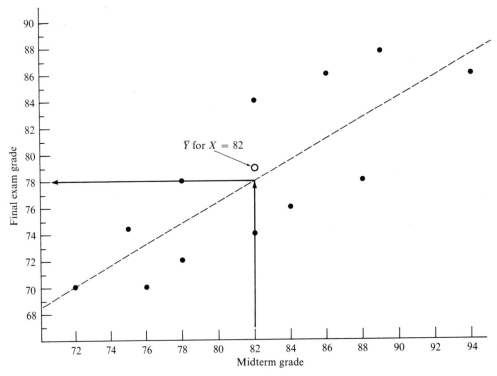

Figure 6.1-1. Scatter diagram for paired midterm and final exam grades.

wonders how he'll do on the final. According to the scatter diagram, two students made 82; the mean of their grades—and hence George's predicted grade—is $(74 + 84)/2 = 79$.

While this prediction method works, it has a serious disadvantage. The prediction is based only on the two Y scores corresponding to $X_i = 82$; the other ten paired scores are ignored. Predictions based on such small samples tend to be unstable—that is, to vary markedly from sample to sample. Prediction can be improved by utilizing all the data rather than a small subset. George notes that the relationship between the midterm and final grades appears to be linear, so he determines the best-fitting linear regression line. It is shown as a dashed line in Figure 6.1-1. To predict his final grade George draws a vertical line from $X_i = 82$ to the regression line and then a horizontal line to the Y axis. His predicted grade is 78.

Predictions based on the regression line take into account all the sample data and hence are more stable than those based on only the mean of the Y scores corresponding to a given X score. Both procedures presuppose that the population represented by the current sample (George's statistics class) doesn't differ from that represented by the earlier sample (last year's class). Obviously, if this assumption isn't tenable, we can have little faith in the prediction. The regression approach also presupposes that the data points have been fitted by the correct regression equation—in our example, the equation for a straight line. Fortunately, the tenability of this assumption is easily checked by looking at the scatter diagram.

6.2 Criterion for the Line of Best Fit

Predicting Y from X

We have frequently referred to the line of best fit without ever defining it—what *is* the line of best fit for data points? Best fit can be defined in a number of ways. It seems reasonable that we should want a line that minimizes some function of the error in predicting Y_i from X_i. A **prediction error**, e_i, *is defined as the difference between the ith person's actual score, Y_i, and the score predicted for that person, Y_i'*—that is, $e_i = Y_i - Y_i'$. Prediction errors are illustrated in Figure 6.2-1. One definition of best fit widely used by mathematicians is based on the *principle of least squares* and may be stated as follows. *The line of best fit is the one that minimizes the sum of the squared prediction errors—that is, the line for which $\sum_{i=1}^{n} e_i^2 = \sum_{i=1}^{n} (Y_i - Y_i')^2$ is a minimum.* We will limit our discussion to linearly related data. For this case the predicted values fall on a straight line, called the *regression line*. The equation for a straight line is $Y_i' = a_{Y \cdot X} + b_{Y \cdot X} X_i$, where Y_i' is the predicted value, $a_{Y \cdot X}$ is the point where the line crosses the Y axis, $b_{Y \cdot X}$ is the slope of the line, and X_i is a value of the independent variable. The subscript $Y \cdot X$ is read "Y given X" and indicates that we are predicting Y from X. According to the least squares criterion, we want values of the constants $a_{Y \cdot X}$ and $b_{Y \cdot X}$ such that $\sum_{i=1}^{n} e_i^2 = \sum_{i=1}^{n} (Y_i - Y_i')^2 = \sum_{i=1}^{n} [Y_i - (a_{Y \cdot X} + b_{Y \cdot X} X_i)]^2$ is as small as it possibly can be.

The method of finding numerical values for $a_{Y \cdot X}$ and $b_{Y \cdot X}$ that minimizes $\sum e_i^2$ utilizes the differential calculus and hence is beyond the scope of this text.[1] We shall

---

[1] McNemar (1969, pp. 133–136) describes the method in a text written for behavioral science students.

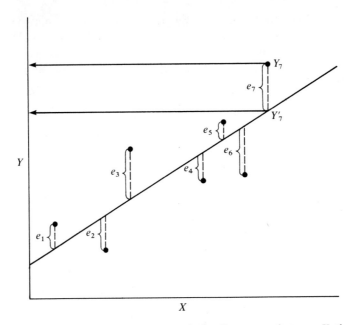

Figure 6.2-1. A prediction error, e_i, is the discrepancy between Y_i, the actual observed score for person i, and Y_i', the predicted score based on the regression line; for example, $e_7 = Y_7 - Y_7'$.

simply report here the formulas for computing their values:

$$a_{Y \cdot X} = \bar{Y} - b\bar{X} \quad \text{and} \quad b_{Y \cdot X} = \frac{\sum_{i=1}^{n}(X_i - \bar{X})(Y_i - \bar{Y})}{\sum_{i=1}^{n}(X_i - \bar{X})^2}.$$

A computational example is given in Table 6.2-1 for the data in Figure 6.1-1. The values of the constants are $a_{Y \cdot X} = 12.154$ and $b_{Y \cdot X} = 0.803$. Hence, the linear equation that minimizes the sum of the squared prediction errors is $Y_i' = 12.154 + 0.803X_i$. According to the equation the line crosses the Y axis at 12.154. In other words, when $X = 0$, $Y' = 12.154$ and the slope of the line is 0.803, which means that as X increases 1 unit Y increases 0.803 unit. To determine the predicted value for, say, $X_i = 82$, we enter the value of X_i in the equation, $Y_i' = 12.154 + 0.803(82)$, and solve for Y_i'. The predicted value is 78.

Alternatively, we can determine predicted values by graphic means as we did in Figure 6.1-1. The first step is to draw the line of best fit. Since a straight line is defined by two points, we can begin by solving for Y_i' when $X_i = 72$ and 94. The corresponding Y_i' values are, respectively, 69.97 and 87.54. Once a line connecting the (X_i, Y_i) points (72, 69.97) and (94, 87.54) has been drawn, it can be used to obtain Y_i' for other values of X_i.

Table 6.2-1. Computation of Least Squares Values for Constants in a Linear Equation (Data from Figure 6.1-1)

(i) Data:

X_i	Y_i	$(X_i - \bar{X})$	$(Y_i - \bar{Y})$	$(X_i - \bar{X})(Y_i - \bar{Y})$	$(X_i - \bar{X})^2$	$(Y_i - \bar{Y})^2$
72	70	−10	−8	80	100	64
75	74	−7	−4	28	49	16
76	70	−6	−8	48	36	64
78	72	−4	−6	24	16	36
78	78	−4	0	0	16	0
82	74	0	−4	0	0	16
82	84	0	6	0	0	36
84	76	2	−2	−4	4	4
86	86	4	8	32	16	64
88	78	6	0	0	36	0
89	88	7	10	70	49	100
94	86	12	8	96	144	64

$\sum X_i = 984 \qquad \sum Y_i = 936$
$\bar{X} = 82 \qquad \bar{Y} = 78$

$\sum (X_i - \bar{X})(Y_i - \bar{Y}) = 374 \qquad \sum (X_i - \bar{X})^2 = 466 \qquad \sum (Y_i - \bar{Y})^2 = 464$

(ii) Computation of $a_{Y \cdot X}$ and $b_{Y \cdot X}$:

$$b_{Y \cdot X} = \frac{\sum_{i=1}^{n}(X_i - \bar{X})(Y_i - \bar{Y})}{\sum_{i=1}^{n}(X_i - \bar{X})^2} = \frac{374}{466} = 0.803$$

$$a_{Y \cdot X} = \bar{Y} - b_{Y \cdot X}\bar{X} = 78 - 0.803(82) = 12.154$$

(iii) Computation of $a_{X \cdot Y}$ and $b_{X \cdot Y}$:

$$b_{X \cdot Y} = \frac{\sum_{i=1}^{n}(X_i - \bar{X})(Y_i - \bar{Y})}{\sum_{i=1}^{n}(Y_i - \bar{Y})^2} = \frac{374}{464} = 0.806$$

$$a_{X \cdot Y} = \bar{X} - b_{X \cdot Y}\bar{Y} = 82 - 0.806(78) = 19.132$$

Predicting X from Y

If the value of Y_i is known, X_i can be predicted from the equation $X_i' = a_{X \cdot Y} + b_{X \cdot Y} Y_i$. The subscript $X \cdot Y$ indicates that X is predicted from Y. It should be noted that $a_{X \cdot Y}$ is different from $a_{Y \cdot X}$ and $b_{X \cdot Y}$ from $b_{Y \cdot X}$, since they apply to different regression lines. The constants of the linear equation for predicting X from Y are given by

$$a_{X \cdot Y} = \bar{X} - b_{X \cdot Y} \bar{Y} \quad \text{and} \quad b_{X \cdot Y} = \frac{\sum_{i=1}^{n} (X_i - \bar{X})(Y_i - \bar{Y})}{\sum_{i=1}^{n} (Y_i - \bar{Y})^2}.$$

The formulas for $a_{X \cdot Y}$ and $b_{X \cdot Y}$ were derived so as to minimize the sum of the squared prediction errors defined by $\sum_{i=1}^{n} e_i^2 = \sum_{i=1}^{n} (X_i - X_i')^2$. These prediction errors are illustrated in Figure 6.2-2. The computation of $a_{X \cdot Y}$ and $b_{X \cdot Y}$ is illustrated in Table 6.2-1. The regression equation is $X_i' = 19.132 + 0.806 Y_i$.

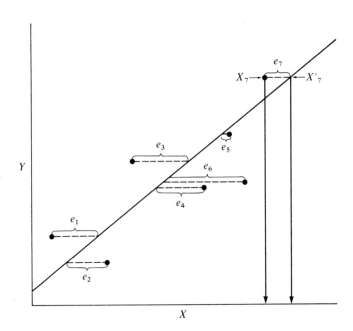

Figure 6.2-2. The error in predicting X_i from Y_i is the discrepancy between X_i, the actual observed value for person i, and X_i', the predicted value based on the regression line; for example, $e_7 = X_7 - X_7'$.

For any set of paired data points we can compute two regression lines—the regression of Y on X given by $Y_i' = a_{Y \cdot X} + b_{Y \cdot X} X_i$ and the regression of X on Y given by $X_i' = a_{X \cdot Y} + b_{X \cdot Y} Y_i$. The two lines are shown in Figure 6.2-3 for the data in Table 6.2-1. There are two lines because in predicting Y from X we want to minimize one error, $\sum (Y_i - Y_i')^2$, but in predicting X from Y we minimize a different error, $\sum (X_i - X_i')^2$.

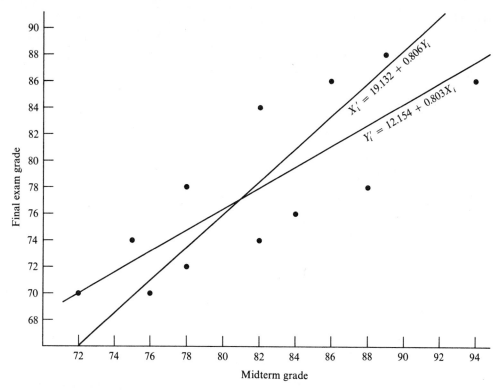

Figure 6.2-3. Regression lines for predicting Y_i from X_i and X_i from Y_i (data from Table 6.2-1).

Relationship between r and the Slopes of the Regression Lines

There are a number of interesting relationships between r and the two regression coefficients $b_{Y \cdot X}$ and $b_{X \cdot Y}$. It is a simple matter to show that $r(S_Y/S_X) = b_{Y \cdot X}$ and $r(S_X/S_Y) = b_{X \cdot Y}$ and that $\pm \sqrt{b_{Y \cdot X} b_{X \cdot Y}} = \pm r$.[2] According to the latter relationship, r is the *geometric mean* of the two regression coefficients. *A **geometric mean** is a special kind of central tendency measure often used for averaging rates of change. It is defined as the nth root of the product of n positive numbers.*

A little algebra is all that is necessary to prove the relationships. One of the formulas for r given in Chapter 5 was

$$r = \frac{\sum (X_i - \bar{X})(Y_i - \bar{Y})/n}{\sqrt{[\sum (X_i - \bar{X})^2/n][\sum (Y_i - \bar{Y})^2/n]}}.$$

The standard deviations of X and Y are $S_X = \sqrt{\sum (X_i - \bar{X})^2/n}$ and $S_Y = \sqrt{\sum (Y_i - \bar{Y})^2/n}$.

[2] The sign of r is positive if both regression coefficients are positive and negative if both coefficients are negative; $b_{Y \cdot X}$ and $b_{X \cdot Y}$ always have the same sign.

Therefore,

$$r\frac{S_Y}{S_X} = \frac{\sum(X_i - \bar{X})(Y_i - \bar{Y})/n}{\sqrt{[\sum(X_i - \bar{X})^2/n][\sum(Y_i - \bar{Y})^2/n]}} \frac{\sqrt{\sum(Y_i - \bar{Y})^2/n}}{\sqrt{\sum(X_i - \bar{X})^2/n}}$$

$$= \frac{\sum(X_i - \bar{X})(Y_i - \bar{Y})}{\sum(X_i - \bar{X})^2} = b_{Y \cdot X}.$$

$$r\frac{S_X}{S_Y} = \frac{\sum(X_i - \bar{X})(Y_i - \bar{Y})/n}{\sqrt{[\sum(X_i - \bar{X})^2/n][\sum(Y_i - \bar{Y})^2/n]}} \frac{\sqrt{\sum(X_i - \bar{X})^2/n}}{\sqrt{\sum(Y_i - \bar{Y})^2/n}}$$

$$= \frac{\sum(X_i - \bar{X})(Y_i - \bar{Y})}{\sum(Y_i - \bar{Y})^2} = b_{X \cdot Y}.$$

Similarly, we can show that $\sqrt{b_{Y \cdot X} b_{X \cdot Y}} = r$:

$$\sqrt{b_{Y \cdot X} b_{X \cdot Y}} = \sqrt{\left[\frac{\sum(X_i - \bar{X})(Y_i - \bar{Y})}{\sum(X_i - \bar{X})^2}\right]\left[\frac{\sum(X_i - \bar{X})(Y_i - \bar{Y})}{\sum(Y_i - \bar{Y})^2}\right]}$$

$$= \frac{\sqrt{[\sum(X_i - \bar{X})(Y_i - \bar{Y})]^2}}{\sqrt{[\sum(X_i - \bar{X})^2][\sum(Y_i - \bar{Y})^2]}}$$

$$= \frac{\sum(X_i - \bar{X})(Y_i - \bar{Y})}{\sqrt{[\sum(X_i - \bar{X})^2][\sum(Y_i - \bar{Y})^2]}} = r.$$

The last formula for r can be written in the form given earlier by multiplying both the numerator and the denominator by $1/n$.

The linear equation for predicting Y_i from X_i can be expressed in terms of $r(S_Y/S_X)$ instead of $b_{Y \cdot X}$:

$$Y'_i = \overbrace{\bar{Y} - r\frac{S_Y}{S_X}\bar{X}}^{a_{Y \cdot X}} + \overbrace{r\frac{S_Y}{S_X}}^{b_{Y \cdot X}} X_i = \bar{Y} + r\frac{S_Y}{S_X}(X_i - \bar{X}).$$

In this form we can see what happens when $r = 0$; we obtain

$$Y'_i = \bar{Y} + 0\frac{S_Y}{S_X}(X_i - \bar{X})$$

$$= \bar{Y}.$$

This means that when the correlation coefficient equals zero, the predicted value of Y is the mean of the Y scores regardless of the X value used to predict Y. In other words, knowing X_i is no help if $r = 0$, since in every case the predicted Y value is $\bar{Y}$.

Review Exercises for Sections 6.1 and 6.2

*1. If Y increases 2 units for every 4 unit increase in X, what is the slope of the regression line of Y on X?

2. If Y decreases 5 units for every 2 unit increase in X, what is the slope of the regression line of Y on X?

*3. In an experiment on sex-typed behavior, a random sample of boys ages 5–8 was given choices among such toys as a football, doll carriage, dump truck, and dishes. The number of sex-appropriate choices for boys at each age are listed in the table.

Age, X	Number of Appropriate Choices, Y	Age, X	Number of Appropriate Choices, Y
7.5	18	7.5	15
7.0	13	5.0	7
5.5	11	5.5	8
8.0	20	6.0	12
6.5	13	8.0	17
6.0	14	7.0	14
5.0	9	6.5	12
8.0	18	5.5	10
6.5	14	5.0	8
6.0	10	7.0	16
7.5	19		

a. Construct a scatter diagram and decide whether the data appear to be linearly related.
b. Compute the values of $a_{Y \cdot X}$ and $b_{Y \cdot X}$ for the line of best fit and draw the line in the scatter diagram. Compute r using the relationship $r = b_{Y \cdot X}(S_X/S_Y)$.
c. Estimate Y for a 6-year-old boy using both the regression equation and the line of best fit in the scatter diagram.

4. For the experiment described in Exercise 3, data were also obtained for a random sample of girls, as listed in the table.

Age, X	Number of Appropriate Choices, Y	Age, X	Number of Appropriate Choices, Y
7.5	10	8.0	14
6.0	11	7.0	11
5.5	10	7.5	13
8.0	15	6.5	9
7.5	14	6.5	11
5.0	6	6.0	10
6.0	8	5.5	8
7.0	12	7.0	10
8.0	12	6.5	13
5.0	7	5.0	9
5.5	9		

a. Construct a scatter diagram and decide whether the data appear to be linearly related.
b. Compute the values of $a_{Y \cdot X}$, $b_{Y \cdot X}$, $a_{X \cdot Y}$, and $b_{X \cdot Y}$ for the two regression equations, and draw the lines of best fit in the scatter diagram.
c. Compute r using the relationship $r = \pm\sqrt{b_{Y \cdot X} b_{X \cdot Y}}$.
d. Estimate Y for a 6-year-old girl and X for a girl who made 11 appropriate choices using the lines of best fit in the scatter diagram.

*5. In what sense are the regression lines in Exercises 3 and 4 best-fitting lines?
6. What characteristics of the line of best fit do $a_{Y \cdot X}$ and $b_{Y \cdot X}$ describe?
7. Distinguish between $b_{Y \cdot X}$ and $b_{X \cdot Y}$.
*8. The formula for $b_{Y \cdot X}$ in Table 6.2-1 is not the most convenient one for computational purposes. Show that it is algebraically equivalent to $[n \sum X_i Y_i - (\sum X_i)(\sum Y_i)]/[n \sum X_i^2 - (\sum X_i)^2]$, which is simpler to compute.
9. If $r = 0$, the predicted Y score for all subjects is the mean of Y. Draw a scatter diagram that illustrates this.
*10. a. If $Y_i' = a + bX_i$ for all i and $a = \bar{Y} - b\bar{X}$, prove that $\sum Y_i' = \sum Y_i$.
   b. If $Y_i' = Y_i$ for all i, what do you know about r?
11. Terms to remember:
   a. Regression analysis
   b. Line of best fit
   c. Prediction error
   d. Principle of least squares

6.3 Another Measure of Ability to Predict: The Standard Error of Estimate

We have seen that our ability to predict Y from X is a function of the degree of correlation between the two variables. The higher the correlation, the more closely the data points cluster around the regression line and the smaller our prediction error. A measure of the prediction error is given by the **standard error of estimate**, $S_{Y \cdot X}$, which is a kind of standard deviation. For comparison purposes the formulas for the standard error of estimate and standard deviation are given below.[3]

$$S_{Y \cdot X} = \sqrt{\frac{\sum (Y_i - Y_i')^2}{n}} \qquad S_Y = \sqrt{\frac{\sum (Y_i - \bar{Y})^2}{n}}$$

In computing $S_{Y \cdot X}$, the deviation $(Y_i - Y_i')$ is from the predicted value or regression line, whereas for S_Y, the deviation is from the mean of Y. The two deviations are illustrated in Figure 6.3-1.

Let's look at $S_{Y \cdot X}$ more closely. The regression line denoted by Y' can be thought of as a kind of mean—a "running mean" giving the predicted value of Y for a particular value of X. Whereas $\bar{Y}$ is the mean of Y, Y' estimates the mean of Y for a particular value of X. The deviation of a Y score from a running mean is a prediction error ($e_i = Y_i - Y_i'$). Viewed in this light, $S_{Y \cdot X}$ (like S_Y) is computed from the sum of squared deviations from means and hence is a standard deviation. However, $S_{Y \cdot X}$ is the standard deviation of predicted errors,[4] while S_Y is the standard deviation of scores; $S_{Y \cdot X}$ can be interpreted in much the same way as a conventional standard deviation.

---

[3] When the population standard error of estimate is estimated from sample data, a better estimator is
$$\hat{\sigma}_{Y \cdot X} = \sqrt{\frac{\sum (Y_i - Y_i')^2}{n - 2}}.$$

[4] See Section 6.7-1 for a mathematical development of this idea.

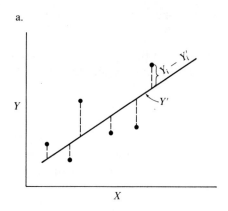 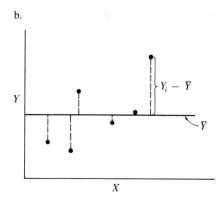

Figure 6.3-1. Comparison of the deviation $Y_i - Y_i'$ used to compute the standard error of estimate (part a) and the deviation $Y_i - \bar{Y}$ used to compute the standard deviation (part b).

Interpretation of $S_{Y \cdot X}$

The larger $S_{Y \cdot X}$, the greater the dispersion of Y scores around the regression line and hence the larger the average prediction error. If the distribution of Y scores at every X score is approximately normal and if all the Y score distributions have the same dispersion, 68.3% of the Y scores will fall within the limits of $Y' \pm S_{Y \cdot X}$. This information is illustrated in Figure 6.3-2. Similarly, 95.4% of the Y scores will fall within the limits $Y' \pm 2S_{Y \cdot X}$, and 99.7% within the limits $Y' \pm 3S_{Y \cdot X}$. This follows from the normal distribution; see Figure 4.4-1.

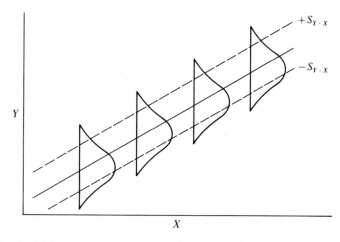

Figure 6.3-2. Illustration of the standard error of estimate. Approximately 68% of the Y scores fall within the limits $Y' \pm S_{Y \cdot X}$ if the distribution of Y scores at every X score is approximately normally distributed and all the Y score distributions have the same dispersion.

6.4 Alternative Formula for $S_{Y \cdot X}$

The formula for the standard error of estimate given in the last section isn't convenient for computing $S_{Y \cdot X}$. An equivalent formula[5] that is much easier to use is

$$S_{Y \cdot X} = S_Y \sqrt{1 - r^2}.$$

The maximum and minimum possible values of $S_{Y \cdot X}$ can be readily ascertained from this formula. The maximum value occurs when $r = 0$, in which case $S_{Y \cdot X} = S_Y$. In other words, if $r = 0$, the dispersion of Y scores around the regression line is as large as the standard deviation of Y. Knowing the X score doesn't reduce the error of prediction. The minimum value occurs when $r = 1$. In this case, $S_{Y \cdot X} = 0$ and there is no dispersion around the regression line and no error in predicting Y from X. To summarize, the upper limit for $S_{Y \cdot X}$ is S_Y and occurs when $r = 0$; the lower limit is zero and occurs when $r = 1$.

Coefficient of Alienation, k

The relationship between $S_{Y \cdot X}$ and r described above leads to a useful relative index of prediction error called the *coefficient of alienation*, denoted by k. As we have seen, when $r = 0$ we have the worst possible prediction situation—the magnitude of our prediction error is as large as it can be so that $S_{Y \cdot X} = S_Y$. As r approaches ± 1, prediction error decreases until when $r = \pm 1$, $S_{Y \cdot X} = 0$. The coefficient of alienation compares $S_{Y \cdot X}$, which reflects the magnitude of predictive error, with S_Y, which reflects the error that would be obtained in the worst possible prediction situation, and this occurs when $S_{Y \cdot X} = S_Y$. The coefficient is simply the ratio $S_{Y \cdot X}/S_Y$.

$$k = \begin{pmatrix} \text{Comparison of predictive} \\ \text{error for a set of data, with the} \\ \text{predictive error in the worst} \\ \text{possible prediction situation} \end{pmatrix} = \frac{S_{Y \cdot X}}{S_Y} = \frac{S_Y \sqrt{1 - r^2}}{S_Y} = \sqrt{1 - r^2}$$

For example, if $r = 0.80$, then $k = \sqrt{1 - (0.80)^2} = 0.60$. This tells us that the standard error of estimate, a measure of our predictive error, is 60% of the size it would be if the correlation were zero. To state it another way, when $r = 0.80$, we have reduced the magnitude of the prediction error by $100\% - 60\% = 40\%$ by taking X into account. The reduction in predictive error is called the **index of forecasting efficiency**, E,[6] and is computed from the formula $E = 1 - k$.

The values of k and E for selected values of r are given in Table 6.4-1. We note from the table that relative predictive error decreases very slowly as r increases from

---

[5] See Section 6.7-2 for the derivation of the formula. When the population standard error of estimate is estimated from sample data, a better estimator is

$$S_{Y \cdot X} = S_Y \sqrt{\frac{n}{n-2}(1 - r^2)}.$$

[6] E is also called the *coefficient of predictive efficiency*.

Table 6.4-1. Values of k and E for Selected Values of r

r	k	E	r	k	E
1.00	0.00	1.00	0.45	0.89	0.11
0.95	0.31	0.69	0.40	0.92	0.08
0.90	0.44	0.56	0.35	0.94	0.06
0.85	0.53	0.47	0.30	0.95	0.05
0.80	0.60	0.40	0.25	0.97	0.03
0.75	0.66	0.34	0.20	0.98	0.02
0.70	0.71	0.29	0.15	0.989	0.011
0.65	0.76	0.24	0.10	0.995	0.005
0.60	0.80	0.20	0.05	0.999	0.001
0.55	0.84	0.16	0.00	1.00	0.00
0.50	0.87	0.13			

0 to 1. For $r = 0.85$ the standard error of estimate is 53% of the size it would be if the correlation were zero. We see also that for high correlations a small increase in r produces a greater reduction in relative predictive error than for low correlations. For example, a 0.05 change in r from 0.85 to 0.90 reduces k by 9%, whereas a 0.05 change in r from 0.20 to 0.25 only reduces k by 1%.

Descriptive Application of $S_{Y \cdot X}$

Although the standard error of estimate is most often used in inferential statistics, a descriptive application that was briefly mentioned in Section 6.3 is illustrative. Suppose an experiment was conducted to determine the relationship between length of time necessary to reach a decision, Y, and the number of alternative choices presented, X. The following data were obtained: $S_X = 1.5$, $S_Y = 12.5$, $\bar{X} = 4.5$, $\bar{Y} = 45$, $r = 0.78$, and $n = 100$. Assume the distribution of Y scores for every X score is approximately normal and all the Y score distributions have the same dispersion. The predicted reaction time for a person presented with a choice from among, say, three alternatives is given according to the regression equation $Y' = \bar{Y} + r(S_Y/S_X)(X_i - \bar{X})$:

$$Y' = 45 + 0.78 \frac{12.5}{1.5}(3 - 4.5)$$
$$= 45 + 6.5(-1.5)$$
$$= 35.25.$$

The standard error of estimate is

$$S_{Y \cdot X} = S_Y\sqrt{1 - r^2}$$
$$= 12.5\sqrt{1 - (0.78)^2}$$
$$= 12.5(0.6258)$$
$$= 7.82.$$

We can conclude that approximately 68% of the subjects had reaction times between

6.4 Alternative Formula for $S_{Y \cdot X}$

43.07 and 27.43, as we see from

$$Y' \pm S_{Y \cdot X} = 35.25 \pm 7.82 = 43.07 \text{ and } 27.43.$$

Similarly, approximately 95% had reaction times between

$$Y' \pm 2S_{Y \cdot X} = 35.25 \pm 2(7.82) = 50.89 \text{ and } 19.61.$$

6.5 Assumptions Associated with Estimate of Prediction Error

When we make predictions using the regression equation $Y_i' = a + bX_i$ we assume only that the relationship between X and Y is linear. If the assumption is tenable, the principle of least squares ensures that $a + bX_i$ provides the best possible fit for the data points. For prediction purposes we don't have to make any assumptions regarding the shape of the X and Y distributions.

The use of the standard error of estimate and k and E involves more stringent assumptions. In addition to the linearity assumption, we must also assume that (1) for any value of X the associated Y scores are approximately normally distributed and (2) the standard deviation of Y scores, $S_{Y \cdot X}$, is the same for all the X scores. The latter assumption is referred to as the *homoscedasticity* assumption.

In making predictions from Y to X the same assumptions are required, but they must be rephrased to reflect the revised roles of X and Y.

Review Exercises for Sections 6.3, 6.4, and 6.5

*12. Chimpanzees were exposed to white noise 8 hours a day for 3 months to determine if it affected their hearing. Ten animals were randomly assigned to the following noise levels: 75 dBA,[7] 85 dBA, 95 dBA, 105 dBA, and 115 dBA.

Animal	Noise Level (dBA)	Hearing Loss (dB at 1000 Hz)	Animal	Noise Level (dBA)	Hearing Loss (dB at 1000 Hz)
1	105	11	6	85	9
2	85	6	7	105	13
3	95	10	8	115	11
4	115	15	9	75	5
5	75	7	10	95	8

a. Compute $S_{Y \cdot X}$ using the formula $S_Y \sqrt{1 - r^2}$.
b. Assuming a large sample in which the distribution of Y scores for every X score is approximately normal and all the distributions have the same dispersion, compute the limits of the interval that will contain 68% of the scores for a noise level of 115 dBA.

[7] The abbreviation *dBA* stands for the sound pressure level in decibels as measured on the A weighting network of a sound-level meter.

c. Compute the value of $S_{Y \cdot X}$ for $r = 0$ and $r = 1$.
d. Compute k and E and interpret.

13. The relationship between birth order and participation in dangerous sports such as football, soccer, and boxing was investigated. College records were screened to obtain four males who were first-born, four who were second-born, and so on. The data in the table were obtained.

Subject	Birth Order	Number of Dangerous Sports	Subject	Birth Order	Number of Dangerous Sports
1	4	1	11	1	0
2	3	1	12	2	0
3	2	0	13	3	2
4	4	2	14	5	1
5	1	0	15	3	1
6	5	2	16	2	1
7	1	0	17	5	2
8	4	1	18	1	1
9	2	1	19	3	1
10	5	3	20	4	2

a. Compute $S_{Y \cdot X}$ using the formula $S_Y \sqrt{1 - r^2}$.
b. Assuming a large sample in which the distribution of Y scores for every X score is approximately normal and all the distributions have the same dispersion, compute the limits that will contain 68% of the scores for fourth-born males.
c. Compute the value of $S_{Y \cdot X}$ for $r = 0$ and $r = 1$.
d. Compute k and E and interpret.

14. In what sense is Y' a mean?
15. How is $S_{Y \cdot X}$ related to the magnitude of prediction error?
16. How is k related to the magnitude of prediction error?
17. Describe the effect of changes in r on the value of k.
18. Compare the assumptions associated with predictions using r, Y', and $S_{Y \cdot X}$.
19. Terms to remember:
    a. Standard error of estimate
    b. Coefficient of alienation
    c. Index of forecasting efficiency (coefficient of predictive efficiency)
    d. Homoscedasticity

6.6 Summary

The term "regression" was first used by Sir Francis Galton to refer to the tendency for short parents to have slightly taller offspring and tall parents to have slightly shorter offspring. We still speak of regression toward the mean, but today the term has a broader meaning. It refers to any analysis of paired data $(X_1, Y_1), (X_2, Y_2), \ldots, (X_n, Y_n)$, where X is the independent variable and Y the dependent variable.

In regression analysis the line of best fit, called the regression line, is used to predict Y from a knowledge of X. The line of best fit according to the least squares principle is the one for which the sum of the squared prediction errors, the discrepancy between the observed value of Y_i and the predicted value is as small as it can possibly be.

If $r = 1$, the value of Y_i can be predicted perfectly from the equation $Y_i' = a + bX_i = \bar{Y} + r(S_Y/S_X)(X_i - \bar{X})$. If $|r| < 1$, there is likely to be some discrepancy between the observed value Y_i and the predicted value Y_i'. A measure of the prediction error is given by the standard error of estimate $S_{Y \cdot X}$, which is a kind of standard deviation of scores around the regression line. The maximum value of $S_{Y \cdot X}$ is the standard deviation of Y, S_Y, and occurs when $r = 0$. The minimum value is zero and occurs when $r = 1$.

The coefficient of alienation, k, is a relative index of predictive error. It is the ratio of $S_{Y \cdot X}$, which reflects the magnitude of predictive error, to S_Y, which reflects the magnitude of predictive error in the worst possible prediction situation—when $S_{Y \cdot X} = S_Y$. The coefficient of alienation is conveniently computed from $k = S_{Y \cdot X}/S_Y = \sqrt{1 - r^2}$. If $r = 60$, then $k = 0.80$, which means that $S_{Y \cdot X}$ is 80% of the size it would be if the correlation between X and Y were zero. A related statistic, the index of forecasting efficiency, $E = 1 - k$, measures the reduction in magnitude of the prediction error over what it would have been had the value of X been ignored in predicting Y.

In predicting Y from X we assume only that the relationship between the variables is linear. Interpretations involving $S_{Y \cdot X}$, k, or E also assume that the distribution of Y scores at every X score is approximately normal and all the Y score distributions have the same dispersion. When prediction involves different samples, as when the performance of one group of students is predicted from that of another, we must also assume the populations represented by the two samples are identical with respect to the relevant characteristics.

†6.7 Technical Notes

6.7-1 Standard Error of Estimate as the Standard Deviation of Prediction Errors

We will show in this section that $S_{Y \cdot X}$ is the standard deviation of prediction errors. We defined a prediction error in Section 6.3 as $e_i = Y_i - Y_i'$. The definitional formula for a standard deviation is $S_X = \sqrt{\sum (X_i - \bar{X})^2/n}$. Substituting e_i for X_i and $\bar{e}$ for $\bar{X}$ in the definitional formula gives

$$S_e = \sqrt{\frac{\sum (e_i - \bar{e})^2}{n}}.$$

The mean of the prediction errors can be shown to equal zero.

$$\bar{e} = \frac{1}{n}\sum e_i$$

(6.7-1-1)
$$= \frac{1}{n}\sum (Y_i - Y_i')$$

† These notes, which are not essential to the text, can be omitted without loss of continuity.

We show in Section 6.2 that $Y_i' = \bar{Y} + r(S_Y/S_X)(X_i - \bar{X})$. Substituting this expression for Y_i' in (6.7-1-1) gives

$$\bar{e} = \frac{1}{n}\sum\left\{Y_i - \left[\bar{Y} + r\frac{S_Y}{S_X}(X_i - \bar{X})\right]\right\}$$

$$= \frac{1}{n}\left[\sum(Y_i - \bar{Y}) - r\frac{S_Y}{S_X}\sum(X_i - \bar{X})\right],$$

but as was shown in Technical Note 3.9-2, $\sum(Y_i - \bar{Y}) = 0$ and $\sum(X_i - \bar{X}) = 0$, so

$$\bar{e} = \frac{1}{n}\left[0 - r\frac{S_Y}{S_X}0\right]$$

$$= 0.$$

Thus,

$$S_e = \sqrt{\frac{\sum(e_i - \bar{e})^2}{n}} = \sqrt{\frac{\sum(e_i^2)}{n}}.$$

Since $e_i = Y_i - Y_i'$, we can write

$$S_e = \sqrt{\frac{\sum(Y_i - Y_i')^2}{n}}.$$

Therefore, the formula $\sqrt{\sum(Y_i - Y_i')^2/n}$ we defined as $S_{Y \cdot X}$ in Section 6.3 is the formula for the standard deviation of prediction errors.

6.7-2 Derivation of Alternative Formula for $S_{Y \cdot X}$

The formula $S_{Y \cdot X} = \sqrt{\sum(Y_i - Y_i')^2/n}$ in Section 6.3 is not convenient for computational purposes. It is much easier to compute $S_{Y \cdot X}$ from $S_Y\sqrt{1 - r^2}$. We show in this section that the two formulas are algebraically equivalent. We begin with the formula given in Section 6.3.

$$S_{Y \cdot X} = \sqrt{\frac{\sum(Y_i - Y_i')^2}{n}}$$

According to Section 6.2, $Y_i' = \bar{Y} + r(S_Y/S_X)(X_i - \bar{X})$. Making the substitution gives

$$S_{Y \cdot X} = \sqrt{\frac{1}{n}\sum\left[Y_i - \left(\bar{Y} + r\frac{S_Y}{S_X}(X_i - \bar{X})\right)\right]^2}$$

$$= \sqrt{\frac{1}{n}\sum\left[(Y_i - \bar{Y}) - r\frac{S_Y}{S_X}(X_i - \bar{X})\right]^2}$$

$$= \sqrt{\frac{1}{n}\sum\left[(Y_i - \bar{Y})^2 - 2r\frac{S_Y}{S_X}(X_i - \bar{X})(Y_i - \bar{Y}) + r^2\frac{S_Y^2}{S_X^2}(X_i - \bar{X})^2\right]}$$

$$= \sqrt{\frac{\sum(Y_i - \bar{Y})^2}{n} - 2r\frac{S_Y}{S_X}\frac{\sum(X_i - \bar{X})(Y_i - \bar{Y})}{n} + r^2\frac{S_Y^2}{S_X^2}\frac{\sum(X_i - \bar{X})^2}{n}}.$$

Multiplying the middle term by (S_Y/S_Y) gives

$$S_{Y \cdot X} = \sqrt{\frac{\sum (Y_i - \bar{Y})^2}{n} - 2rS_Y^2 \frac{\sum (X_i - \bar{X})(Y_i - \bar{Y})}{nS_X S_Y} + r^2 \frac{S_Y^2}{S_X^2} \frac{\sum (X_i - \bar{X})^2}{n}},$$

but $\sum (Y_i - \bar{Y})^2/n = S_Y^2$, $\sum (X_i - \bar{X})(Y_i - \bar{Y})/nS_X S_Y = r$, and $\sum (X_i - \bar{X})^2/n = S_X^2$. Substituting for these terms gives

$$\begin{aligned} S_{Y \cdot X} &= \sqrt{S_Y^2 - 2r^2 S_Y^2 + r^2 S_Y^2} \\ &= \sqrt{S_Y^2 - r^2 S_Y^2} \\ &= \sqrt{S_Y^2(1 - r^2)} \\ &= S_Y \sqrt{1 - r^2}. \end{aligned}$$

6.7-3 Partition of S_Y^2/S_Y^2 into r^2 and k^2

In Section 5.4 we stated that $S_Y^2/S_Y^2 = r^2 + k^2$. To prove this we begin with the identity,

$$(Y_i - \bar{Y}) = (Y_i - Y_i' + Y_i' - \bar{Y}),$$

where Y_i is the score for individual i, Y_i' is the individual's predicted score, and $\bar{Y}$ is the mean of all the Y scores. Summing over all the scores and rearranging the terms gives

$$\sum_{i=1}^{n} (Y_i - \bar{Y}) = \sum_{i=1}^{n} [(Y_i - Y_i') + (Y_i' - \bar{Y})].$$

Squaring both sides of the equation and distributing the summation leads to

$$\begin{aligned} \sum_{i=1}^{n} (Y_i - \bar{Y})^2 &= \sum_{i=1}^{n} [(Y_i - Y_i') + (Y_i' - \bar{Y})]^2 \\ &= \sum_{i=1}^{n} [(Y_i - Y_i')^2 + 2(Y_i - Y_i')(Y_i' - \bar{Y}) + (Y_i' - \bar{Y})^2] \\ &= \sum_{i=1}^{n} (Y_i - Y_i')^2 + 2 \sum_{i=1}^{n} (Y_i - Y_i')(Y_i' - \bar{Y}) + \sum_{i=1}^{n} (Y_i' - \bar{Y})^2. \end{aligned}$$

We can show that the middle term on the right is equal to zero. We saw in Section 6.2 that $Y_i' = \bar{Y} + r(S_Y/S_X)(X_i - \bar{X})$. Making the substitution for Y_i' in the middle term gives

$$\begin{aligned} 2 \sum_{i=1}^{n} (Y_i - Y_i')(Y_i' - \bar{Y}) \\ &= 2 \sum_{i=1}^{n} \left\{ \left[Y_i - \bar{Y} - r\frac{S_Y}{S_X}(X_i - \bar{X}) \right] \left[\bar{Y} + r\frac{S_Y}{S_X}(X_i - \bar{X}) - \bar{Y} \right] \right\} \\ &= 2 \sum_{i=1}^{n} \left\{ \left[(Y_i - \bar{Y}) - r\frac{S_Y}{S_X}(X_i - \bar{X}) \right] \left[r\frac{S_Y}{S_X}(X_i - \bar{X}) \right] \right\} \\ &= 2 \sum_{i=1}^{n} \left[r\frac{S_Y}{S_X}(X_i - \bar{X})(Y_i - \bar{Y}) - r^2 \frac{S_Y^2}{S_X^2}(X_i - \bar{X})^2 \right] \\ &= 2r \frac{S_Y}{S_X} \sum_{i=1}^{n} (X_i - \bar{X})(Y_i - \bar{Y}) - 2r^2 \frac{S_Y^2}{S_X^2} \sum_{i=1}^{n} (X_i - \bar{X})^2. \end{aligned}$$

(6.7-3-1)

Since $r = [\sum_{i=1}^{n}(X_i - \bar{X})(Y_i - \bar{Y})]/nS_XS_Y$, $nrS_XS_Y = \sum_{i=1}^{n}(X_i - \bar{X})(Y_i - \bar{Y})$, and since $S_X^2 = \sum_{i=1}^{n}(X_i - \bar{X})^2/n$, $nS_X^2 = \sum_{i=1}^{n}(X_i - \bar{X})^2$. Making these substitutions in 6.7-3-1, we have

$$2\sum_{i=1}^{n}(Y_i - Y_i')(Y_i' - \bar{Y}) = 2r\frac{S_Y}{S_X}nrS_XS_Y - 2r^2\frac{S_Y^2}{S_X^2}nS_X^2$$
$$= 2nr^2S_Y^2 - 2nr^2S_Y^2 = 0.$$

Therefore,

(6.7-3-2) $$\sum_{i=1}^{n}(Y_i - \bar{Y})^2 = \sum_{i=1}^{n}(Y_i - Y_i')^2 + \sum_{i=1}^{n}(Y_i' - \bar{Y})^2.$$

According to Technical Note 6.7-2,

(6.7-3-3) $$nS_{Y \cdot X}^2 = \sum_{i=1}^{n}(Y_i - Y_i')^2 = nS_Y^2(1 - r^2).$$

As we saw in Section 6.2, $Y_i' = \bar{Y} + r(S_Y/S_X)(X_i - \bar{X})$. Hence, $(Y_i' - \bar{Y}) = r(S_Y/S_X)(X_i - \bar{X})$. Squaring both sides and summing over the n scores gives

$$\sum_{i=1}^{n}(Y_i' - \bar{Y})^2 = r^2\frac{S_Y^2}{S_X^2}\sum_{i=1}^{n}(X_i - \bar{X})^2$$
$$= r^2\frac{S_Y^2}{S_X^2}nS_X^2$$

(6.7-3-4) $$= nr^2S_Y^2.$$

Substituting equations (6.7-3-3) and (6.7-3-4) in (6.7-3-2) gives

$$\sum_{i=1}^{n}(Y_i - \bar{Y})^2 = nS_Y^2(1 - r^2) + nr^2S_Y^2.$$

Dividing both sides of the equation by nS_Y^2 gives

$$\frac{\sum_{i=1}^{n}(Y_i - \bar{Y})^2}{nS_Y^2} = \frac{nS_Y^2(1 - r^2)}{nS_Y^2} + \frac{nr^2S_Y^2}{nS_Y^2}.$$

Since $nS_Y^2 = \sum_{i=1}^{n}(Y_i - \bar{Y})^2$, we can write $S_Y^2/S_Y^2 = (1 - r^2) + r^2$. In Section 5.4 we used the symbol k^2 to represent $1 - r^2$. Thus,

$$\frac{S_Y^2}{S_Y^2} = k^2 + r^2.$$

We have just shown that the total variance, S_Y^2, expressed as a proportion is equal to the coefficient of nondetermination, k^2, plus the coefficient of determination, r^2.

7
Probability

7.1 Introduction to Probability
   The Subjective–Personalistic View
   The Classical or Logical View
   The Empirical Relative-Frequency View
   Review Exercises for Section 7.1
7.2 Basic Concepts
   Simple and Compound Events
   Graphing Simple Events
   Formal Properties of Probability
   Review Exercises for Section 7.2
7.3 Probability of Combined Events
   Addition Rule of Probability
   Addition Rule for Mutually Exclusive Events
   Addition Rule for Three Events
   Multiplication Rule of Probability
   Multiplication Rule for Statistically Independent Events
   Multiplication Rules for Three Events
   Common Errors in Applying the Rules of Probability
   Review Exercises for Section 7.3
7.4 Summary

7.1 Introduction to Probability

Everyone has some intuitive notion of what probability is. However, its definition is a topic for continuing debate among mathematicians. Three views of probability will be described: (1) the subjective–personalistic view, (2) the classical or logical view, and (3) the empirical relative–frequency view. Fortunately, the different views are not incompatible.

The Subjective–Personalistic View

*According to the **subjective view**, probability is a measure of the strength of one's expectation that an event will occur.* For example, one might assert "Chances are I'll pass statistics" or "I think I'll go home this weekend." Such assertions express a degree of belief concerning an event with an outcome that is for the moment unknown. Subjective probabilities affect our lives since they enter into our decision-making process. For most people the subjective probability of being struck by a car while crossing the street is low, so they proceed as if the event won't happen. But if a person's subjective probability is high enough, all suitable preparations for the event's occurrence will be made.

Although our behavior is influenced by subjective probability, there are difficulties in incorporating it into a formal decision-making process. Equally knowledgeable individuals often disagree on the probability that should be assigned to an event. We find that some people's subjective probabilities follow closely the rules of probability theory described later, but other people's don't. Hence, a subjective probability can't be considered apart from the person holding it. The measurement of subjective probability poses another problem, although behavioral scientists are beginning to find solutions to this problem. In spite of the problems, a formal approach to decision-making that utilizes subjective probability has been developed and is popular in economics and business management and is beginning to find acceptance in behavioral research. This approach, called *Bayesian inference*,[1] enables an experimenter to make decisions about some true state of affairs using not only sample data but also any prior information that is available, either from previous samples or simply in the form of informed opinions or beliefs. The approach represents an extension of classical inferential methods. Bayesian methods, however, emphasize the steady accumulation and utilization of information from many sources. Furthermore, they lead to different ways of interpreting data, but not necessarily different conclusions about data. The body of classical statistical methods introduced in Chapters 11–15 is pretty well mapped out; this isn't true for Bayesian methods. The next decade will see many new developments.

The Classical or Logical View

Suppose we want to know the probability of rolling a 2 with a fair die. We reason that since a fair die is symmetrical and dynamically balanced, all six faces are equally likely. Of the six possible events, only one is a 2, and therefore the probability of rolling

---

[1] Bayesian inference is named for the Reverend Thomas Bayes (1702–1761), whose theorem laid the groundwork for the approach. Application of Bayes' theorem is described in Section 8.2. For a fuller discussion see Hays (1973, chap. 19), McGee (1971, chaps. 10 and 11), and Novick and Jackson (1974).

a 2 is $p(2) = \frac{1}{6}$. According to the **classical view**, the probability of an event, say A, is given by the number of events favoring A, n_A, divided by the total number of equally likely events, n_S.[2] Thus, $p(A) = n_A/n_S$. The value of $p(A)$ is always a number between 0 and 1 inclusive, because the number of events favoring A can never exceed the total number of events—that is, $n_A \leq n_S$.

The classical view of probability is based on logical analysis. We reason, for example, that when a fair coin is tossed there are two possible outcomes, a head or a tail, and that the outcomes are equally likely. It follows that the probability of a head is $p(H) = n_H/n_S = \frac{1}{2}$. The probabilities $\frac{1}{2}$ for a head and $\frac{1}{6}$ for a 2 were arrived at by logical analysis of dynamic, event-generating processes. In effect, we developed a mathematical model based on a postulate and logic. We postulated that certain events are equally likely and deduced the consequences. If our logic is correct, the deductions $p(H) = \frac{1}{2}$ and $p(2) = \frac{1}{6}$ are formally correct, although they may not correspond to empirical results since for any particular coin or die the postulate may be incorrect. Experience has demonstrated that the classical view generates probability estimates that closely approximate empirical probabilities for certain fairly simple event-generating processes such as coin tossing. Consequently, this approach is useful for practical problems.

The Empirical Relative-Frequency View

A third view of probability can be adopted for event-generating processes that can be repeated without changing their characteristics. Probability according to this view is estimated from experience—by performing an experiment and determining the ratio of the number of events of interest to the total number of events. This leads to our final definition of probability. According to the **empirical relative-frequency view**, the probability of event A, $p(A)$, is a number approached by the ratio n_A/n as the total number of observations, n, approaches infinity. For example, in a simple experiment such as tossing a coin, the probability of a head can be estimated by making many tosses of the coin and recording the outcomes. If a head is obtained 12 times in 20 tosses, the best estimate of the probability of heads is $n_A/n = \frac{12}{20} = 0.6$. If a head is obtained 120 times in 200 tosses, our confidence in the estimate $120/200 = 0.6$ is even greater. We assume that the sample estimate n_A/n moves closer and closer to some "true probability" as n approaches infinity and thus we have greater confidence in larger samples.

The empirical view of probability is useful and intuitively simple, but it too has certain difficulties. It is meaningful to speak of the probability of rain tomorrow or the probability of getting an A on Tuesday's quiz; however, there is only one tomorrow and only one such quiz. The interpretation of probability as the number approached by n_A/n as the number of tomorrows approaches infinity is unconvincing.

We must conclude that none of the views of probability is completely adequate. Since they are all useful and they are not incompatible, they coexist amicably in the mathematician's bag of conceptual tools. In the discussion that follows we will rely most on the classical and empirical views.

[2] The letter S, denoting a sample space, will be defined in Section 7.2.

Review Exercises for Section 7.1

1. Why is subjective probability difficult to incorporate into a formal decision-making process?
2. What distinguishes Bayesian inference from traditional classical methods?
*3. (a) According to the classical view, what is the probability of observing an odd number on the toss of a die? (b) What assumptions were required to arrive at the answer?
4. (a) What is the probability of drawing the king of hearts from a well-shuffled deck of 52 cards? (b) What assumptions were required to arrive at the answer?
*5. (a) According to the relative-frequency view, what is the probability that a head will occur on the next toss of a fair coin if a head appeared on 52 of the last 100 tosses? (b) According to the logical view, what is the probability that a head will occur?

7.2 Basic Concepts

For behavioral scientists and educators, probability theory is a means to an end. It is the vehicle for making inferences about the characteristics of populations by observing samples drawn from the populations. Sample data are obtained by observing events in nature or experimenting in the laboratory. We will denote either procedure by the term *experiment*. In particular, we are interested in experiments with outcomes that can't be predicted with certainty. For example, will desensitization therapy result in more symptom relief than symbolic modeling therapy? Will a rat turn right or left in a T maze?

Simple and Compound Events

One of the simplest experiments we can perform is tossing a die and observing the number that appears on the upper face. Some of the possible outcomes are the following.

1. Event A—observe an odd number.
2. Event B—observe an even number.
3. Event C—observe a number less than 4.
4. Event E_1—observe a 1.
5. Event E_2—observe a 2.
6. Event E_3—observe a 3.
7. Event E_4—observe a 4.
8. Event E_5—observe a 5.
9. Event E_6—observe a 6.

An *event* is an observable happening. Events A, B, and C are called *compound* events because they can be decomposed into simpler events. For example, event A is the occurrence of one of the simpler events E_1, E_3, or E_5. Events $E_1, \ldots, E_6$ are called *simple* events because they can't be decomposed. An experiment in which two coins are tossed has four possible simple events.

	Coin 1	Coin 2
Event E_1	Head	Head
Event E_2	Head	Tail
Event E_3	Tail	Head
Event E_4	Tail	Tail

The event—observe two like coins—is a compound event because it can be decomposed into the simpler events E_1 and E_4. A single trial of an experiment will result in one and only one simple event. A list of simple events provides a breakdown of all possible outcomes of the experiment.

Graphing Simple Events

It is convenient to represent the simple events in an experiment by a graph called a Euler diagram.[3] To do this we assign to each simple event a point called a *sample point*. The symbol E_i is used to denote both the ith simple event and its sample point. *The set of all sample points is called the* **sample space** *and is denoted by the letter S*. A Euler diagram representing the sample space for the die-tossing experiment is shown in Figure 7.2-1.

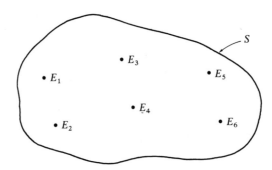

Figure 7.2-1. Euler diagram for die-tossing experiment. The set of all sample points $E_1, \ldots, E_6$ defines the sample space S of the experiment.

Compound events are represented in the diagram by encircling the sample points in that event. For example, earlier we defined event A as observing an odd number on the toss of a die and event C as a number less than 4. The two events are represented in Figure 7.2-2 by two subsets of the sample points.

The probability of event A according to the classical view is $p(A) = n_A/n_S = \frac{3}{6}$; the probability of event C is $p(C) = n_C/n_S = \frac{3}{6}$. By examining the sample space we can also determine the probability for combined events. What is the probability that when a die is tossed the outcome will be an odd number *and* less than 4? We could observe an odd number and a number less than 4, events A and C, if either E_1 or E_3 occurred. Hence, the probability that the outcome will represent both events A and C is $\frac{2}{6} = \frac{1}{3}$.

---

[3] The diagram was developed by Leonhard Euler (1707–1783), a Swiss mathematician.

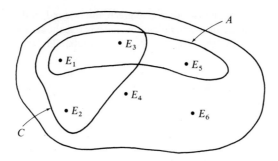

Figure 7.2-2. Euler diagram for event A, observing an odd number, and event C, observing a number less than 4.

We could observe an odd number *or* a number less than 4, event A or C or both A and C in four ways: if E_1, E_2, E_3, or E_5 occurred. Hence, the probability of A or C or both A and C is $\frac{4}{6} = \frac{2}{3}$. Formal rules for computing the probabilities of combined events are presented in Section 7.3, but first we will describe three properties of probabilities.

Formal Properties of Probability

Probability theory can be thought of as a system of definitions and operations pertaining to a sample space. According to the classical view described in Section 7.1, the probability of event A is the ratio of the number of sample points that are examples of A to the total number of sample points, provided all sample points are equally likely. For the die-tossing experiment represented in Figure 7.2-1, $p(E_1) = p(E_2) = \cdots = p(E_6) = n_{E_i}/n_S = \frac{1}{6}$. This follows from the assumption that all six faces are equally likely—$n_{E_i} = 1$ for the $i = 1, \ldots, 6$ sample points—and the fact that there are six sample points in the sample space—$n_S = 6$. To each point in the sample space we can assign a number called the probability of E_i such that (1) $0 \leq p(E_i) \leq 1$ for all i, (2) $\sum_{i=1}^{n_S} p(E_i) = 1$, and (3) $p(S) = 1$. In other words, (1) the probability assigned is a number greater than or equal to 0 and less than or equal to 1, (2) the sum of the probabilities over the sample space equals 1, and (3) the probability of the sure event, S, is always 1.

Review Exercises for Section 7.2

*6. An experiment consists of tossing three fair coins. (a) Represent the sample space by a Euler diagram, and encircle the sample points corresponding to observing two heads, event A, and observing at least one head, event B. (b) What is the probability of event A? (c) What is the probability of event B?

7. A class contains six psychology (P) majors, one sociology (S) major, and three history (H) majors. Assume that no students have double majors. (a) Represent the sample space by a Euler diagram. (b) If a student is selected at random, what is the probability that he or she will be a psychology major? (c) What is the probability that the student selected will be a psychology or a sociology major?

8. An experiment consists of tossing two dice, one green and one red, and recording the outcome. (a) Represent the sample space by a Euler diagram, and encircle the sample points corresponding to observing a 7 as the sum of the dice. (b) What is the probability that the sum of two dice is 7? (c) What is the probability that the sum of two dice is less than 5?

*9. Determine the probability (a) that a man chosen randomly from a group of ten men is a psychologist if the group contains three psychologists (b) of winning a car if you bought six raffle tickets and 10,000 tickets were sold.

10. A fair die is rolled once. You win $5 if the outcome is even, event A, or if it is divisible by 3, event B. (a) Represent the sample space by a Euler diagram and encircle events A and B. (b) What is the probability of winning the $5?

11. The following are properties of probabilities. In your own words state what each property means. (a) $0 \le p(E_i) \le 1$, for all i. (b) $\sum_{i=1}^{n_S} p(E_i) = 1$. (c) $p(S) = 1$.

12. Terms to remember:
    a. Experiment
    b. Simple and compound events
    c. Euler diagram
    d. Sample point
    e. Sample space

7.3 Probability of Combined Events

This section describes rules for determining the probabilities of combined events. For example, we might want to know the probability that the outcome of an experiment will be event A or event B or both A and B. We denote this probability by $p(A \text{ or } B)$.[4] Alternatively, we might want to know the probability that the outcome will be both A and B. We denote this probability by $p(A \text{ and } B)$.[5]

The *union* of two events A and B is the set of elements that belong to A or to B or to both. The probability of the union of two events, $p(A \text{ or } B)$, is computed by using the addition rule of probability. The *intersection* of two events A and B is the set of elements that belong to both A and B. Its probability, $p(A \text{ and } B)$, is computed by the multiplication rule.

Addition Rule of Probability

The **addition rule** states that the probability of the union A or B is equal to $p(A \text{ or } B) = p(A) + p(B) - p(A \text{ and } B)$. For example, let event A be an even number when a die is tossed and event B a number less than 5. The events are represented in Figure 7.3-1. Their probabilities are determined by counting sample points; $p(A) = n_A/n_S = \frac{3}{6} = \frac{1}{2}$ and $p(B) = n_B/n_S = \frac{4}{6} = \frac{2}{3}$. The probability of event A and B is the ratio of the number of sample points that are examples of both A and B to the total number of sample points. In symbols $p(A \text{ and } B) = n_{A \text{ and } B}/n_S = \frac{2}{6} = \frac{1}{3}$, since there are two simple events in both A and B and six in the sample space. Given this information, $p(A \text{ or } B) = \frac{1}{2} + \frac{2}{3} - \frac{1}{3} = \frac{5}{6}$. The probability of observing an even number or a number less than

[4] Some texts use the Boolean algebraic symbol ∪ in place of *or*.
[5] Some texts use the Boolean algebraic symbol ∩ in place of *and*.

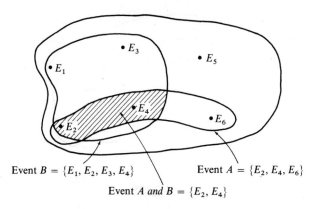

Event B = {E_1, E_2, E_3, E_4}   Event A = {E_2, E_4, E_6}
Event A and B = {E_2, E_4}

Figure 7.3-1. Euler diagram for event A, observing an even number, and event B, observing a number less than 5. The intersection of A and B is the shaded area.

5 is $\frac{5}{6}$. The value $p(A \text{ and } B) = \frac{1}{3}$ is subtracted from $p(A) + p(B)$ to avoid counting the simple events E_2 and E_4 twice, since they are contained in event A and event B.

The information contained in Figure 7.3-1 can be presented in a table, as shown in Table 7.3-1. This mode of presentation is easier to interpret, especially when the number of events exceeds two.

Table 7.3-1. Tabular Presentation of Information in Figure 7.3-1

		Event A	**Event** Not A	
Event	B	A and B = {E_2, E_4}	Not A and B = {E_1, E_3}	B = {E_1, E_2, E_3, E_4}
	Not B	A and Not B = {E_6}	Not A and Not B = {E_5}	Not B = {E_5, E_6}
		A = {E_2, E_4, E_6}	Not A = {E_1, E_3, E_5}	

Addition Rule for Mutually Exclusive Events

Two events may contain no sample points in common, in which case the events are said to be *mutually exclusive*. For example, consider the events observing an even number on the toss of a die, event A, and observing an odd number, event B. As shown in Figure 7.3-2, the intersection A and B contains no sample points, and hence A and B are mutually exclusive.

For mutually exclusive events the addition rule is simplified because $p(A \text{ and } B) = 0$. The rule is

$$p(A \text{ or } B) = p(A) + p(B).$$

The probability of observing an even number or odd number is $p(A \text{ or } B) = \frac{3}{6} + \frac{3}{6} = 1$.

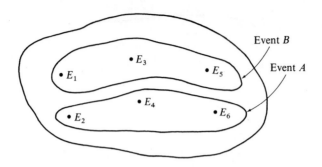

Figure 7.3-2. Euler diagram for event *A*, observing an even number, and event *B*, observing an odd number. Since the intersection *A* and *B* contains no sample points, the events are mutually exclusive.

Since the probability is 1, we know that when a die is tossed one of the events must occur. *Events for which the probability of their union equals 1 are called* **collectively exhaustive** *or simply* **exhaustive**.

Addition Rule for Three Events

The addition rule can be extended to three or more events. For three events *A*, *B*, and *C*, the probability of *A* or *B* or *C* is given by

$$p(A \text{ or } B \text{ or } C) = p(A) + p(B) + p(C) - p(A \text{ and } B) - p(A \text{ and } C)$$
$$- p(B \text{ and } C) + p(A \text{ and } B \text{ and } C).$$

The rationale for the formula is apparent from the Euler diagram in Figure 7.3-3. If the events are mutually exclusive, the formula simplifies to

$$p(A \text{ or } B \text{ or } C) = p(A) + p(B) + p(C).$$

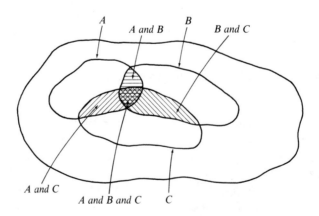

Figure 7.3-3. Euler diagram illustrating terms in the formula $p(A \text{ or } B \text{ or } C) = p(A) + p(B) + p(C) - p(A \text{ and } B) - p(A \text{ and } C) - p(B \text{ and } C) + p(A \text{ and } B \text{ and } C)$.

Multiplication Rule of Probability

The multiplication rule is used to compute the probability of the joint occurrence, or intersection, of two or more events. For example, suppose 100 psychology majors have been classified according to sex and class level. The number of students in each category is given in Table 7.3-2. If a student is selected by lottery, what is the probability that the student will be both a female and a lowerclassperson? The multiplication rule lets us determine the probability that the student selected will be in the intersection *female and lowerclassperson*—that is, both a female and a lowerclassperson. This information is different from that given by the addition rule, which tells us the probability that the student selected will be a female or a lowerclassperson or a female-lowerclassperson.

Table 7.3-2. Number of Psychology Majors by Sex and Class Level

	Lowerclassperson, L	Upperclassperson, U	Marginal Total
Women, W	10 $p(W \text{ and } L) = n_{W \text{ and } L}/n_S$ $= \frac{10}{100}$ $= 0.10$	20	30 $p(W) = n_W/n_S$ $= \frac{30}{100}$ $= 0.30$
Men, M	40	30	70 $p(M) = n_M/n_S$ $= \frac{70}{100}$ $= 0.70$
Marginal Total	50 $p(L) = n_L/n_S$ $= \frac{50}{100}$ $= 0.50$	50 $p(U) = n_U/n_S$ $= \frac{50}{100}$ $= 0.50$	100

Before presenting the multiplication rule we need to discuss the concept of conditional probability. Two events are often related so that the probability of one event depends upon whether the other has or has not occurred. Consider the events: your roommate reports he feels bad, event A, and his temperature is 103, event B. The two events are obviously related since the probability of an elevated temperature, $p(B)$, is much higher if a person feels bad than if he feels good. We refer to this type of relationship as a *conditional probability*. The **conditional probability** of B given that A has occurred is denoted by $p(B|A)$ and is equal to $p(B|A) = p(A \text{ and } B)/p(A)$. Similarly, the conditional probability of A given that B has occurred is $p(A|B) = p(A \text{ and } B)/p(B)$. The vertical line | is read "given," or "given that."

The calculation of conditional probability will be illustrated for the data in Table 7.3-2. The probability that a student selected by lottery is a woman given that you know the student is a lowerclassperson is $p(W|L) = p(W \text{ and } L)/p(L) = (\frac{10}{100})/(\frac{50}{100}) = 0.20$. We see that the condition of being a lowerclassperson limits the outcome to the left column of Table 7.3-2, which is a smaller sample space of size 50. The probability of

7.3
Probability of Combined Events

selecting a woman is a subset of this smaller sample space, namely, 10 events out of 50. The probability of selecting a woman in the absence of information about class level is $p(W) = \frac{30}{100} = 0.30$. The events W and L are related since a knowledge of one event affects the probability of the other.

The multiplication rule can now be stated. *Given two events A and B, the probability of obtaining both A and B jointly is the product of the probability of obtaining one event times the conditional probability of the other event. Stated symbolically, the joint probability p(A and B) is given by*

$$p(A \text{ and } B) = p(A)p(B|A)$$
$$= p(B)p(A|B).$$

For the events defined in Table 7.3-2, the probability of selecting a student who is both a female and a lowerclassperson is

$$p(W \text{ and } L) = p(W)p(L|W) = p(W)[p(W \text{ and } L)/p(W)]$$
$$= 0.30(0.10/0.30) = 0.10$$
$$= p(L)p(W|L) = p(L)[p(W \text{ and } L)/p(L)]$$
$$= 0.50(0.10/0.50) = 0.10.$$

The multiplication rule may seem unnecessarily complicated since if the intersection (A and B) is known, the rule is not needed. However, sometimes only a *marginal probability*, $p(A)$ or $p(B)$, and a conditional probability, $p(A|B)$ or $p(B|A)$, are known. For example, suppose we want to know the probability of drawing two aces from a 52 card deck. On the first draw, $p(ace) = \frac{4}{52}$. If an ace is drawn on the first draw and not replaced in the deck, the conditional probability of drawing an ace on the second draw is $p(ace \text{ on second draw}|ace \text{ on first draw}) = \frac{3}{51}$. The probability of drawing two aces is

$$p(\text{two aces}) = p(ace)p(ace \text{ on second draw}|ace \text{ on first draw})$$
$$= \left(\frac{4}{52}\right)\left(\frac{3}{51}\right) \simeq 0.0045.$$

Multiplication Rule for Statistically Independent Events

*Two events A and B are **statistically independent** if the probability of one event occurring is unaffected by the occurrence of the other. Stated symbolically, A and B are statistically independent if $p(A|B) = p(A)$. If this equality holds, it must also be true that*

$$p(B|A) = p(B).$$

The events $p(W)$ and $p(L)$ in Table 7.3-2 are not statistically independent since $p(W|L) = 0.20$ is not equal to $p(W) = 0.30$. We can easily construct an example in which the events are independent. Consider an experiment in which a fair coin and a die are tossed. Since the coin can land in one of two ways, H or T, and the die, in one of six ways, $1, \ldots, 6$, the possible outcomes are $H1, T1, H2, T2, \ldots, H6, T6$. The sample space for the experiment is shown in Figure 7.3-4. Let event A be a head and B be a 5. The probabilities required to demonstrate independence of A and B are $p(A) = n_A/n_S = \frac{6}{12} = \frac{1}{2}$ and $p(A|B) = (n_{A \text{ and } B}/n_S)/(n_B/n_S) = n_{A \text{ and } B}/n_B = \frac{1}{2}$. Since $p(A) =$

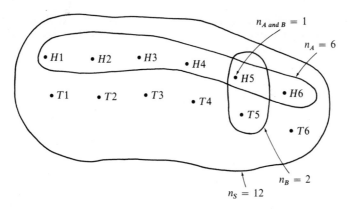

Figure 7.3-4. Euler diagram for event A, observing a head, and event B, observing a 5, when a coin and die are tossed.

$p(A|B)$, the events are statistically independent; this agrees with our intuition that what happens on the toss of a die can in no way affect the outcome of tossing a coin.

For statistically independent events the multiplication rule is simplified because $p(A|B) = p(A)$ and $p(B|A) = p(B)$. The rule becomes

$$p(A \text{ and } B) = p(A)p(B).$$

As we saw above, the probability of observing a head and a 5 are independent; hence, the probability of their joint occurrence is $p(A \text{ and } B) = (\frac{1}{2})(\frac{1}{6}) = \frac{1}{12}$.

Multiplication Rules for Three Events

If we want to find the joint probability of three events, the multiplication rule is

$$p(A \text{ and } B \text{ and } C) = p(A)p(B|A)p(C|A \text{ and } B).$$

To illustrate, suppose we have a box containing five marbles, each a different color: blue, green, red, yellow, and orange. Three marbles are selected from the box at random, one at a time. Once drawn, a marble is not replaced. The procedure is called *sampling without replacement*. Consider the following events.

> A—first marble is blue
> B—second marble is green
> C—third marble is red

According to the classical view, $p(A) = n_A/n_S = \frac{1}{5}$. If a blue marble is selected first, then four remain—one of which is green. The probability of selecting a green marble given that the first is blue is $p(B|A) = \frac{1}{4}$. Similarly, the probability of selecting a red marble given that a blue and green one have been selected is $p(C|A \text{ and } B) = \frac{1}{3}$. The joint probability for A, B, and C is $p(A \text{ and } B \text{ and } C) = \frac{1}{5} \times \frac{1}{4} \times \frac{1}{3} = \frac{1}{60}$.

If the marbles sampled from the box are replaced (*sampling with replacement*), the multiplication rule for three events is $p(A \text{ and } B \text{ and } C) = p(A)p(B)p(C)$. This follows since the selection of the first marble does not decrease the number of sample points in the sample space, and hence $p(B|A) = p(B)$ and $p(C) = p(C|A \text{ and } B)$. If we had replaced each marble after it was drawn from the box, the joint probability for events A, B, and C would be $p(A \text{ and } B \text{ and } C) = \frac{1}{5} \times \frac{1}{5} \times \frac{1}{5} = \frac{1}{125}$.

Common Errors in Applying the Rules of Probability

The probability rules described in this section are often incorrectly used. Some of the more common errors are the following.

1. Using the addition rule $p(A \text{ or } B) = p(A) + p(B)$ for events that are not mutually exclusive. For example, let event A be the classification "psychology major" and event B, "biology major." If $p(A) = 0.20$ and $p(B) = 0.15$, one might conclude that the probability of a student being either a psychology or a biology major is $p(A \text{ or } B) = (0.20) + (0.15) = 0.35$. This is incorrect because some students have a double major and these students have been counted twice, once in computing $p(A)$ and once in computing $p(B)$. If $p(A \text{ and } B) = 0.03$, the correct probability is given by $p(A \text{ or } B) = p(A) + p(B) - p(A \text{ and } B) = 0.20 + 0.15 - 0.03 = 0.32$.
2. Using the addition rule when the multiplication rule should be used, and vice versa. For example, the probability of observing a 3, event A, or a 5, event B, on the toss of a die is given by $p(A \text{ or } B) = p(A) + p(B) = \frac{1}{6} + \frac{1}{6} = \frac{2}{6}$ and not by $p(A \text{ and } B) = p(A)p(B) = (\frac{1}{6})(\frac{1}{6}) = \frac{1}{36}$.
3. Using the multiplication rule $p(A \text{ and } B) = p(A)p(B)$ when the events are not statistically independent. Suppose the probability of seeing an advertisement for a product, event A, is 0.40 and the probability of buying the product, event B, is 0.30. If the dependency between A and B is ignored, the probability of both seeing an advertisement and buying the product is $p(A \text{ and } B) = (0.40)(0.30) = 0.12$. The correct probability takes into account the conditional probability of buying the product given that the ad has been seen, $p(B|A) = 0.50$, so that $p(A \text{ and } B) = p(A)p(B|A) = (0.40)(0.50) = 0.20$.

Review Exercises for Section 7.3

*13. A standard deck contains 52 cards, 10 number cards of each suit (counting the ace as a 1) and three face cards of each suit. If a card is drawn from the deck at random, what is the probability that it will be (a) an ace? (b) a heart? (c) an ace or a heart or both? (d) a heart or a spade? (e) a face card? (f) a card less than 5?

**14. Events A, B, and C are mutually exclusive and exhaustive, each having a probability of $\frac{1}{3}$. Determine (a) $p(A \text{ or } B)$, (b) $p(A \text{ or } B \text{ or } C)$, *(c) $p(\text{not } A)$, and (d) $p[\text{not}(A \text{ or } B)]$.

*15. Events A and B are independent; $p(A) = 0.6$ and $p(B) = 0.8$. What is the probability (a) that both will occur? (b) that neither will occur? (c) that one or the other or both will occur?

*16. Recent highway accident statistics show that 10% of all automobile accidents and half of all fatal automobile accidents are caused by drunken drivers. Four in 1000 reported accidents are fatal. (a) Fill in the table with the appropriate probabilities. (b) What is the joint probability that a fatal accident is caused by a drunken driver?

Probability		Fatal, F	Nonfatal, Not F	
	Drunken Driver, D	$p(D \text{ and } F) =$	$p(D \text{ and Not } F) =$	$p(D) =$
	Other Cause, O	$p(O \text{ and } F) =$	$p(O \text{ and Not } F) =$	$p(O) =$
		$p(F) =$	$p(\text{Not } F) =$	

*17. You ask your roommate to mail a letter. The probability that she will mail it is 0.98. The probability that the post office will fail to deliver it, given it was mailed, is 0.15. What is the probability that the letter will be mailed and the post office will fail to deliver it?

18. For the data in the table determine if the events "attend college" and "male" are statistically independent.

	Attend College		
	Yes	No	
Male	0.30	0.20	0.50
Female	0.10	0.40	0.50
	0.40	0.60	1.00

19. Data were obtained on the incidence of rheumatic disease and the presence of grimacing in schizophrenic patients. In a sample of 1942 patients, 6% had a known history of rheumatic disease, 21.8% grimaced, and 1.8% had a history of both rheumatic disease and grimacing. (a) Fill in the table with the appropriate probabilities. (b) What is the probability of grimacing given a history of rheumatic disease? (c) Are grimacing and rheumatic disease statistically independent?

	Grimacing	No Grimacing
History of Rheumatic Disease		
No History of Rheumatic Disease		

20. A smoker has ten pipes, three of which are meerschaums. Of his six curved-stem pipes, two are meerschaums. He asks his son to bring him a curved-stem meerschaum. Since the boy doesn't know a meerschaum from other pipes, he picks up a curved-stem pipe at random. (a) Fill in the table with the appropriate probabilities. (b) What is the probability that the son picked the right pipe?

	Meerschaum, M	Other Kind of Pipe, O	
Curved Stem, C	$p(C$ and $M) =$	$p(C$ and $O) =$	$p(C) =$
Straight Stem, S	$p(S$ and $M) =$	$p(S$ and $O) =$	$p(S) =$
	$p(M) =$	$p(O) =$	

**21. One hundred students are enrolled in a university course. Fifty are males (M) and 50 are females (F). Of the 100 students, 60 are undergraduates (U) and 40 are graduate students (G). Of the 100, 20 are both males and undergraduates. If a student is selected at random from the class, compute the following probabilities. *(a) $p(F$ and $U)$, (b) $p(F$ and $G)$, (c) $p(M$ and $G)$, (d) $p(M|U)$, (e) $p(F|G)$. [Hint: $p(U) = p(M$ and $U) + p(F$ and $U)$.]

22. A family of three goes to a photography studio to have a group picture made. They consider a picture good only if everyone in it looks good. Suppose the probability that any one person looks good is 0.5. Assume independence. (a) What is the probability that the picture will be good? (b) If two pictures are made, what is the probability that at least one of them will be good? (c) How many pictures must be made to have a 75% chance of getting a good one?

23. A survey of 100 students taking courses in algebra (A), history (H), and psychology (P) revealed the following enrollments: $A = 65$, $H = 37$, $P = 59$, A and $H = 17$, A and $P = 44$, H and $P = 14$. (a) Represent the sample space by a Euler diagram. If a student is selected at random, what is the probability that he or she is taking (b) all three subjects? (c) algebra but not psychology? (d) history but not algebra? (e) psychology but not algebra or history?

24. A drawer contains six intelligence tests (I), four personality tests (P), and five aptitude tests (A). If three tests are drawn randomly from the drawer, what is the probability that the tests are drawn in the order I–P–A if each test (a) is replaced in the drawer after it is drawn? (b) is not replaced after it is drawn?

25. Terms to remember:
    a. Union
    b. Intersection
    c. Addition rule
    d. Mutually exclusive events
    e. Exhaustive events
    f. Multiplication rule
    g. Conditional probability
    h. Marginal probability
    i. Statistical independence
    j. Sampling with or without replacement

7.4 Summary

Probability is an abstract mathematical concept that can be defined in a number of ways. The three most useful views of probability are the subjective–personalistic view, the classical or logical view, and the empirical relative-frequency view.

Our interest in probability is pragmatic; we want to make statements about the likelihood of observing various outcomes in experiments. An experiment is any well-defined act or process that leads to an outcome. An outcome is either a compound event that can be decomposed into simpler events, such as observing an even number on the toss of a die, or a simple event that can't be decomposed. If we assign to each simple event a point called a sample point, the possible outcomes of an experiment can be represented by a Euler diagram. The set of all sample points is called the sample space S.

Regardless of one's view, probability is based on a system of definitions and operations pertaining to a sample space. If S is the sample space for an experiment, and n_S is the number of sample points in S, we can associate with each event E_i a real number called the probability of E_i or $p(E_i)$, satisfying the following properties.

1. $0 \leq p(E_i) \leq 1$, for all i
2. $\sum_{i=1}^{n_S} p(E_i) = 1$
3. $p(S) = 1$

These properties describe probabilities, but they don't tell us how to compute them. If we adopt the classical view, the probability of an event A is computed from the formula $p(A) = n_A/n_S$, where n_A is the number of events favoring A and n_S is the total number of equally likely events in the sample space S. This view of probability is based on logical analysis. We reason that an experiment has n_S possible outcomes, that the outcomes are equally likely, and that n_A of the outcomes favor A. If our reasoning is correct, the value we compute for $p(A)$ will agree closely with that based on the relative-frequency view.

According to the relative-frequency view, the probability of event A is the number approached by n_A/n as the total number of observations, n, approaches infinity. The estimate n_A/n is based on experience, since it is computed for a sample from the population of possible experiments. On the average, the larger the sample the closer the estimate is to the true probability.

Probabilities for combined events can be computed by the addition rule and the multiplication rule. The addition rule states that the probability that an event will be A or B or both is $p(A \text{ or } B) = p(A) + p(B) - p(A \text{ and } B)$. For mutually exclusive events, $p(A \text{ and } B) = 0$ and the rule simplifies to $p(A \text{ or } B) = p(A) + p(B)$. The multiplication rule states that the probability that an event will be both A and B is $p(A \text{ and } B) = p(A)p(B|A) = p(B)p(A|B)$. For statistically independent events, $p(B|A) = p(B)$ and $p(A|B) = p(A)$, and the rule simplifies to $p(A \text{ and } B) = p(A)p(B)$.

8

More about Probability

8.1 Counting Simple Events
Fundamental Counting Rule
Permutation of n Objects Taken n at a Time, $_nP_n$
Permutation of n Objects Taken r at a Time, $_nP_r$
Permutation of n Objects When Some of the Objects Are Alike, $_nP_{r_1, r_2, \ldots, r_k}$
Combination of n Objects Taken r at a Time, $_nC_r$
Review Exercises for Section 8.1
8.2 Revising Probabilities Using Bayes' Theorem
Bayes' Theorem
Controversial Issues
Review Exercises for Section 8.2
8.3 Summary

8.1 Counting Simple Events

It is usually a lot of trouble to list all the simple events in an experiment. Even a small experiment, such as recording the outcome of tossing three dice, has a large sample space—in this case $6 \times 6 \times 6 = 216$ sample points. Fortunately, it is not necessary to enumerate simple events to compute probabilities. The required information can be determined using the counting rules discussed in this section.

Fundamental Counting Rule[1]

Suppose an event can occur in n_1 ways and a second in n_2 ways, and each of the first event's n_1 ways can be followed by any of the second's n_2 ways. Then, according to the **fundamental counting rule**, event 1 followed by event 2 can occur in $n_1 n_2$ ways. To illustrate, suppose we toss a coin and then a die. The number of possible outcomes of the experiment is

$$n_1 n_2 = (2)(6) = 12,$$

since a coin can land heads or tails ($n_1 = 2$) and a die has six faces ($n_2 = 6$). The simple events are enumerated on the tree diagram of Figure 8.1-1.

The fundamental counting rule can be extended to $k > 2$ events. If there are k events (event 1 having n_1 outcomes, followed by event 2 having n_2 outcomes, and so forth), the outcome can occur in $n_1 n_2 \cdots n_k$ ways. For example, the number of possible outcomes of tossing three dice and a coin is $6 \times 6 \times 6 \times 2 = 432$.

Permutation of n Objects Taken n at a Time, $_nP_n$

Suppose we have n different objects and we want to find the number of different ordered sequences in which the objects can be arranged. In how many ordered sequences can the letters A, B, and C be arranged? The answer is six: ABC, ACB, BAC, BCA, CAB, CBA. Arranging n objects is equivalent to putting them into a big box with n ordered compartments. The first slot can be filled in any of n ways,

1	2	3	$\cdots$	n
n ways	$n-1$ ways	$n-2$ ways		1 way

which uses up one of the objects; the second slot can be filled in any of $n - 1$ ways, ..., and the last slot, in only one way. Applying the fundamental counting rule, the number of ordered arrangements of n objects is the product $n(n - 1)(n - 2) \cdots 1$. The quantity $n(n - 1)(n - 2) \cdots 1$ is denoted by the symbol $n!$, which is read "n factorial."

An ordered sequence of n objects taken all together is called a **permutation** of the objects. The total number of such permutations, denoted by $_nP_n$, is given by

$$_nP_n = n! = n(n-1)(n-2) \cdots 1.$$

The symbol $_nP_n$ is read "the permutation of n objects taken n at a time."

[1] Also called the *multiplication principle*.

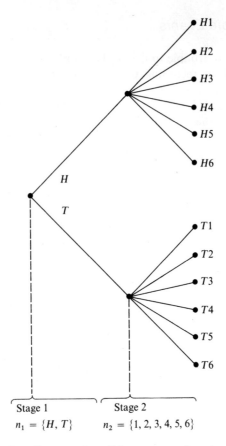

Figure 8.1-1. Tree diagram of possible outcomes of tossing a coin and then a die.

Suppose we are doing a taste preference experiment in which a panel of ten experts rates five nondairy coffee creamers. In how many ordered sequences can coffee prepared with the creamers be presented? The answer is $5! = 5(4)(3)(2)(1) = 120$. Finding ten experts willing to sit through the 120 sequences is probably impossible. However, to control for sequence effects the experiment can be performed by presenting the coffee in 12 of the sequences to one expert, 12 different sequences to another expert, and so on.

Permutation of n Objects Taken r at a Time, $_nP_r$

The number of permutations of n different objects taken r at a time, where $r \leq n$, is denoted[2] by $_nP_r$ and is equal to $n(n - 1)(n - 2) \cdots (n - r + 1)$. For example, the number of ordered sequences of five letters, A, B, C, D, E, taken three at a time is $_5P_3 = 5(5 - 1)(5 - 3 + 1) = 5(4)(3) = 60$. The rationale behind the formula is as follows.

More about Probability

[2] Also denoted by P_r^n, $P(n, r)$, and $(n)_r$.

Consider the box

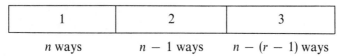

with $r = 3$ ordered components. The first slot can be filled in any of $n = 5$ ways and the second in $n - 1 = 4$ ways. When we come to the $r = $ 3rd slot, we have used $r - 1 = 2$ of the n letters so that $n - (r - 1) = n - r + 1 = 3$ letters are left to fill the last position. According to the fundamental counting rule, the number of ordered sequences is the product of $n(n - 1)(n - r + 1)$. Therefore, the number of ordered sequences of the five letters taken three at a time is $5(4)(3) = 60$.

The right side of the formula $_nP_r = n(n - 1)(n - 2) \cdots (n - r + 1)$ is often written $n!/(n - r)!$.[3] To see that the expressions are equivalent, note that the next term in $n(n - 1)(n - 2) \cdots (n - r + 1)$ is $(n - r)$, followed by $(n - r - 1) \cdots (3)(2)(1)$. The terms $(n - r)(n - r - 1) \cdots (3)(2)(1)$ can be denoted by $(n - r)!$. If we multiply $n(n - 1) \cdots (n - r + 1)$ by $(n - r)!/(n - r)!$, we obtain the more familiar formula $_nP_r = n!/(n - r)!$ since $[n(n - 1) \cdots (n - r + 1)](n - r)! = n!$.

The taste preference experiment described earlier could be performed using the method of paired comparisons. In this method an expert sips first one and then a second cup of coffee prepared with two of the creamers and indicates a preference. This is repeated until each creamer has been compared twice with every other creamer, once in the first cup sipped and once in the second cup. In how many ordered sequences can five creamers be presented two at a time? The answer is given by

$$_5P_2 = 5(5 - 2 + 1) = 5(4) = 20$$

or

$$_5P_2 = \frac{5!}{(5 - 2)!} = \frac{(5)(4)(3)(2)(1)}{(3)(2)(1)} = 20.$$

The method of paired comparisons would require each expert to make 20 judgments to compare each creamer with every other creamer.

Permutation of n Objects When Some of the Objects Are Alike, $_nP_{r_1,r_2,\ldots,r_k}$

Suppose we have n objects that can be divided into k subsets such that each subset contains like objects, but the objects in any given subset are different from those in the other subsets. Let r_1 be the number of like objects of one kind and r_2 be the number of like objects of another kind, and so on for the k subsets, with $r_1 + r_2 + \cdots + r_k = n$. The number of permutations of n objects in which $r_1, r_2, \ldots, r_k$ are alike is denoted by $_nP_{r_1,r_2,\ldots,r_k}$ and is equal to $n!/(r_1!r_2!\cdots r_k!)$. For example, the number of ordered sequences of the ten letters in PSYCHOLOGY is

$$_{10}P_{1,1,2,1,1,2,1,1} = \frac{10!}{1!1!2!1!1!2!1!1!} = 907,200.$$

[3] In computations involving $(n - r)!$, remember that $1! = 1$, and by definition $0! = 1$.

Combination of n Objects Taken r at a Time, $_nC_r$

Sometimes we are not interested in the number of ordered sequences or permutations of n objects taken r at a time, but instead in the number of different sets of r objects that can be selected from n when order is ignored. This is referred to as the **combination** of n objects taken r at a time and is denoted[4] by $_nC_r$. The formula for $_nC_r$ is

$$_nC_r = \frac{n!}{r!(n-r)!}.$$

The rationale for the formula is as follows. Consider four letters, A, B, C, and D taken two at a time. The number of ordered sequences is $_4P_2 = 4!/(4-2)! = 12$. But suppose we don't want to distinguish AB from BA, CB from BC, and so on. We note that any sequence of $r = 2$ objects can be permuted in $r! = 2(1) = 2$ ways. If we want to ignore the order of the r objects in $_nP_r$, we can divide $_nP_r$ by $r!$ This gives

$$_nC_r = \frac{_nP_r}{r!} = \frac{\frac{n!}{(n-r)!}}{r!} = \frac{n!}{r!(n-r)!}.$$

The number of different sets of $r = 2$ letters that can be selected from $n = 4$ letters, A, B, C, D, is

$$_4C_2 = \frac{4!}{2!(4-2)!} = \frac{4(3)(2)(1)}{2(1)[2(1)]} = 6.$$

The six sets are as follows: AB, AC, AD, BC, BD, CD. Because the order of letters in a pair is of no interest, the six could just as well have been written: BA, CA, AD, CB, BD, CD.

The combination of n objects taken r at a time will be used in Chapter 9 to develop the binomial distribution, which describes the possible outcomes of a particular kind of experiment.

Review Exercises for Section 8.1

*1. Determine the number of possible outcomes for the following. (a) Three coins are tossed. (b) Four dice are tossed. (c) A coin and a die are tossed.
2. If there are three candidates for governor and five for mayor, in how many ways can the two offices be filled?
3. The Greasy Spoon menu offers a choice of five appetizers, four salads, eight entrees, seven vegetables, and nine desserts. If a meal consists of one each, in how many ways can you select a dinner?
*4. The four Russian novels *War and Peace*, *Anna Karenina*, *Crime and Punishment*, and *The Brothers Karamazov* are to be placed on a shelf. In how many ordered sequences can the books be arranged?
5. Two different psychology books, four different statistics books, and three different sociology books are to be arranged on a shelf. (a) In how many ordered sequences can the books

[4] Also denoted by C_r^n, $C(n, r)$, and $\binom{n}{r}$.

be arranged? (b) If the books in each subject must be kept together, how many ordered sequences are possible?
*6. In how many different ways can ten people be seated on a bench if only four seats are available?
7. How many ordered arrangements can be made for the letters in the word *statistics*?
8. If on a statistics examination consisting of 12 questions a student may omit five, in how many ways can the student select the problems to answer?
9. (a) In how many ways can six people be seated in a row at the head banquet table? (b) Suppose the people are to be seated in pairs at separate tables; in how many ways can they be seated if we consider the arrangement AB to be different from BA? In how many ways if we consider AB to be equivalent to BA?
*10. Given nine areas from which to choose, in how many ways can a student select (a) a major–minor area? (b) a major and first and second minors? (c) a major and two minors if it is not necessary to designate the order of the minors?
11. How many different committees of three men and four women can be formed from eight men and six women?
12. Terms to remember:
   a. Permutation of n objects
   b. Combination of n objects

†8.2 Revising Probabilities Using Bayes' Theorem

You charge out the door, late for class, but stop short. The sky is overcast—do you need a raincoat? A moment ago your expectation of rain was low; seeing the clouds changed that. We constantly revise our expectations as new information becomes available. In this section we describe a principle for formally incorporating information into decision-making. The principle is called Bayes' theorem after the Reverend Thomas Bayes (1702–1761), who proposed it.[5] The theorem is used to compute the conditional probability of an event E_i, given that a particular datum D_j has been observed, for example, the probability of wearing a raincoat given a menacing sky. Similarly, the theorem can be used to compute the conditional probability that a particular state of nature E_i exists, given a particular datum D_j, for example, the probability that a person has tuberculosis given a positive X ray.

Bayes' Theorem[6]

Consider an experiment involving several mutually exclusive events, $E_1, E_2, \ldots, E_i, \ldots, E_n$. The events could be "wear a raincoat," E_1, or "don't wear a raincoat," E_2. Suppose the probability of each event $p(E_1), p(E_2), \ldots, p(E_n)$ is known. These are called *prior probabilities*, because they represent the probability of the event before certain data are known. The possible data of the experiment, $D_1, D_2, \ldots, D_j, \ldots, D_k$, and the

---

† This section can be omitted without loss of continuity.
[5] Bayes apparently had some doubts about the validity of the application of the theorem and accordingly withheld its publication. It was published posthumously in 1763.
[6] For a detailed discussion of the use of Bayes' theorem in psychology, see Edwards, Lindman, and Savage (1963).

events are statistically dependent. The data could be "sunny day," D_1, "cloudy sky," $D_2, \ldots,$ "menacing sky," D_k. Also suppose that for any datum D_j, we know either the joint probability $p(E_i \text{ and } D_j)$ or the conditional probabilities $p(D_j|E_1), \ldots, p(D_j|E_n)$. After a datum D_j has been observed, we want to know the conditional probability $p(E_i|D_j)$—that is, the probability of E_i, given the datum D_j. For example, what is the probability of wearing a raincoat given a menacing sky. This conditional probability is called a *posterior probability*, since it represents the probability of an event after some datum is known. The probability can be computed from

$$p(E_i|D_j) = \frac{p(E_i \text{ and } D_j)}{p(D_j)}$$

if $p(E_i \text{ and } D_j)$ and $p(D_j) \neq 0$ are known. If the two probabilities aren't known, $p(E_i|D_j)$ can be computed from Bayes' theorem:

$$p(E_i|D_j) = \frac{p(E_i)p(D_j|E_i)}{p(E_1)p(D_j|E_1) + \cdots + p(E_n)p(D_j|E_n)}.$$

The following example illustrates the use of Bayes' theorem. Suppose a university is considering using an advanced-placement test to assign students to the first or second biology course. This year all students requesting advanced placement to the second course were admitted. They all took the placement test, but it was not scored until the students had completed the course. The following data were obtained.

Proportion of students who passed the second course:

$$p(E_1) = 0.60 \text{ (prior probability)}$$

Proportion of students who failed the second course:

$$p(E_2) = 0.40 \text{ (prior probability)}$$

Proportion of students who passed the placement test given that they passed the course:

$$p(D_1|E_1) = 0.80 \text{ (conditional probability)}$$

Proportion of students who passed the placement test given that they failed the course:

$$p(D_1|E_2) = 0.40 \text{ (conditional probability)}$$

Assuming that students next year are similar to those this year, should passing the placement test be a requirement for admission to the second biology course? The probability of passing the second course without utilizing the test information is 0.60. We can compute the posterior probability of passing the second course given that the placement test was passed from

$$p(E_1|D_1) = \frac{p(E_1)p(D_1|E_1)}{p(E_1)p(D_1|E_1) + p(E_2)p(D_1|E_2)}$$

$$= \frac{(0.60)(0.80)}{(0.60)(0.80) + (0.40)(0.40)} = \frac{0.48}{0.48 + 0.16} = 0.75.$$

Use of the placement test as a screening device seems warranted, since the posterior probability based on the outcome of the test, 0.75, differs markedly from the prior probability, 0.60.

A more complete summary of the placement test and course data is presented in Table 8.2-1. From this table it can be seen that the formulas for computing posterior probability—$p(E_1 \text{ and } D_1)/p(D_1)$ and $p(E_1)p(D_1|E_1)/[p(E_1)p(D_1|E_1) + p(E_2)p(D_1|E_2)]$—are equivalent. Consider first the numerator $p(E_1 \text{ and } D_1)$. By definition the conditional probability $p(D_1|E_1) = p(E_1 \text{ and } D_1)/p(E_1)$, and hence $p(E_1 \text{ and } D_1) = p(E_1)p(D_1|E_1)$. To show that the denominators are equivalent, we note from Table 8.2-1 that $p(D_1) = p(E_1 \text{ and } D_1) + p(E_2 \text{ and } D_1)$. Again by definition, $p(E_1 \text{ and } D_1) = p(E_1)p(D_1|E_1)$ and $p(E_2 \text{ and } D_1) = p(E_2)p(D_1|E_2)$. Therefore, $p(D_1) = p(E_1)p(D_1|E_1) + p(E_2)p(D_1|E_2)$.

Table 8.2-1. Joint Probability Table

	Pass Course, E_1	Fail Course, E_2	
Pass Test, D_1	$p(E_1 \text{ and } D_1) = 0.48$	$p(E_2 \text{ and } D_1) = 0.16$	$p(D_1) = 0.64$
Fail Test, D_2	$p(E_1 \text{ and } D_2) = 0.12$	$p(E_2 \text{ and } D_2) = 0.24$	$p(D_2) = 0.46$
	$p(E_1) = 0.60$	$p(E_2) = 0.40$	

Controversial Issues

Bayes' theorem is a convenient mathematical rule for computing a conditional probability, more specifically the posterior probability $p(event|datum)$. In the previous examples the prior probabilities—wearing a raincoat and passing the second biology course—were empirical relative frequencies. The controversy surrounding Bayes' theorem concerns (1) the use of subjective probabilities in place of empirical probabilities, (2) the mixing of the two kinds of probabilities in computing posterior probabilities, and (3) the problem of measuring subjective probabilities. The increasing use of Bayesian methods indicates that for many investigators the controversy has been resolved.

Review Exercises for Section 8.2

*13. The incidence of having cancer among men aged 45–55 is 2 in 100, $p(E_1) = 0.02$. Of those with cancer, 99% will show a positive X ray indicating cancer, $p(D_1|E_1) = 0.99$, while of those who don't have cancer, 3 in 1000 will show a positive test, $p(D_1|E_2) = 0.003$. (a) What is the probability that a man 45–55 years old will have a positive X ray test? (b) What is the probability that a man 45–55 years old who has a positive X ray has cancer?

14. Students who read their statistics texts and work the problems in the review exercises have a 0.98 probability of passing the course. Those who don't have a 0.15 probability of passing. If 70% of students do the reading and work the problems, (a) what is the probability that a student who passed did the reading and problems? (b) What is the probability of passing the course?

*15. A life insurance company plans to use an aptitude test in selecting salespeople. For the past 2 years the test has been administered to all applicants but not used in making hiring decisions. A partial summary of the job performance and test scores of those hired is given in the table. The conditional probability of scoring 70 or above, given satisfactory job performance, is 0.90.

For those whose performance was unsatisfactory the probability is 0.30. (a) Fill in the missing values in the table. (b) What is the posterior probability that an applicant will be satisfactory given a test score of 70 or above? (c) What is the posterior probability that an applicant will be satisfactory given a test score below 70?

		Job Performance		
		Satisfactory, E_1	Unsatisfactory, E_2	
Test Score	70 or Above, D_1	$p(D_1$ and $E_1) =$	$p(D_1$ and $E_2) =$	$p(D_1) =$
	Below 70, D_2	$p(D_2$ and $E_1) =$	$p(D_2$ and $E_2) =$	$p(D_2) =$
		$p(E_1) = 0.50$	$p(E_2) = 0.50$	

16. A large sample of males and females was examined for color-blindness. A partial summary of the data is given in the table. The conditional probability of being a male, given some color deficiency, is 0.857; for those with normal color vision the probability is 0.511. (a) Fill in the missing values in the table. (b) What is the posterior probability of color-blindness given that the person is a male? (c) What is the posterior probability of color-blindness given that the person is a female?

	Normal Color Vision, E_1	Color-Blind, E_2	
Male, D_1	$p(D_1$ and $E_1) =$	$p(D_1$ and $E_2) =$	$p(D_1) =$
Female, D_2	$p(D_2$ and $E_1) =$	$p(D_2$ and $E_2) =$	$p(D_2) =$
	$p(E_1) = 0.958$	$p(E_2) = 0.042$	

17. Terms to remember:
    Prior and posterior probabilities

8.3 Summary

This chapter has presented rules developed to simplify the determination of the number of simple events in an experiment. The key rules are as follows.
1. Fundamental counting rule. If there are k events, event 1 followed by event 2, ..., followed by the kth event, the outcome can occur in $n_1 n_2 \cdots n_k$ ways.
2. Permutation of n objects taken n at a time. The number of ordered sequences of n objects taken all together is $_nP_n = n! = n(n-1)(n-2)\cdots 1$.
3. Permutation of n objects taken r at a time. The number of ordered sequences of n objects taken r at a time is $_nP_r = n!/(n-r)!$.
4. Permutation of n objects when some of the objects are alike. The number of ordered sequences of n objects in which $r_1, r_2, \ldots, r_k$ are alike is $_nP_{r_1, r_2, \ldots, r_k} = n!/(r_1! r_2! \cdots r_k!)$.
5. Combination of n objects taken r at a time. The number of sequences, ignoring order, of n objects taken r at a time is $_nC_r = n!/[r!(n-r)!]$.

The subjective–personalistic view of probability is often used with Bayes' theorem. Suppose you entertain two mutually exclusive and exhaustive hypotheses, E_1 and E_2. Before doing an experiment your subjective probability of E_1 being true is 0.5; that is, you feel there is a 50:50 chance of E_1 being true. You conduct the experiment and observe datum D_1. How does the datum affect your prior subjective probability that E_1 is true? The impact of D_1 on $p(E_1)$ is given by

$$p(E_1|D_1) = \frac{p(E_1)p(D_1|E_1)}{p(E_1)p(D_1|E_1) + p(E_2)p(D_1|E_2)}.$$

Before the experiment, the plausibility of E_1 is $p(E_1)$; after the datum is known, the plausibility of E_1 is the posterior probability $p(E_1|D_1)$. Bayes' theorem is a convenient mathematical rule for converting a prior probability into a posterior probability that reflects accumulated evidence.

9

Random Variables and Probability Distributions

9.1 Random Sampling
  Defining the Population
  Sampling with or without Replacement
  Random Sampling Procedures
  Review Exercises for Section 9.1
9.2 Random Variables and Their Distributions
  Random Variables
  Distribution of a Discrete Random Variable
  Expected Value of a Discrete Random Variable
  Expected Value of a Continuous Random Variable
  Standard Deviation of a Discrete Random Variable
  Review Exercises for Section 9.2
9.3 Binomial Distribution
  Bernoulli Trial
  Binomial Distribution
  Expected Value and Standard Deviation of Binomial Distribution
  Multinomial and Hypergeometric Distributions
  Review Exercises for Section 9.3
9.4 Summary

9.1 Random Sampling

Inferential statistics are used in reasoning from a sample to the population. Some samples provide a sound basis for this process, while others don't. The difference lies in the method by which the samples are selected. *The method of drawing samples from a population so that every possible sample of a particular size has the same probability of being selected is called* **random sampling** *and the resulting sample a* **random sample**. As the definition indicates, randomness is a property of the procedure rather than of the particular sample obtained. By the term "random sample" we mean simply a sample produced by a random sampling procedure. Any sampling method based on haphazard or purposeless choices such as utilizing volunteers, students enrolled in a psychology course, or every tenth name in an alphabetical listing is called *nonrandom sampling*. The resulting samples, unlike random samples, don't provide a sound basis for deducing the properties of populations.

As we will see, the inferential procedures described in subsequent chapters assume either random sampling or random assignment of subjects to the various conditions of an experiment. There is no guarantee that a particular random sample will resemble the population, but in the long run, random samples are more likely to do so than nonrandom samples. Random assignment ensures that systematic bias isn't introduced, as it would be, for example, if the best subjects were unwittingly assigned to the experimental conditions that are expected to be superior.

Defining the Population

The first step in drawing a random sample is to identify the population. *A* **population** *is the collection of all objects or observations having one or more specified characteristics.* The population is identified when we specify the common characteristics, for example, this year's freshmen at Oregon State University or the outcomes of tossing a die for eternity. *A single object or observation is called an* **element** *of the population.* The elements of the population can be *finite*[1] (or limited) in number, as this year's freshmen at Oregon State, or *infinite* in number, as the outcomes of tossing a die for eternity.

In practice, it is difficult to obtain a random sample from large populations like residents of a city or students at a university. There are two obstacles—obtaining an accurate list of the population elements and securing their participation once they have been selected. Some cities have lists of their residents, but unfortunately the information isn't updated frequently. Telephone directories are more current, but exclude certain segments of society more often than others. Both introduce systematic bias into the experiment. An investigator faced with the choice may prefer to redefine the population to fit the more current list. Instead of all city residents, the population is defined as all families in the telephone directory.

---

[1] The probability of drawing a particular sample from a finite population is given by $_rC_r/_nC_r$ (see Section 8.1), where r denotes the sample size and n, the population size. For example, the probability for $r = 2$ and $n = 100$ is $\{2!/[2!(2-2)!]\}/\{100!/[2!(100-2)!]\} = 1/19{,}800$.

Sampling with or without Replacement

*After identifying the population one must decide whether to sample with replacement or without replacement. In sampling **with replacement**, a sampled element is returned to the population so it is available to be drawn again; in sampling **without replacement**, the element is not replaced and hence can only be drawn once.*[2] Sampling with replacement is rarely appropriate for the kinds of problems investigated in the behavioral sciences and education, since the sampled elements may be significantly and permanently altered by participating in the experiment. For example, once a child has learned an arithmetic unit, that child is no longer a naive learner with respect to the unit; once tissue has been surgically removed, it can't be removed again should the organism happen to be sampled a second time.

Random Sampling Procedures

A variety of procedures can be used to draw a random sample. If the population is finite, each element can be identified on a slip of paper, the slips placed in a container, thoroughly mixed, and drawn blindly from the container. If sampling with replacement is used, the identity of an element selected is noted and the slip is returned to the container; it is then available to be drawn again. The blind drawing-of-slips procedure seems simple enough, but it isn't always random—as witness the December 1969 draft lottery. More slips containing birthdates in the last 6 months of the year were drawn early than were dates in the first 6 months, much to the dismay of Virgos, Libras, Scorpios, and Sagittarians. The result was attributed to placing the slips in the bowl in chronological order and failing to shake them up thoroughly. Slips for the later months were the last ones in the bowl and the first ones drawn.

Another technique for drawing a random sample is to flip a coin or spin a roulette wheel with the outcome of the random device determining whether an element is or isn't included in the sample. This procedure is practical for selecting a small sample, but becomes tedious for larger ones.

Most researchers prefer to use a table of random numbers to draw their sample. Random number tables like Appendix Table D.1 were prepared so that integers from 0 to 9 occur with about equal frequency and appear in the table in a random fashion. The digits are usually in groups of two or more to make them easier to read, but the grouping has no other significance.

Suppose we want a random sample of 30 speech therapy majors. A printout listing 273 majors constituting the population is obtained from the computer center, and the students are numbered serially from 001 to 273. We turn to Table D.1 and note that it has two pages with 50 rows and 25 columns each. To decide where to begin in the table we close our eyes and drop our pencil on the table. Suppose it lands on the second page with the point closest to the first number in row 21 and column 13. The numbers reading from left to right are 22 00 20 35 55.... We let the first number, 2, identify the table page we begin on, the next two digits, 20, the row, and the next two

---

[2] The number of samples of size r that can be drawn without replacement from a population of size n is $_nC_r$. The number of samples with replacement is $n_1 n_2 \cdots n_r$.

digits, 02, the column. Having previously decided to read the numbers from left to right, although any sequence can be used, we proceed to draw our sample. Numbers are read in groups of three, until we obtain 30 numbers between 001 and 273, inclusive. To sample without replacement, we ignore numbers after their first appearance. The students corresponding to the 30 numbers compose the sample.

In sampling from a list with many pages, such as a telephone directory or a student directory, it isn't necessary to number each population element if the number of names on each page is about the same. Instead, each page and each position on the page is numbered. To select an element, two numbers are drawn from a random number table; the first identifies the directory page and the second, the position of the element on the page. Another procedure, called *systematic sampling*, for sampling from a list is sometimes used. It involves sampling every nth element, say, every 20th person, in the list. In spite of the simplicity of this procedure it can't be recommended, because it doesn't satisfy the definition of random sampling.

Review Exercises for Section 9.1

*1. List the steps involved in drawing a random sample.
2. What advantages do random samples have over nonrandom samples?
**3. *(a) How many samples of size 5 can be drawn without replacement from a population of size 50? (b) How many samples of size 5 can be drawn with replacement from a population of size 50?
*4. A sample of four supermarkets is to be selected from a total of eight in a small town. (a) How many different random samples without replacement can be drawn? (b) What is the probability that a given sample will be selected? (c) How many random samples with replacement can be drawn?
5. A sample of five students is to be selected from a class of ten students. (a) How many different random samples without replacement can be drawn? (b) What is the probability that a given sample will be selected? (c) How many random samples with replacement can be drawn?
6. Use a random number table to draw two random samples of ten students from the following population. For one use sampling with replacement and for the other sampling without replacement. (b) Describe in detail how you used the table.

Terry	Ed	Bob	Mac	Maureen	Jay
Jack	Vena	Roger	Les	Myrtie	Lanelle
Bill	Larry	Tommy	Bud	John	Nancy

7. Terms to remember:
   a. Random sampling
   b. Sampling with or without replacement
   c. Systematic sampling

9.2 Random Variables and Their Distributions

Random Variables

In rolling a pair of dice we can observe the total number of dots; in tossing a coin three times the total number of heads; in observing a naive rat in a three-choice T maze the total number of incorrect turns. The variable, number of dots or number of heads or number of incorrect turns, is called a *random variable* because the value of the variable for a particular experiment is determined by chance. In the dice example, the random variable, number of dots, can assume values of 2, ..., 12; in the T maze example, the random variable can assume values of 0, 1, 2, or 3 errors. Random variables are usually denoted by capital letters toward the end of the alphabet, such as X, Y, or Z. It helps to think of a random variable as the name for the number associated with the outcome of a random experiment before the experiment is performed. Performing the experiment converts the random variable into some specific number.

You may be wondering, why the fancy name? How does a random variable differ from just a plain old variable? We may contrast the two kinds of variables as follows.

1. The variable X is a name for any one of a set of permissible values.
2. The random variable X is a name for any one of a set of permissible numerical values of a random experiment.

Let's pursue the meaning of a random variable a bit further. In Section 7.2 we saw that all the possible outcomes of a random experiment can be represented by points in a sample space. A random variable associates one and only one numerical value with each point; hence, in the language of the mathematician, a random variable is a *function*. To understand this, recall from algebra that a function consists of two sets of elements and a rule that assigns to each element in the first set one and only one element in the second set. The definition of a function is quite general; $\{(a, 1), (b, 5), (c, 6)\}$ is a function, as are $\{$(Jack, tall), (Bill, medium), (Terry, medium)$\}$ and $\{$(No errors, 0), (One error, 1), (Two errors, 2), (Three errors, 3)$\}$. To state it more simply, a function is a set of ordered pairs of elements, no two of which have the same first element. If the second element of a pair is a number, the function is said to be numerically valued. A random variable associates one and only one number with each point in a sample space; thus, it is a numerically valued function defined over a sample space. Most readers will find the following definition easier to remember. A random variable is a numerical quantity with a value determined by the outcome of a random experiment.

Random variables are classified according to the nature of the numbers they can assume. *A random variable is* **discrete** *if its range can assume only a finite number of values or an infinite number of values that is countable*; for example, family size, number of dates per week, or scores on a test. *A random variable is* **continuous** *if its range is uncountably infinite;* for example, temperature in Monterey, California, duration of a kiss, or height. We must distinguish between the values the random variable can assume and those yielded by our measuring instruments. A thermometer is usually calibrated in $1°$ steps, a stop watch in 0.1 sec, and a ruler in $\frac{1}{16}$ in. Consequently, our measurement of continuous random variables is always approximate.

Distribution of a Discrete Random Variable

We learned in Chapter 2 that a frequency distribution associates a frequency with each value or class interval of a variable. *A similar representation that associates a probability with each value of a random variable is called a* **probability distribution**.

Table 9.2-1. Probability Distribution for Outcome of Tossing a Die

Possible Values, r, of the Random Variable X	$p(X = r)$
1	1/6
2	1/6
3	1/6
4	1/6
5	1/6
6	1/6

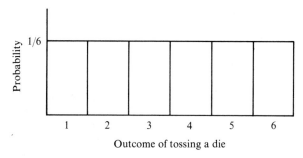

Figure 9.2-1. Histogram for probability distribution in Table 9.2-1.

A probability distribution for an experiment of tossing a die is shown in Table 9.2-1, and a graph in which probability is represented by the height of rectangles is shown in Figure 9.2-1. The distribution is said to be uniform, since each value of the random variable has the same probability. Notice that the probabilities sum to 1, because the events $X = 1, \ldots, 6$ are mutually exclusive and collectively exhaustive.

Consider next the three-choice T maze experiment mentioned earlier. Suppose the correct series of turns is *right, left, right* (R, L, R). We know from the fundamental counting rule in Section 8.1 that a rat can traverse the maze in $2 \times 2 \times 2 = 8$ ways, since three right–left choices must be made. The eight ways and the number of errors associated with each are listed in the table. The probability of making 0, 1, 2, or 3 errors, the random variable, can be computed by $p(X = r) = n_r/n_S$, where n_r is the number

	Number of Errors
R, L, R	0
R, R, R	1
R, L, L	1
L, L, R	1
R, R, L	2
L, R, R	2
L, L, L	2
L, R, L	3

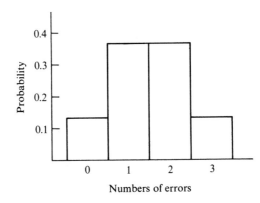

Table 9.2-2. Probability Distribution for Number of Errors in a Three-Choice T Maze

Possible Values, r, of Random Variable X	$p(X = r)$
0	0.125
1	0.375
2	0.375
3	0.125

Figure 9.2-2. Histogram for probability distribution in Table 9.2-2.

of events (maze routes) favoring r errors and n_S is the number of events. For example, the probability of one error is $\frac{3}{8} = 0.375$. The probability distribution is given in Table 9.2-2 and a graph of the distribution in Figure 9.2-2. The table and figure can be used to answer any question about the probability associated with the random variable. For instance, the probability that X is odd: $p(X = 1, 3) = 0.375 + 0.125 = 0.5$. Or the probability that X is less than 3: $p(X < 3) = 0.125 + 0.375 + 0.375 = 0.875$.

A probability distribution is similar to a frequency distribution in that it associates a probability with each value of a variable, while the frequency distribution associates a frequency with each value of the variable. We saw in Chapter 3 that the arithmetic mean is often used to describe the central tendency of a frequency distribution. A similar index of the central tendency of a probability distribution is called an *expected value*.[3] We now turn to the subject of how to compute expected values.

Expected Value of a Discrete Random Variable

If an extremely large number of naive rats were to run the three-choice T maze, how many errors on the average would we expect them to make? Stated more formally, what is the *expected value* of that random variable? If X is a discrete random variable

---

[3] The terms *expected value* and *expectation* are synonymous.

that assumes values $X_1, X_2, \ldots, X_n$ with probabilities $p(X_1), p(X_2), \ldots, p(X_n)$, then the expected value of X denoted by $E(X)$ is defined as[4]

$$E(X) = p(X_1)X_1 + p(X_2)X_2 + \cdots + p(X_n)X_n = \sum_{i=1}^{n} p(X_i)X_i,$$

where $p(X_1) + p(X_2) + \cdots + p(X_n) = 1$. For the T maze example, $E(X)$ for the values in Table 9.2-2 is

$$E(X) = 0.125(0) + 0.375(1) + 0.375(2) + 0.125(3) = 1.5.$$

Thus, we would expect a rat to make on the average 1.5 errors in the maze. Note the similarity between the formula for $E(X)$ and that for the mean of a grouped frequency distribution:

$$\bar{X} = \frac{f_1}{n} X_1 + \frac{f_2}{n} X_2 + \cdots + \frac{f_k}{n} X_k.$$

Here, X is a discrete variable that assumes values $X_1, X_2, \ldots, X_k$ with frequencies $f_1, f_2, \ldots, f_k$, where $f_1 + f_2 + \cdots + f_k = n$; $\bar{X}$ and $E(X)$ differ in that $\bar{X}$ is the mean of a sample defined by its frequency distribution; $E(X)$ is the mean of a theoretical population defined by its probability distribution.

Originally, the expected value concept was used in games of chance to tell a player what the long-run average loss or gain per play would be. Consider the popular casino game of roulette. A player places a bet, the wheel is spun, and the ball is set in motion. The ball can drop into one of 38 slots; 36 slots are numbered from 1 to 36 with half red and half black and two green slots are numbered 0 and 00. Suppose a player places $1 on number 7. If the ball drops into the 7 slot, the player receives a $35 payoff; otherwise, the bet is lost. We can calculate the player's expected winnings as shown in Table 9.2-3. Thus, if the player keeps making $1 bets indefinitely, an average loss of 5.3¢ per bet will be incurred. On any given gamble, the player stands to either win $35 or lose only $1. What the player may choose to ignore is that, on the average, $35 is won in only 1 out of 38 gambles, while $1 is lost in 37 out of 38.

Table 9.2-3. Expected Value of a Bet

Possible Winnings, X_i	$p(X_i)$	$p(X_i)X_i$
+$35	$\frac{1}{38}$	$\frac{1}{38}(\$35) = \$\frac{35}{38}$
−$1	$\frac{37}{38}$	$\frac{37}{38}(-\$1) = -\$\frac{37}{38}$
	$E(X) = \sum_{i=1}^{n} p(X_i)X_i =$	$-\$\frac{2}{38} = -0.053$

The term *expected value* is misleading in one sense because $E(X)$ is often not one of the possible outcomes of an experiment. In the T maze example, $E(X) = 1.5$, but the possible values of the random variable are 0, 1, 2, or 3 errors. Similarly, the gambler can win $35 or lose $1 on any given play, although $E(X) = -5.3$¢. In both examples, $E(X)$ is an average result and in this respect it is like a sample mean, $\bar{X}$.

[4] This definition of $E(X)$ doesn't commit us to a particular view of probability, since the $p(X_i)$'s can be subjective, logical, or empirical.

Expected Value of a Continuous Random Variable

Computing the expected value of a discrete random variable is fairly simple because we need only multiply probabilities by random variable values and sum the products—that is, $E(X) = \sum_{i=1}^{n} p(X_i)X_i$. The continuous random variable case is more complicated because the variable can assume an infinite number of values. The probability that a continuous random variable X has a particular value is zero.[5] Consequently, instead of referring to the probability that X has a particular value, we refer to the probability that X lies in an interval between two values of the random variable. This notion is illustrated in Figure 9.2-3. The expected value of a continuous random variable X is the sum of the products formed by multiplying each value that X can assume by the height of the probability distribution curve above that value of X. Since X can assume an infinite number of values, its expected value is not actually computed by physically multiplying each X by the height of the curve at X but instead by means of the integral calculus.[6] As we will see, tables for most random variables of interest have been prepared; these simplify the calculation of the probability that X lies in an interval.

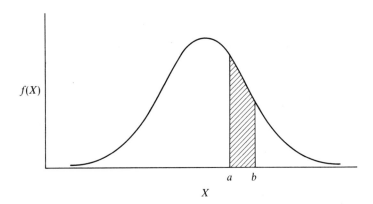

Figure 9.2-3. The probability that X will assume a value between a and b is equal to the area under the curve between those two points. For many random variables, tables are available that simplify determining the area between two points (see Section 10.1).

Standard Deviation of a Discrete Random Variable

In Chapter 4 we learned that the standard deviation is a useful measure of dispersion. For a sample defined by a grouped frequency distribution it is given by $S = \sqrt{\sum f_j(X_j - \bar{X})^2/n}$. A similar measure of dispersion for a theoretical population defined

---

[5] This may not be obvious. For a discussion of this point, see Hays (1973, pp. 119–123).
[6] For those familiar with the calculus, the expected value is $E(X) = \int_{X_{min}}^{X_{max}} Xf(X)\,dX$, where X_{min} is the smallest value of X and X_{max} is the largest value.

by the probability distribution of a discrete random variable is given by
$$\sigma = \sqrt{E\{[X - E(X)]^2\}} = \sqrt{\sum p(X_i)[X_i - E(X)]^2}.$$
We note from the formula on the left that σ is the square root of the expected value of a squared deviation, $[X - E(X)]^2$. We compute this expected value in the same way as we did for $E(X)$, where we multiplied each value of X_i by its probability. To compute $E\{[X - E(X)]^2\}$, we multiply each $[X_i - E(X)]^2$ by its probability $p(X_i)$ and sum the products. The computation of σ is illustrated for the T maze data in Table 9.2-2: $E(X) = 1.5$ and the standard deviation is

$$\sigma = \sqrt{0.125(0 - 1.5)^2 + 0.375(1 - 1.5)^2 + 0.375(2 - 1.5)^2 + 0.125(3 - 1.5)^2}$$
$$= 0.866.$$

The symbol σ is used instead of S, since this standard deviation is a population parameter. The value $\sigma = 0.866$ together with $E(X) = 1.5$ provide a useful summary of the theoretical population.

Review Exercises for Section 9.2

8. Distinguish between a variable and a random variable.
**9. *(a) Construct a probability distribution for a four-choice T maze. Assume the correct series of turns is *right, right, left, right*. (b) Graph the probability distribution.
*10. Let the random variable X be the number of cars per household. Suppose that in Waco, Texas, X has the probability distribution listed in the table. If a household is selected at random, com-

X	0	1	2	3	4	5
$f(X)$	0.16	0.54	0.23	0.05	0.01	0.01

pute the following.
a. $p(X \le 2)$ b. $p(X \ge 3)$ c. $p(1 \le X \le 2)$
d. $E(X)$ e. σ

11. Let the random variable X be the number of children in a family. Suppose that X has the probability distribution listed in the table. If a family is selected at random, compute the following.

X	0	1	2	3	4	5	6	7
$f(X)$	0.40	0.18	0.15	0.11	0.09	0.05	0.01	0.01

a. $p(X = 0)$ b. $p(X \ge 4)$ c. $p(X < 3)$
d. $p(2 \le X \le 5)$ e. $E(X)$ f. σ

12. How does an expected value differ from the mean of a frequency distribution?
*13. What is the maximum you would be willing to pay to enter a game in which you can win $30 with probability 0.6 and $10 with probability 0.4?
14. If it rains, a fortune-teller loses $6 per day; if it is fair, she earns $45 per day. Assume the probability of rain is 0.3. What are her expected earnings per day?
15. The random variable X has the probability distribution listed in the table.

X	0	1	2	3	4
$f(X)$	0	$\frac{2}{5}$	$\frac{1}{5}$	$\frac{1}{5}$	$\frac{1}{5}$

(a) Compute $E(X)$.   (b) Compute σ.

16. Terms to remember:
    a. Discrete random variable
    b. Continuous random variable
    c. Probability distribution
    d. Expected value

9.3 Binomial Distribution

Bernoulli Trial

Many experiments have only two possible outcomes: a new drug is effective or it isn't, an animal takes the correct turn or the wrong turn, a job is given to an applicant or it isn't. These experiments have much in common with the simpler coin-tossing experiment. In each, the random variable is discrete and can assume only two values, often denoted "success" and "failure." Flipping a coin once and noting whether it landed heads or tails or randomly sampling one person from a population of former students and noting whether he or she graduated is called a *Bernoulli trial* or *Bernoulli experiment*.[7] Our interest is usually in several Bernoulli trials. We toss a coin n times and note the number of heads, or we randomly sample n persons and note the number of graduates. The probability that on any given trial we will observe a success is denoted by p and the probability of a failure by q. Since the two outcomes are mutually exclusive and exhaustive, $p + q = 1$. When there are n Bernoulli trials, the random variable of interest is the number of successes; its value can range from 0 to n.

The characteristics of a Bernoulli trial are as follows.

1. A trial can result in one of two outcomes.
2. The probability of a success remains constant from trial to trial.
3. The outcomes of successive trials are independent.

Very few real-life situations will perfectly satisfy the requirements. Strictly speaking, the last two are satisfied only when sampling is done with replacement or from an infinite population. In most research, sampling is done without replacement from a finite population. This practical departure from the ideal is of little consequence as long as the population is large relative to the sample size.

Below we describe a binomial distribution in which the random variable is a sum—the number of successes observed on $n \geq 2$ Bernoulli trials. *A probability distribution where the random variable is a statistic based on the results of $n \geq 2$ trials is given a special name—**sampling distribution***. For convenience, we will examine a relatively simple binomial distribution here and defer discussion of the special properties of sampling distributions to Chapter 10.

The binomial distribution will be encountered repeatedly in subsequent chapters. It is the theoretical model for a variety of statistics, as we will see in Sections 12.4, 14.2, 15.4, 17.2, 17.3, and 17.4.

---

[7] After James Bernoulli (1654–1705), who discussed such trials in his *Ars Conjectandi* (1713).

Binomial Distribution

The number of successes observed on $n \geq 2$ identical Bernoulli trials is called a **binomial random variable** and its probability distribution is called a **binomial distribution**.[8] Suppose we toss a fair coin five times. The probability of observing exactly r heads (successes) in n tosses is given by the function rule

$$p(X = r) = {}_nC_r p^r q^{n-r},$$

where $p(X = r)$ is the probability that the random variable X equals r heads, ${}_nC_r$ is the combination of n objects taken r at a time,[9] p is the probability of success (a head), and $q = 1 - p$. For example, the probability that the random variable X equals four heads is

$$p(X = 4) = {}_5C_4 \left(\frac{1}{2}\right)^4 \left(\frac{1}{2}\right)^{5-4} = \frac{5!}{4!(5-4)!} \left(\frac{1}{2}\right)^4 \left(\frac{1}{2}\right)^1 = \frac{5}{32}.$$

The complete probability distribution is given in Table 9.3-1 and a graph of the distribution in Figure 9.3-1. The probability that X equals or exceeds some value or that it

Table 9.3-1. Binomial Distribution for $n = 5$ and $p = \frac{1}{2}$

Number of Heads, r	0	1	2	3	4	5
$p(X = r)$	$\frac{1}{32}$	$\frac{5}{32}$	$\frac{10}{32}$	$\frac{10}{32}$	$\frac{5}{32}$	$\frac{1}{32}$

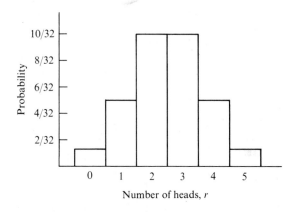

Figure 9.3-1. Histogram for binomial distribution in Table 9.3-1.

---

[8] So named because the probabilities associated with the distribution can be obtained by raising a binomial (an algebraic expression containing two terms) to the nth power. For example,

$$(p + q)^n = p^n + np^{n-1}q + \frac{n(n-1)}{2(1)} p^{n-2} q^2 + \cdots + q^n,$$

where p is the probability of success, $q = 1 - p$, and n is the number of Bernoulli trials. The first term, p^n, gives the probability of n successes, the second term, $np^{n-1}q$ the probability of $n - 1$ successes, and so on.

[9] Discussed in Section 8.1.

lies in a given interval can be obtained by combining probabilities from the table or figure. For example, the probability of obtaining four or more heads in five tosses of a fair coin is

$$p(X \geq 4) = p(X = 4) + p(X = 5) = \frac{5}{32} + \frac{1}{32} = \frac{6}{32}.$$

The probability distribution of a binomial random variable is completely specified by n, the number of trials, and the parameter p, the probability of success. When $p < 0.5$ a graph of the probability distribution is positively skewed, for $p = 0.5$ it is symmetrical, and for $p > 0.5$ it is negatively skewed. As n increases, the shape of the distribution approaches more and more closely that of the normal bell-shaped distribution. It should be apparent that the binomial distribution is actually a family of distributions, one for each set of p and n values. The thread that binds the distributions into a family is their common function rule, $p(X = r) = {}_nC_r p^r q^{n-r}$.[10] The following example illustrates another member of the binomial family.

Suppose we are interested in the probability that more than half of a random sample of six patients will show improvement following treatment. Let the probability of improvement for any patient equal 0.7. The probability of observing exactly six successes in six patients is given by

$$p(X = 6) = {}_6C_6(0.7)^6(0.3)^0 = \frac{6!}{6!(6-6)!}(0.70)^6(0.30)^0 = 0.118.$$

Table 9.3-2. Distribution Showing Probability of Improvement following Treatment

Number Improved, r	0	1	2	3	4	5	6
$p(X = r)$	0.001	0.008	0.059	0.185	0.324	0.302	0.118

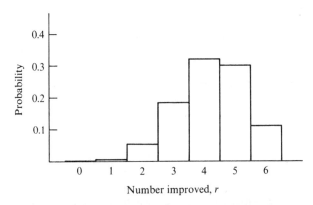

Figure 9.3-2. Histogram for probability that patients will show improvement following treatment.

[10] Other examples of families of discrete probability distributions are the uniform distribution, multinomial distribution, hypergeometric distribution, Poisson distribution, and negative binomial distribution. The first three distributions are discussed in this text. For a discussion of the latter two distributions see Hays (1973, chap. 5), Stilson (1966, chap. 10), and Walpole (1968, chap. 6).

The complete probability distribution is given in Table 9.3-2 and graphed in Figure 9.3-2. The probability that in a random sample of six patients more than half will show improvement is given by

$$p(X \geq 4) = p(X = 4) + p(X = 5) + p(X = 6) = 0.324 + 0.302 + 0.118 = 0.744.$$

Expected Value and Standard Deviation of Binomial Distribution

The expected value of a discrete random variable can always be computed from $E(X) = \sum_{i=1}^{n} p(X_i)X_i$. For a binomial random variable there is a simpler formula for computing the expected value of X (number of successes):[11]

$$E(X) = np,$$

where n is the number of trials and p is the probability of a success on any trial. For the probability distribution in Table 9.3-2, the expected number of patients showing improvement is $E(X) = 6(0.7) = 4.2$. The same result is obtained using the longer formula $E(X) = \sum_{i=0}^{6} p(X_i)X_i = 0.001(0) + 0.008(1) + \cdots + 0.118(6) = 4.2$.

The standard deviation of a binomial distribution is given by $\sigma = \sqrt{npq}$. For the probability distribution in Table 9.3-2 the standard deviation is

$$\sigma = \sqrt{6(0.7)(0.3)} = 1.12.$$

Multinomial and Hypergeometric Distributions

As we have seen, the binomial distribution is the appropriate model when (1) randomly sampling from a population with elements that belong to one of two classes and (2) the probability of obtaining an element in a class remains constant from trial to trial, as when sampling with replacement or from an infinite population. In subsequent chapters we will refer to two other models that apply when one or both of these conditions aren't satisfied. These models—the multinomial and hypergeometric distribution—are described below.

Consider first an experiment in which a trial results in an outcome from one of $k \geq 2$ classes and the probabilities associated with the classes remain constant. Such an experiment is called a *multinomial experiment* and its associated probability distribution is called a *multinomial distribution*.[12] If n observations are made independently and at random from $k \geq 2$ mutually exclusive and exhaustive classes with probabilities $p_1, p_2, \ldots, p_k$, then the probability that exactly n_1 will be of kind 1, n_2 will be of kind 2, ..., and n_k will be of kind k, where $n_1 + n_2 + \cdots + n_k = n$, is given by the **multinomial function rule**:

$$p(X_1 = n_1 \text{ and } X_2 = n_2 \text{ and } \cdots \text{ and } X_k = n_k) = \frac{n!}{n_1! n_2! \cdots n_k!} (p_1)^{n_1}(p_2)^{n_2} \cdots (p_k)^{n_k}.$$

When $k = 2$, this rule is the same as the binomial function rule because then $n_1 = r$, $n_2 = n - r$, $p_1 = p$, and $p_2 = 1 - p_1 = q$. A multinomial experiment is simply an extension of a binomial experiment and applies when there are two or more classes.

---

[11] A derivation of the formula is given by Hays (1973, pp. 227–228).
[12] So named because the probabilities associated with the distribution can be obtained by raising a multinomial (an algebraic expression containing three or more terms) to the nth power.

The computation of probabilities using the multinomial rule is not too difficult when n and the number of categories is small. Suppose we want to know the probability of drawing 2 red, 1 white, and 0 blue marbles from a box that contains 4 red, 2 white, and 2 blue marbles. If we randomly draw one marble at a time and replace it before drawing the next, the probability is given by

$$p(X_R = 2 \text{ and } X_W = 1 \text{ and } X_B = 0) = \frac{n!}{n_R! n_W! n_B!} (p_R)^{n_R}(p_W)^{n_W}(p_B)^{n_B}$$

$$= \frac{3!}{2!1!0!} (0.50)^2 (0.25)^1 (0.25)^0$$

$$= 0.188.$$

Another computational example is given in Section 17.6-2. When n and the number of categories is large, the multinomial function rule requires a prohibitive amount of computation. As we will see in Chapter 17, this problem can be circumvented through the use of the chi-square approximation of the multinomial.

Suppose that instead of replacing marbles after they are drawn from the box, we sample without replacement. What is the probability of drawing 2 red, 1 white, and 0 blue marbles? Now the probabilities associated with drawing the different colored marbles will change for each draw. This experiment is called a *hypergeometric experiment* and its associated probability distribution is called a *hypergeometric distribution*. If n observations are drawn at random and without replacement from a finite population containing a total of t elements that are divided into k mutually exclusive and exhaustive classes with t_1 in class 1, t_2 in class 2, ..., t_k in class k, then the probability that exactly n_1 will be of kind 1, n_2 will be of kind 2, ..., and n_k will be of kind k, where $n_1 + n_2 + \cdots + n_k = n$ and $t_1 + t_2 + \cdots + t_k = t$, is given by the **hypergeometric function rule**:

$$p(X_1 = n_1 \text{ and } X_2 = n_2 \text{ and } \cdots \text{ and } X_k = n_k) = \frac{(_{t_1}C_{n_1})(_{t_2}C_{n_2}) \cdots (_{t_k}C_{n_k})}{_t C_n}.$$

The probability for our marble experiment is given by

$$p(X_R = 2 \text{ and } X_W = 1 \text{ and } X_B = 0) = \frac{(_4C_2)(_2C_1)(_2C_0)}{_8C_3}$$

$$= \frac{\{4!/[2!(4-2)!]\}\{2!/[1!(2-1)!]\}\{2!/[0!(2-0)!]\}}{\{8!/[3!(8-3)!]\}}$$

$$= 0.214.$$

This probability is different from the 0.188 given by the multinomial rule. The two rules give similar results when the population is extremely large.

Review Exercises for Section 9.3

17. Compare a Bernoulli random variable with a binomial random variable.
18. Interpret the statement $p(X = 3) = 0.2$.
*19. Let the random variable X be the number of males in a random sample of size two taken from a population that contains 60% males and 40% females. (a) Determine the probability of the sample containing 0, 1, or 2 males. (b) Graph the probability distribution. (c) Compute $E(X)$ and σ.

20. Suppose that 20% of eligible voters in a given city voted in the last election. A random sample of ten eligible voters is obtained to investigate reasons for the poor turnout. (a) If X is the number of people who didn't vote, determine the probability distribution for X. (b) Compute $E(X)$ and σ.

*21. Thirty percent of elementary students in a school system have a reading ability below the national standard for their grade level. (a) If ten children are selected at random, what is the probability that no more than one will be functioning below grade level? (b) Compute $E(X)$ and σ.

22. Ten percent of patients fail to improve after being placed on medication. (a) If five patients are selected at random, what is the probability that two or more will not show improvement? (b) Compute $E(X)$ and σ.

*23. Of 800 families with five children each, how many would you expect to have (a) 3 girls? (b) 5 boys? (c) either 2 or 3 girls? Assume equal probabilities for girls and boys.

24. What is the probability of guessing correctly at least 6 of 10 answers on a true–false examination?

*25. Suppose a box contains 50% nickels, 30% dimes, and 20% quarters. If 10 coins are drawn at random and with replacement, what is the probability of drawing 6 nickels, 3 dimes, and 1 quarter?

*26. Suppose that in Exercise 25 the box contained 20 coins and 10 are drawn at random and without replacement. What is the probability of drawing 6 nickels, 3 dimes, and 1 quarter?

27. A graduate admissions committee has 20 doctoral applicants, but they can only admit 10. What is the probability that the 10 selected will include the 5 most qualified applicants? (*Hint*: $t = 20$, $t_1 = 5$, and $t_2 = 15$.)

28. Terms to remember:
   a. Bernoulli trial
   b. Binomial random variable
   c. Multinomial experiment
   d. Hypergeometric experiment

9.4 Summary

Some kind of random procedure should be a part of all research with samples; most often it takes the form of random sampling from a population or random assignment of subjects to experimental conditions. Randomness is a property of a procedure rather than of a sample. Any procedure for drawing samples from a population so that every possible sample of a particular size has the same probability of being selected is called random sampling and the resulting sample is called a random sample.

A random variable is a numerical quantity with values determined by the outcomes of a random experiment. A table showing the possible values of a random variable and the associated probabilities is called a probability distribution. Probability distributions and the frequency distributions discussed in Chapter 2 are similar; each associates a number with each of the possible values of a variable. However, for a frequency distribution, the number is a frequency; for a probability distribution, it is a probability. This reflects a fundamental difference between them. A frequency distribution describes a set of data that has actually been observed; it is empirical. A probability distribution describes data that might be observed under certain well-specified conditions; hence, it is hypothetical or theoretical. Probability distributions are used in inferential statistics as models of how random variables are expected to behave. If empirical data deviate appreciably from the predictions of a model, doubt is cast on the correctness of the model and/or its assumptions. For example, if five fair coins are tossed, according to the binomial model we should observe five heads on the average once in every 32 tosses. If instead of observing five heads once we observe them in ten of the first 32 tosses, we would begin to question the assumption that the coin is fair.

The central tendency of a theoretical population defined by a probability distribution can be described in the same way as the central tendency of a sample—by a mean. The mean of random variable values for a theoretical population is called an *expected value*, and is given by $E(X) = \sum p(X_i)X_i$.

An experiment is called a Bernoulli trial if (1) its random variable has only two possible outcomes, denoted "success" and "failure"; (2) the probability of a success remains constant from trial to trial; and (3) the outcomes of successive trials are independent. The probability distribution of a Bernoulli random variable could hardly be simpler, since it represents the possible outcomes of a single trial. Consider next a series of Bernoulli trials. The number of successes in n identical Bernoulli trials is a discrete random variable that can assume integer values from 0 to n. Its distribution is called a binomial distribution. The binomial distribution is one of the more useful models of how a discrete random variable should behave. Two other useful models are the multinomial and hypergeometric distributions. The multinomial distribution is an extension of the binomial model for the case in which a trial results in an outcome from one of $k \geq 2$ classes. If elements are sampled at random but without replacement from a finite population with $k \geq 2$ classes, the results are described by the hypergeometric distribution.

10

Normal Distribution and Sampling Distributions

10.1 Normal Distribution
  Characteristics of the Normal Distribution
  Converting Scores to Standard Scores
  Finding Areas under the Normal Curve
  Finding Scores when the Area Is Known
  Normal Approximation to the Binomial Distribution
  Review Exercises for Section 10.1
10.2 Interpreting Scores in Terms of z Scores and Percentile Ranks
  Standard Score
  Percentile Rank
  Relative Advantages of z Scores and Percentile Ranks
  Other Kinds of Standard Scores
  Comparing Performance on Different Tests
  Review Exercises for Section 10.2
10.3 Sampling Distributions
  Looking Ahead
  Introduction to Sampling Distributions
  Sampling Distribution of the Mean
  Central Limit Theorem
  Standard Error of a Statistic
  Two Properties of Good Estimators
  Test Statistics
  Review Exercises for Section 10.3
10.4 Summary
10.5 Technical Note
  Demonstration Showing that $E(\hat{\sigma})^2 = \sigma^2$ and $E(\sigma_{est}^2) = \sigma^2$ but $E(S)^2 \neq \sigma^2$

10.1 Normal Distribution

The normal distribution is the most important probability distribution in statistics. It is important partly because so many variables in science and nature have probability distributions that closely resemble it. For example, people's heights and weights are approximately normally distributed, as are intelligence, mechanical aptitude, introversion, and most other psychological attributes. The normal distribution is important also because it is a convenient model for estimating other theoretical distribution probabilities. We will see later that it provides an excellent approximation to the binomial distribution when the number of trials is large. It will become apparent that the normal distribution is a useful model, but this hardly accounts for its eminent position in statistical theory. To understand this we must consider the distribution of a sample statistic such as the mean. Suppose that from a population we drew 100 random samples of size n (where n is fairly large), computed the mean of each sample, and constructed a histogram for the sample means. We would find that the resulting graph closely resembled the normal distribution. This might not surprise us if the sampled population were normally distributed, but the striking thing is that the resemblance holds regardless of the population's shape if n is sufficiently large. The tendency for the distribution of a sample statistic to approximate the normal form as n increases plays a key role in inferential statistics; more will be said about this when we discuss the central limit theorem in Section 10.3.

Serendipity has produced many breakthroughs in science, and one of them is the normal distribution. Abraham de Moivre (1667–1754), a mathematics tutor, was search-

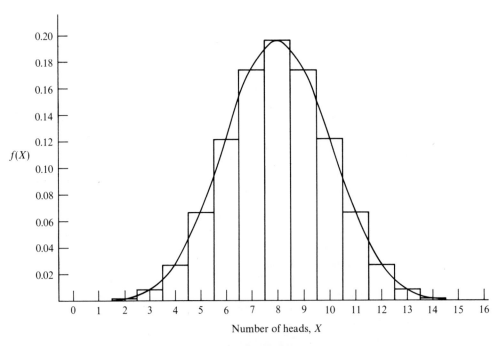

Figure 10.1-1. Comparison of the histogram for the probability distribution of 16 fair coins and the normal curve.

ing for a shortcut method of computing probabilities for binomial random variables and in the process derived the function rule for the ubiquitous normal distribution. If we toss ten coins it doesn't take too much effort to compute the probability of observing 0 heads, 1 head, and so on. But suppose we toss 100 coins. The amount of work necessary to calculate the probabilities associated with 100 heads is prohibitive.

Consider the graph in Figure 10.1-1 of the probability distribution for 16 fair coins. If we superimpose the graph of a normal distribution on the histogram, it provides a fairly good fit. The fit would be even better if we had graphed the distribution for 50 coins. If the number of coins were increased indefinitely, the number of bars in the histogram would increase and their outline would eventually coincide with that of the normal curve. De Moivre derived the function rule giving the height of the curve for any value of a random variable.

Characteristics of the Normal Distribution

A random variable X is said to be normally distributed if its probability distribution is given by the normal function rule. *The function rule for the **normal distribution** is*

$$f(X) = \frac{1}{\sigma\sqrt{2\pi}} e^{-(X-\mu)^2/(2\sigma^2)},$$

where $f(X)$ is the height of the curve at X, π is approximately 3.142, e (the base of the system of natural logarithms) is approximately 2.718, and μ and σ identify a particular normal distribution in the family of normal distributions. Fortunately, we don't have to use the rule to determine areas under the curve between various values of X; as we will see, this can be determined from Appendix Table D.2.

The normal curve is shaped like a bell. Since it is unimodal and symmetrical, its mean, median, and mode have the same value and that value corresponds to the highest point on the curve. The mean plus and minus the standard deviation, as shown in Figure 10.1-2, defines the inflection points of the curve—that is, the points at which the curve

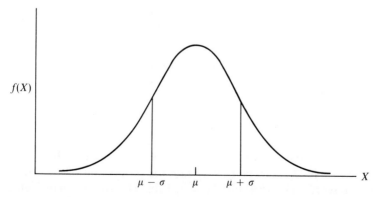

Figure 10.1-2. Graph of the normal curve. The inflection points occur at $\mu + \sigma$ and $\mu - \sigma$.

changes from being concave to convex or vice versa. Although they are not shown in the figure, the tails of the curve extend indefinitely in both directions, never quite touching the horizontal axis. The total area under the curve is equal to one.

Converting Scores to Standard Scores

There are as many normal curves as there are possible values of μ and σ, the parameters that identify a particular curve. To avoid having to develop an infinite number of tables, statisticians have made one particular curve the standard. It has a mean equal to zero and a standard deviation equal to one and is called the *standard normal curve*. This is the curve with areas that are tabulated in Table D.2. Random variable values for this curve are called *standard scores* and are denoted by z.

If a random variable doesn't happen to have $\mu = 0$ and $\sigma = 1$, its values can be transformed to standard scores in order to use Table D.2. This is accomplished by the formula

$$z = \frac{X - \mu}{\sigma},$$

where X is a random variable value, μ is the population mean, and σ is its standard deviation. A comparable formula for a sample is

$$z = \frac{X - \bar{X}}{S},$$

where X is a random variable value, $\bar{X}$ is the sample mean, and S is the sample standard deviation. Suppose the mean of a random variable is 100 and its standard deviation is 15. The z score corresponding to $X = 130$ is $z = (130 - 100)/15 = 2$. A z score transformation alters the mean and standard deviation of a random variable, but not the relative location of scores in the distribution. For example, $X = 130$ is two standard deviations above the mean of 100, since $130 = 100 + 2(15)$. Similarly, the transformed value $z = 2$ is also two standard deviations above the mean of 0, since $2 = 0 + 2(1) = 2$. If we were to graph the distribution of the random variable and the distribution of the transformed variable, we would find that they are identical in shape although they differ in central tendency and dispersion. Transforming scores to standard scores doesn't change the shape of the distribution or the relative position of scores—only the mean and standard deviation.

For approximately normal distributions most z scores are between -3 and $+3$. As we learned in Chapter 4, 99.73% of the area under the normal curve lies within ± 3 standard deviations of the mean.

Finding Areas under the Normal Curve

If a random variable is approximately normally distributed, the standard normal distribution (see Table D.2) can be used to find the proportion of the total area falling between any two scores. The areas tabulated in Table D.2 are shown in Figures 10.1-3(a) and (b).

a.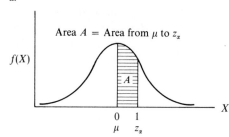

b.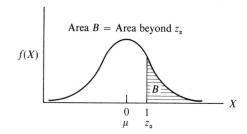

c.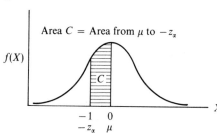

d.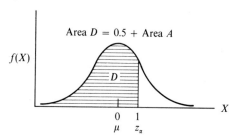

e.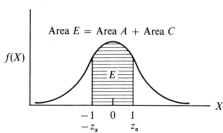

f.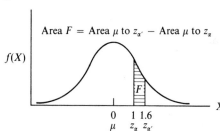

Figure 10.1-3. Illustrations of the areas of the standard normal distribution. Areas A and B are given in Table D.2. A standard score is denoted by z_α, where α indicates the proportion of the standard normal distribution that falls beyond the score. In considering area D, recall that the mean divides the total area in half, so that 0.5 falls above the mean and 0.5 below the mean.

1. *Area between μ and a score above it (area A)*

    Suppose the distribution of college students' IQs is approximately normal with $\mu = 115$ and $\sigma = 15$, and you want to know the proportion of students with IQs between the mean and 130. The first step is to convert 130 into a standard score: $z = (130 - 115)/15 = 1$. According to Table D.2, the proportion of the area from μ to $z = 1$ is 0.3413; thus, approximately 34% of students have IQs between 115 and 130. This area is shown as area A in Figure 10.1-3(a).

10.1 Normal Distribution

2. *Smaller area in the tail (area B)*
   The proportion of students with IQs above 130, which corresponds to a standard score of 1, is shown as area *B* in Figure 10.1-3(b). It is equal to 0.1587. Since the normal distribution is symmetrical, area *A* + area *B* = 0.5. Standard scores are sometimes denoted by *z* and a subscript that indicates the proportion of the normal distribution that falls beyond the score. For example, the standard score of 1 is denoted by $z_{0.1587}$. The symbol z_α denotes the standard score beyond which α proportion of the normal distribution falls.
3. *Area between μ and a score below it (area C)*
   To determine the proportion of the total area from μ to a score below the mean, say 100, we first convert the score to a *z* score: $z = (100 - 115)/15 = -1$. Table D.2 only gives areas for positive *z* scores, but since the distribution is symmetrical, the area from μ to $z = -1$ is the same as that from μ to $z = +1$. Thus, area *C* is obtained by ignoring the negative sign and looking up the *z* score in area *A*. The area from μ to $z = -1$ is 0.3413.
4. *Larger area in the tail (area D)*
   To find area *D* for a score of, say, 130 (with a *z* score of 1), we find area *A* and add 0.5 to it. For example, area $D = 0.5 + 0.3413 = 0.8413$. A student with an IQ of 130 scores above approximately 84% of college students.
5. *Area between scores on opposite sides of the mean (area E)*
   To find the proportion of the total area between two scores on opposite sides of the mean, add areas *A* and *C*. For example, if the scores are 130 and 100, the *z* scores are 1 and -1. The sum of areas *A* and *C* is $0.3413 + 0.3413 = 0.6826$.
6. *Area between scores on the same side of the mean (area F)*
   Suppose we want to determine the proportion of the total area between scores of 130 and 139. We first transform the scores to *z* scores: $(130 - 115)/15 = 1$ and $(139 - 115)/15 = 1.6$. Area *A* for $z = 1$ is 0.3413 and for $z = 1.6$ is 0.4452. Area *F* for these *z* values is given by (area μ to $z = 1.6$) − (area μ to $z = 1$) = $0.4452 - 0.3413 = 0.1039$.

Finding Scores when the Area Is Known

A different kind of problem arises when we have a percentile rank in mind or know the relative size of the area above or below a point in a distribution, and we want to determine the untransformed score corresponding to that rank or point. If we know the size of the area, we can determine from Table D.2 the *z* score that marks the boundary of the area. In the previous examples we knew *X*, μ, and σ, and solved for *z* using the formula $z = (X - \mu)/\sigma$. If we know *z*, μ, and σ, it is a simple matter to solve for *X*. A little algebra is all that is needed to express the formula in the desired form.

$$z = \frac{(X - \mu)}{\sigma}$$

$$\sigma z = X - \mu$$

$$X = \mu + \sigma z$$

Suppose we want to know the IQ score corresponding to the 80th percentile rank. We know that 0.80 of the area under the normal curve falls below the score and that $0.80 - 0.50 = 0.30$ of the area falls between μ and the *z* score. To find the *z* score corresponding to the area between μ and *z* of 0.30, we look in column 2 of Table D.2 until we locate 0.30. The corresponding *z* score is approximately 0.84. Knowing that

$\mu = 115$ and that $\sigma = 15$, we have all the information necessary to solve for X in the formula $X = \mu + \sigma z$. Substituting in the formula, we obtain $115 + 15(0.84) = 127.6$. Thus, a score of 127.6 corresponds to the 80th percentile rank.

To take one more example, suppose we want to know the IQ score corresponding to the 40th percentile rank. We know that the score is below the mean and that 0.40 of the area lies beyond (below) the score. To find the z score corresponding to the area beyond 0.40, we look in column 3 of Table D.2 until we locate 0.40 and find that $-z \simeq -0.25$. Remember that the sign of z scores below the mean is negative. Substituting in the formula $X = \mu + \sigma z$ gives $X = 115 + 15(-0.25) = 111.25$, the score corresponding to the 40th percentile rank.

Normal Approximation to the Binomial Distribution

The normal distribution function rule was originally derived by de Moivre to estimate binomial distribution probabilities when the number of trials, n, is large. As we will see, the approximation is excellent even when n is small.

Consider an experiment in which a fair coin is tossed five times. The random variable of interest is the number of heads. A graph of the distribution for $n = 5$ and $p = 0.5$ is given in Figure 10.1-4. A normal distribution has been superimposed on the graph. The probability of observing four or more heads can be computed from Table 9.3-1 (page 183): $p(X \geq 4) = \frac{5}{32} + \frac{1}{32} = \frac{6}{32} = 0.1875$.

The probability can be estimated using the normal distribution table by finding the area including and to the right of four heads. Since the normal distribution is continuous, we must think of four heads as occupying a class interval from 3.5 to 4.5; the lower limit of the class interval is 3.5. To find the area above the lower limit of four heads,

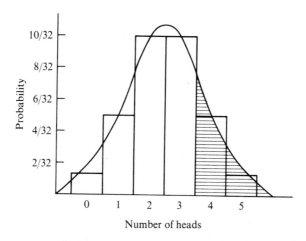

Figure 10.1-4. Histogram for binomial distribution with $n = 5$ and $p = 0.5$. A normal distribution is superimposed over the histogram. The normal curve area corresponding to the probability of observing four or more heads is represented by the shaded area.

we first convert 3.5 to a z score. Recall from Section 9.3 that the mean and standard deviation of a binomial distribution are given by, respectively, $E(X) = np$ and $\sigma = \sqrt{npq}$. For our example, $E(X) = 5(0.5) = 2.5$ and $\sigma = \sqrt{5(0.5)(0.5)} = 1.118$. The z score is

$$z = \frac{X - E(X)}{\sigma} = \frac{3.5 - 2.5}{1.118} = 0.894.$$

According to Table D.2, the area beyond $z = 0.894$ is 0.1857. This is close to the exact value of 0.1875 computed from the binomial distribution. So we see that, although the normal distribution approximation was not intended to be used for such a small n, it yielded a value quite close to the exact probability.

Review Exercises for Section 10.1

1. a. Why is the normal distribution so important in statistics?
   b. How does a standard normal curve differ from other normal curves?
*2. Which of the following variables do you think approximate the normal distribution? For those that you don't think are normal, sketch the form of the distribution you would expect. (a) Amount of coffee per cup dispensed by a vending machine, (b) Extroversion scores of college students, (c) Incomes of families in the United States, (d) Errors in reading voltages on a meter, (e) Ages of residents in Normal, Ohio, and (f) Time students arrive for class.
**3. A set of scores has a mean of 20 and a standard deviation of 5. Transform the following to z scores.
   *a. 30   *b. 12   *c. 15
   d. 0   e. 40   f. 20
   g. 25   h. 10   i. 18
**4. If z is a normally distributed random variable with $\mu = 0$ and $\sigma = 1$, determine the percentage of the area under the standard normal curve for the following.
   *a. Above $z = 1.5$ *b. Below $z = -2$
   *c. From μ to $z = 3$ d. Between $z = 1$ and $z = -2$
   *e. Between $z = 1$ and $z = 2$ f. Between $z = -2$ and $z = -3$
   g. From μ to $z = -1$ h. Between $z = -1$ and $z = 1.5$
**5. Determine the percentage of the area of the standard normal distribution that falls between $\mu - k\sigma$ and $\mu + k\sigma$, where k is equal to the following.
   *a. 1.0   *b. 1.64   *c. 1.96
   d. 2.58   e. 0.5   f. 3.30
   g. 2.33   h. 0.25   i. 0.67
**6. Compute the score corresponding to each of the following z scores. Assume the original distribution had a mean of 150 and a standard deviation of 20.
   *a. 2.0   *b. -1.5   *c. 3.1
   d. 0   e. 1   f. -2.1
   g. 2.5   h. -1   i. 1.8
**7. Find the z score such that at least the following proportion of the area under the standard normal distribution falls above it.
   *a. 0.50   *b. 0.05   *c. 0.40
   d. 0.70   e. 0.95   f. 0.16
   g. 0.025   h. 0.84   i. 0.01

8. In the general population, Stanford–Binet IQs are nearly normally distributed, with a mean of 100 and a standard deviation of 16. (a) What is the probability that a randomly selected person will have an IQ between 100 and 124? (b) What percentage of the population will have IQs above 132?

*9. Junior college grade point averages (GPAs) have $\mu = 2.8$ and $\sigma = 0.24$. A university is considering raising its minimum entrance score from 2.2 to 2.5. If GPA is normally distributed, how will the proposed change affect the percentage of students eligible to enter the university from junior colleges?

10. "Grading on the curve" means assigning grades according to the normal distribution. The mean of a test is 50 with a standard deviation of 10. If 10% of the class receives A, what is the lowest score that receives an A?

11. The time from conception to birth in humans is approximately normally distributed, with a mean of 280.5 days and a standard deviation of 8.4 days. In a paternity case it was proved that the time from the alleged conception to the birth of a $6\frac{1}{2}$ lb baby was at least 306 days. (a) Compute the percentage of women having this or a longer gestation time. (b) Discuss the significance of the evidence.

*12. Use the normal approximation to the binomial distribution to determine the probability of guessing correctly (a) at least 12 of 20 answers on a true–false examination, and (b) at least 24 of 40 answers.

*13. Suppose 10% of physicians' diagnoses at a clinic are incorrect. Use the normal approximation to the binomial distribution to determine the probability that of 400 diagnoses (a) at most 30 will be incorrect, (b) between 30 and 50 will be incorrect, (c) more than 50 will be incorrect.

14. Terms to remember:
    a. Inflection point
    b. Standard normal curve
    c. Standard (z) score

10.2 Interpreting Scores in Terms of z Scores and Percentile Ranks

Your roommate announces that she got a 62 on the midterm. Not knowing whether to rejoice with her or sympathize, you ask "What was the class average?" Forty-one, she replies. You press further "What was the range?" The lowest score was 22 and the highest was 62. A celebration is in order.

This example illustrates a problem in interpreting scores. A score by itself is uninterpretable; a frame of reference is needed to know whether a score is good or bad. The frame of reference in the example was provided by the central tendency of the distribution and its dispersion. The score became interpretable when it was related to the performance of other students.

Standard Score

It would be convenient to have one number that would provide all the information necessary to interpret a score instead of having to relate it to the mean and the standard deviation or perhaps the range. We have already discussed two such numbers that can be used to interpret a score: standard score and percentile. *A **standard score** is a single number that expresses the value of a score relative to the mean and the standard deviation of its distribution.* For example, if $\bar{X} = 50$ and $S = 10$, a score of 70 is

$$z = \frac{X - \bar{X}}{S} = \frac{70 - 50}{10} = \frac{20}{10} = 2,$$

which is two standard deviations above the mean. A standard score tells us the location of a score relative to the mean in standard deviation units.

Percentile Rank

The second kind of number that can be used to interpret a score is percentile rank, which is discussed in Chapter 4. *The **percentile rank** of a score indicates the percentage of the scores of the distribution that fall below that score.* For example, if a score has a percentile rank of 80, we know that 80% of scores fall below it and 20% fall above. The range of the transformed scale is from the 0th percentile rank to the 100th percentile rank, and the median is the 50th percentile rank. Transforming a score to a percentile rank locates it on a scale from 0 to 100 and indicates the percentage of scores below. Thus, as in the case of a standard score, a single number, the percentile rank, is sufficient to interpret a score.

Because percentile ranks are familiar to most people, they are widely used in presenting psychological test scores. Standard scores, on the other hand, are less familiar but, as we will see, possess a number of advantages over percentiles.

Relative Advantages of z Scores and Percentile Ranks

Consider the distribution of IQ scores in Figure 10.2-1(a); it is slightly negatively skewed. A graph of the percentile ranks corresponding to scores in Figure 10.2-1(a) is shown in part (b) of the figure. The percentile rank graph has a rectangular shape. We can see from the figure that the transformation of scores to percentile ranks has altered four characteristics of the distribution: (1) central tendency, (2) dispersion, (3) skewness, and (4) kurtosis. The only characteristic that isn't changed by the transformation is the rank order of scores within the distribution. In addition, we see that the 10-point difference between, for example, the 50th and 60th percentiles corresponds to a relatively small difference between IQ scores, but a 10-point difference between the 80th and 90th percentiles corresponds to a large difference between IQs. To put it another way, there is a greater difference in intellectual functioning between two individuals at the 80th and 90th percentiles than between individuals at the 50th and 60th percentiles. Thus, the interpretation of a 10-point difference between percentile ranks depends on where the difference is on the 0–100 scale. This problem doesn't occur with standard scores.

As we have seen, transforming scores to percentile ranks alters four characteristics of the distribution; a standard score transformation alters only two characteristics—central tendency and dispersion. Standard scores have the added advantage that they

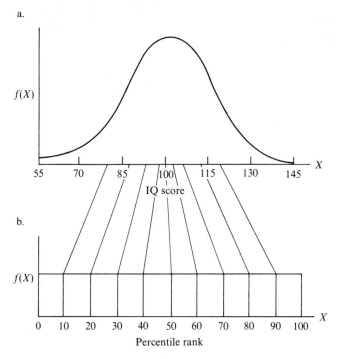

Figure 10.2-1. (a) Graph of distribution of IQ scores. (b) Graph of distribution of percentile ranks. The transformation of scores to percentile ranks alters the shape of the distribution.

can be manipulated arithmetically. For these reasons, those who use or develop psychological tests prefer standard scores over percentile ranks even though they are less familiar to the average person.

Other Kinds of Standard Scores

The standard scores we have described range approximately from -3 to $+3$ and have a mean of zero and a standard deviation of one. It is a minor inconvenience to have to deal with negative numbers, and fortunately this can be avoided. If a sufficiently large constant is added to each z score, all the z scores will be positive with a new mean equal to the constant. Similarly, if each z score is multiplied by a constant, the standard deviation is changed from one to the value of the constant. The formula

$$z' = \frac{(X - \bar{X})}{S} S' + \bar{X}'$$

is used to change the mean and standard deviation of z scores to any desired values, where z' is the transformed standard score, S' is the value of the desired standard deviation, and $\bar{X}'$ is the value of the desired mean.

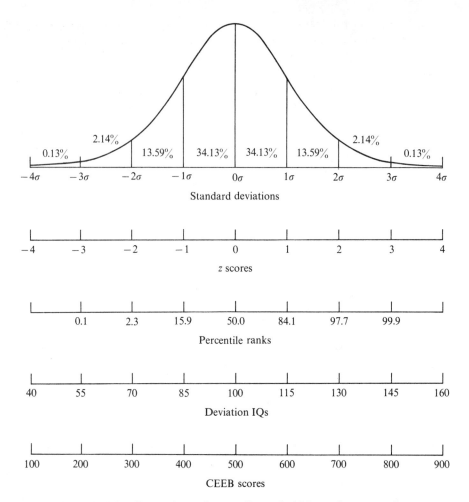

Figure 10.2-2. Comparison of percentiles and widely used systems of standard scores.

A surprising number of test scores are actually transformed z scores. Most IQ scores are really z scores that have been multiplied by 15 and then had 100 added to the product. The resulting z' scores have $\bar{X} = 100$ and $S = 15$. Other examples of common transformed standard scores are shown in Figure 10.2-2.

Comparing Performance on Different Tests

Randy got a raw score of 68 in arithmetic and a 42 in English. He seems to be doing better in arithmetic than in English, but is he really? His teacher converts the class's arithmetic and English scores to standard scores with a mean of 50 and a standard deviation of 10. A different picture of Randy's performance emerges; his arithmetic z score is 40, 1 standard deviation below the class mean, but his English score is 65, 1.5 standard deviations above the mean. So Randy is much better in English than in arithmetic, relative to others in his class. As this example illustrates, z scores are useful for determining an individual's strengths and weaknesses—that is, for making intra-individual comparisons. They permit us to compare performance on different tasks that are measured on different scales, as were Randy's arithmetic and English tests. For the comparisons to be meaningful it is necessary only that for both variables the z scores be based on the same or equivalent reference groups. Reference groups are equivalent with respect to a variable if their distributions have essentially the same mean, standard deviation, and shape. It wouldn't have been possible to compare Randy's z scores if he had been in an accelerated arithmetic class and a remedial English class.

Review Exercises for Section 10.2

15. The statement "Fred got a 29 on the quiz" is uninterpretable. Discuss.
16. Compare the relative merits of standard scores and percentile ranks for interpreting scores.
*17. Suppose three tests were given in your statistics course. The class means and standard deviations were as listed in the table. Your scores on tests 1, 2, and 3 were, respectively, 72, 61, and 63.

Test	μ	σ
1	56	11
2	44	17
3	53	8

On which test did you do your best and on which did you do your worst?

18. On a mechanical aptitude test, Bill scored 110 and Vena scored 85. The national mean for men is 104 with a standard deviation of 20. The comparable norms for women are 70 and 30. Which of the two did better?
*19. Your statistics professor returned the midterm exam and said the mean was 82 and the standard deviation was 14. The top 15% of test scores received an A. Your score was 99. Did you get an A?
*20. Suppose the mean of a test was 22 and the standard deviation was 5. Transform a score of 18 to standard scores with the following means and standard deviations.
  *a. $\bar{X} = 100, S = 15$
  *b. $\bar{X} = 50, S = 10$
  *c. $\bar{X} = 10, S = 2$
  d. $\bar{X} = 100, S = 16$
  e. $\bar{X} = 500, S = 100$
  f. $\bar{X} = 80, S = 10$

10.3 Sampling Distributions

Looking Ahead

So far we have covered descriptive statistics, probability, and probability distributions. These topics provide the necessary background for moving on to inferential statistics, the subject of the second half of this book. Inferential statistics are procedures for using sample data to make inferences about one or more population parameters. Two kinds of procedures are categorized under inferential statistics—estimation and hypothesis testing.

The term *estimation* is used in statistics in much the same way it is used in everyday language. A student might estimate that the mean grade point average of members of a sailing club is 2.9 or that it is between 2.7 and 3.1. We call the first type of estimate a *point estimate* because the one number representing the estimate may be associated with a point on a line. The second type, involving two numbers, is called an *interval estimate* because the two numbers and associated points define an interval on a line. *An **estimator** is a rule, usually in the form of a formula such as $\sum X_i/n$, that tells us how to calculate an estimate of a population parameter using sample information. The estimate is the numerical value that results from applying the rule to a sample.* The value of a point estimate will vary from one random sample to the next; hence, the value for a particular sample is likely to differ from the population parameter. Interval estimation is used in conjunction with point estimation to specify an interval that has a high likelihood of containing the parameter of interest. These procedures are discussed in Chapter 15.

The other approach to statistical inference, *hypothesis testing*, is similar in many respects to the scientific method. The scientist observes nature, formulates a hypothesis, and then proceeds to test the hypothesis by comparing its predictions with data. Similarly, hypothesis testing begins with a question about nature that leads to a hypothesis regarding the value of one or more population parameters. The experimenter obtains a sample from the population and compares the sample value to the hypothesized value. If the sample value is inconsistent with the hypothesis, the hypothesis is rejected; otherwise it isn't rejected. These procedures are discussed in Chapters 11–14.

In summary, estimation is concerned with getting a reasonable idea of the value of a parameter. Hypothesis testing is concerned with deciding whether a hypothesis is or isn't true. In estimation, the result is a number or an interval bounded by two numbers. In hypothesis testing, the result is a decision about a hypothesis.

Before turning to hypothesis testing, which is the subject of Chapter 11, we will lay a little more groundwork for statistical inference.

Introduction to Sampling Distributions

As we have learned, inferential statistics is concerned with reasoning from a sample to the population—from the particular to the general. Such reasoning is based on a knowledge of the sample-to-sample variability of a statistic—that is, on its sampling behavior. Before data have been collected we can speak of a sample statistic such as $\bar{X}$ in terms of probability. Its value is yet to be determined and will depend on which score values happen to be randomly selected from the population. Thus, at this stage of an investigation a sample statistic is a random variable, since it is computed from score values obtained by random sampling. *Like any random variable, a sample statistic has a probability distribution that gives the probability associated with each value of the statistic over all possible samples of the same size that could be drawn from the population. The distribution of a statistic is called a **sampling distribution** to distinguish it from a probability distribution for, say, a score value.* Sampling distributions play a key role in statistical inference, since they describe the sample-to-sample variability of statistics computed from random samples. In subsequent chapters we will use sampling distributions (1) to determine the tenability of the hypothesis that a population parameter is equal to a particular value and (2) to specify a range of values that has a high likelihood of including the parameter.

Sampling Distribution of the Mean

Some of the important characteristics of a sampling distribution will be introduced by an example that, though obviously unrealistic, has the virtue of allowing a concrete approach to the topic. Our discussion will focus on the sampling distribution of a mean, but the ideas we will develop apply to any sampling distribution. Suppose we have a discrete uniform (rectangular) population consisting of the four scores 1, 2, 3, and 4. A graph of the population is shown in Figure 10.3-1. The mean of the population is $\mu = \sum X_i/n = (1 + 2 + 3 + 4)/4 = 2.5$, and its standard deviation is

$$\sigma = \sqrt{\sum (X_i - \mu)^2/n}$$
$$= \sqrt{[(1 - 2.5)^2 + (2 - 2.5)^2 + (3 - 2.5)^2 + (4 - 2.5)^2]/4}$$
$$= 1.118.$$

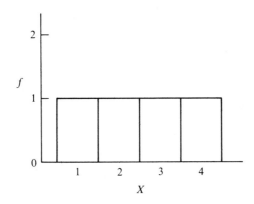

Figure 10.3-1. Histogram of a discrete uniform population.

If we draw all possible samples of size two with replacement, there are 16 different samples that can be drawn (see Table 10.3-1). This follows since the first element can be drawn in any one of four ways and the second in any one of four ways, making a total of $4 \times 4 = 16$ samples. The probability of drawing a particular sample is, according to the multiplication rule, $(\frac{1}{4})(\frac{1}{4}) = \frac{1}{16}$. The 16 equally likely samples and their means are given in Table 10.3-1. As shown in the table, the mean of the 16 means, denoted by $\mu_{\bar{X}}$, is equal to 2.5; the standard deviation of the means, denoted by $\sigma_{\bar{X}}$, is equal to 0.791. A chart depicting the sampling procedure along with a graph of the sampling distribution of the mean is presented in Figure 10.3-2, p. 205.

Table 10.3-1. All Possible Samples of Size Two from the Population in Figure 10.3-1

(i) Data:

Sample Number	Sample Values	$\bar{X}$	Sample Number	Sample Values	$\bar{X}$
1	1, 1	1.0	9	2, 3	2.5
2	1, 2	1.5	10	3, 2	2.5
3	2, 1	1.5	11	2, 4	3.0
4	1, 3	2.0	12	4, 2	3.0
5	3, 1	2.0	13	3, 3	3.0
6	1, 4	2.5	14	3, 4	3.5
7	4, 1	2.5	15	4, 3	3.5
8	2, 2	2.0	16	4, 4	4.0

(ii) Mean and standard deviation of means:

$$\mu_{\bar{X}} = \frac{\sum_{i=1}^{n} \bar{X}_i}{n} = \frac{1.0 + 1.5 + \cdots + 4.0}{16} = \frac{40}{16} = 2.5$$

$$\sigma_{\bar{X}} = \sqrt{\frac{\sum_{i=1}^{n}(\bar{X}_i - \mu)^2}{n}} = \sqrt{\frac{(1.0 - 2.5)^2 + (1.5 - 2.5)^2 + \cdots + (4.0 - 2.5)^2}{16}} = 0.791$$

Three characteristics of the sampling distribution are especially significant.

1. The distribution of the sample means doesn't resemble the original population (which was rectangular in appearance), but instead resembles the normal distribution. We could show that if the sample size were increased from $n = 2$ to $n = 3, 4, \ldots$, the number of possible $\bar{X}$ values would increase and the distribution of $\bar{X}$'s would resemble more and more closely the normal distribution.
2. The mean of the 16 sample means, $\mu_{\bar{X}} = 2.5$, equals the mean of the four score values in the population, $\mu = 2.5$.
3. The standard deviation of the 16 sample means, $\sigma_{\bar{X}} = 0.791$, is equal to the standard deviation of the four scores in the population divided by the square root of the sample size—that is, $\sigma_{\bar{X}} = \sigma/\sqrt{n} = 1.118/\sqrt{2} = 0.791$.

Several implications of the third point are easily overlooked. It says in effect that we can compute the standard deviation of sample means in two ways—from $\sqrt{\sum(\bar{X}_i - \mu)^2/n}$ or $\sigma/\sqrt{n}$. Since in practical situations the distribution of sample means isn't available, we rely on a knowledge of, or an estimate of, the population standard deviation and the formula $\sigma/\sqrt{n}$ in computing $\sigma_{\bar{X}}$. The formula $\sigma_{\bar{X}} = \sigma/\sqrt{n}$ also gives us a reason for having greater confidence in large samples. We know that the standard deviation of the population (σ) is a constant. Therefore, it follows from $\sigma_{\bar{X}} = \sigma/\sqrt{n}$ that as n (the sample size) increases, $\sigma_{\bar{X}}$ (the dispersion of sample means) decreases and hence the closer a randomly selected mean is likely to be to μ. In general, the larger the sample size, the more probable it is that the sample mean comes arbitrarily close to the population

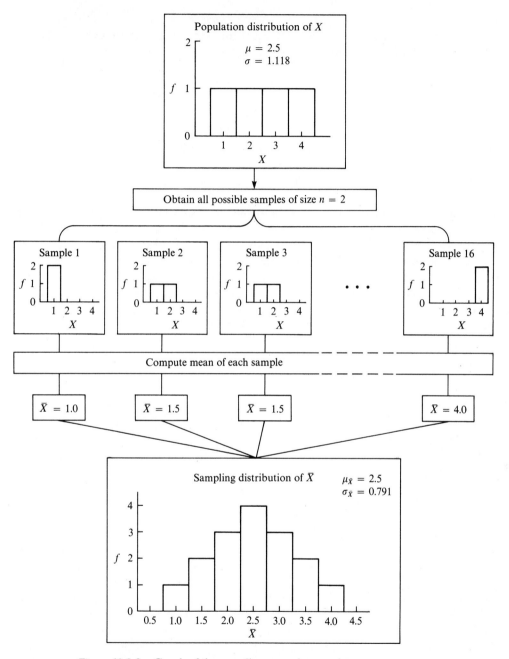

Figure 10.3-2. Graph of the sampling procedure used to construct a sampling distribution for samples of size $n = 2$ from a discrete, uniform population. Note that $\mu_{\bar{X}} = \mu = 2.5$ and that $\sigma_{\bar{X}} = \sqrt{\sum_{i=1}^{n}(\bar{X}_i - \mu)^2/n} = 0.791$ which is equal to $\sigma/\sqrt{n} = 1.118/\sqrt{2} = 0.791$.

mean. This fact, often called the *law of large numbers*, is one justification for using random samples to learn about populations. If the sample is large enough, the sample information is likely to be very accurate.

Central Limit Theorem

The characteristics of the sampling distribution of a mean are succinctly stated in the *central limit theorem*, one of the most important theorems in statistics. In one form the theorem states that *if random samples are selected from a population with mean μ and finite standard deviation σ, as the sample size n increases the distribution of $\bar{X}$ approaches a normal distribution with mean μ and standard deviation $\sigma/\sqrt{n}$.* Probably the most significant point is that regardless of the shape of the sampled population, the means of sufficiently large samples will be nearly normally distributed. Just how large is sufficiently large? This depends on the shape of the sampled population; the more a population departs from the normal form, the larger n must be. For most populations encountered in the behavioral sciences and education, a sample size of 100 is sufficient to produce a nearly normal sampling distribution of $\bar{X}$. The tendency for the sampling distributions of statistics to approach the normal distribution as n increases helps to explain why the normal distribution is so important in statistics.

Standard Error of a Statistic

We have used the term *standard deviation* to refer to a measure of dispersion both for scores in a frequency distribution and statistics in a sampling distribution. To avoid confusion in the future we will use the term *standard error* to denote the latter measure. The symbol for a standard error always includes a subscript indicating the statistic to which it applies, for example, $\sigma_{\bar{X}}$, σ_S, and σ_r. There are as many standard errors as there are sample statistics, but they are all interpreted analogously to a standard deviation.

Two Properties of Good Estimators

As we have discussed, the value of sample statistics varies from one random sample to the next. Consequently, it is unlikely that a given statistic will equal the population parameter it is used to estimate. This is a little frustrating, but it is something we have to live with. However, we can require that the mean of the distribution of estimates yielded by an estimator must equal the parameter it estimates and that the estimates must vary from one random sample to the next as little as possible. Statistics that satisfy these two intuitively reasonable requirements are said to be unbiased estimators and minimum variance estimators. *More formally, an estimator $\hat{\theta}$ is an* **unbiased estimator** *for the parameter θ if $E(\hat{\theta}) = \theta$. An estimator $\hat{\theta}$ is a* **minimum variance estimator** *for the parameter θ if the variance of $\hat{\theta}$ (Var $\hat{\theta}$) is smaller than that for any other possible unbiased estimator of θ.* The sample mean is a good estimator for μ, because it satisfies both these requirements; $E(\bar{X}) = \mu$ and Var($\bar{X}$) is a minimum. It can be shown that the sample median is also an unbiased estimator for μ for normally distributed populations, but it is not a minimum variance estimator. This can be seen by comparing the variance error of the median with that for the mean: $\mathrm{Var}(Mdn) = 1.57\sigma^2/n > \mathrm{Var}(\bar{X}) = \sigma^2/n$. This confirms what we stated in Section 3.5, namely, that the sample mean is more stable than the sample median. The sample variance S^2 isn't a good estimator for σ^2, since $E(S^2) \neq \sigma^2$. Instead, we use $\hat{\sigma}^2 = \sum (X_i - \bar{X})^2/(n-1)$ to estimate σ^2, because $E(\hat{\sigma}^2) = \sigma^2$.[1]

Test Statistics

The statistics we have presented thus far—$\bar{X}$, Mdn, S, and so on—are useful for describing samples. If they are computed from a random sample, they can also be used to estimate population parameters, although, as we have seen, some are better for this purpose than others. In subsequent chapters we will introduce a new kind of statistic that is used for testing hypotheses about the values of population parameters. These statistics are called *test statistics*. Consider a test statistic used to test a hypothesis about a population mean μ. Its formula is

$$z = \frac{\bar{X} - \mu_0}{\sigma_{\bar{X}}},$$

where $\bar{X}$ is the mean of a random sample, μ_0 is the hypothesized value of the population mean, and $\sigma_{\bar{X}}$ is the standard error of the mean. If the sampled population is normal or the sample size is sufficiently large, it is possible to specify the sampling distribution of the z test statistic—it is the standard normal distribution. The similarity in appearance between this z test statistic and a z score ($z = (X - \mu)/\sigma$) is obvious. In either case, z is obtained by subtracting the mean (μ or the hypothesized value of the mean, μ_0) from the statistic and dividing this difference by the standard deviation of the statistic (σ in the case of X, and $\sigma_{\bar{X}}$ in the case of $\bar{X}$). Other test statistics that will be introduced in later chapters include $t = (\bar{X} - \mu_0)/\hat{\sigma}_{\bar{X}}$ (also used to test a hypothesis about μ), $\chi^2 = (n-1)\hat{\sigma}^2/\sigma_0^2$ (used to test a hypothesis about σ^2), and $F = \hat{\sigma}_1^2/\hat{\sigma}_2^2$ (used to test the hypothesis that $\sigma_1^2 = \sigma_2^2$).

Review Exercises for Section 10.3

21. Distinguish a sampling distribution from a sample (frequency) distribution.
*22. A population consists of four scores: 0, 1, 2, and 3. (a) List the $(4)(4) = 16$ samples of size two that can be drawn with replacement from the population. (b) Compute the mean and standard error of the mean using the formulas $\mu = \sum X_i/n$ and $\sigma_{\bar{X}} = \sigma/\sqrt{n}$, where $\sigma = \sqrt{\sum(X_i - \mu)^2/n}$. (c) Compute the mean and standard error of the mean using the formulas $\mu_{\bar{X}} = \sum \bar{X}_i/n$ and $\sigma_{\bar{X}} = \sqrt{\sum(\bar{X}_i - \mu)^2/n}$.
23. For the population in Exercise 22, (a) list the $_4C_2 = 6$ samples of size two that can be drawn without replacement. (b) Compute the mean and standard error of the mean using the formulas $\mu = \sum X_i/n$ and $\sigma_{\bar{X}} = \sigma/\sqrt{n}$, where $\sigma = \sqrt{\sum(X_i - \mu)^2/n}$. (c) Compute the mean and standard

---

[1] A demonstration of this is given in Technical Note 10.5-1.

error of the mean using the formulas $\mu_{\bar{X}} = \sum \bar{X}_i/n$ and $\sigma_{\bar{X}} = \sqrt{[\sum (\bar{X}_i - \mu)^2]/n}$. The means computed from the two formulas should be equal, but the standard error computed from $\sigma/\sqrt{n}$ overestimates the true value, since it assumes an infinite population and/or sampling with replacement. A correction for a finite population or when sampling without replacement can be made: $\sigma_{\bar{X}} = (\sigma/\sqrt{n})\sqrt{(n_p - n)/(n_p - 1)}$, where n_p is the number of scores in the population and n is the number in the sample. (d) Apply the correction to $\sigma/\sqrt{n}$.

24. Distinguish a standard error from a standard deviation.
25. How is the dispersion of the sampling distribution of $\bar{X}$ related to σ and n?
*26. A sample of size n is to be drawn from a population with a mean of 100 and a standard deviation of 10. Complete the table.

	n	$\sigma_{\bar{X}}$		n	$\sigma_{\bar{X}}$
*a.	2		e.	32	
*b.	4		f.	64	
c.	8		g.	128	
d.	16		h.	256	

*27. The registrar claims the mean IQ of students at a university (μ_0) is 120, with a standard deviation (σ) of 10. You obtain a random sample of 25 students and find their mean ($\bar{X}$) is 115. What is the probability of obtaining a mean of 115 or lower if the true mean is 120? (*Hint*: Transform $\bar{X}$ to a z score and use the standard normal distribution to find the area below 115.)

*28. An elevator has a maximum safe load of 1638 lb. If mens' weights are approximately normally distributed with a mean of 165 lb and a standard deviation of 15 lb, what is the probability that nine men (whose weights can be assumed to be independent) will overload the elevator?

29. Terms to remember:
    a. Point estimate
    b. Interval estimate
    c. Estimator
    d. Sampling distribution
    e. Law of large numbers
    f. Central limit theorem
    g. Standard error
    h. Unbiased estimator
    i. Minimum variance estimator
    j. Test statistic

10.4 Summary

Two theoretical distributions provide a bridge between descriptive and inferential statistics—a probability distribution and its close relative, a sampling distribution. A probability distribution associates a probability with each value of a random variable. If the random variable is some function of two or more population elements, say, a mean or a standard deviation, a probability distribution is called a *sampling distribution*.

The normal distribution is the most widely applicable theoretical model in statistics. It provides an excellent approximation to the binomial distribution and other theoretical distributions with probabilities that are laborious to work out when n is large. In addition, it serves as a model for the many variables in science and nature that are approximately normally distributed. But its most important use is as a model for the sampling distribution of statistics based on large n's. According to the central limit theorem, as the sample size n increases, the distribution of $\bar{X}$'s from random samples approaches a normal distribution with mean μ and standard deviation $\sigma/\sqrt{n}$, regardless of the shape of the original population.

The normal distribution is actually a family of distributions, one for each possible combination of μ and σ. The distribution with $\mu = 0$ and $\sigma = 1$ is called the *standard normal curve* and has the areas under the curve given in Table D.2. To use the standard normal table a score is transformed into a standard (z) score by the formula $z = (X - \mu)/\sigma$. The transformation doesn't affect the shape of the original distribution but does change its mean and standard deviation to 0 and 1, respectively. Standard scores are widely used for reporting psychological test scores, since they contain all the information necessary to interpret a score.

An important new measure of dispersion was introduced in this chapter—the standard error, which is the standard deviation of a statistic. It is a random variable that has been computed from two or more population elements. The standard error describes the dispersion of a statistic over all possible samples of the same size. It is denoted by σ with a subscript identifying the statistic, for example, $\sigma_{\bar{X}}$ denotes the standard error of a mean. In the following chapters we will see how the elements—standard error, sampling distribution, and test statistic—are used in inferential statistics.

†10.5 Technical Note

10.5-1 Demonstration Showing that $E(\hat{\sigma})^2 = \sigma^2$ and $E(\hat{\sigma}_{est}^2) = \sigma^2$ but $E(S)^2 \neq \sigma^2$

In Section 10.3 (p. 203) we drew all possible samples of size two with replacement from a finite population in order to show that the standard error of the mean, $\sigma_{\bar{X}}$, is equal to the standard deviation of the population divided by the square root of the sample size, $\sigma/\sqrt{n}$. In this technical note we will use the same sampling procedure and data to show that $E(\hat{\sigma})^2 = \sigma^2$ and $E(S^2) \neq \sigma^2$, which means that $\hat{\sigma}^2$ is an unbiased estimator for the parameter σ^2 but S^2 is a biased estimator. The values of $\hat{\sigma}^2$ and S^2 are shown in Table 10.5-1 and are based on the population in Figure 10.3-1 and the random samples in Table 10.3-1. For the moment we will ignore σ_{est}^2 in column 6 of Table 10.5-1. Since $\hat{\sigma}^2$ is a discrete random variable that assumes values $\hat{\sigma}_1^2, \hat{\sigma}_2^2, \ldots, \hat{\sigma}_n^2$ with probabilities $p(\hat{\sigma}_1^2), p(\hat{\sigma}_2^2), \ldots, p(\hat{\sigma}_n^2)$, the expected value of $\hat{\sigma}^2$ is given by $E(\hat{\sigma}^2) = \sum_{i=1}^{n} p(\hat{\sigma}_i^2) \hat{\sigma}_i^2$ (see pp. 178–179). Similarly, the expected value of S^2 is given by $E(S^2) = \sum_{i=1}^{n} p(S_i^2) S_i^2$. It can be seen from the computations in Table 10.5-1 (part ii) that $\hat{\sigma}^2$ is an unbiased estimator for σ^2 since $E(\hat{\sigma}^2) = 1.25 = \sigma^2$; however, $E(S^2) = 0.625 < \sigma^2$, which means S^2 is a biased estimator for σ^2. Thus, dividing $\Sigma(X_i - \bar{X})^2$ by $n - 1$ instead of by n provides an unbiased estimator of the population variance.

---

† This technical note can be skipped without loss of continuity.

Table 10.5-1. Computation of $\hat{\sigma}^2$, S^2, and σ^2_{est} for All Possible Samples of Size Two from the Population in Figure 10.3-1 (For this population, $\mu = 2.5$ and $\sigma^2 = 1.25$.)

(i) Data:

(1) Sample Number	(2) Sample Values	(3) $\bar{X}$	(4) $\hat{\sigma}^2 = \dfrac{\sum_{i=1}^{n}(X_i - \bar{X})^2}{n-1}$	(5) $S^2 = \dfrac{\sum_{i=1}^{n}(X_i - \bar{X})^2}{n}$	(6) $\sigma^2_{est} = \dfrac{\sum_{i=1}^{n}(X_i - \mu)^2}{n}$
1	1, 1	1.0	0.0	0.00	2.25
2	1, 2	1.5	0.5	0.25	1.25
3	2, 1	1.5	0.5	0.25	1.25
4	1, 3	2.0	2.0	1.00	1.25
5	3, 1	2.0	2.0	1.00	1.25
6	1, 4	2.5	4.5	2.25	2.25
7	4, 1	2.5	4.5	2.25	2.25
8	2, 2	2.0	0.0	0.00	0.25
9	2, 3	2.5	0.5	0.25	0.25
10	3, 2	2.5	0.5	0.25	0.25
11	2, 4	3.0	2.0	1.00	1.25
12	4, 2	3.0	2.0	1.00	1.25
13	3, 3	3.0	0.0	0.00	0.25
14	3, 4	3.5	0.5	0.25	1.25
15	4, 3	3.5	0.5	0.25	1.25
16	4, 4	4.0	0.0	0.00	2.25

(ii) Computation of expected value:

$$p(\hat{\sigma}_i^2) = p(S_i^2) = p(\sigma^2_{est\,i}) = \tfrac{1}{16} = 0.0625$$

$$E(\hat{\sigma}^2) = p(\hat{\sigma}_1^2)\hat{\sigma}_1^2 + p(\hat{\sigma}_2^2)\hat{\sigma}_2^2 + \cdots + p(\hat{\sigma}_n^2)\hat{\sigma}_n^2$$
$$= 0.0625(0) + 0.0625(0.5) + \cdots + 0.0625(0) = 1.25$$

$$E(S^2) = p(S_1^2)S_1^2 + p(S_2^2)S_2^2 + \cdots + p(S_n^2)S_n^2$$
$$= 0.0625(0) + 0.0625(0.25) + \cdots + 0.0625(0) = 0.625$$

$$E(\sigma^2_{est}) = p(\sigma^2_{est\,1})\sigma^2_{est\,1} + p(\sigma^2_{est\,2})\sigma^2_{est\,2} + \cdots + p(\sigma^2_{est\,n})\sigma^2_{est\,n}$$
$$= 0.0625(2.25) + 0.0625(1.25) + \cdots + 0.0625(2.25) = 1.25$$

(iii) Computation of variance:

$$\text{Var}(\hat{\sigma}^2) = \sum_{i=1}^{n} p(\hat{\sigma}_i^2)[\hat{\sigma}_i^2 - E(\hat{\sigma}^2)]^2$$
$$= 0.0625(0 - 1.25)^2 + 0.0625(0.5 - 1.25)^2 + \cdots + 0.0625(0 - 1.25)^2$$
$$= 2.0625$$

$$\text{Var}(\sigma^2_{est}) = \sum_{i=1}^{n} p(\sigma^2_{est\,i})[\sigma^2_{est\,i} - E(\sigma^2_{est})]^2$$
$$= 0.0625(2.25 - 1.25)^2 + 0.0625(1.25 - 1.25)^2 + \cdots + 0.0625(2.25 - 1.25)^2$$
$$= 0.5000$$

A second unbiased estimator for the population variance is σ_{est}^2, where $\sum (X_i - \mu)^2$ is divided by n. According to Table 10.5-1 (part ii), $E(\sigma_{est}^2) = 1.25 = \sigma^2$, which means that σ_{est}^2, like $\hat{\sigma}^2$, is an unbiased estimator. It is also a better estimator for σ^2 than $\hat{\sigma}^2$, since σ_{est}^2 varies less from sample to sample. This is shown in part iii of the table: $\text{Var}(\sigma_{est}^2) = 0.5000 < \text{Var}(\hat{\sigma}^2) = 2.0625$.

To compute σ_{est}^2 we need to know μ, one of the parameters of the population. Since μ is rarely known in real-life situations, we rely instead on $\hat{\sigma}^2$. In computing $\hat{\sigma}^2$ we use $\bar{X}$ in place of μ; in effect, $\bar{X}$ is used to estimate the unknown population parameter. As a consequence, $\sum (X_i - \bar{X})^2$ must be divided by $n - 1$ (1 is the number of parameters estimated in the computation) instead of by n to obtain an unbiased estimator of σ^2.

11

Statistical Inference: One Sample

11.1 Introduction to Hypothesis Testing
  Scientific Hypotheses
  Why Statistical Inference?
  Statistical Hypotheses
  Hypothesis Testing and the Method of Indirect Proof
  Rejection or Nonrejection of H_0—What Does It Mean?
  The Role of Logic in Evaluating a Scientific Hypothesis
  Review Exercises for Section 11.1
11.2 Hypothesis Testing
  Step 1: Stating the Statistical Hypotheses
  Step 2: Specifying the Test Statistic
  Step 3: Specifying n and the Sampling Distribution
  Step 4: Specifying α
  Step 5: Making a Decision
  Review Exercises for Section 11.2
11.3 One-Sample z Test for μ When σ^2 Is Known
  Review Exercises for Section 11.3
11.4 More about Hypothesis Testing
  One-Tailed and Two-Tailed Tests
  Exact versus Inexact Hypotheses
  Type I and Type II Errors
  Determining the n Required to Achieve an Acceptable α, $1 - \beta$, and $\mu - \mu_0$
  More about Type I and Type II Errors
  Review Exercises for Section 11.4
11.5 Summary
11.6 Technical Note
  Derivation of Formula for Estimating n

11.1 Introduction to Hypothesis Testing

Evaluating the effectiveness of a new teaching technology or assessing consumer acceptance of a seaweed substitute for meat involves making a decision on the basis of incomplete information. The experimenter's information is usually incomplete because it is impossible or impractical to observe all the people in the population of interest. Fortunately, there are procedures for making rational decisions about nature that use a sample containing only a small portion of the elements in the population. These procedures, called *statistical inference*, are the subject of this and subsequent chapters.

There are several approaches to decision-making that use information from a sample, but we will limit our discussion to classical statistical inference,[1] which evolved from the work of R. A. Fisher and, more directly, Jerzy Neyman and Egon Pearson.[2] Two complementary topics are categorized under classical statistical inference—hypothesis testing and interval estimation. We will describe hypothesis testing procedures in Chapters 11–14 and take up interval estimation in Chapter 15.

Scientific Hypotheses

People are by nature inquisitive. We ask questions, develop hunches, and sometimes put our hunches to the test. Over the years a formalized procedure for testing hunches has evolved—the scientific method. It involves (1) observing nature, (2) asking questions, (3) formulating hypotheses, (4) conducting experiments, and (5) developing theories and laws. Let's examine in detail the third characteristic, formulating hypotheses.

*A **scientific hypothesis** is a testable supposition that is tentatively adopted to account for certain facts and to guide in the investigation of others. It is a statement about nature that requires verification.* Some examples of scientific hypotheses are the following: The child-rearing practices of parents affect the personalities of their offspring. Student radicals have higher IQs than the average college student. Cigarette smoking is associated with high blood pressure. Children who feel insecure engage in overt aggression more frequently than children who feel secure. These hypotheses have three characteristics in common with all scientific hypotheses: (1) They are intelligent, informed guesses about phenomena of interest. (2) They can be stated in the *if–then* form of an implication, for example, "*if* John smokes, *then* he will show signs of high blood pressure." (3) Their truth or falsity can be determined by observation and experimentation. Many interesting hypotheses don't qualify as scientific hypotheses because they aren't testable by recourse to experience. Such questions as "Can three or more angels sit on the head of a pin?" and "Does life exist in more than one galaxy in the universe?" can't be investigated because no procedures presently exist for observing angels or other galaxies. This doesn't mean that the question concerning the existence of life in other galaxies can never be investigated. Indeed, with continuing advances in space science it is likely that this question eventually will be answered.

---

[1] Other less widely used approaches are the likelihood function and Bayesian decision theory. For a discussion of these approaches see Kirk (1972a, chap. 10).

[2] The major contributions of Fisher, Neyman, and Pearson to statistical theory are summarized by Dudycha and Dudycha (1972, pp. 21–24).

Why Statistical Inference?

We have said that statistical inference is a form of reasoning whereby rational decisions about states of nature can be made on the basis of incomplete information. Rational decisions often can be made without resorting to statistical inference, as when a scientific hypothesis concerns some limited phenomenon that is directly observable, for example, "this rat is running." The truth or falsity of the hypothesis can be determined by observing the rat. Many scientific hypotheses, on the other hand, refer to phenomena that can't be directly observed—that is, to populations with elements that are so numerous it is impossible or impractical to view them all, for example, "all rats run under condition X." It is impossible to observe the entire population of rats under condition X. Likewise, it is impossible to observe all parents rearing their children, all student radicals, all smokers, or all insecure children. In such cases, recourse to statistical inference is necessary to make a reasonable decision concerning the probable truth or falsity of the scientific hypothesis.

Statistical Hypotheses

Scientific hypotheses are statements about phenomena of nature and man, and are normally stated in fairly general terms—at least in the initial stages of an inquiry. As we have seen, if a scientific hypothesis can't be evaluated directly by observing all members of a population, it may be possible to evaluate the hypothesis indirectly by statistical inference. To do this, an experimenter must express a scientific hypothesis in the form of a *statistical hypothesis*. *A **statistical hypothesis** is a statement about one or more parameters of a population distribution that requires verification.* For example, $\mu > 115$ is a statistical hypothesis; it states that the population mean is greater than 115. Another statistical hypothesis can be formulated that states the mean is equal to or less than 115—that is, $\mu \leq 115$. These hypotheses, $\mu \leq 115$ and $\mu > 115$, are mutually exclusive and exhaustive; if one is true, the other must be false. They are examples, respectively, of the *null hypothesis* H_0 and the *alternative hypothesis* H_1. The null hypothesis is the one whose tenability is actually tested. If on the basis of this test the null hypothesis is rejected, only the alternative hypothesis remains tenable. According to convention, the alternative hypothesis is formulated so that it corresponds to the experimenter's hunch.[3] The process of choosing between H_0 and H_1 is called *hypothesis testing*.

An example may help to explicate the nature of H_0 and H_1. Consider the scientific hypothesis that campus radicals are more intelligent than the average college student. Let's assume the population mean of college students' IQs, denoted by μ_0, is known to equal 115. (This information could come from college entrance examinations.) The population mean of campus radicals' IQs, denoted by μ, is unknown but can be estimated from a random sample. The null hypothesis, which is contrary to what we believe to be true about μ, is $H_0: \mu \leq 115$. This is read "the null hypothesis states that the population mean of campus radicals' IQs is equal to or less than 115." The alterna-

---

[3] The merits of this convention have been extensively debated by Binder (1963), Edwards (1965), Grant (1962), Wilson and Miller (1964), and Wilson, Miller, and Lower (1967). Articles setting forth the basic issues in the controversy are contained in Kirk (1972a, chap. 4).

tive hypothesis is $H_1: \mu > 115$. We have followed the convention of equating the alternative hypothesis with the situation we believe to be true—that student radicals have higher IQs than the average college student. In this example the scientific hypothesis and its negation are expressed as two mutually exclusive and exhaustive statistical hypotheses concerning the value of μ, the population mean of campus radicals.

Sometimes an experimenter's scientific hypothesis as originally formulated is identical in meaning to the alternative statistical hypothesis. This is most likely to occur in research areas that have been extensively investigated. In such cases it is necessary only to formulate an appropriate null hypothesis.

Hypothesis Testing and the Method of Indirect Proof

The reader may marvel at the circuitous procedure whereby an experimenter tests a null hypothesis that is believed to be untrue in the hope of rejecting it and thereby accepting the alternative hypothesis that is believed to be true. On reflection, you may recall a similar technique taught in plane geometry and algebra—the method of indirect proof. This method consists of listing all possible answers or solutions to a problem and showing that all save one are contrary to known fact and lead to an absurdity. By a process of elimination the one that doesn't lead to an absurdity must be true. The success of the method of indirect proof depends upon listing all possibilities and finding a contradiction for all but one. The comparable procedure in hypothesis testing consists of formulating H_0 and H_1 so that they exhaust all possibilities concerning a population parameter. A sample is obtained from the population and an appropriate statistic is computed from the sample. If it is highly improbable that the obtained value of the statistic would have occurred if the null hypothesis were true, then the null hypothesis must be considered a poor prediction of the parameter and should be rejected in favor of the alternative hypothesis.

There is one important difference between the method of indirect proof and hypothesis testing. In indirect proof a possibility is rejected only if it is found to lead to a contradiction to known fact. In hypothesis testing the null hypothesis is rejected if the obtained value of a sample statistic would have been unlikely if the null hypothesis were indeed true. Therefore, hypothesis testing, unlike indirect proof, doesn't provide incontrovertible proof, since the null hypothesis is rejected because of the occurrence of an event that is improbable rather than impossible.

Rejection or Nonrejection of H_0—What Does It Mean?

If the null hypothesis isn't rejected, what conclusion can be drawn? Is the null hypothesis true? Not necessarily, since there are always two possibilities when the null hypothesis isn't rejected: (1) H_0 is true and shouldn't be rejected and (2) H_0 is false and should be rejected but the particular sample that was obtained belies this fact or the experimental methodology is not sufficiently sensitive to detect the true situation. An experimental methodology can lack sensitivity because the size of the sample is too small, the measuring procedure results in large random and/or systematic errors, and so on. Even if the sample size, measurement precision, and so forth are known to be adequate, there is always the possibility that one has obtained an unrepresentative sample. We know, for example, that a fair coin will, on occasion, produce ten or even 100 consecutive heads. If the null hypothesis isn't rejected, the experimenter has two options: state that he or she failed to reject the null hypothesis, in which case it remains credible; or suspend judgment about the null and scientific hypotheses pending completion of a new, improved experiment.

On the other hand, what does it mean if the null hypothesis is rejected? The experimenter can conclude that the alternative and scientific hypotheses are probably true. Here, too, there is always the possibility that one's sample isn't representative, but, as we will see later, the probability of erroneously rejecting a true null hypothesis is determined by the experimenter and can be made as small as desired.

The Role of Logic in Evaluating a Scientific Hypothesis

Above, we have dealt with the evaluation of statistical hypotheses. We turn now to the experimenter's ultimate objective—evaluation of the scientific hypothesis. This evaluation involves a chain of deductive and inductive logic that begins and ends with the scientific hypothesis. The chain is diagrammed in Figure 11.1-1. First, by means of deductive logic the scientific hypothesis and its negation are expressed as two mutually exclusive and exhaustive statistical hypotheses that make predictions concerning a population parameter. These predictions, denoted by H_0 and H_1, are made about the population mean, median, variance, and so on. If, as is usually the case, all the elements in the population can't be observed, a random sample is obtained from the population. The sample provides an estimate of the unknown population parameter.

*The process of deciding whether to reject the null hypothesis is called a **statistical test**. The decision is based on (1) a test statistic computed from a random sample from the population, (2) hypothesis testing conventions, and (3) a decision rule.* These are described in subsequent sections. The outcome of the statistical test is the basis for the final link in the chain shown in Figure 11.1-1, an inductive inference concerning the probable truth or falsity of the scientific hypothesis.

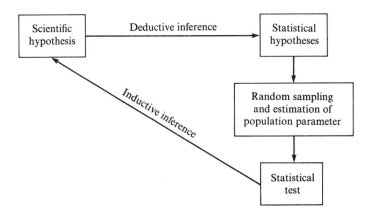

Figure 11.1-1. The evaluation of a scientific hypothesis using deductive and inductive logic.

Logic therefore plays a key role in hypothesis testing. It is the basis for arriving at both the statistical hypothesis that is tested and the final decision regarding the scientific hypothesis. If errors occur in the deductive or inductive links in the chain of logic, the statistical hypothesis subjected to test may have little or no bearing on the original scientific hypothesis, and/or the inference concerning the scientific hypothesis may be incorrect. Consider the scientific hypothesis that cigarette smoking is associated with high blood pressure. If this hypothesis is true, a measure of central tendency such as mean blood pressure should be higher for the population of smokers than for non-smokers. The statistical hypotheses are

$$H_0: \mu_1 - \mu_2 \leq 0$$
$$H_1: \mu_1 - \mu_2 > 0,$$

where μ_1 and μ_2 designate unknown population means for smokers and nonsmokers, respectively. The null hypothesis H_0 states in effect that the mean blood pressure of smokers is equal to or less than that of nonsmokers; H_1 states that the mean blood pressure of smokers is greater than that of nonsmokers. These hypotheses follow logically from the original scientific hypothesis. Suppose the experimenter formulated the statistical hypotheses in terms of population variances, for example,

$$H_0: \sigma_1^2 - \sigma_2^2 \leq 0$$
$$H_1: \sigma_1^2 - \sigma_2^2 > 0,$$

where σ_1^2 and σ_2^2 denote the population variances of smokers and nonsmokers, respectively. A statistical test of this null hypothesis, which states in effect that the variance of blood pressure for the population of smokers is equal to or less than the variance for nonsmokers, would have little bearing on the original scientific hypothesis. However, it would be relevant if the experimenter were interested in determining whether the two populations differed in dispersion.

The reader should not infer that for any scientific hypothesis there is only one pertinent null hypothesis. A null hypothesis stating that the correlation between cigarette smoking and blood pressure is zero bears more directly on the scientific hypothesis than the one involving population means. If cigarette smoking is associated with high blood pressure, there should be a positive correlation, $\rho > 0$, between cigarette consumption and blood pressure. The statistical hypotheses are

$$H_0: \rho \leq 0$$
$$H_1: \rho > 0,$$

where ρ denotes the population correlation coefficient for cigarette consumption and blood pressure. So we see that both creativity and deductive skill are required to formulate pertinent statistical hypotheses.

Review Exercises for Section 11.1

*1. Which of the following are scientific hypotheses?
  a. Right-handed people tend to be taller than left-handed people.
  b. Behavior therapy is more effective than hypnosis in helping smokers kick the habit.
  c. Most clairvoyant people are able to communicate with beings from outer space.
  d. Rats are likely to fixate an incorrect response if it is followed by an intense noxious stimulus.

2. Why might it be necessary to use the techniques of statistical inference in evaluating a scientific hypothesis?
*3. Which of the following are examples of statistical hypotheses?
   a. H_0: $\mu = 100$ b. H_0: $S^2 \leq 50$
   c. H_1: $\rho > 0$ d. H_0: $\bar{X} \geq 100$
   e. H_1: $\sigma^2 > 0$ f. H_1: $\bar{X} < 15$
   g. H_0: $\mu \geq 60$ h. H_0: $r = 0$
   i. H_0: $\sigma^2 = 225$ j. H_0: $\rho = 0$
4. a. According to convention, which statistical hypothesis corresponds to the experimenter's scientific hunch?
   b. Which is the hypothesis that actually is tested?
5. Why might an experimenter fail to reject H_0?
6. If H_0 is rejected, what does this imply about the experimental methodology?
*7. Assume an experimenter has a hunch that insecure children engage in overt aggression more frequently than children who feel secure. Let μ and μ_0 denote the mean daily number of aggressive acts, respectively, of insecure and secure children, where it is known that $\mu_0 = 8$. State H_0 and H_1 for the experiment.
8. Terms to remember:
   a. Scientific hypothesis
   b. Statistical inference
   c. Statistical hypothesis
   d. Null hypothesis
   e. Alternative hypothesis
   f. Statistical test

11.2 Hypothesis Testing

We will now describe the procedures for testing statistical hypotheses. For the sake of clarity they are stated in five steps. This should not suggest that hypothesis testing is a formal or a rigid procedure—it isn't. However, as the experimenter plans research, each of the items in the following steps must be considered.

Step 1. State the null and alternative hypotheses.
Step 2. Specify the test statistic.
Step 3. Specify the size n of the sample to be obtained and make assumptions that permit specification of the sampling distribution of the test statistic given that H_0 is true.
Step 4. Specify an acceptable risk of rejecting H_0 when it is true.
Step 5. Obtain a random sample of size n from the population, compute the test statistic, and make a decision about the null and alternative hypotheses.

Rejection of H_0 leads to the inductive inference that the scientific hypothesis is true. These five steps and the conventions they summarize require explication.

Step 1: Stating the Statistical Hypotheses

Let's assume we are interested in testing the scientific hypothesis "student radicals have higher IQs than the average college student." The corresponding statistical hypothesis is $\mu > \mu_0$, where μ denotes the unknown population mean of student radicals and μ_0 denotes the population mean of college students. Assume also that μ_0 and the population standard deviation, σ, of college students are known to equal 115 and 15, respectively. The first step is to state the null and alternative hypotheses,

$$H_0: \quad \mu \leq 115$$
$$H_1: \quad \mu > 115,$$

where μ_0 has been replaced by 115, the known population mean of college students. As written, the null hypothesis is inexact because it states a whole range of possible values for the population mean. However, one exact value is specified—$\mu = 115$—and that is the value actually tested. If the null hypothesis $\mu = 115$ can be rejected, then the hypothesis $\mu < 115$ is automatically rejected. Obviously, if $\mu = 115$ is considered improbable because the sample statistic exceeds 115, then any $\mu < 115$ would be considered even less probable.

Step 2: Specifying the Test Statistic

A relatively small number of test statistics are used to evaluate hypotheses about population parameters. The principal ones are denoted by z, t, χ^2, and F. A test statistic is called a *z statistic* if its sampling distribution is the normal distribution; a test statistic is called a *t statistic* if its sampling distribution is a t distribution, and so forth. As we will see, the choice of a test statistic is determined by (1) the hypothesis to be tested, (2) the information about the population that is known, and (3) the assumptions about the population that appear to be tenable. Which test statistic should be used to test the hypothesis $H_0: \mu \leq 115$? Since the hypothesis concerns the mean of a single population, the population standard deviation is known, and the population is assumed to be approximately normal, the appropriate test statistic is

$$z = \frac{\bar{X} - \mu_0}{\sigma/\sqrt{n}}.$$

Step 3: Specifying n and the Sampling Distribution

In specifying the sample size, we must keep in mind that to use the standard normal distribution Table D.2, the sampling distribution of our z test statistic must be normal. The sampling distribution of z will be normal if the sampling distribution of $\bar{X}$ is normal, since z is a linear transformation of $\bar{X}$; that is, $z = a + b\bar{X}$ where $a = -\mu_0/(\sigma/\sqrt{n})$ and $b = 1/(\sigma/\sqrt{n})$. The sampling distribution of $\bar{X}$ will be normal if the population distribution of X is normal or if the sample is fairly large, say $n \geq 100$. This follows from the central limit theorem discussed in Section 10.3. In the case of campus radicals' IQs, the population distribution is probably nearly normal, since college students' IQs are approximately normally distributed, but to be on the safe side we will rely on a large sample and obtain an n of 100. Later, we will describe a more rational way to determine the value of n. For practical reasons (limitations on the availability of research funds and time, and the difficulty of securing subjects) we don't want to specify a larger n than is needed.

We can specify the sampling distributions of the random variables $\bar{X}$ and z given that H_0 is true on the basis of the information (1) $\mu_0 = 115$, (2) $\sigma = 15$, and (3) $n = 100$; and the assumptions (1)[4] $\mu = 115$ and (2) the population distribution of X is approximately normal. The sampling distribution of $\bar{X}$ will be normal with mean equal to 115 and standard error equal to $\sigma/\sqrt{n} = 15/\sqrt{100} = 1.5$. Because z is simply a linear transformation of $\bar{X}$, the sampling distribution of z will also be normal with (as we saw in Chapter 10) mean equal to zero and standard deviation equal to one. We have now carried out the third step in testing the null hypothesis; we have specified the sample size, n, and sampling distribution of the z test statistic.

Step 4: Specifying α

If we decided that $\mu > 115$ when in fact $\mu \leq 115$, we would make a decision error. The fourth step is to specify an acceptable risk of making this kind of error—that is, rejecting H_0 when it is true. We will touch on this subject here and return to it later.

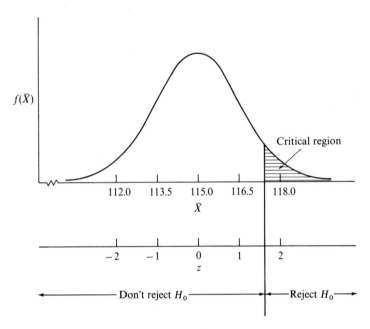

Figure 11.2-1. Sampling distribution of $\bar{X}$ and z given that H_0 is true. The critical region, which corresponds in this example to the upper 0.05 portion of the sampling distribution, defines values of $\bar{X}$ and z that are improbable if the null hypothesis, $H_0: \mu \leq 115$, is true. Hence, if the z test statistic falls in the critical region, the null hypothesis is rejected. The value of z that cuts off the upper 0.05 portion can be found in the normal distribution Table D.2 and is 1.645. It can be shown that the corresponding $\bar{X}$ value is given by $\mu_0 + z\sigma_{\bar{X}} = 115 + 1.645(1.5) = 117.47$.

---

[4] Recall that this "assumption" is the null hypothesis to be tested and hopefully rejected.

Considering the sample-to-sample variability of random variables, we wouldn't expect the mean of a single random sample, $\bar{X}$, to exactly equal the predicted value, μ_0, although $\mu = \mu_0$. We would probably be willing to attribute a small discrepancy between $\bar{X}$ and μ_0 to chance, but if the discrepancy is large enough, we would be inclined to believe that μ_0 is incorrect and the null hypothesis should be rejected. According to hypothesis-testing convention, a large discrepancy is one that would occur by chance five times or fewer in 100 if the null hypothesis is true. To state it another way, given that H_0 is true, a discrepancy between $\bar{X}$ and μ_0 with probability less than or equal to 0.05 is considered to be sufficient evidence for believing that $\mu \neq \mu_0$. Therefore, such a discrepancy leads to rejection of the null hypothesis.

A probability of 0.05 is by convention the largest risk an experimenter is willing to take of rejecting a true null hypothesis—declaring, for example, that $\mu > 115$ when in fact $\mu \leq 115$. Such a probability, called a *level of significance*, is designated by the Greek letter α. For $\alpha = 0.05$ and $H_1: \mu > 115$, the region for rejecting H_0, called the *critical region*, is shown in Figure 11.2-1. The location and size of the critical region are determined, respectively, by H_1 and α.

A decision to adopt the 0.05, 0.01, or any other level of significance is generally based on hypothesis-testing conventions, which have evolved over the past 50 years. Unfortunately, these conventions don't always lead to decisions that are optimal for an experimenter's purposes. We will return to the problem of selecting a level of significance in Section 11.4.

Step 5: Making a Decision

The fifth step in testing a statistical hypothesis is to obtain a sample from the population of interest, compute the test statistic, and make a decision. The decision rule is to reject H_0 if the test statistic falls in the critical region; otherwise, don't reject H_0.

Review Exercises for Section 11.2

9. For the past several years the mean arithmetic-achievement score for a population of ninth-grade students has been 45. After participating in an experimental teaching program a random sample of 100 students had a mean score of 50 and a standard deviation of 15. (a) List the five steps you would follow in testing the hypothesis that the new program is superior to the old program, and supply the required information. (b) State the decision rule.
10. For the data in Exercise 9 draw the sampling distribution associated with the null hypothesis and indicate the regions that lead to rejection and nonrejection of the null hypothesis.
*11. Which of the following statistical hypotheses will actually be tested?

$$H_0: \mu \leq 15$$
$$H_1: \mu > 15$$

12. What determines the size of the critical region and its location?
13. Can you think of some reasons why an experimenter should always specify H_0, H_1, α, and n before collecting data?

14. Terms to remember:
    a. Level of significance
    b. Critical region

11.3 One-Sample z Test for μ When σ^2 Is Known

We will now illustrate the use of

$$z = \frac{\bar{X} - \mu_0}{\sigma/\sqrt{n}}$$

in testing a hypothesis about a population mean. Recall that $\bar{X}$ is the mean of a random sample from the population of interest, μ_0 is the mean specified in the null hypothesis, σ is the standard deviation of the population, and n is the number of elements in the sample used to compute $\bar{X}$. To use this formula we must know the population variance. In Chapter 12 we will treat the case in which σ^2 is unknown but can be estimated from sample data.

Let's assume that a random sample of 100 campus radicals has been obtained from the population of radicals at the Big Ten universities and that the mean IQ of this sample is 117. How improbable is a sample mean of 117 if the population mean is 115? Would it occur five or fewer times in 100 by chance? Stated another way, does the sample statistic $\bar{X} = 117$ fall in the critical region, which for our example is the upper 5% of the sampling distribution? To answer this question we will first transform the random variable $\bar{X}$ into a z random variable with sampling distribution (if H_0 is true) identical to the standard normal distribution listed in Appendix Table D.2. If z falls in the upper 5% of the sampling distribution of z, we know that $\bar{X}$ also falls in the upper 5% of the sampling distribution of $\bar{X}$.[5] The linear transformation of $\bar{X}$ into z is a convenience that enables us to use the standard normal distribution table. The decision rule and the steps to be followed in testing the hypothesis are as follows.

Step 1. State the statistical hypotheses: $H_0: \mu \leq 115$
$H_1: \mu > 115.$

Step 2. Specify the test statistic: $z = \dfrac{\bar{X} - \mu}{\sigma/\sqrt{n}}.$

Step 3. Specify the sample size: $n = 100$;
and the sampling distribution: standard normal distribution because σ^2 is known and the population distribution of X is approximately normal in form.

Step 4. Specify the level of significance: $\alpha = 0.05.$

Step 5. Obtain a random sample of size n, compute z, and make a decision.

Decision rule: Reject H_0 if z falls in the upper 5% of the sampling distribution of z; otherwise, don't reject H_0.

[5] Figure 11.2-1 illustrates the correspondence between the sampling distributions of $\bar{X}$ and z.

The z statistic for our example is

$$z = \frac{\bar{X} - \mu_0}{\sigma/\sqrt{n}} = \frac{117 - 115}{15/\sqrt{100}} = \frac{2}{1.5} = 1.33.$$

According to the standard normal distribution Table D.2, the value of z that cuts off the upper 0.05 region of the sampling distribution is 1.645—that is, $z_{0.05} = 1.645$. The subscript 0.05 denotes the proportion of the sampling distribution that falls above the z value. Since $z = 1.33$ is less than $z_{0.05} = 1.645$, the observed z falls short of the upper 0.05 critical region. This is illustrated in Figure 11.3-1. According to our decision rule, we fail to reject H_0 and therefore conclude that our sample data don't indicate that campus radicals at the Big Ten universities have higher IQs than the average college student. Two points need to be emphasized. First, we haven't proven that the null hypothesis is true—only that the evidence doesn't warrant its rejection. Second, our conclusion has been restricted to the population from which we sampled, namely, the Big Ten universities.

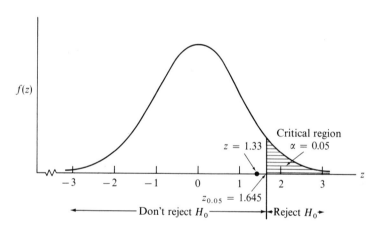

Figure 11.3-1. Sampling distribution of z under the null hypothesis. Since $z = 1.33$ falls short of the critical region, H_0 is not rejected.

Review Exercises for Section 11.3

*15. Under what conditions is the sampling distribution of $z = (\bar{X} - \mu_0)/\sigma_{\bar{X}}$ the same as the standard normal distribution?

16. Assume the Pd (Psychopathic deviate) scale of the Minnesota Multiphasic Personality Inventory has been given to a random sample of 30 men classified as habitual criminals. The experimenter wants to test the hypothesis that habitual criminals have higher Pd scores than noncriminals. The latter population is known to be normally distributed, with mean and standard deviation equal to 50 and 10, respectively. List the five steps you would follow in testing the scientific hypothesis. Let $\alpha = 0.05$.

*17. Assume the data in the table have been obtained for the habitual criminals in Exercise 16. (a) Compute a z statistic for these data. (b) What conclusion can be drawn about the scientific hypothesis in Exercise 16?

11.3
One-Sample z Test for μ when σ^2 Is Known

Subject	Pd Score	Subject	Pd Score
1	50	16	55
2	51	17	56
3	54	18	48
4	55	19	45
5	25	20	41
6	61	21	82
7	64	22	65
8	55	23	67
9	55	24	75
10	52	25	40
11	71	26	61
12	57	27	35
13	59	28	56
14	54	29	56
15	55	30	55

18. If $\alpha = 0.001$ in Exercise 16, what conclusion would have been drawn about the scientific hypothesis?

19. One of the prison guards confessed that for a lark he filled out the Pd scale and used a prisoner's name, subject number 22. (a) Recompute the z statistic for the data in Exercise 16 eliminating subject 22's score. (b) What conclusion can be drawn about the scientific hypothesis?

11.4 More about Hypothesis Testing

In Section 11.2 we described the steps used in testing a hypothesis; we illustrated these steps in Section 11.3 by means of a one-sample z test. We now turn to several new concepts that round out our discussion of hypothesis testing.

One-Tailed and Two-Tailed Tests

A statistical test for which the critical region is in either the upper or lower tail of the sampling distribution is called a **one-tailed test**. If the critical region is in both the upper and lower tails of the sampling distribution, the statistical test is called a **two-tailed test**.

A one-tailed test is used whenever the experimenter makes a *directional* prediction concerning the phenomenon of interest, for example, that student radicals have higher IQs than the average college student. We know from Section 11.3 that the statistical hypotheses corresponding to this scientific hypothesis are

$$H_0: \mu \leq \mu_0$$
$$H_1: \mu > \mu_0.$$

The region for rejecting the null hypothesis is shown in Figure 11.3-1. If the scientific hypothesis stated that student radicals have lower IQs than the average college student, the following statistical hypotheses would be appropriate.

$$H_0: \mu \geq \mu_0$$
$$H_1: \mu < \mu_0$$

The region for rejecting this null hypothesis is shown in Figure 11.4-1(a). To be significant, an observed z test statistic would have to be less than $-z_{0.05} = -1.645$.

Often, we don't have sufficient information to make a directional prediction about a population parameter; we simply believe the parameter is not equal to the value specified by the null hypothesis. This situation calls for a two-tailed, or *nondirectional*, test. The statistical hypotheses for a two-tailed test have the following form.

$$H_0: \mu = \mu_0$$
$$H_1: \mu \neq \mu_0$$

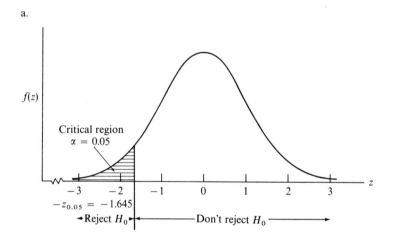

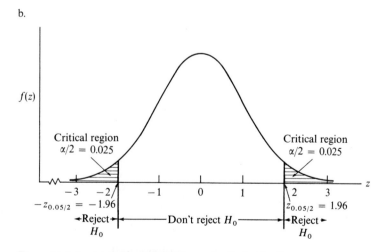

Figure 11.4-1. (a) Critical region for one-tailed test; $H_0: \mu \geq \mu_0$; $H_1: \mu < \mu_0$; $\alpha = 0.05$. (b) Critical regions for two-tailed test; $H_0: \mu = \mu_0$; $H_1: \mu \neq \mu_0$; $\alpha = 0.025 + 0.025 = 0.05$.

For a two-tailed test the region for rejecting the null hypothesis lies in both the upper and lower tails of the sampling distribution. The two-tailed critical region is shown in Figure 11.4-1(b).

In summary, a one-tailed, or directional, hypothesis is called for when the experimenter's original hunch is expressed in such terms as "more than," "less than," "increased," or "decreased." Such a hunch indicates that the experimenter has quite a bit of knowledge about the research area. The knowledge could come from previous research, a pilot study, or perhaps theory. If the experimenter is interested only in determining whether something happened or whether there is a difference without specifying the direction of the difference, a two-tailed, or nondirectional, hypothesis should be used. Generally, significance tests in the behavioral sciences are nondirectional, because most experimenters lack the information necessary to make directional predictions.

How does the choice of a one- or two-tailed test affect the probability of rejecting a false null hypothesis? We will answer this question by means of an illustration. Assume that $\alpha = 0.05$ and the following hypotheses have been advanced.

$$H_0: \mu = \mu_0$$
$$H_1: \mu \neq \mu_0$$

If the z statistic falls in either the upper or lower 0.025 region of the sampling distribution, the result is said to be significant at the 0.05 level of significance because $0.025 + 0.025 = 0.05$. The values of z that cut off the upper and lower $0.05/2 = 0.025$ region are $-z_{0.05/2} = -1.96$ and $z_{0.05/2} = 1.96$. An observed z test statistic is significant at the 0.05 level if its value is greater than 1.96 or less than -1.96; or, more simply, if its absolute value $|z|$ is greater than 1.96. Now consider the hypotheses

$$H_0: \mu \geq \mu_0$$
$$H_1: \mu < \mu_0;$$

again $\alpha = 0.05$. If the z statistic falls in the appropriate tail of the sampling distribution, the result is said to be significant at the 0.05 level of significance. The critical regions for the two cases are shown in Figure 11.4-1. From an inspection of this figure it should be apparent that the size of the difference $\bar{X} - \mu_0$ necessary to reach the critical region for a two-tailed test is larger than that required for a one-tailed test. Consequently, an experimenter is less likely to reject a false null hypothesis with a two-tailed test than with a one-tailed test. The term *power* refers to the probability of rejecting a false null hypothesis. A one-tailed test is more powerful than a two-tailed test if the experimenter's hunch about the true difference $\mu - \mu_0$ is correct—that is, if the alternative hypothesis places the critical region in the correct tail of the sampling distribution. If the directional hunch is incorrect, the rejection region will be in the wrong tail and the experimenter will most certainly fail to reject the null hypothesis even though it is false. An experimenter is rewarded for making a correct directional prediction and penalized for making an incorrect directional prediction. In the absence of sufficient information for using a one-tailed test, the experimenter should play it safe and use a two-tailed test.

In one kind of research situation it is customary to use a one-tailed test. Consider a manufacturer who is evaluating a modified version of a product. If the modified version is better than the old product, the manufacturer will begin producing the modified version; otherwise, production of the old product will continue. In this example only one research outcome will lead to a product change—the case in which the modified version is superior. In this and similar research situations, where an experimenter is interested only in a change in one direction, a one-tailed test is appropriate.[6]

Exact versus Inexact Hypotheses

A statistical hypothesis can be either exact or inexact. The hypothesis $H_0: \mu = 100$ is an exact hypothesis. The hypothesis $H_0: \mu \leq 100$ is inexact, because instead of specifying a single value for μ it specifies a range of values—those equal to or less than 100.

If the null hypothesis is exact, the alternative hypothesis can be either exact or inexact; for example,

$H_0: \quad \mu = 100$ (exact H_0)
$H_1: \quad \mu = 115$ (exact H_1)

or

$H_0: \quad \mu = 100$ (exact H_0)
$H_1: \quad \mu \neq 100$ (inexact H_1).

If the null hypothesis is inexact, the alternative hypothesis must also be inexact; for example,

$H_0: \mu \leq 100$ (inexact H_0)
$H_1: \mu > 100$ (inexact H_1)

or

$H_0: \mu \geq 100$ (inexact H_0)
$H_1: \mu < 100$ (inexact H_1).

In the behavioral sciences inexact alternative hypotheses are the rule rather than the exception because it is rarely possible to make a precise numerical prediction about the experimental outcome.

As we have seen from the above examples, null hypotheses can be either exact or inexact, but the hypothesis testing procedure is the same in both cases. If the null hypothesis is inexact, say $\mu \geq \mu_0$, it still embodies at least one exact statement, namely $\mu = \mu_0$, and this is the hypothesis that is actually tested. An examination of Figure 11.4-1(a) should convince you that if the null hypothesis $\mu = \mu_0$ is rejected at $\alpha = 0.05$, any null hypothesis for which $\mu > \mu_0$ would be automatically rejected.

Type I and Type II Errors

When the null hypothesis is tested, an experimenter's decision will be correct or incorrect. *An incorrect decision can be made in two ways. The experimenter can reject the null hypothesis when it is true; this is called a **type I error**. Or the experimenter can fail to reject the null hypothesis when it is false; this is called a **type II error**. Likewise, a correct decision can be made in two ways. If the null hypothesis is true and the experimenter does not reject it, a **correct acceptance** has been made; if the null hypothesis is false and the experimenter rejects it, a **correct rejection** has been made.* The two kinds of correct decisions and the two kinds of errors are summarized in Table 11.4-1.

[6] For an in-depth discussion of the issues involved in choosing between one- and two-tailed tests see Kirk (1972a, chap. 8).

Table 11.4-1. Decision Outcomes Categorized

		True Situation	
		H_0 True	H_0 False
Experimenter's Decision	Fail to Reject H_0	Correct acceptance Probability = $1 - \alpha$	Type II error Probability = β
	Reject H_0	Type I error Probability = α	Correct rejection Probability = $1 - \beta$

The probability of making a type I error is determined by the experimenter when the level of significance, α, is specified. If $\alpha = 0.05$, the probability of making a type I error is 0.05. The level of significance also determines the probability of a correct acceptance of a true null hypothesis, since this probability is equal to $1 - \alpha$.

The probability, symbolized by β, of making a type II error and the probability of making a correct rejection, which is equal to $1 - \beta$, are determined by a number of variables: (1) the level of significance, (2) the size of the sample, (3) the size of the population standard deviation, (4) the magnitude of the difference between μ and μ_0, and (5) whether a one- or two-tailed test is used. *The probability of making a correct rejection, $1 - \beta$, is called the* **power** *of a statistical test.* To compute β and power it is necessary either to know μ or to specify a value of μ that would be of interest. The latter approach is usually necessary, since in any practical situation μ is unknown.

Perhaps an example will help clarify the meaning of α, $1 - \alpha$, β, and $1 - \beta$. In Section 11.3 we tested the hypothesis that student radicals have higher IQs than the average college student, and we failed to reject the null hypothesis. Suppose for purposes of exposition that the sample mean equals 117.47 instead of 117. In this case, the z

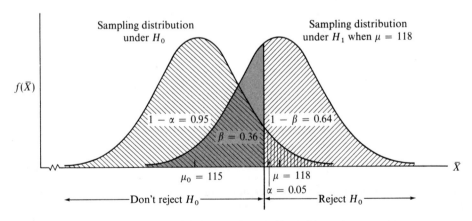

Figure 11.4-2. Regions corresponding to probabilities of making a type I error (α) and a type II error (β). The size and location of the region corresponding to a type I error are determined by α and H_1, respectively. If for a given n, σ, H_0, and true H_1, α is made smaller, the probability of making a type II error is increased.

statistic is

$$z = \frac{\bar{X} - \mu_0}{\sigma/\sqrt{n}} = \frac{117.47 - 115}{15/\sqrt{100}} = 1.647.$$

Also assume that the population mean of student radicals' IQs is equal to 118. This information is illustrated in Figure 11.4-2. Two sampling distributions are shown in this figure, one associated with the null hypothesis and the other with the true IQ of campus radicals. The region corresponding to a type II error (labeled β) is determined from

$$z = \frac{\bar{X} - \mu}{\sigma/\sqrt{n}} = \frac{117.47 - 118}{15/\sqrt{100}} = \frac{-0.53}{1.50} = -0.35,$$

where μ, the true population mean, is substituted for μ_0 in the z formula. According to the standard normal distribution Table D.2, the area below a z of -0.35 is 0.36. Thus, if the true parameter is 118, the probability of making a type II error (β) is equal to 0.36 and the probability of making a correct rejection (power) is equal to $1 - \beta = 1 - 0.36 = 0.64$. A power of 0.64 is considerably less than the minimum usually considered acceptable, which is 0.80.[7] Table 11.4-2 summarizes the probabilities associated with the possible decision outcomes when $\mu = 115$ and when $\mu = 118$.

Table 11.4-2. Probabilities Associated with the Decision Process

		True Situation	
		$\mu = 115$	$\mu = 118$
Experimenter's Decision	$\mu \leq 115$	Correct acceptance $1 - \alpha = 0.95$	Type II error $\beta = 0.36$
	$\mu > 115$	Type I error $\alpha = 0.05$	Correct rejection $1 - \beta = 0.64$

For the example we have described, the probability of making a correct decision is larger when the null hypothesis is true (Probability = 0.95) than when the null hypothesis is false (Probability = 0.64). It is also apparent that the probability of making a type I error (0.05) is much smaller than the probability of making a type II error (0.36). In most research situations the experimenter follows the convention of setting $\alpha = 0.05$ or $\alpha = 0.01$. This convention of choosing a small numerical value for α is based on the notion that a type I error is very bad and should be avoided. Unfortunately, as the probability of a type I error is made smaller and smaller, the probability of a type II error increases and vice versa. This can be seen from an examination of Figure 11.4-2; if the vertical line cutting off the upper α region is moved to the

---

[7] The selection of 0.80 as the minimum acceptable power is a convenient rule of thumb and reflects the view that type I errors are more serious than type II errors. Consider the value of the ratio p(Type II error)$/p$(Type I error) when the conventional 0.05 level of significance is adopted and power is equal to 0.80. Under these conditions the ratio is $0.20/0.05 = 4$. The probability of a type II error is four times larger than that for a type I error. An experimenter who adopts $\alpha = 0.05$ and $1 - \beta = 0.80$ is saying in effect that a type I error is considered to be four times more serious than a type II error.

right or to the left, the region designated β is made, respectively, larger or smaller. In our example the decision rule, reject H_0 if z falls in the upper 0.05 region of the sampling distribution, is weighted in favor of deciding that the population mean is less than or equal to 115 rather than greater than 115 because the associated probabilities are 0.95 and 0.64, respectively.

Determining the n Required to Achieve an Acceptable α, $1 - \beta$, and $\mu - \mu_0$

As Figure 11.4-2 shows, power can be increased by adopting a large numerical value for α, but this increases the probability of making a type I error and may therefore be undesirable. Often the simplest way to increase the power of a statistical test is to increase the sample size. Until now we haven't said very much about specifying n except that it should be large enough—but not too large. There is a more rational approach to specifying sample size. The factors we have been discussing (α, $1 - \beta$, σ, n, and $\mu - \mu_0$) are interrelated; if we specify any four of them, the fifth is determined. The appropriate size of n can be estimated, therefore, if we know or specify acceptable values for α, $1 - \beta$, σ, and $\mu - \mu_0$. The student radical experiment, where $\alpha = 0.05$, $\sigma = 15$, and $\mu_0 = 115$, will be used to illustrate the procedure. Suppose we want the power of the experiment to be 0.80. That leaves one factor, $\mu - \mu_0$, unspecified. There is no way of knowing $\mu - \mu_0$ without measuring all the student radicals in the population. However, we can specify the minimum IQ difference between μ and μ_0 that we would be interested in finding—if in fact $\mu \neq \mu_0$. Suppose this difference is 3 IQ points; then $\mu - \mu_0 = 118 - 115 = 3$. By specifying that $\mu - \mu_0 = 3$ we are saying that any difference less than 3 points is too small to be of practical interest. The formula for determining the sample size[8] is

$$n = \frac{(z_\alpha - z_\beta)^2}{(\mu - \mu_0)^2/\sigma^2},$$

where z_α is the value of z that cuts off the α region of the sampling distribution of μ_0 and z_β is the value that cuts off the β region of the sampling distribution of μ. These regions are shown in Figure 11.4-2. If the alternative hypothesis is nondirectional, z_α is replaced by $z_{\alpha/2}$. For our example, $z_{0.05} = 1.645$ and $-z_{0.20} = -0.84$. Substituting in the formula gives

$$n = \frac{[1.645 - (-0.84)]^2}{(118 - 115)^2/(15)^2} = 154.4,$$

which when rounded to the next larger integer value is 155. Thus, the minimum sample size required to detect a 3 point IQ difference with $\alpha = 0.05$ and power equal to 0.80 is 155. This sample size gives the experimenter a fighting chance of detecting a difference considered worth finding. We can be more concrete. Suppose we repeated the experiment 100 times using random samples of 155 student radicals. If the null hypothesis

---

[8] The derivation of the formula is given in Technical Note 11.6-1.

is true, we would expect to reject it five times (a type I error) and to fail to reject it 95 times (a correct acceptance). If, however, the null hypothesis is false and the true difference between μ and μ_0 is 3 IQ points, we would expect to reject the null hypothesis 80 times (a correct rejection) and to fail to reject it 20 times (a type II error).

Suppose we consider type I and type II errors to be equally serious. If we set both errors at 0.05, the required sample size is

$$n = \frac{[1.645 - (-1.645)]^2}{(118 - 115)^2/(15)^2} \simeq 271.$$

To increase the power from 0.80 to 0.95, other things being equal, we have to increase the sample size from 155 to 271—a 75% increase. Research always involves a series of tradeoffs, as this example illustrates for power and sample size. Another tradeoff, as we will see, involves n and $\mu - \mu_0$; the larger $\mu - \mu_0$ is, the smaller is the n required to reject the null hypothesis.

In many experiments the dependent variable is a new untried measure or one with which the experimenter has had little experience. For example, one may have developed a new test of assertiveness for which there are no norms or a new apparatus for measuring complex reaction time. In such cases it is difficult if not impossible to specify the minimum difference $\mu - \mu_0$ that would be worth detecting. Fortunately, there is an alternative procedure for estimating n that doesn't require either an estimate of $\mu - \mu_0$ or a knowledge of σ. In Section 10.2 we saw that a standard score $(X - \mu)/\sigma$ is a pure number free of the original unit of measurement. If the difference we want to detect is divided by the population standard deviation, $(\mu - \mu_0)/\sigma$, the resulting score is similar to a standard score. Cohen (1969, pp. 18–25) calls this relative measure an *effect size* and denotes it by the symbol d. It expresses the magnitude of the difference $\mu - \mu_0$ in standard deviation units. The value of d for our student radical experiment is

$$d = \frac{\mu - \mu_0}{\sigma} = \frac{118 - 115}{15} = \frac{3}{15} = 0.2.$$

Hence, the difference we wanted to detect in that experiment was 0.2 of a standard deviation. Cohen refers to a d value of 0.2 as a *small-effect size*. A medium-effect size is one for which $d = 0.5$, and a large-effect size is one for which $d = 0.8$. Using Cohen's rule of thumb, a medium-effect size for the IQ data is one for which $\mu - \mu_0 = 122.5 - 115 = 7.5$, since

$$d = \frac{7.5}{15} = 0.5.$$

Similarly, a large-effect size corresponds to $\mu - \mu_0 = 127 - 115 = 12$, since

$$d = \frac{12}{15} = 0.8.$$

The specification of the minimum difference between μ and μ_0 that one wants to detect in terms of effect size simplifies the formula for estimating n. If d is substituted for $(\mu - \mu_0)/\sigma$, the formula becomes

$$n = \frac{(z_\alpha - z_\beta)^2}{(\mu - \mu_0)^2/\sigma^2} = \frac{(z_\alpha - z_\beta)^2}{d^2}.$$

To compute n, all we have to specify is (1) d, (2) the probability of a type I error, and (3) power. Using Cohen's rule of thumb concerning the interpretation of $d = 0.2, 0.5$, and 0.8, the n's necessary to detect small-, medium-, and large-effect sizes for the student radical experiment are, respectively,

$$n = \frac{[1.645 - (-0.84)]^2}{(0.2)^2} \simeq 155,$$

$$n = \frac{[1.645 - (-0.84)]^2}{(0.5)^2} \simeq 25, \text{ and}$$

$$n = \frac{[1.645 - (-0.84)]^2}{(0.8)^2} \simeq 10.$$

As in the earlier computation of n, the probability of a type I error is 0.05 ($z_{0.05} = 1.645$) and the power is equal to 0.80 ($-z_{0.20} = -0.84$).

It is obvious that one's sample size can be too small, resulting in insufficient power. But n can also be too large, resulting in wasted time and resources. An experimenter can avoid these problems by using the formulas just described to make a rational choice of sample size. This procedure has two other less obvious benefits: it focuses attention on the interrelationships among n, α, $1 - \beta$, σ, and $\mu - \mu_0$; and it forces the experimenter to think about the size of the difference $\mu - \mu_0$ that would be worth detecting. Critics of hypothesis testing procedures have observed that the population mean is rarely equal to the exact value specified by the null hypothesis, and hence by obtaining a large enough sample virtually any H_0 can be rejected.[9] For this reason it is important to distinguish between statistical significance, which leads to the decision that $\mu \neq \mu_0$, and practical significance, which means that the difference $\mu - \mu_0$ is large enough to be useful in the real world. By estimating the n required to reject a difference of interest, an experimenter increases the chances of detecting practical differences. One of the challenges of research is to design an experiment that (1) has adequate power for detecting a meaningful or practical difference between μ and μ_0, (2) uses minimum resources, n, (3) provides adequate protection against making a type I error, and (4) minimizes the effects of extraneous variables.

More about Type I and Type II Errors

In many research situations the cost of committing a type I error can be large relative to that of a type II error. It is a serious matter to commit a type I error by deciding that a new medication is more effective than conventional therapies in halting the production of cancer cells and therefore can be used in place of other medical procedures. On the other hand, falsely deciding that the new medication is not more effective, a type II error, would result in withholding the medication from the public and lead to further research. The further research would eventually demonstrate the effectiveness of the new medication. In this example a type I error is more costly and should be avoided more than a type II error. The use of the 0.01, or even the 0.001,

---

[9] For a discussion of this issue see Bakan (1966), Nunnally (1960), and Rozeboom (1960).

level of significance seems warranted. However, in other research situations that don't involve life and death, a type I error may be less costly than a type II error. For example, an experimenter who makes a type II error may discontinue a promising line of research whereas a type I error would lead to further exploration into a blind alley. Faced with these two alternatives many experimenters would set the level of significance equal to 0.05 or even 0.20, preferring to make a type I error rather than a type II error.

It is apparent that the loss functions associated with type I and type II errors must be known before a rational choice of α can be made. Unfortunately, experimenters in the behavioral sciences are generally unable to specify the losses associated with the two kinds of errors, and therein lies the problem. The problem is resolved by using the conventional but arbitrary 0.05 or 0.01 level of significance.

Hopefully, this discussion has dispelled the magical aura that surrounds the 0.05 and 0.01 levels of significance—their use in hypothesis testing is simply a convention. A test of significance yields the probability of committing an error in rejecting the null hypothesis. It embodies no information concerning the importance or usefulness of the result obtained. It is just one bit of information used in making a decision concerning a scientific hypothesis.

Review Exercises for Section 11.4

**20. For each of the following statistical hypotheses, sketch the sampling distribution, designate the critical region(s), and indicate their size.

*a. H_0: $\mu = 60$
   H_1: $\mu \neq 60$
   $\alpha = 0.01$

b. H_0: $\mu \leq 100$
   H_1: $\mu > 100$
   $\alpha = 0.05$

c. H_0: $\mu \geq 25$
   H_1: $\mu < 25$
   $\alpha = 0.005$

21. Under what condition is a one-tailed test less powerful than a two-tailed test?
*22. Which of the hypotheses in Exercise 20 are exact and which are inexact?
23. Indicate the type of error or correct decision for each of the following.
    a. A true null hypothesis was rejected.
    b. The experimenter failed to reject a false null hypothesis.
    c. The null hypothesis is false and the experimenter rejected it.
    d. The experimenter did not reject a true null hypothesis.
    e. A false null hypothesis was rejected.
    f. The experimenter rejected the null hypothesis when he or she should have failed to reject it.
**24. The calculation of power was illustrated in this section by means of the campus radicals' IQ data. We saw that if $\bar{X}$ is equal to 117.47 and the campus radicals' mean IQ is really equal to 118.0, the probability of correctly rejecting the null hypothesis is only 0.64.
    *a. If a sample of $n = 150$ instead of $n = 100$ had been obtained, what would the power have been?
    *b. How many subjects would be required to achieve a power of 0.90 in part a? Assume that μ_0 is equal to 115.
    c. If a sample of $n = 100$ had been obtained but $\mu = 118.5$ instead of 118.0, what would the power have been?
    d. How many subjects would be required to achieve a power of 0.80 in part c?
25. Set up a table that summarizes the probabilities associated with the four possible decision outcomes in Exercise 24a.
26. List the ways in which an experimenter can increase the power of experimental methodology. What are their relative merits?

27. Terms to remember:
    a. One-tailed test
    b. Two-tailed test
    c. Power
    d. Exact statistical hypothesis
    e. Inexact statistical hypothesis
    f. Type I error
    g. Type II error
    h. Correct acceptance
    i. Correct rejection

11.5 Summary

Hypothesis testing procedures, one form of statistical inference, use sample data in making a decision about a scientific hypothesis when it is impossible or impractical to observe all the elements in the population. The main features of hypothesis testing are as follows. An experimenter formulates from a scientific hypothesis two mutually exclusive and exhaustive statistical hypotheses—the null hypothesis, H_0, and the alternative hypothesis, H_1. The null hypothesis is a testable prediction about one or more parameters of a population distribution. A test of H_0 consists of determining whether the obtained value of a sample statistic would be improbable if H_0 is true. If the statistic would be improbable, then H_0 is a poor prediction and should be rejected in favor of H_1. The criterion for what constitutes an improbable value is specified prior to the test. It is expressed as a probability and is called a level of significance, α. If, for example, the probability associated with the value of a sample statistic is equal to or less than, say, 0.05, the sample value is said to be significant or to deviate significantly from the parameter specified in H_0.

How does one determine the probability associated with a sample statistic? This is where things begin to get complicated. An experimenter must specify H_0, H_1, a test statistic, n, and α. The population standard deviation, σ, must be known; a random sample from the population must be obtained; and, with good cause, the population distribution must be assumed to be normal. If the latter assumption is not tenable, the sample n must be fairly large. The above is sufficient to enable the experimenter to (1) completely specify the sampling distribution of the sample statistic given that H_0 is true, (2) specify the critical region for rejecting H_0, and (3) transform the sample statistic $\bar{X}$ into a z test statistic with a sampling distribution that is normal with $\mu = 0$ and $\sigma = 1$. Now, to answer the question we just posed: the probability associated with z, the transformed sample statistic, can be determined from a table of the normal distribution. This probability, and the sign of z if the test is one-tailed, determines whether z is in the critical region for rejecting H_0, in other words, if z is statistically significant.

There is a tendency among experimenters to impart surplus meaning to the term *statistical significance*. All it really means is that a result has been obtained that is improbable if the null hypothesis is true. Statistical significance doesn't connote importance or usefulness, and it shouldn't be confused with practical significance. In the simplest terms, a significant result is one for which chance is an unlikely explanation.

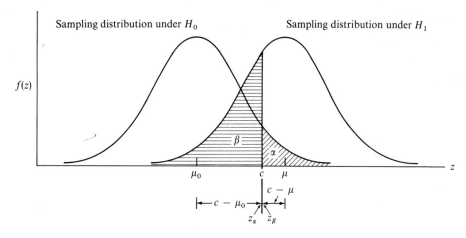

Figure 11.6-1. Sampling distributions under H_0 and H_1. The ordinate at c cuts off the α and β regions.

†11.6 Technical Note

11.6-1 Derivation of Formula for Estimating n

Consider Figure 11.6-1 in which the ordinate at c cuts off the α and β regions as shown. The size of the α region corresponds to the probability of making a type I error; the size of the β region corresponds to the probability of making a type II error. The distance from μ_0 to c and the distance from μ to c can be expressed as z scores by

$$z_\alpha = \frac{c - \mu_0}{\sigma/\sqrt{n}} \quad \text{and} \quad z_\beta = \frac{c - \mu}{\sigma/\sqrt{n}}.$$

Rearranging terms for both z scores, we obtain

$$c - \mu_0 = \frac{z_\alpha \sigma}{\sqrt{n}} \quad \text{and} \quad c - \mu = \frac{z_\beta \sigma}{\sqrt{n}}$$

$$c = \frac{z_\alpha \sigma}{\sqrt{n}} + \mu_0 \quad \quad c = \frac{z_\beta \sigma}{\sqrt{n}} + \mu.$$

Since both quantities are equal to c, they are equal to each other, so we can write

$$\frac{z_\alpha \sigma}{\sqrt{n}} + \mu_0 = \frac{z_\beta \sigma}{\sqrt{n}} + \mu$$

$$\frac{z_\alpha \sigma - z_\beta \sigma}{\sqrt{n}} = \mu - \mu_0.$$

† This technical note can be skipped without loss of continuity.

Squaring both sides of the equation,
$$\frac{(z_\alpha - z_\beta)^2 \sigma^2}{n} = (\mu - \mu_0)^2,$$
and solving for n, we obtain
$$n = \frac{(z_\alpha - z_\beta)^2}{(\mu - \mu_0)^2/\sigma^2}.$$
If effect size, d, is defined as $(\mu - \mu_0)/\sigma$, the formula for n becomes
$$n = \frac{(z_\alpha - z_\beta)^2}{d^2}.$$

12

Statistical Inference: Other One-Sample Test Statistics

12.1 Introduction to Other One-Sample Test Statistics
12.2 One-Sample t Test for μ when σ^2 Is Unknown
   Computational Example for One-Sample t Test
   Some Experimental Design Considerations
   Statistical Significance and Practical Significance
   Determining the n Required to Achieve an Acceptable α, $1 - \beta$, and $(\mu - \mu_0)/\sigma$
   More about the Normality Assumption
   Review Exercises for Section 12.2
12.3 One-Sample Chi-Square Test for a Population Variance
   Computational Example for One-Sample Chi-Square Test
   More about the Normality Assumption for χ^2
   Review Exercises for Section 12.3
12.4 One-Sample z Test for a Population Proportion
   Computational Example for One-Sample z Test
   Correction for Continuity
   Computational Procedure when the Sample Size Is Equal to or Greater than 10% of the Population
   Review Exercises for Section 12.4
12.5 One-Sample t and z Tests for a Population Correlation
   t Test for $H_0: \rho = 0$
   Computational Example for t Test
   z Test for Correlations Other than Zero
   Computational Example for z Test
   Review Exercises for Section 12.5
12.6 Summary

12.1 Introduction to Other One-Sample Test Statistics

This chapter could just as well be entitled "Theme with Variations"—it applies the five-step hypothesis testing format introduced in the last chapter to several new test statistics. In Chapter 11 procedures for testing hypotheses about μ when σ^2 is known were described. These same procedures, with slight modifications, are used in testing hypotheses about μ when σ^2 is unknown. And they can be used to make decisions about other population parameters such as variance (σ^2), proportion (p), and correlation (ρ). The statistics that accomplish this and associated null hypotheses are (1) t for $H_0: \mu = \mu_0$, (2) χ^2 for $H_0: \sigma^2 = \sigma_0^2$, (3) z for $H_0: p = p_0$, (4) t for $H_0: \rho = 0$, and (5) z for $H_0: \rho = \rho_0$.

12.2 One-Sample t Test for μ when σ^2 Is Unknown

When an experimenter tests a hypothesis about μ the value of σ^2 is rarely known, and as we have seen, σ^2 is required to compute the z statistic. *The t statistic does not require a knowledge of σ^2.* Instead, it requires only an estimate of σ^2 computed from a random sample from an approximately normally distributed population. The formula for the one-sample t statistic is

$$t = \frac{\bar{X} - \mu_0}{\hat{\sigma}/\sqrt{n}},$$

where $\bar{X}$ is the mean of a random sample, μ_0 is the mean specified in the null hypothesis, n is the number of elements in the sample used to compute $\bar{X}$, and $\hat{\sigma}$ is an estimator of the population standard deviation. This estimator is given by

$$\hat{\sigma} = \sqrt{\frac{\sum(X_i - \bar{X})^2}{n-1}} \text{ or } \sqrt{\frac{\sum X_i^2 - (\sum X_i)^2/n}{n-1}}.$$

A caret (or hat) over σ signifies that the statistic is a sample estimator of the population parameter. The denominator of the t statistic, $\hat{\sigma}/\sqrt{n}$, is called the *standard error of a mean* and is denoted by $\hat{\sigma}_{\bar{X}}$.

The sampling distribution of t was derived by William Sealey Gossett, who published under the pseudonym *Student*; hence, the distribution is often referred to as *Student's t distribution*. The shape of the t distribution is similar to that of the z distribution—it is symmetrical about a mean of zero. However, its variance depends on sample size, or more specifically, degrees of freedom. To discuss this we must understand the concept of degrees of freedom, abbreviated df or v. The term comes from the physical sciences, where it refers to the number of planes or directions in which an object is free to move. In statistics the term *degrees of freedom* refers to the number of scores with values that are free to vary, as we will now see. Consider a sample of size three with mean equal to five—that is, $(X_1 + X_2 + X_3)/3 = 5$. If we arbitrarily specify that $X_1 = 4$ and $X_2 = 5$, then X_3 must equal 6 since $(4 + 5 + 6)/3 = 15/3 = 5$. Given the statement $\bar{X} = 5$, we are free to assign any values to $n - 1 = 2$ of the scores, but having done so, the value of the nth score is determined. Thus, the number of degrees of freedom associated with $\bar{X}$ is $n - 1$. Similarly, the number of degrees of freedom associated with $\hat{\sigma} = \sqrt{\sum (X_i - \bar{X})^2/(n - 1)}$ is $n - 1$. This follows because once $n - 1$ of the n deviations $(X_i - \bar{X})$ have been arbitrarily specified, the nth deviation is not free to vary, since $\sum (X_i - \bar{X})$ must equal zero (as shown in Technical Note 3.9-2). The number of degrees of freedom for the one-sample t test is equal to $n - 1$.

It can be shown that for $v > 2$ the variance of the t sampling distribution is

$$\text{Var}(t) = \frac{v}{v - 2},$$

where $v = n - 1$. According to the formula, if random samples of size $n = 5$ are obtained from a population, the variance of the resulting t sampling distribution is

$$\text{Var}(t) = \frac{v}{v - 2} = \frac{4}{2} = 2.$$

When $v = 4$ the variance is 2, which is twice the variance of the z sampling distribution. As the number of degrees of freedom increases the variance of the sampling distribution of t approaches more and more closely that of the standard normal distribution. For example, when $n = 30$,

$$\text{Var}(t) = \frac{29}{29 - 2} = 1.0741,$$

which differs only slightly from the variance of the standard normal distribution. Because the two sampling distributions are so similar for samples equal to or larger than 30, an n of 30 is often taken as a dividing point between large and small samples.

The t distribution is actually a family of distributions with a shape that is dependent on the degrees of freedom. A comparison of three members of the t family and the standard normal distribution is shown in Figure 12.2-1. As this figure illustrates, the t and z distributions are alike in that both have a mean of zero, are symmetrical, and are unimodal. The distributions differ when $v < \infty$ in that the distribution of t is more leptokurtic and has a larger variance.

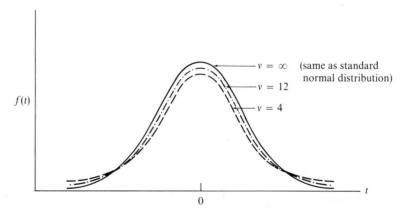

Figure 12.2-1. Graph of the distribution of t for 4, 12, and ∞ degrees of freedom. The t distribution for $v = \infty$ is identical to the z distribution.

The t and z test statistics look alike, but a difference can be seen on close inspection.

$$z = \frac{\bar{X} - \mu_0}{\sigma/\sqrt{n}} = \frac{(\text{Random variable}) - (\text{Constant})}{(\text{Constant})}$$

$$t = \frac{\bar{X} - \mu_0}{\hat{\sigma}/\sqrt{n}} = \frac{(\text{Random variable}) - (\text{Constant})}{(\text{Random variable})}$$

The z statistic is the ratio of a random variable to a constant; t is the ratio of two random variables. When $\nu < \infty$ both $\bar{X}$ and $\hat{\sigma}$ vary from sample to sample; therefore, even though $\bar{X} - \mu_0$ is normally distributed, the ratio $(\bar{X} - \mu_0)/(\hat{\sigma}/\sqrt{n})$ is not normally distributed.

The t test statistic can be used to test hypotheses of the form

$$H_0: \mu \geq \mu_0 \qquad H_0: \mu \leq \mu_0 \qquad H_0: \mu = \mu_0$$
$$H_1: \mu < \mu_0 \qquad H_1: \mu > \mu_0 \qquad H_1: \mu \neq \mu_0,$$

where μ denotes the unknown population mean and μ_0 denotes the hypothesized value of the population mean. The assumptions associated with using t to test these hypotheses are that (1) the population distribution of X is normal in form, (2) the X's are a random sample from the population of interest, (3) the population variance isn't known, and (4) the null hypothesis is true, which is the hypothesis the experimenter hopes to reject. The null hypothesis is rejected if an observed t falls in the critical region of its sampling distribution. The critical value of t that cuts off the upper α region of the sampling distribution for ν degrees of freedom is given in Appendix Table D.3 and is denoted by $t_{\alpha,\nu}$; the value that cuts off the lower α region is denoted by $-t_{\alpha,\nu}$. If α is to be divided equally between the two tails, as in performing a two-tailed test, the critical values are denoted by $t_{\alpha/2,\nu}$ and $-t_{\alpha/2,\nu}$. To illustrate, suppose an experimenter wants to test the one-tailed hypothesis $H_0: \mu \leq 50$ versus $H_1: \mu > 50$, and suppose α is set at 0.05 and $\nu = 30$. According to Table D.3 the value of t that cuts off the upper 0.05 region is 1.697. The null hypothesis would be rejected if $t \geq 1.697$. For $H_0: \mu \geq 50$ and $H_1: \mu < 50$, the null hypothesis would be rejected if $t \leq -1.697$. The critical value for a two-tailed hypothesis is 2.042, in which case the null hypothesis would be rejected if $|t| \geq 2.042$.

Computational Example for One-Sample t Test

Consider the scientific hypothesis that a new class registration procedure at Idle-On-In College will reduce the time required for a student to register. Over the past several years the mean time required to register has been 3.1 hr. A trial run to test the new procedure was conducted with a random sample of 21 undergraduates. The decision rule and the steps to be followed in testing the null hypothesis are as follows.

Step 1. State the statistical hypotheses: $H_0: \mu \geq 3.1$
$H_1: \mu < 3.1$.

Step 2. Specify the test statistic: $t = \dfrac{\bar{X} - \mu_0}{\hat{\sigma}/\sqrt{n}}$.

Step 3. Specify the sample size: $n = 21$;
and the sampling distribution: t distribution with $v = n - 1$, because σ^2 is unknown and must be estimated and we assume the population distribution of X is approximately normal in form.

Step 4. Specify the significance level: $\alpha = 0.05$.

Step 5. Obtain a random sample of size n, compute t, and make a decision.
Decision rule: Reject H_0 if t falls in the lower 5% of the sampling distribution of t; otherwise, don't reject H_0.

The data for the trial run are shown in Table 12.2-1. According to Table D.3, a t of -1.725 with $n - 1 = 20$ degrees of freedom cuts off the lower 0.05 region of the

Table 12.2-1. Registration Time Data

(i) Data:

Student	Registration Time, X_i (Hours)	X_i^2	Student	Registration Time, X_i (Hours)	X_i^2
1	2.9	8.41	12	3.0	9.00
2	2.7	7.29	13	2.8	7.84
3	2.8	7.84	14	3.1	9.61
4	3.0	9.00	15	2.9	8.41
5	2.6	6.76	16	2.7	7.29
6	2.9	8.41	17	3.2	10.24
7	3.1	9.61	18	2.8	7.84
8	2.9	8.41	19	2.8	7.84
9	3.0	9.00	20	3.1	9.61
10	2.7	7.29	21	3.0	9.00
11	2.9	8.41		$\sum X_i = 60.9$	$\sum X_i^2 = 177.11$

(ii) Computation:

$$\bar{X} = \frac{\sum X_i}{n} = \frac{60.9}{21} = 2.9$$

$$\hat{\sigma} = \sqrt{\frac{\sum X_i^2 - (\sum X_i)^2/n}{n-1}} = \sqrt{\frac{177.11 - (60.9)^2/21}{21-1}} = \sqrt{\frac{0.50}{20}} = 0.158$$

$$t = \frac{\bar{X} - \mu_0}{\hat{\sigma}/\sqrt{n}} = \frac{2.9 - 3.1}{0.158/\sqrt{21}} = \frac{-0.2}{0.034} = -5.882$$

$$v = n - 1 = 21 - 1 = 20$$

$$-t_{0.05, 20} = -1.725$$

sampling distribution—that is, $-t_{0.05, 20} = -1.725$. Since $-5.882 < -1.725$, the null hypothesis is rejected and the experimenter concludes that registration can be completed in less time with the new procedure than with the old procedure.

Some Experimental Design Considerations

Note that there are alternative explanations for why the sample of 21 students completed registration in less time with the new procedure. The school administrators would like to believe that the new procedure is extremely efficient and, if adopted for all students, would cut down registration time. But what are some alternative explanations? Since they were selected for the trial run, the 21 students may have felt they should make a special effort to complete registration quickly—an effort they wouldn't make once the new procedure was adopted and they were no longer under scrutiny. It is also possible that the personnel assisting in registration were more alert and tried to expedite the registration because they too were under scrutiny and because the procedure was a break from the usual routine. Other explanations could be advanced, and unless these explanations can be ruled out, the administrators may be disappointed if they adopt the new procedure. Once the novelty of the new procedure wears off it may be no better or even poorer than the old registration procedure.

Designing an experiment with an outcome that can be unambiguously interpreted requires careful planning. *It is customary in behavioral science research to use one or more* **control groups**; *these contain subjects that are not subjected to the treatment in order to provide data on the effects of extraneous variables that affect the interpretation of the experiment.*[1] For example, the design of our registration experiment could be improved by drawing a sample of 42 students, half to be assigned to the new procedure and half to the old. This would provide data on the effects of being specially selected to participate in the trial run.

Statistical Significance and Practical Significance

Let us assume for purposes of discussion that results obtained in the trial run would also be obtained if the new registration procedure were adopted for all students. The difference in registration time between the procedures was $2.9 - 3.1 = -0.2$ hr, or -12 min. Before adopting the new procedure the school administrators would have to decide whether the statistically significant difference in registration time is significant from a practical point of view. Is a savings of 12 min worth the cost of changing the registration procedures? If the change involves hiring new personnel, scrapping costly registration forms, or extensively modifying physical facilities, the administrators might decide the change isn't worth it.

Statistical significance is concerned with whether a result could have occurred by chance; it says nothing about the practical significance of the result, which must be decided on the basis of other considerations.

[1] For an excellent discussion of these procedures see Campbell (1957). This article is reprinted in Kirk (1972a).

Determining the *n* Required to Achieve an Acceptable α,
$1 - \beta$, and $(\mu - \mu_0)/\sigma$

Procedures for making a rational choice of *n* were discussed in Section 11.4. Similar procedures can be used with *t*, but the computations are difficult to perform. If the difference one is interested in detecting is expressed as an effect size, *d*, the required *n* can be looked up in Appendix Table D.9.[2] Cohen (1969, pp. 18–25) refers to a *d* value of 0.2 as a small-effect size; *d* values of 0.5 and 0.8 are called, respectively, medium- and large-effect sizes. To use Table D.9, it is necessary to specify *d*, α, $1 - \beta$, and whether the test is directional or nondirectional. In the class registration experiment, suppose we wanted to detect a large-effect size ($d = 0.8$), and we wanted α to equal 0.05 and $1 - \beta$ to equal 0.80. The required sample size according to Table D.9 is 12, which is approximately half the number used in the example.

More about the Normality Assumption

One of the assumptions of the *t* test is that the population distribution is normal in form, in which case the sampling distribution of $\bar{X}$ is normal for any size *n*. We know from the central limit theorem that the sampling distribution of $\bar{X}$ tends to the normal form even if *X*, the underlying variable, is not normal. The larger the sample size *n*, the better the approximation to the normal form. Thus, if *n* is large, the *t* test statistic can be used even though the population distribution is not normal. But what if *n* is small and the population distribution is not normal? In this case the sampling distribution of $\bar{X}$ may deviate significantly from the normal form, particularly if the population distribution is markedly skewed. We can find some comfort in the fact that the *t* test is not very sensitive to departure from normality; that is, it is *robust* with respect to nonnormality (Boneau, 1960). Thus, the actual probability of a type I error will be fairly close to the nominal or specified probability of a type I error even when the population distribution is not normal and *n* is small.

Research has also indicated that power and the probability of a type II error are not significantly affected by lack of normality of the population distribution, provided *n* is large. However, when *n* is small an experimenter may fail to reject a false null hypothesis due to lack of power resulting from nonnormality of the population distribution.

Review Exercises for Section 12.2

1. List the similarities and differences between the *t* and *z* tests.
2. Suppose several first-grade teachers have complained that their classes this year are unusually slow in learning to read. The school principal has asked you to determine if the children are below average in intelligence—have a mean IQ below 100. Since there are 362 first-grade children, it is not feasible to give an individual intelligence test to each of them. Instead, you administer the Wechsler Intelligence Scale for Children to a random sample of 16 children. List the steps you would follow in testing the scientific hypothesis. Let $\alpha = 0.05$.

[2] More extensive tables are provided by Cohen (1969), Dixon and Massey (1957), and Minium (1970).

*3. Assume the data in the table have been obtained for the children in Exercise 2. (a) Compute a t statistic for these data and make a decision about the scientific hypothesis. (b) Use Table D.9 to estimate the sample size needed to detect a large-effect size for $\alpha = 0.05$ and $1 - \beta = 0.80$.

Child	IQ	Child	IQ
1	89	9	86
2	96	10	88
3	86	11	92
4	92	12	101
5	78	13	87
6	110	14	93
7	82	15	97
8	69	16	74

4. Make a frequency distribution for the data in Exercise 3. Use ten class intervals, with a class interval size of five. From a visual inspection of the distribution, is it reasonable to assume the population distribution is normal in form?

5. A random sample of 65 freshman college students was selected to participate in a new look–say teaching program designed to increase reading speed in French. The final exam consisted of a French passage that the students translated. The time required for each student to complete the translation was recorded. The sample statistics were $\bar{X} = 302$ sec and $\hat{\sigma} = 56$ sec. According to departmental records, the mean for students in conventional classes was 320 sec. Let $\alpha = 0.05$.
   a. List the steps you would use in testing the scientific hypothesis that the look–say program resulted in a decrease in time required to translate the French passage.
   b. Compute a t statistic and make a decision about the scientific hypothesis.
   c. How could the design of the experiment be improved?
   d. Is the sample adequate to detect a medium-effect size if a power of 0.95 is desired?

6. Terms to remember:
   a. Degrees of freedom
   b. Robust

12.3 One-Sample Chi-Square Test for a Population Variance

Most scientific hypotheses are concerned with central tendency, but hypotheses concerning population variance, skewness, kurtosis, and correlation also can be of interest. In this section we describe a chi-square (χ^2) statistic for evaluating a hypothesis about a single population variance. *The chi-square test statistic is given by*

$$\chi^2 = \frac{(n-1)\hat{\sigma}^2}{\sigma_0^2},$$

where n is the number of elements in a random sample from the population, $\hat{\sigma}^2$ is an unbiased estimator of the population variance, and σ_0^2 is the value of the population variance specified in the null hypothesis.

The chi-square statistic can be used to test hypotheses such as the following.

H_0: $\sigma^2 = \sigma_0^2$	H_0: $\sigma^2 \leq \sigma_0^2$	H_0: $\sigma^2 \geq \sigma_0^2$
H_1: $\sigma^2 \neq \sigma_0^2$	H_1: $\sigma^2 > \sigma_0^2$	H_1: $\sigma^2 < \sigma_0^2$

The assumptions associated with using the chi-square statistic to test these hypotheses are that (1) the population distribution of X is normal in form, (2) the X's are a random sample from the population of interest, and (3) the null hypothesis is true, which is the hypothesis that the experimenter hopes to reject.

The sampling distribution of chi square was derived in 1876 by F. R. Helmert; in 1900 it was used by Karl Pearson to test hypotheses. The chi-square distribution, like the t distribution, is actually a family of distributions with a shape that is dependent on degrees of freedom. The number of degrees of freedom, v, for χ^2 is equal to $n - 1$. Unlike the z and t distributions, the chi-square distribution is positively skewed for small degrees of freedom, but as v increases the distribution approaches the normal form with mean and variance, respectively,

$$E(\chi_v^2) = v \quad \text{and} \quad \text{Var}(\chi_v^2) = 2v.$$

Since χ^2 is a squared quantity, it can range over only nonnegative numbers, zero to positive infinity, whereas t and z can range over all real numbers. The chi-square distributions for several different degrees of freedom are shown in Figure 12.3-1. The critical value of χ^2 that cuts off the upper α region of the sampling distribution for v degrees of freedom is given in Appendix Table D.4 and is denoted by $\chi_{\alpha,v}^2$; the value that cuts off the lower α region is denoted by $\chi_{1-\alpha,v}^2$. If α is to be divided equally between the two tails, as in performing a two-tailed test, the critical values are denoted by $\chi_{\alpha/2,v}^2$ and $\chi_{1-\alpha/2,v}^2$.

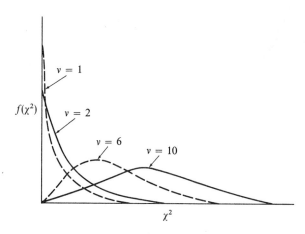

Figure 12.3-1. Chi-square distributions for different degrees of freedom.

Computational Example for One-Sample Chi-Square Test

It has been shown that students who take a driver education course in high school receive on the average higher scores on the state licensing test than do students who haven't taken the course. One might also expect less variability among the scores of students who have taken driver education than among students who haven't.

We can make an inference regarding this scientific hypothesis by means of the one-sample chi-square test. Let us assume that the variance on the licensing test for high school seniors who haven't taken the driver education course is known to equal 324. The decision rule and the steps to be followed in testing the null hypothesis are as follows.

Step 1. State the statistical hypotheses: $H_0: \sigma^2 \geq 324$
$H_1: \sigma^2 < 324$.

Step 2. Specify the test statistic: $\chi^2 = \dfrac{(n-1)\hat{\sigma}^2}{\sigma_0^2}$.

Step 3. Specify the sample size: $n = 16$;
and the sampling distribution: chi-square distribution with $v = n - 1$, because the population distribution of X is assumed to be normal in form.

Step 4. Specify the significance level: $\alpha = 0.05$.

Step 5. Obtain a random sample of size n, compute χ^2, and make a decision.

Decision rule: Reject H_0 if χ^2 falls in the lower 5% of the sampling distribution of χ^2; otherwise, don't reject H_0.

Assume that driver license scores for a random sample of 16 high school seniors who have taken the driving course have been obtained from the Bureaucracy of Motor Vehicles. The data are given in Table 12.3-1. For this sample the variance is equal to

Table 12.3-1. Driver License Test Scores for a Random Sample of High School Seniors Who Have Completed a Driving Course

(i) Data:

Student Number	Score, X_i	X_i^2	Student Number	Score, X_i	X_i^2
1	70	4900	9	57	3249
2	68	4624	10	61	3721
3	81	6561	11	72	5184
4	95	9025	12	78	6084
5	63	3969	13	71	5041
6	69	4761	14	82	6724
7	47	2209	15	87	7569
8	79	6241	16	59	3481
				$\sum X_i = 1139$	$\sum X_i^2 = 83{,}343$

(ii) Computation:

$$\hat{\sigma}^2 = \frac{\sum X_i^2 - (\sum X_i)^2/n}{n-1} = \frac{83{,}343 - (1139)^2/16}{16 - 1} = 150.6958$$

$$\chi^2 = \frac{(n-1)\hat{\sigma}^2}{\sigma_0^2} = \frac{(16-1)150.6958}{(18)^2} = \frac{2260.437}{324.000} = 6.977$$

$v = n - 1 = 16 - 1 = 15$

$\chi^2_{0.95,15} = 7.261$

150.7. Is this variance small enough to warrant the conclusion that the population parameter is not greater than or equal to 324?

According to the chi-square Table D.4, a χ^2 of 7.261 cuts off the lower 0.05 region of the sampling distribution—that is, $\chi^2_{0.95,15} = 7.261$. Since $6.977 < 7.261$, the null hypothesis is rejected and the experimenter concludes that the population variance of driver license scores is smaller for students who have had a driving course in high school than for students who haven't had the course.

More about the Normality Assumption for χ^2

How important is the assumption that the population distribution of X is normal in form? In Section 12.2 we learned that the t test is robust with respect to departures from normality. Unfortunately, this isn't true for this chi-square test. If the sampling distribution of χ^2 is to have the same form as the chi-square distribution, the population distribution of X must be normal in form. Deviations from the normal form can result in large discrepancies between the actual and nominal type I and II errors.

Review Exercises for Section 12.3

7. List the similarities and differences between the χ^2 and t sampling distributions.
*8. For the data in Exercise 3 (page 244), test the hypothesis $H_0: \sigma^2 = (15)^2$. Let $\alpha = 0.10$.
9. The mean number of movies attended per month by 14-year-old boys in the United States is 4.8, with a standard deviation of 1.3. According to juvenile court records in Houston, Texas, the corresponding statistics for a random sample of 31 boys who appeared in court are $\bar{X} = 4.9$ and $\hat{\sigma} = 1.6$. Evaluate the hypothesis that the dispersion of the movie attendance distribution for boys appearing in the Houston juvenile court is different from that for the nation at large. Let $\alpha = 0.02$.
10. List the assumptions involved in testing the null hypothesis in Exercise 9.

12.4 One-Sample z Test for a Population Proportion

An experimenter is often interested in testing a hypothesis about a population proportion. For example, an opinion pollster may want to know if a majority of the voters favor a certain candidate, an automobile manufacturer may want to know if at least 0.70 of new car buyers are willing to pay $150 for a safety device, or the United States Marine Corps may want to know if at least 0.35 of its volunteers plan to reenlist. In each of these examples there is a large number of occasions, or independent trials, in which one of two outcomes can occur, and the probabilities associated with the two outcomes remain constant from trial to trial. For convenience the outcomes are designated "success" and "failure," with probabilities p and $1 - p = q$, respectively.

We saw in Section 9.3 that a Bernoulli trial results in success with probability p or failure with probability q and that the probability of exactly r successes in n independent trials for the binomial random variable X is given by

$$p(X = r) = {}_nC_r p^r q^{n-r}.$$

If n is very small, this binomial function rule can be used to determine the probabilities associated with values of the random variable X—that is, the probabilities that various numbers of successes will occur. If, however, n is large, this procedure is not computationally feasible. In this case, the normal distribution approximation to the binomial distribution can be used. This approximation is excellent if n is large and $p = 0.5$; as n becomes smaller and/or p approaches either 0 or 1 the approximation becomes poorer. As a rule of thumb, if np_0 (the sample size multiplied by the population proportion specified in the null hypothesis) and $n(1 - p_0)$ are both greater than 5, the normal approximation is satisfactory.

The z statistic for testing the null hypothesis about a population proportion is given by

$$z = \frac{\hat{p} - p_0}{\sqrt{p_0 q_0 / n}} \quad \text{or} \quad z = \frac{X - np_0}{\sqrt{np_0 q_0}},$$

where $\hat{p}$ is the sample estimator of the population proportion, p_0 is the hypothesized value of the population proportion specified in the null hypothesis, q_0 is equal to $1 - p_0$, n is the size of the sample used to compute $\hat{p}$, and X is equal to $n\hat{p}$ (the number of elements in the success category). The z statistic can be used to test hypotheses such as the following.

$$\begin{array}{ccc} H_0: p = p_0 & H_0: p \leq p_0 & H_0: p \geq p_0 \\ H_1: p \neq p_0 & H_1: p > p_0 & H_1: p < p_0 \end{array}$$

Here, p is the population proportion.

Computational Example for One-Sample z Test

Suppose the League for Better Housing has made a survey to determine whether the proportion of substandard dwelling units in a large city has changed since the last census 5 years ago. At that time 0.30 of the dwelling units were classified as substandard. A random sample of 900 dwelling units is surveyed and the proportion of substandard units is found to be 0.34. The decision rule and the steps to be followed in testing the null hypothesis are as follows.

Step 1. State the statistical hypotheses: $H_0: p = 0.30$
$H_1: p \neq 0.30$.

Step 2. Specify the test statistic: $z = \dfrac{\hat{p} - p_0}{\sqrt{p_0 q_0 / n}}.$

Step 3. Specify the sample size:[3] $n = 900$;
and the sampling distribution: standard normal distribution, because z is a satisfactory approximation to the binomial distribution since $np_0 = 270$ and $n(1 - p_0) = 630$ are both greater than 5.

[3] Cohen (1969, chap. 6) provides tables for estimating the sample size required for various values of α, $1 - \beta$, and effect size.

Step 4. Specify the significance level: $\alpha = 0.05$.
Step 5. Obtain a random sample of size n, compute z, and make a decision.
Decision rule: Reject H_0 if z falls in either the lower 2.5% or the upper 2.5% of the sampling distribution of z; otherwise, don't reject H_0.

For a $\hat{p} = 0.34$, the proportion of substandard dwelling units in the last survey, the z statistic is

$$z = \frac{0.34 - 0.30}{\sqrt{(0.30)(0.70)/900}} = \frac{0.04}{\sqrt{0.0002333}} = 2.62.$$

According to Appendix Table D.2, z's of $+1.96$ and -1.96, respectively, cut off the upper and lower 0.025 regions of the sampling distribution. Since $2.62 > 1.96$, the null hypothesis is rejected and it is concluded that the housing situation hasn't improved but rather has deteriorated.

Correction for Continuity

Many variables, such as the number of substandard dwelling units, have a discrete probability distribution. *When the continuous normal distribution is used to estimate probabilities for a discrete distribution and n is small, it is desirable to apply a **correction for continuity***. In making a correction for continuity we treat a number, say $X = 6$, as representing the interval 5.5–6.5; that is, a number is considered to have an actual lower limit equal to $X -$ (Unit of measurement)/2 and an actual upper limit equal to $X +$ (Unit of measurement)/2. *The addition or subtraction of $\frac{1}{2}$ the unit of measurement is the correction for continuity.* The rule for applying the correction is simple. Add $\frac{1}{2}$ the unit of measurement to X in the formula $z = (X - np_0)/\sqrt{np_0 q_0}$ if $X < np_0$; if $X > np_0$, subtract $\frac{1}{2}$ the unit of measurement from X.

If n is large, the use of the continuity correction has little effect on the z test and can be dispensed with. The correction has little effect on the z statistic for the housing survey data, as the following computation shows.

$$z = \frac{(X - \frac{1}{2}) - np_0}{\sqrt{np_0 q_0}} = \frac{(306 - 0.5) - 900(0.30)}{\sqrt{900(0.30)(0.70)}} = \frac{35.5}{\sqrt{189}} = 2.58.$$

Computational Procedure when the Sample Size Is Equal to or Greater than 10% of the Population

If the population is finite and sampling is carried out without replacement, the random variable X is distributed as the hypergeometric distribution (see Section 9.3) rather than as the binomial distribution. For this situation the z statistic should be modified if the ratio $n/n_{Pop} \geq 0.10$, where n is the number of elements in the sample

12.4
One-Sample z Test for a Population Proportion

and n_{Pop} is the number of elements in the population. The modified z statistic is

$$z = \frac{\hat{p} - p_0}{\sqrt{(p_0 q_0/n)[(n_{Pop} - n)/(n_{Pop} - 1)]}}.$$

The denominator adjustment $(n_{Pop} - n)/(n_{Pop} - 1)$ can be ignored if $n/n_{Pop} < 0.10$.

Review Exercises for Section 12.4

*11. If $p_0 = 0.20$, how large should n be in order to use the normal approximation to the binomial distribution?

*12. Suppose you are interested in testing babies' color preferences. On each of $n = 20$ trials you offer a baby a choice between two balls—one red and one green. The baby chooses a red ball on 12 of the 20 trials. Can you conclude that the baby has a preference for red over green? Use the correction for continuity. Let $\alpha = 0.05$.

13. The probability of recovery for schizophrenic patients after receiving 6 months of conventional therapy at Happyfarm Hospital was 0.60. A token economy program was introduced for a random sample of 30 schizophrenic patients. At the end of the 6 months' trial period, 21 patients had recovered. Test the hypothesis that the recovery probability for the token economy program would be higher than that for the conventional therapy. Let $\alpha = 0.05$.

14. Sketch the sampling distribution for z in Exercise 13 and label the critical region.

15. Assume that in Exercise 13 there were only 90 schizophrenic patients in the hospital. Test the null hypothesis using the z statistic for the finite population case.

16. Term to remember:
Correction for continuity

12.5 One-Sample t and z Tests for a Population Correlation

t Test for $H_0: \rho = 0$

In many correlation problems the hypotheses of interest are

$$H_0: \rho = 0$$
$$H_1: \rho \neq 0.$$

A sample correlation coefficient, r, can differ from zero due to chance sampling variability. Thus, just because $r \neq 0$ it can't be concluded that the population correlation parameter, ρ, differs from zero. In this section we describe a t test that helps an experimenter decide on the basis of sample data whether ρ is or isn't equal to zero.

Computational Example for t Test

Consider an experimenter who is interested in determining whether there is a correlation between college grades and income 10 years after graduation. Assume grade point averages and income data for a random sample of 62 male graduates of Lollipop

Day-Care College have been obtained. The product-moment correlation between grade point average and income for this sample is 0.16. How likely is it that a sample correlation coefficient of this size would have been obtained if the correlation between income and grades is really equal to zero?

A test of the hypothesis $H_0: \rho = 0$ is given by the t statistic

$$t = \frac{r\sqrt{n-2}}{\sqrt{1-r^2}}$$

with $n - 2$ degrees of freedom. The sampling distribution of r can be regarded as approximately normal when ρ is close to zero; for other values of ρ, the sampling distribution of r tends to be very skewed and the t statistic shouldn't be used.

The t test of the hypothesis $\rho = 0$ assumes that the population of paired X and Y scores is *bivariate normal* in form. A **bivariate normal population** is one in which the distributions of X and Y are normal, the relationship between X and Y is linear, and the distribution of Y for any value of X is normal with variance that does not depend on the X value selected and vice versa.

The decision rule and the steps to be followed in testing the hypothesis are as follows.

Step 1. State the statistical hypotheses: $H_0: \rho = 0$
$H_1: \rho \neq 0$.

Step 2. Specify the test statistic:
$$t = \frac{r\sqrt{n-2}}{\sqrt{1-r^2}}.$$

Step 3. Specify the sample size:[4] $n = 62$;
and the sampling distribution: t distribution with $v = n - 2$, because the population is assumed to be bivariate normal in form.

Step 4. Specify the significance level: $\alpha = 0.05$.

Step 5. Obtain a random sample of size n, compute t, and make a decision.
Decision rule: Reject H_0 if t falls in either the upper or lower 2.5% of the sampling distribution of t; otherwise, don't reject H_0.

Since $r = 0.16$ for the random sample of $n = 62$ male graduates, the t statistic is

$$t = \frac{0.16\sqrt{62-2}}{\sqrt{1-(0.16)^2}} = \frac{1.239}{0.987} = 1.26,$$

with $v = 62 - 2 = 60$. According to Appendix Table D.3, $t_{0.05/2, 60} = 2.00$. Since 1.26 < 2.00, the null hypothesis isn't rejected. The data do not warrant the conclusion that the population correlation coefficient is not equal to zero.

[4] Cohen (1969, chap. 4) provides tables for estimating the sample size required for various values of α, $1 - \beta$, and effect size.

There is a much simpler way to test the null hypothesis when it specifies that $\rho = 0$. The critical values of r and r_s necessary to reject the null hypothesis are tabulated in Tables D.6 and D.7, respectively. If $|r|$ is equal to or greater than the critical value in the table, the hypothesis that $\rho = 0$ can be rejected.

†z Test for Correlations Other than Zero

The t test just described can be used to test the null hypothesis when the predicted value equals zero. But what if the experimenter wants to test a hypothesis in which the predicted value is equal to some number other than zero, for example, $\rho = 0.50$? A procedure due to R. A. Fisher can be used in such cases. It uses a particular function of r rather than r; the function is called the *Fisher r-to-Z' transformation*. By means of Table D.8, r can be easily transformed into Fisher's Z'. This table gives for each value of r the corresponding Fisher Z' statistic. For example, if $r = 0.50$, $Z' = 0.549$. Fisher showed that the sampling distribution of Z' is approximately normal in form if (1) ρ is not too close to 1 or -1, (2) the population is bivariate normal in form, and (3) the sample size is at least moderately large, say $n > 50$.

The one-sample z statistic used to test hypotheses about ρ is given by

$$z = \frac{Z' - Z_0'}{\sqrt{1/(n-3)}},$$

where Z' is the transformed sample r, Z_0' is the transformed value of ρ_0 specified by the null hypothesis, and n is the size of the sample used to estimate the population correlation parameter.

Computational Example for z Test

Assume an experimenter believes that the population correlation between two variables is greater than 0.25. A random sample of 83 subjects from the population of interest has been obtained and r is found to equal 0.47. The decision rule and the steps to be followed in testing the hypothesis are as follows.

Step 1. State the statistical hypotheses: H_0: $\rho \leq 0.25$
H_1: $\rho > 0.25$.

Step 2. Specify the test statistic:
$$z = \frac{Z' - Z_0'}{\sqrt{1/(n-3)}}.$$

Step 3. Specify the sample size: $n = 83$;
and the sampling distribution: z distribution, because the population is assumed to be bivariate normal in form and the parameter specified in H_0 is not zero.

† This subsection can be omitted without loss of continuity.

Step 4. Specify the significance level: $\alpha = 0.05$.
Step 5. Obtain a random sample of size n, compute z, and make a decision.
Decision rule: Reject H_0 if z falls in the upper 5% of the sampling distribution of z; otherwise, don't reject H_0.

The test statistic is given by

$$z = \frac{Z' - Z_0'}{\sqrt{1/(n-3)}} = \frac{0.5101 - 0.2550}{\sqrt{1/(83-3)}} = \frac{0.2551}{\sqrt{0.0125}} = 2.28.$$

According to Table D.2, a z of 1.64 cuts off the upper 0.05 region of the sampling distribution of z—that is, $z_{0.05} = 1.64$. Since $2.28 > 1.64$, we reject the null hypothesis and conclude that the population correlation coefficient is greater than 0.25.

Review Exercises for Section 12.5

*17. Assume $r = 0.24$ has been computed for a random sample of 26 subjects. Test the hypothesis $H_0: \rho = 0$. Let $\alpha = 0.05$.

**18. Convert r into Z'.
   *a. $r = 0.46$ *b. $r = -0.23$
   c. $r = -0.96$ d. $r = 0.15$

**19. Convert Z' into r.
   *a. $Z' = 0.549$ *b. $Z' = -0.203$
   c. $Z' = 0.121$ d. $Z' = -1.256$

20. Assume $r = 0.63$ has been computed for a random sample of 96 subjects. Test the hypothesis $H_0: \rho \leq 0.50$. Let $\alpha = 0.05$.

*21. a. If r equalled 0.48 in Exercise 20, would it be necessary to compute a test of significance?
    b. Why?

22. The sampling distribution of r is not likely to be normal when ρ deviates appreciably from zero. From what you know about r, why is this true?

23. Terms to remember:
    a. Bivariate normal
    b. Fisher r-to-Z' transformation

12.6 Summary

We have covered a lot of ground in Chapters 11 and 12—the basic concepts of hypothesis testing and a variety of test statistics for the one-sample case. In spite of the fact that the test statistics have different formulas and are used to test hypotheses about different parameters, they all use the same five steps in arriving at a decision about a hypothesis. We will see that the logic underlying hypothesis testing procedures described in these chapters generalizes to the two-sample case and to more complex decision-making situations.

The test statistics for the one-sample case are presented in Table 12.6-1, along with a summary of the parameters they are used to test, their formulas, and their assumptions.

Table 12.6-1. Summary of One-Sample Test Statistics

Chapter Section	Statistic	Statistical Hypotheses	Formula	Assumptions
11.3	z	$H_0: \mu = \mu_0$ $H_1: \mu \neq \mu_0$	$z = \dfrac{\bar{X} - \mu_0}{\sigma/\sqrt{n}}$	1. Random sampling 2. Normality 3. Variance is known
12.2	t	$H_0: \mu = \mu_0$ $H_1: \mu \neq \mu_0$	$t = \dfrac{\bar{X} - \mu_0}{\hat{\sigma}/\sqrt{n}}$	1. Random sampling 2. Normality 3. Variance is unknown
12.3	χ^2	$H_0: \sigma^2 = \sigma_0^2$ $H_1: \sigma^2 \neq \sigma_0^2$	$\chi^2 = \dfrac{(n-1)\hat{\sigma}^2}{\sigma_0^2}$	1. Random sampling 2. Normality
12.4	z	$H_0: p = p_0$ $H_1: p \neq p_0$	$z = \dfrac{\hat{p} - p_0}{\sqrt{p_0 q_0 / n}}$ $z = \dfrac{X - np_0}{\sqrt{np_0 q_0}}$	1. Random sampling 2. Binomial distribution 3. $np_0 > 5$, $n(1 - p_0) > 5$
12.4	z	$H_0: p = p_0$ $H_1: p \neq p_0$	$z = \dfrac{\hat{p} - p_0}{\sqrt{(p_0 q_0 / n)[(n_{Pop} - n)/(n_{Pop} - 1)]}}$	1. Random sampling 2. Hypergeometric distribution 3. $np_0 > 5$, $n(1 - p_0) > 5$ 4. $n/n_{Pop} \geq 0.10$
12.5	t	$H_0: \rho = 0$ $H_1: \rho \neq 0$	$t = \dfrac{r\sqrt{n-2}}{\sqrt{1 - r^2}}$	1. Random sampling 2. Bivariate normality
12.5	z	$H_0: \rho = \rho_0$ $H_1: \rho \neq \rho_0$	$z = \dfrac{Z' - Z_0'}{\sqrt{1/(n - 3)}}$	1. Random sampling 2. Bivariate normality 3. Moderately large sample

13

Statistical Inference: Two Samples

13.1 Introduction to Hypothesis Testing for Two Samples
13.2 Two-Sample z Test for $\mu_1 - \mu_2$
  Review Exercises for Section 13.2
13.3 Two Randomization Strategies: Random Sampling and Random Assignment
  The Strategy of Random Sampling
  The Strategy of Random Assignment
  Advantages and Disadvantages of the Two Research Strategies
  Review Exercises for Section 13.3
13.4 Two-Sample t Test for $\mu_1 - \mu_2$
  Computational Example for Two-Sample t Test (Independent Samples)
  Review Exercises for Section 13.4
13.5 The z and t Tests for $\mu_1 - \mu_2$ Using Dependent Samples
  Introduction to Dependent Samples
  z Test for Dependent Samples
  t Test for Dependent Samples
  Computational Example for Two-Sample t Test (Dependent Samples)
  Determining the n Required to Achieve an Acceptable α, $1 - \beta$, and $(\mu_1 - \mu_2)/\sigma$
  Group Matching: A Research Strategy to be Avoided
  Review Exercises for Section 13.5
13.6 Summary

13.1 Introduction to Hypothesis Testing for Two Samples

Are men able to withstand weightlessness better than women? Do disadvantaged children learn more quickly in a contingency management classroom than in the traditional classroom? Do people who jog have fewer heart attacks than those who don't? Is one antilitter slogan more effective than another? Each of these questions involves a comparison of two population distributions. Population distributions can differ in central tendency, dispersion, skewness, and kurtosis. Most questions in the behavioral sciences and education are concerned with central tendency and, more specifically, with whether the means of two populations differ.

We learned in Chapter 11 that scientific hypotheses often involve predictions about populations with elements that are so numerous it is impossible to view them all (all men and women in a weightless environment, all disadvantaged school children, all joggers and nonjoggers) or about phenomena that can't be directly observed (the effectiveness of two antilitter slogans). In such cases we can obtain random samples from the populations, and by using hypothesis testing procedures, we can make inferences as to whether the means, variances, and so on, of the populations differ. These procedures are a straightforward extension of those for the one-sample case described in Chapters 11 and 12.

13.2 Two-Sample z Test for $\mu_1 - \mu_2$

We will now describe a z test statistic for determining whether the means of two populations differ. The statistic can be used to test any of the following null hypotheses.

$H_0: \mu_1 - \mu_2 = \delta_0$ versus $H_1: \mu_1 - \mu_2 \neq \delta_0$
$H_0: \mu_1 - \mu_2 \leq \delta_0$ versus $H_1: \mu_1 - \mu_2 > \delta_0$
$H_0: \mu_1 - \mu_2 \geq \delta_0$ versus $H_1: \mu_1 - \mu_2 < \delta_0$

Here, δ_0 is the predicted difference between the population means. Usually, an experimenter is interested in testing the hypothesis that two population means are equal, in which case $\delta_0 = 0$.

Consider populations with unknown means μ_1 and μ_2. Assume the variances of the populations, σ_1^2 and σ_2^2, are known but are not necessarily equal and random samples of size n_1 and n_2 have been obtained from the respective populations. If the population distributions of X_1 and X_2 are normal in form or if the samples are sufficiently large, the sampling distribution of the difference between sample means, $\bar{X}_1 - \bar{X}_2$, is normal. It can be shown that the expectation of the difference between sample means $E(\bar{X}_1 - \bar{X}_2)$ is equal to $\mu_1 - \mu_2$ and that the variance of the difference between sample means $\sigma^2_{\bar{X}_1 - \bar{X}_2}$ is equal to $\sigma_1^2/n_1 + \sigma_2^2/n_2$. The statistic $\sigma_{\bar{X}_1 - \bar{X}_2}$ is called the *standard error of the difference between two means*. The z statistic for testing the null hypothesis is given by

$$z = \frac{(\bar{X}_1 - \bar{X}_2) - \delta_0}{\sigma_{\bar{X}_1 - \bar{X}_2}} = \frac{(\bar{X}_1 - \bar{X}_2) - \delta_0}{\sqrt{\sigma_1^2/n_1 + \sigma_2^2/n_2}}.$$

If the null hypothesis specifies that $\delta_0 = 0$, the formula for z simplifies to

$$z = \frac{\bar{X}_1 - \bar{X}_2}{\sqrt{\sigma_1^2/n_1 + \sigma_2^2/n_2}}.$$

When the null hypothesis is true, the sampling distribution of z is approximately normal[1] in form with mean equal to zero, $E(z) = 0$, and variance equal to 1, $\sigma_z^2 = 1$. Hence, the sampling distribution of z has approximately the same shape, mean, and standard deviation as the standard normal distribution with values listed in Table D.2. These similarities enable us to use the standard normal distribution in determining whether an obtained z, our transformed difference between sample means, lies in the critical region of the sampling distribution of z.

In the preceding discussion we assumed that σ_1^2 and σ_2^2 were somehow known so that we could compute the standard error of the difference between means. In the real world such omniscience is rare; the fact of the matter is that population variances are seldom known. We can, however, compute unbiased estimates of the population variances from the samples used to estimate the population means. When the population variances are estimated, the appropriate test statistic is t. But before describing this statistic we will turn briefly to an issue concerning the design of experiments: selection of the appropriate randomization procedure.

Review Exercises for Section 13.2

1. The null hypothesis is sometimes written $H_0: \mu_1 = \mu_2$. What does this indicate about δ_0?
2. Under what conditions does the sampling distribution of $z = [(\bar{X}_1 - \bar{X}_2) - \delta_0]/\sigma_{\bar{X}_1 - \bar{X}_2}$ approximate the standard normal distribution?
3. An experimenter is interested in testing the hypothesis that college freshmen who are on probation have lower academic aptitude scores than those not on probation. Random samples of $n_1 = 50$ and $n_2 = 50$ are obtained from the populations of probationers and nonprobationers, respectively. The populations are known to be normally distributed with $\sigma_1 = 15$ and $\sigma_2 = 15$. List the five steps you would follow in testing the null hypothesis and state the decision rule. Let $\alpha = 0.05$.
*4. Suppose that for Exercise 3, $\bar{X}_1 = 112$, $\bar{X}_2 = 116$, and α has been set at 0.05. Compute the test statistic and make a decision.
5. Discuss the statement "The absolute magnitude of the z test statistic is indicative of the importance or practical significance of the difference between two sample means."

13.3 Two Randomization Strategies: Random Sampling and Random Assignment

Two randomization strategies can be used in investigating scientific hypotheses. An experimenter can obtain random samples from two existing populations of interest or can randomly assign elements of a population to two experimental conditions. In rare cases

---

[1] In practice, z can never be normally distributed because variables do not range from $-\infty$ to $+\infty$.

the two methods can be combined—that is, sample randomly from a population and assign elements of the sample randomly to the experimental conditions. The choice of strategy affects an experimenter's conclusions.

The Strategy of Random Sampling

Consider the scientific hypothesis that males who jog have fewer heart attacks than those who don't. The statistical hypotheses are

$$H_0: \mu_1 - \mu_2 \geq 0$$
$$H_1: \mu_1 - \mu_2 < 0,$$

where μ_1 and μ_2 designate the mean number of heart attacks of the populations of joggers and nonjoggers, respectively. The alternative hypothesis, which corresponds to the experimenter's hunch, states that the mean number of heart attacks is smaller for joggers than for nonjoggers. To test the null hypothesis, an experimenter could obtain a random sample of 70-year-old men who have jogged regularly since they were 40 and a second sample of men the same age who have never jogged. Suppose the mean number of heart attacks is $\bar{X}_1 = 0.2$ for the joggers and $\bar{X}_2 = 1.1$ for the nonjoggers and that the difference between the means, $0.2 - 1.1 = -0.9$, is significant at the 0.01 level. It can be concluded that the population of joggers has fewer attacks than the nonjoggers, and hence the scientific hypothesis is supported.

Can the experimenter conclude that the difference between the population means is due to jogging per se? Unfortunately, the answer is no, because in all likelihood the two populations of men differ in other ways besides jogging. More than likely, males who jog are concerned about health and about staying in good physical shape. Joggers are probably less often obese, have better muscle tone, and have more healthful diets than nonjoggers. If our experimenter had sampled from populations of obese and nonobese males, or from males with good and poor muscle tone, or from males who are and aren't diet conscious, a significant difference in mean number of heart attacks would probably have been found.

The Strategy of Random Assignment

Suppose a population of 40-year-old prisoners at Oops Penitentiary is available and it is possible to exercise some control over their lives for a period of 30 years. The prisoners are randomly assigned to one of two groups, which we will call the experimental and the control groups. Those assigned to the experimental group participate in a jogging program for 30 years; those in the control group don't participate in the jogging program. Suppose that at the end of 30 years the mean number of heart attacks for those in the experimental and control groups are, respectively, $\bar{X}_1 = 0.3$ and $\bar{X}_2 = 1.4$, and that the difference, $0.3 - 1.4 = -1.1$, is significant at the 0.01 level. As in the previous experiment, the scientific hypothesis is supported.

Can the experimenter conclude that the difference between the experimental and control groups is due to jogging per se? Again the answer is no. What have we accomplished by using random assignment? This procedure ensures that idiosyncratic characteristics of the subjects, such as their interest in staying in shape, their tendency toward obesity, the adequacy of their diets, and so forth, will be randomly distributed over the two groups and therefore will not selectively bias the dependent measure for either group. To state it another way, random assignment ensures the comparability of the experimental and control groups, since prior to the experiment they should differ no more than would be expected by chance. If at the conclusion of the experiment there is a significant difference between them in the incidence of heart attacks, the experimenter can be confident that the difference is due to events that occurred after the experiment began rather than to idiosyncratic characteristics of the subjects that existed prior to the experiment. And if during the experiment all conditions except the independent variable of jogging are held constant, differences between the groups in number of heart attacks must be due to jogging per se. Unfortunately, it is often difficult and sometimes impossible to verify that all conditions except the independent variable have been held constant.

Advantages and Disadvantages of the Two Research Strategies

Most experiments in the behavioral sciences and education are designed to establish **concomitant relationships** *rather than* **causal relationships**. *To establish that an independent variable X causes an effect Y, it is necessary to demonstrate that X is both necessary and sufficient for the occurrence of Y. To establish a concomitant relationship, it is only necessary to demonstrate that the occurrence or nonoccurrence of one event is accompanied by the occurrence or nonoccurrence of the other event.* The distinction between causal and concomitant relationships is important because it affects the way in which an experimenter interprets his or her results. Neither the random sampling nor the random assignment experiments just described have established that jogging per se results in fewer heart attacks—a causal relationship—but they have established that males who jog have, on the average, fewer attacks than nonjoggers—a concomitant relationship.

The strategy of drawing random samples from two existing populations that are known to differ in X can't be used to establish causality, because the two populations may also differ on other variables. One or more of the other variables could be responsible for the effect Y. An experimenter samples randomly from two existing populations so conclusions can be generalized to the populations. In many research situations, most notably opinion polling, the discovery of a concomitant relationship is sufficient for the experimenter's purposes.

In the behavioral sciences and education most experimenters have neither the time nor the resources to obtain random samples. In the rare cases in which random samples are obtained, the populations are often so narrowly defined that they are of little interest. For example, human subjects are frequently randomly sampled from a population of sophomore students enrolled at a college, or from volunteers, and so forth. Experimenters who work with animal subjects generally do not attempt to obtain random samples.

The second strategy of randomly assigning subjects to experimental conditions can be used to establish the existence of a causal relationship if all conditions except the independent variable can be held constant. This is a big if since the requirement is difficult to satisfy in the behavioral sciences and education.

Random assignment of a large number of subjects to the experimental and control conditions equalizes the effects of extraneous subject differences across experimental conditions and allows the experimenter to rule out prior differences as an explanation for results. This is obviously an important advantage of this research strategy. A disadvantage is that the conclusions apply only to the population of subjects used in the experiment—in our example, prisoners at Oops Penitentiary. To the extent that these prisoners resemble all those in the United States, the findings can be logically generalized to the larger population. It should be emphasized that any generalization to populations not actually sampled is an exercise in logic rather than statistical inference.[2] In this example it seems unlikely that the resemblance is great enough to permit such a generalization.

If an experimenter wants to generalize findings to some population and also ensure the comparability of subjects in the experimental and control groups, the two research strategies can be combined. A random sample of subjects from the population of interest can be obtained and then these subjects randomly assigned to the two conditions. This combined strategy can obviously not be used when it is necessary to sample from two populations that differ with respect to the independent variable, for example, populations of joggers and nonjoggers. Such populations are referred to as *intact populations*.

A final point: If an experimenter wants to use statistical inference, the experimental design must include some form of randomization. Which randomization procedure is used will depend upon the objectives of the experiment.

Review Exercises for Section 13.3

6. An experimenter in Conception, Iowa, wished to determine whether there is a relationship between children's IQs and their mothers' ages when they were born. Using school records, a list was compiled of 10-year-olds whose mothers were over 35 at parturition and a second list was compiled of 10-year-olds whose mothers were 20 or under at parturition. The experimenter randomly sampled 50 children from each list and administered the Stanford–Binet intelligence test to them. The IQs were found to be considerably higher for the children of older mothers, and the difference was significant beyond the 0.001 level. The experimenter concluded that a woman should postpone childbearing until later life to ensure a high IQ for her offspring. (a) Comment on the appropriateness of the experimenter's conclusion. (b) List some alternative explanations for the observed difference in IQs.
7. a. In Exercise 6, which sampling strategy was used?
   b. Would this strategy enable the experimenter to establish a causal relationship between the IQs of children and the ages of their mothers at parturition?
*8. In Exercise 6, what does the fact that the test statistic was significant at the 0.001 level tell you about the magnitude of the difference between the population means?
9. What condition in the random assignment strategy must be satisfied to establish a causal relationship between the independent and dependent variables?
10. What are the advantages and disadvantages of random sampling and random assignment?
**11. For each of the following research topics, indicate the research strategy that seems most appropriate. Justify your choice.

[2] This point is thoroughly discussed by Edgington (1966). This article is reprinted in Kirk (1972a).

*a. Relative resistance to extinction of a bar-pressing response acquired under 100% reinforcement and 50% reinforcement.
b. Difference between adult males and females in the incidence of alcohol use.
c. Relationship between grades in college and number of hours studied per week.
d. Difference in reaction time to the onset of a light versus the onset of a tone.
12. Terms to remember:
a. Concomitant relationship
b. Causal relationship
c. Intact populations

13.4 Two-Sample t Test for $\mu_1 - \mu_2$

If an experimenter wants to test the null hypothesis that the difference between two population means is equal to some value, say zero, in all likelihood the population variances, σ_1^2 and σ_2^2, will not be known. Hence, the z test statistic described in Section 13.2 can't be computed. The solution to this problem is to estimate the population variances from sample data and use a t test instead of the z test. For the t test, one assumes the population variances are unknown but equal and random samples of size n_1 and n_2 have been obtained from the respective populations. If the population distributions of X_1 and X_2 are normal in form or if the samples are sufficiently large, the sampling distribution of the difference between sample means, $\bar{X}_1 - \bar{X}_2$, is normal in form. It can be shown that the expectation of the difference between sample means, $E(\bar{X}_1 - \bar{X}_2)$, is equal to $\mu_1 - \mu_2$ and that an unbiased estimator of the variance of the difference between sample means, $\hat{\sigma}^2_{\bar{X}_1 - \bar{X}_2}$, is $\hat{\sigma}^2_{Pooled}/n_1 + \hat{\sigma}^2_{Pooled}/n_2$, where $\hat{\sigma}^2_{Pooled} = [(n_1 - 1)\hat{\sigma}_1^2 + (n_2 - 1)\hat{\sigma}_2^2]/[(n_1 - 1) + (n_2 - 1)]$. The t test statistic is given by

$$t = \frac{(\bar{X}_1 - \bar{X}_2) - \delta_0}{\hat{\sigma}_{\bar{X}_1 - \bar{X}_2}} = \frac{(\bar{X}_1 - \bar{X}_2) - \delta_0}{\sqrt{(\hat{\sigma}^2_{Pooled}/n_1) + (\hat{\sigma}^2_{Pooled}/n_2)}} = \frac{(\bar{X}_1 - \bar{X}_2) - \delta_0}{\sqrt{\hat{\sigma}^2_{Pooled}(1/n_1 + 1/n_2)}},$$

where δ_0 is the predicted difference between the population means, which is usually equal to zero. The number of degrees of freedom, v, for the t statistic is equal to $n_1 + n_2 - 2$. Sample 1 contributes $n_1 - 1$ degrees of freedom, 1 less than the number of independent observations, and, likewise, sample 2 contributes $n_2 - 1$ degrees of freedom.

The use of $\hat{\sigma}^2_{Pooled}$ in the formula for t requires a word of explanation. We assumed that the variances of populations 1 and 2 are equal; hence, $\hat{\sigma}_1^2$ and $\hat{\sigma}_2^2$ are both estimators for the same parameter, σ^2. Whenever two independent estimators of σ^2 are available, a pooled estimator is likely to be better than either of the sample estimators taken alone; $\hat{\sigma}^2_{Pooled}$ is simply a weighted mean of $\hat{\sigma}_1^2$ and $\hat{\sigma}_2^2$, where the weights are the respective v's. This can be seen from the equation

$$\hat{\sigma}^2_{Pooled} = \frac{(n_1 - 1)\hat{\sigma}_1^2 + (n_2 - 1)\hat{\sigma}_2^2}{(n_1 - 1) + (n_2 - 1)}.$$

As in the case of the two-sample z test statistic, t can be used to test any of the following null hypotheses.

$H_0: \mu_1 - \mu_2 = \delta_0$ versus $H_1: \mu_1 - \mu_2 \neq \delta_0$
$H_0: \mu_1 - \mu_2 \leq \delta_0$ versus $H_1: \mu_1 - \mu_2 > \delta_0$
$H_0: \mu_1 - \mu_2 \geq \delta_0$ versus $H_1: \mu_1 - \mu_2 < \delta_0$

Computational Example for Two-Sample t Test (Independent Samples)

Let's suppose that a student in an experimental psychology course is investigating the hypothesis that distributed practice is superior to massed practice in developing skill on a mirror tracing task. The task requires a subject to trace a star pattern on a sheet of paper with his nonpreferred hand; he can see himself tracing the pattern only by looking in a mirror. Forty students, an entire introductory psychology class, are randomly assigned to the practice conditions, with the restriction that 20 students are assigned to each condition. Although the subjects are assigned to the conditions without replacement (once selected a subject is no longer available for another assignment), the samples can be regarded for all practical purposes as independent. Subjects in the distributed condition have a 3 min rest period at the end of each practice trial. Subjects in the massed condition have only a 5 sec pause at the end of each trial—just long enough to permit the experimenter to place a new sheet of paper in the tracing apparatus. Both groups receive 15 practice trials. Since the groups may differ in fatigue at the conclusion of practice, the dependent variable is measured the following day. The subjects are given two warm-up trials; the dependent variable is the time required to trace the star pattern on the next three trials. The decision rule and the steps to be followed in testing the null hypothesis are as follows.

Step 1. State the statistical hypotheses: $H_0: \mu_1 - \mu_2 \geq 0$
$H_1: \mu_1 - \mu_2 < 0$,
where μ_1 and μ_2 denote the population means, respectively, for the distributed and massed conditions.

Step 2. Specify the test statistic: $t = \dfrac{\bar{X}_1 - \bar{X}_2}{\hat{\sigma}_{\bar{X}_1 - \bar{X}_2}}$.

Step 3. Specify the sample sizes:[3] $n_1 = 20$ and $n_2 = 20$;
and the sampling distribution: t distribution, because the population variances are estimated from sample data and the population distributions of X_1 and X_2 are approximately normal in form.

Step 4. Specify the significance level: $\alpha = 0.05$.

Step 5. Obtain random samples of size n_1 and n_2, compute t, and make a decision.
Decision rule: Reject H_0 if t falls in the lower 0.05 portion of the sampling distribution of t; otherwise, don't reject H_0.

The data for the experiment are shown in Table 13.4-1. According to Table D.3, a t of -1.69 cuts off the lower 0.05 region of the sampling distribution—that is, $-t_{0.05, 38} = -1.69$. Since $-1.46 > -1.69$, the null hypothesis is not rejected. The test doesn't warrant the inference that distributed practice leads to better performance on the tracing task than massed practice.

[3] Table D.9 can be used to estimate the sample sizes required for various values of α, $1 - \beta$, and effect size, $(\mu_1 - \mu_2)/\sigma$. The concept of effect size is discussed in Section 11.4; use of the table is discussed in Section 12.2.

Table 13.4-1. Mirror Tracing Data

(i) Data:

	Distributed Practice		Massed Practice
Student	Time, X_1 (Seconds)	Student	Time, X_2 (Seconds)
1	17	21	19
2	18	22	20
3	16	23	22
4	18	24	24
5	12	25	10
6	20	26	25
7	18	27	20
8	20	28	22
9	20	29	21
10	22	30	23
11	20	31	20
12	10	32	10
13	8	33	12
14	12	34	14
15	16	35	12
16	16	36	20
17	18	37	22
18	20	38	24
19	18	39	23
20	21	40	17
$n_1 = 20$	$\sum X_1 = 340$	$n_2 = 20$	$\sum X_2 = 380$
	$\sum X_1^2 = 6054$		$\sum X_2^2 = 7662$

(ii) Computation of preliminary statistics:

$$\bar{X}_1 = \frac{\sum X_1}{n_1} = \frac{340}{20} = 17 \qquad \bar{X}_2 = \frac{\sum X_2}{n_2} = \frac{380}{20} = 19$$

$$\hat{\sigma}_1^2 = \frac{n_1 \sum X_1^2 - (\sum X_1)^2}{n_1(n_1 - 1)} \qquad \hat{\sigma}_2^2 = \frac{n_2 \sum X_2^2 - (\sum X_2)^2}{n_2(n_2 - 1)}$$

$$= \frac{20(6054) - (340)^2}{20(20 - 1)} \qquad = \frac{20(7662) - (380)^2}{20(20 - 1)}$$

$$= 14.421 \qquad = 23.263$$

$$\hat{\sigma}_{Pooled}^2 = \frac{(n_1 - 1)\hat{\sigma}_1^2 + (n_2 - 1)\hat{\sigma}_2^2}{(n_1 - 1) + (n_2 - 1)} = \frac{(20 - 1)(14.421) + (20 - 1)(23.263)}{(20 - 1) + (20 - 1)}$$

$$= 18.842$$

(iii) Computation of t:

$$t = \frac{\bar{X}_1 - \bar{X}_2}{\sqrt{\hat{\sigma}_{Pooled}^2(1/n_1 + 1/n_2)}} = \frac{17 - 19}{\sqrt{18.842(1/20 + 1/20)}} = \frac{-2}{1.373} = -1.46$$

$$-t_{0.05, 38} = -1.69$$

13.4
Two-Sample *t* Test
for $\mu_1 - \mu_2$

Review Exercises for Section 13.4

13. Discuss the meaning of the following statement: $t = (\bar{X}_1 - \bar{X}_2)/\sqrt{\hat{\sigma}^2_{Pooled}(1/n_1 + 1/n_2)}$ is the ratio of two random variables, but $z = (\bar{X}_1 - \bar{X}_2)/\sqrt{\sigma^2/n_1 + \sigma^2/n_2}$ is the ratio of a random variable to a constant.

*14. Under what condition is it appropriate to pool $\hat{\sigma}_1^2$ and $\hat{\sigma}_2^2$ in estimating $\hat{\sigma}_{\bar{X}_1 - \bar{X}_2}$?

*15. It has been reported that employment interviewers spend more time talking to applicants who are hired than to applicants who are rejected. To determine whether this is true for college students seeking summer employment through the University Placement Center, an experimenter posing as an applicant accompanied a random sample of referees to their job interviews. A record of the duration and outcome of $n = 49$ interviews was kept. Compute a t test statistic and make a decision about the experimenter's hypothesis. Let $\alpha = 0.05$.

Duration of Interview (Minutes)

Hired		Rejected	
30	23	19	17
21	24	18	18
24	26	22	19
25	27	13	22
29	24	15	15
24	22	18	19
23	25	17	17
24	26	20	20
28	23	18	18
25	24	19	17
24	27	23	
19	26	12	
25	25	18	

16. A college dean believed that car ownership among students leads to lower grades. To test this hypothesis, she obtained a random sample of student car owners and nonowners and looked up their grades. Compute a t test statistic for the data in the table and make a decision about the dean's hypothesis. Let $\alpha = 0.05$.

Grade Point Averages

Students Owning Cars			Students Not Owning Cars		
2.6	2.5	2.4	2.7	2.9	3.0
2.4	2.6	2.5	2.9	2.5	2.9
2.9	2.8	2.8	2.6	3.1	2.7
2.6	2.7	2.6	2.8	2.8	3.2
2.7	3.0	2.5	3.0	2.9	2.9
2.2	2.3	2.6	2.8	3.0	3.0

17. In Exercise 16, the dean decided to prohibit freshmen from bringing cars to campus. (a) Do you think this action was justified by the data? (b) What other kinds of data about car owners and nonowners would be useful in helping the dean arrive at a rational car policy?

13.5 The z and t Tests for $\mu_1 - \mu_2$ Using Dependent Samples

Introduction to Dependent Samples

The tests we have described so far have required the use of independent samples in which the selection of elements in one sample is not affected by the selection of elements in the other. Samples are independent if, for example, an experimenter samples randomly from two populations. If it is feasible to do so, an experimenter may prefer to use dependent samples because this results in a more powerful test of a false null hypothesis. Dependent samples can be obtained by any of the following research procedures.

1. Observing subjects under both the experimental condition and the control condition—that is, obtaining *repeated measures* on the subjects.
2. Matching each subject in the experimental condition with a subject in the control condition on some variable that is correlated with the dependent variable. This is called *subject matching*.
3. Obtaining sets of identical twins or litter mates and assigning one member of the pair randomly to the experimental condition and the other member to the control condition.
4. Obtaining pairs of subjects who are matched by mutual selection, for example, husband and wife pairs or business partners.

Let us consider these in more detail. The first procedure, observing a set of subjects under two conditions, can only be used with independent variables that have relatively short-duration effects. The nature of the independent variable should be such that the effects of one condition dissipate before the subject is observed under the other condition. Otherwise, the second dependent measure will reflect the cumulative effects of two conditions rather than only the effects of the second condition. There is no such restriction, of course, when carryover effects such as learning or fatigue are the principal interest of the experimenter. If repeated measures are obtained, the order of presentation of the two conditions should be randomized independently for each subject if this is at all possible. It is customary to restrict randomization so that half the subjects receive one condition first while the other half receive the other condition first.

The remaining three procedures for obtaining dependent samples involve forming pairs of subjects who are matched on some basis. In subject matching, a matching variable is used to pair up otherwise unrelated subjects; this matching variable must correlate highly with the dependent variable. For example, IQ and ability to learn verbal material are highly correlated; hence, subjects can be assigned to pairs so that members of each pair have similar IQs and therefore similar verbal learning abilities. The higher the correlation between the matching variable and the dependent variable, the more effective the matching. If identical twins or litter mates are used, it can be assumed that subjects within a pair are matched with respect to genetic characteristics. The aptitudes and abilities of identical twins, fraternal twins to some extent, and even siblings are more similar than those of unrelated subjects. When subjects are matched by mutual selection, the experimenter must always ascertain that the subjects within pairs are in fact more similar with respect to the dependent variable than are unmatched subjects. Knowing a husband's attitudes on certain issues, for example, may provide considerable information about his wife's attitudes on the issues, and vice versa. However, knowing the husband's mechanical aptitude or his recreational interests may provide no information about his wife's aptitude or interests.

z Test for Dependent Samples

You probably wonder what difference it makes whether samples are dependent or independent. If the same subjects are observed twice or if subjects in one sample are paired with subjects in another sample, the outcomes for each pair are not statistically independent. This doesn't affect the expectation of the difference between sample means; the expectation $E(\bar{X}_1 - \bar{X}_2)$ is equal to $\mu_1 - \mu_2$. However, dependence within pairs affects the variance of the difference between means. Let us consider first the case in which σ_1^2 and σ_2^2 are known. The variance of $\bar{X}_1 - \bar{X}_2$ when samples are dependent is[4]

$$\sigma^2_{\bar{X}_1 - \bar{X}_2} = \sigma^2_{\bar{X}_1} + \sigma^2_{\bar{X}_2} - 2\rho_{12}\sigma_{\bar{X}_1}\sigma_{\bar{X}_2},$$

where $\sigma^2_{\bar{X}_1} = \sigma^2/n_1$, $\sigma^2_{\bar{X}_2} = \sigma^2/n_2$, and ρ_{12} is the product-moment correlation between X_1 and X_2. The corresponding formula for independent samples given in Section 13.2 is

$$\sigma^2_{\bar{X}_1 - \bar{X}_2} = \sigma^2_{\bar{X}_1} + \sigma^2_{\bar{X}_2}.$$

A comparison of these formulas indicates that the larger the positive correlation between X_1 and X_2, the smaller is the variance error for dependent samples relative to that for independent samples. Hence, if $\rho > 0$, an experimenter will underestimate z if

$$z = \frac{(\bar{X}_1 - \bar{X}_2) - \delta_0}{\sqrt{\sigma^2_{\bar{X}_1} + \sigma^2_{\bar{X}_2}}}$$

is used instead of

$$z = \frac{(\bar{X}_1 - \bar{X}_2) - \delta_0}{\sqrt{\sigma^2_{\bar{X}_1} + \sigma^2_{\bar{X}_2} - 2\rho_{12}\sigma_{\bar{X}_1}\sigma_{\bar{X}_2}}}.$$

t Test for Dependent Samples

We will now consider the t test for dependent samples; t is the appropriate statistic for testing the null hypothesis that $\mu_1 - \mu_2 = \delta_0$ if σ_1^2 and σ_2^2 are unknown but equal and if the population distributions are normal in form or the samples are sufficiently large. The standard error of t defined in Section 13.4 must be modified if X_1 and X_2 are dependent. A modification comparable to that described for z can be made. However, a much simpler computational procedure is available when dependent samples are used—a method that doesn't require the computation of a correlation coefficient. In this procedure the two-sample problem is converted into a one-sample problem by analyzing D, the difference between pairs of scores, instead of X_1 and X_2. The null

---

[4] The derivation of this formula is similar to that for S^2_{X-Y} mentioned in the Review Exercises for Section 5.3.

hypothesis for this analysis is

$$H_0: \mu_D = \delta_0,$$

where μ_D is the mean of differences between paired scores. This null hypothesis is equivalent to $H_0: \mu_1 - \mu_2 = \delta_0$, since $\mu_D = \mu_1 - \mu_2$.

The t test statistic for dependent samples using the difference score approach is

$$t = \frac{\bar{X}_D}{\hat{\sigma}_{\bar{X}_D}} = \frac{\dfrac{\sum D_i}{n}}{\sqrt{\dfrac{[n \sum D_i^2 - (\sum D_i)^2]}{n^2}}\bigg/\sqrt{n-1}},$$

where $\bar{X}_D$ is the mean of difference scores, $\hat{\sigma}_{\bar{X}_D}$ is used to estimate the standard error of the mean of difference scores, $D_i = X_{i1} - X_{i2}$ for the ith pair of subjects, and n is the number of pairs of subjects. The number of degrees of freedom, v, for this test statistic is equal to $n - 1$. On reflection, it seems reasonable that $v = n - 1$. Recall that $\sum (X_i - \bar{X}) = 0$. Also, it can be shown that $\sum (D_i - \bar{X}_D) = 0$. Thus, when any of the $n - 1$ deviations $D_i - \bar{X}_D$ are known, the remaining deviation is also known, since the $n - 1$ deviations plus the remaining deviation must sum to zero.

Computational Example for Two-Sample t Test (Dependent Samples)

The scientific hypothesis for the mirror tracing task described in Section 13.4 could have been investigated using matched subjects. Suppose subjects are tested on the mirror tracing task using their preferred hand. The time required to trace the star pattern on the last three of five trials is used to form pairs of subjects having comparable tracing times and hence similar motor skills. One subject of each pair is randomly assigned to the distributed practice condition and the other subject to the massed condition. Following the matching procedure, the mirror tracing experiment is carried out as described previously. Data for the experiment are shown in Table 13.5-1. According to Table D.3, a t of -1.729 with $v = 20 - 1 = 19$ cuts off the lower 0.05 region of the sampling distribution. Since $-3.04 < -1.729$, the null hypothesis is rejected and it is concluded that distributed practice leads to better performance on the task than massed practice. Of course, this inference applies only to the population of subjects in the experiment and to the particular practice conditions and task that were used.

Has the experimenter gained anything by using matched subjects? To answer this question we must compare the results of using independent samples with those for dependent samples. This comparison can be made from Tables 13.4-1 and 13.5-1, which illustrate the t computational procedures for the same set of data. The null hypothesis is rejected for the dependent samples analysis but not for the independent samples analysis. The respective t test statistics are $-2/0.657 = -3.04$ and $-2/1.373 = -1.46$. An examination of the data for the two practice conditions suggests that they are positively correlated; the product-moment correlation coefficient r is actually 0.793. This example illustrates an important principle: whenever the correlation between

Table 13.5-1. Mirror Tracing Data (Dependent Samples)

(i) Data:

Student Pair, i	Distributed Practice Time, X_{i1} (Seconds)	Massed Practice Time, X_{i2} (Seconds)	Difference, $D_i = X_{i1} - X_{i2}$
1	17	19	−2
2	18	20	−2
3	16	22	−6
4	18	24	−6
5	12	10	2
6	20	25	−5
7	18	20	−2
8	20	22	−2
9	20	21	−1
10	22	23	−1
11	20	20	0
12	10	10	0
13	8	12	−4
14	12	14	−2
15	16	12	4
16	16	20	−4
17	18	22	−4
18	20	24	−4
19	18	23	−5
20	21	17	4
$n = 20$			$\sum D_i = -40$
			$\sum D_i^2 = 244$

(ii) Computation of preliminary statistics:

$$\bar{X}_D = \frac{\sum D_i}{n} = \frac{-40}{20} = -2$$

$$\hat{\sigma}_{\bar{X}_D} = \frac{\sqrt{\left[n \sum_{i=1}^{n} D_i^2 - \left(\sum_{i=1}^{n} D_i\right)^2\right]/n^2}}{\sqrt{n-1}} = \frac{\sqrt{[20(244) - (-40)^2]/(20)^2}}{\sqrt{20-1}} = \frac{2.8636}{4.3589} = 0.657$$

(iii) Computation of t:

$$t = \frac{\bar{X}_D}{\hat{\sigma}_{\bar{X}_D}} = \frac{-2}{0.657} = -3.04$$

$$-t_{0.05, 19} = -1.729$$

samples is positive, the t for dependent samples is larger than the t for independent samples. Hence, the use of dependent samples results in a more powerful test of a false null hypothesis. This statement must be qualified. The number of degrees of freedom for the independent t test, $n_1 + n_2 - 2 = 38$, is larger than that for the dependent t test, $n - 1 = 19$. The values of t that cut off the critical region are $-t_{0.05, 38} = -1.69$ and $-t_{0.05, 19} = -1.73$. Now for the qualification: In order for a t test with dependent samples to be more powerful than a t test with independent samples, the correlation between dependent samples must be large enough to more than compensate for the smaller degrees of freedom and for the larger absolute value of t required for significance.

There are several assumptions associated with the t statistic for dependent samples.

1. The population distribution of differences, $D_i = X_{i1} - X_{i2}$, is approximately normal in form or the number of pairs of differences is large enough that the sampling distribution of $\bar{D} = \sum D_i/n$ is normal in form.
2. The population variance of the mean of the difference scores, $\sigma^2_{\bar{X}_D}$, is unknown.
3. The pairs of matched subjects are a random sample from a population of matched subjects and/or subjects in a pair are randomly assigned to experimental and control conditions.

Determining the n Required to Achieve an Acceptable α, $1 - \beta$, and $(\mu_1 - \mu_2)/\sigma$

Table D.9 can be used to make a rational choice of sample size for the t test with dependent samples (see Sections 11.4 and 12.2). In addition to α, $1 - \beta$, and d, it is necessary to specify ρ, the correlation between the two populations. Since ρ is rarely known, it is necessary to estimate it based on previous research or informed judgment. Consider the mirror tracing task described in Section 13.4. Suppose we wanted to detect a medium-effect size ($d = 0.5$), and we wanted $\alpha = 0.05$ and $1 - \beta = 0.80$. If we estimate that the population correlation between the distributed and massed practice conditions is approximately 0.70, the required n according to Table D.9 is 16.

Group Matching: A Research Strategy to Be Avoided

A procedure called **group matching** *is sometimes seen in the literature. It involves matching samples on one or more relevant characteristics so that the means and variances of the samples are approximately equal. No attempt is made to match individuals in one sample with those in another sample.*

Group matching is often used in *ex post facto* experiments instead of individual matching. In an *ex post facto* experiment the independent variable has occurred prior to the experiment. Thus, the independent variable is not under an experimenter's control, but records or other information are used to construct two samples that differ with respect to the independent variable. The experimenter can then determine whether the two constructed samples differ on the dependent variable. For example, an experimenter might be interested in determining whether there is a difference in amount of community service (the dependent variable) between females who did or did not participate in girl scouting (the independent variable). Scout records can be used to identify those who were scouts. In all likelihood, two constructed samples—those women who did and those who did not participate in scouting—differ on a variety of variables besides the independent variable. Group matching is used in an attempt to equate the constructed samples on extraneous variables such as IQ, school advancement, socioeconomic background, and so on. High school records, for example, can be used to adjust the membership of the constructed samples so that the sample means and variances are identical on the extraneous variables.

Unfortunately, there are several problems inherent in the group matching procedure.[5] Although it results in dependent samples, the t statistic for dependent samples can't be used because $D_i = X_{i1} - X_{i2}$ can't be computed. This follows since group means and variances are matched, but individual subjects aren't. The only "solution" to this dilemma is to use the t statistic for independent samples. This is a poor solution because (1) group matching restricts the ordinary variation between treatment means that is expected on the basis of random sampling and (2) the denominator of the t statistic for independent samples, $\sqrt{\hat{\sigma}^2_{Pooled}(1/n_1 + 1/n_2)}$, overestimates the standard error of the difference between means when the samples are dependent. Hence, the t statistic for independent samples gives a value that is too small, resulting in a less powerful test of a false null hypothesis. An important experimental design principle emerges from this discussion—the sampling, randomization, and control procedures used in an experiment must be reflected in the statistical analysis and interpretation of data. If this is not possible, presumed refinements in the design of an experiment such as group matching should not be used.

Review Exercises for Section 13.5

18. The use of repeated measures isn't appropriate with some kinds of independent variables; list three variables with which you think it would be appropriate and three with which it wouldn't.
*19. If repeated measures are obtained, what restriction is customarily placed on the order of presentation of the conditions in the experiment?
20. (a) List three matching variables that you believe could be used to form pairs of subjects in a learning experiment using nonsense syllables. (b) Which matching variable do you think would have the highest correlation with number of trials required to learn nonsense syllables?
*21. (a) How is the size of the correlation between dependent samples related to the size of the standard error of the difference between means? (b) How is the size of the correlation between dependent samples related to the probability of rejecting a false null hypothesis?
*22. Before and after seeing a film about marijuana, 16 subjects completed a questionnaire designed to assess their attitudes toward legalization of the drug. (a) For the data in the table, compute a t test statistic and decide whether viewing the film can be expected to result in more favorable attitudes toward legalization of marijuana. Let $\alpha = 0.05$.

Favorableness of Attitude

Subject Pair	Before	After	Subject Pair	Before	After
1	13	16	9	19	20
2	16	18	10	16	18
3	10	12	11	15	18
4	14	18	12	14	15
5	15	18	13	12	12
6	12	15	14	13	17
7	11	12	15	14	16
8	18	20	16	15	17

[5] For an in-depth discussion of these problems see Boneau and Pennypacker (1961). This article is reproduced in Kirk (1972).

(b) Use Appendix Table D.9 to estimate the total number of subjects required to reject a false null hypothesis for $\alpha = 0.05$, $1 - \beta = 0.95$, $d = 0.80$ and $\rho = 0.70$.

23. It is well known that increasing room illumination up to some level increases reading speed. A random sample of 14 sixth-grade students read standardized passages under two levels of ambient room illumination, 5 foot-candles and 15 foot-candles. The order in which the conditions were presented was randomized independently for each subject, with the restriction that the conditions were presented first or second equally often. The reading sessions were separated by an interval of 2 hr. (a) For the data in the table, compute a t test statistic and decide whether reading speed is faster under the 15 foot-candle condition than under the 5 foot-candle condition. Let $\alpha = 0.05$. (b) Use Table D.9 to estimate the total number of subjects required to reject a false null hypothesis for $\alpha = 0.05$, $1 - \beta = 0.90$, $d = 0.80$, and $\rho = 0.60$.

Reading Speed (Words/Minute)

Subject Pair	5 Foot-Candles	15 Foot-Candles	Subject Pair	5 Foot-Candles	15 Foot-Candles
1	88	92	8	90	92
2	92	91	9	84	88
3	86	88	10	82	88
4	84	89	11	86	84
5	90	95	12	84	87
6	86	86	13	86	89
7	88	95	14	86	87

24. If the correlation between matched samples equals zero, the t test for dependent samples will be less powerful than the t test for independent samples. Explain why this is true.
25. Why should group matching be avoided?
26. Terms to remember:
    a. Independent samples
    b. Dependent samples
    c. Group matching
    d. *Ex post facto* experiment

13.6 Summary

Two-sample tests for means are described in this chapter. The associated z and t test statistics are presented within the now familiar five-step hypothesis testing format.

Two important topics related to the design of experiments are also discussed. The first concerns the relative merits of the two randomization strategies—random sampling of elements from two populations versus random assignment of elements to experimental and control conditions. An experimenter's research objectives determine whether one or the other procedure is sufficient or both procedures are required. Remember that an experiment must contain some randomization procedure to justify using statistical inference.

The other topic related to the design of experiments concerns the use of independent samples versus dependent samples. It is to an experimenter's advantage to use dependent samples whenever the nature of the independent variable permits it. Matching subjects on some variable that correlates positively with the dependent variable or

Table 13.6-1. Summary of Two-Sample Test Statistics

Chapter Section	Statistic	Statistical Hypotheses	Formula	Assumptions
13.2	z	$H_0: \mu_1 - \mu_2 = \delta_0$ $H_1: \mu_1 - \mu_2 \neq \delta_0$	$z = \dfrac{(\bar{X}_1 - \bar{X}_2) - \delta_0}{\sqrt{\sigma_1^2/n_1 + \sigma_2^2/n_2}}$	1. Random sampling or random assignment 2. Normality 3. Variances known but not necessarily equal 4. Independence between samples
13.4	t	$H_0: \mu_1 - \mu_2 = \delta_0$ $H_1: \mu_1 - \mu_2 \neq \delta_0$	$t = \dfrac{(\bar{X}_1 - \bar{X}_2) - \delta_0}{\sqrt{\hat{\sigma}_{Pooled}^2 (1/n_1 + 1/n_2)}}$ $v = n_1 + n_2 - 2$	1. Random sampling or random assignment 2. Normality 3. Variances unknown but assumed equal 4. Independence between samples
13.5	z	$H_0: \mu_1 - \mu_2 = \delta_0$ $H_1: \mu_1 - \mu_2 \neq \delta_0$	$z = \dfrac{(\bar{X}_1 - \bar{X}_2) - \delta_0}{\sqrt{\sigma_{\bar{X}_1}^2 + \sigma_{\bar{X}_2}^2 - 2\rho_{12}\sigma_{\bar{X}_1}\sigma_{\bar{X}_2}}}$	1. Random sampling or random assignment 2. Normality 3. Variances known 4. Dependence between samples
13.5	t	$H_0: \mu_1 - \mu_2 = \delta_0$ $H_1: \mu_1 - \mu_2 \neq \delta_0$	$t = \dfrac{\sum_{i=1}^{n} D_i / n}{\sqrt{\left[n\sum_{i=1}^{n} D_i^2 - \left(\sum_{i=1}^{n} D_i\right)^2\right] / n^2} / \sqrt{n-1}}$ $v = n - 1$	1. Random sampling or random assignment 2. Normality 3. Variances unknown 4. Dependence between samples

observing the same subjects under both the experimental and control conditions results in a more powerful test of a false null hypothesis than obtaining independent samples. However, the use of group matching instead of individual matching is not recommended because the presumed refinement cannot be taken into account in the statistical analysis. This suggests an important general principle—the sampling, randomization, and control procedures used in an experiment must be reflected in the statistical analysis and interpretation.

The test statistics we have described in this chapter are presented in Table 13.6-1, along with a summary of the parameters they are used to test, their formulas, and their assumptions.

14

Statistical Inference: Other Two-Sample Test Statistics

14.1 Two-Sample Tests for Equality of Variances
    F Test for Equality of Variances (Independent Samples)
    Computational Example for F Test with Independent Samples
    t Test for Equality of Variances (Dependent Samples)
    Review Exercises for Section 14.1
14.2 Two-Sample Tests for $p_1 - p_2$
    z Test for $p_1 = p_2$ (Independent Samples)
    z Test for $p_1 - p_2 = \delta_0$, where $\delta_0 \neq 0$ (Independent Samples)
    z Test for $p_1 = p_2$ (Dependent Samples)
    Computational Example for z Test with Dependent Samples
    Review Exercises for Section 14.2
14.3 Summary

14.1 Two-Sample Tests for Equality of Variances

F Test for Equality of Variances (Independent Samples)

Frequently, an experimenter's research interest focuses on whether two populations differ in dispersion. An experimenter might want to know if placing disadvantaged children in a contingency management classroom results in less variability in their English-achievement scores than does placing them in a traditional classroom. Or he might want to test one of the assumptions of the *t* test—that two unknown population variances are equal.[1]

An **F statistic** for testing $H_0: \sigma_1^2 = \sigma_2^2$ versus $H_1: \sigma_1^2 \neq \sigma_2^2$ is

$$F = \frac{\hat{\sigma}_1^2}{\hat{\sigma}_2^2},$$

where $\hat{\sigma}_1^2$ and $\hat{\sigma}_2^2$ are unbiased estimators of the population variances.[2] The degrees of freedom for the numerator and denominator are, respectively, $v_1 = n_1 - 1$ and $v_2 = n_2 - 1$. The sampling distribution of the *F* statistic was derived by R. A. Fisher in 1924 and given the name *F* in his honor by G. W. Snedecor. The *F* distribution, like the *t* and chi-square distributions, is actually a family of distributions with a shape that is dependent on degrees of freedom. It is positively skewed, as shown in Figure 14.1-1, but it approaches the normal distribution shape for very large values of v_1 and v_2. Since *F* is a ratio of nonnegative numbers, it can take values only from zero to positive infinity; *F* values around 1 are expected if the null hypothesis that $\sigma_1^2 = \sigma_2^2$ is true. The assumptions associated with using the *F* statistic to test this hypothesis are (1) that the population distributions of X_1 and X_2 are independent and normal, (2) that the two collections of *X*'s are random samples from the populations of interest, and (3) that the population variances are equal, which is the hypothesis that the experimenter hopes to reject. This *F* test, unlike the *t* test, is not robust with respect to violation of the normality assumption. Hence, unless the assumption is fulfilled, the probability of making a type I error will not be equal to the preselected value of α.

The critical value of *F* that cuts off the upper α region of the sampling distribution for v_1 and v_2 degrees of freedom is given in Appendix Table D.5 and is denoted by $F_{\alpha; v_1, v_2}$. To use the table, we locate the column corresponding to the numerator degrees of freedom (v_1) along the top of the table and the row corresponding to the denominator degrees of freedom (v_2) along the side. The column–row intersection gives the critical values of *F*. The critical value that cuts off the lower α region is denoted by $F_{1-\alpha; v_1, v_2}$. These values are not given in the table, but they can be determined from $F_{\alpha; v_1, v_2}$, since

$$F_{1-\alpha; v_1, v_2} = \frac{1}{F_{\alpha; v_2, v_1}}.$$

---

[1] Some books recommend always testing the assumption of equality of variances before performing a *t* test for $\mu_1 - \mu_2 = \delta_0$. If this advice is followed, it should be borne in mind that the *t* test is robust with respect to violation of the assumptions of normalcy. However, the *F* test for $\sigma_1^2 = \sigma_2^2$ described in this section is almost as sensitive to nonnormality as it is to nonequality of variances. Hence, an experimenter may be dissuaded from using a *t* test when it is actually appropriate.

[2] This statistic is based on the definition of *F* as the ratio of two independent chi-square random variables, each divided by its degrees of freedom: $F = \chi_{v_1}^2/v_1 / \chi_{v_2}^2/v_2$. According to Section 12.3, $\chi_v^2 = v\hat{\sigma}^2/\sigma_0^2$ and $\chi_v^2/v = \hat{\sigma}^2/\sigma_0^2$; hence, $F = \hat{\sigma}_1^2/\sigma_1^2 / \hat{\sigma}_2^2/\sigma_2^2 = \sigma_2^2 \hat{\sigma}_1^2 / \sigma_1^2 \hat{\sigma}_2^2$. If $H_0: \sigma_1^2 = \sigma_2^2$ is true, then $F = \hat{\sigma}_1^2/\hat{\sigma}_2^2$.

That is, the value of F in the lower portion of the F distribution can be found by computing the reciprocal of the corresponding value in the upper portion, with degrees of freedom for numerator and denominator reversed.

In testing the one-tailed hypothesis $\sigma_1^2 \leq \sigma_2^2$ versus $\sigma_1^2 > \sigma_2^2$, the null hypothesis is rejected if $F \geq F_{\alpha;v_1,v_2}$. For $\sigma_1^2 \geq \sigma_2^2$ versus $\sigma_1^2 < \sigma_2^2$, the null hypothesis is rejected if $F \leq F_{1-\alpha;v_1,v_2}$. For a two-tailed test in which α is divided equally between the two tails of the F distribution, the critical values that cut off the upper and lower $\alpha/2$ regions are denoted by $F_{\alpha/2;v_1,v_2}$ and $F_{1-\alpha/2;v_1,v_2}$, respectively. For this test, the null hypothesis is rejected if $F_{1-\alpha/2;v_1,v_2} \geq F \geq F_{\alpha/2;v_1,v_2}$. An experimenter can avoid having to compute the critical values in the lower tail, $F_{1-\alpha;v_1,v_2}$ and $F_{1-\alpha/2;v_1,v_2}$, by always placing the larger variance in the numerator of $F = \sigma_1^2/\sigma_2^2$.

Computational Example for F Test with Independent Samples

Suppose that 50 disadvantaged children are randomly assigned to contingency management and traditional classrooms, with the restriction that 25 children are placed in each classroom. At the end of the school year an English-achievement test is administered to the two samples. The steps to be followed in testing the null hypothesis that the population dispersion is smaller for children in the contingency management classroom and the decision rule are as follows.

Step 1. State the statistical hypotheses: $H_0: \sigma_1^2 \geq \sigma_2^2$
$H_1: \sigma_1^2 < \sigma_2^2$,
where σ_1^2 and σ_2^2 denote the population variances, respectively, for the contingency management and traditional classrooms.

Step 2. Specify the test statistic: $F = \dfrac{\hat{\sigma}_1^2}{\hat{\sigma}_2^2}$.

Step 3. Specify the sample sizes: $n_1 = 25$ and $n_2 = 25$;
and the sampling distribution: F distribution, because the population distributions of X_1 and X_2 are independent and approximately normal in form.

Step 4. Specify the significance level: $\alpha = 0.05$.

Step 5. Obtain random samples of size n_1 and n_2, compute F, and make a decision.
Decision rule: Reject H_0 if F falls in the lower 0.05 portion of the sampling distribution of F; otherwise, don't reject H_0.

Assume that unbiased estimates of the population variances are $\hat{\sigma}_1^2 = 64$ and $\hat{\sigma}_2^2 = 196$, where $\hat{\sigma}_1^2$ and $\hat{\sigma}_2^2$ are computed from

$$\hat{\sigma}^2 = \dfrac{n \sum X_i^2 - (\sum X_i)^2}{n(n-1)}.$$

The F test statistic is

$$F = \dfrac{64}{196} = 0.33.$$

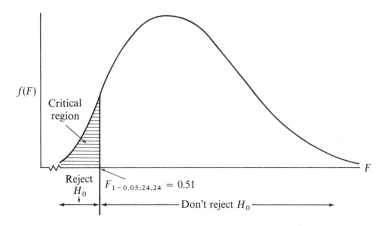

Figure 14.1-1. Sampling distribution of F given that H_0 is true. Since the observed $F = 0.33$ falls in the critical region, H_0 is rejected.

The value of F that cuts off the upper α proportion of the sampling distribution of F is given in Table D.5. The value that cuts off the lower α proportion can be determined from the F table by means of the relation $F_{1-\alpha;v_1,v_2} = 1/F_{\alpha;v_2,v_1}$. Thus, $F_{1-0.05;24,24} = 1/1.98 = 0.51$. The critical region for rejecting H_0 is shown in Figure 14.1-1. Since the observed F falls in the critical region, $0.33 < 0.51$, the null hypothesis is rejected. It is concluded that placing disadvantaged children in a contingency management classroom results in a smaller variance in English-achievement scores than does placing them in a traditional classroom.

t Test for Equality of Variances (Dependent Samples)

When the variances to be compared arise from dependent samples, for example, matched or repeated-measures samples, the F test statistic is not appropriate. Instead, a t test statistic should be used. The formula for t is

$$t = \frac{\hat{\sigma}_1^2 - \hat{\sigma}_2^2}{\sqrt{[4\hat{\sigma}_1^2 \hat{\sigma}_2^2/(n-2)](1 - r_{12}^2)}}$$

with degrees of freedom equal to $n - 2$, where n is the number of pairs of scores and r_{12} is the product-moment correlation coefficient.

To illustrate, assume an arithmetic-achievement test is given to a random sample of $n = 26$ sixth-grade students at the beginning and end of the school year. The experimenter hypothesizes that sixth-grade students become more homogeneous in arithmetic achievement during the year. Suppose achievement test variances at the beginning and end of the year are, respectively, $\hat{\sigma}_1^2 = 256$ and $\hat{\sigma}_2^2 = 100$, with $n = 26$ and $r = 0.60$. A test of $H_0: \sigma_1^2 \leq \sigma_2^2$ versus $H_1: \sigma_1^2 > \sigma_2^2$ is given by

$$t = \frac{256 - 100}{\sqrt{[4(256)(100)/(26-2)][1 - (0.60)^2]}} = \frac{156}{52.256} = 2.99$$

with $v = 26 - 2 = 24$. According to Appendix Table D.3, a t of 1.711 cuts off the upper 0.05 region of the sampling distribution of t. Since $t > t_{0.05, 24}$, the null hypothesis is rejected and the experimenter infers that the population of sixth-grade students is more homogeneous in arithmetic achievement at the end of the school year than they were at the beginning.

Review Exercises for Section 14.1

*1. Can $F = \hat{\sigma}_1^2/\hat{\sigma}_2^2$ be used to test hypotheses of the form $H_0: \hat{\sigma}_1^2 - \hat{\sigma}_2^2 = \delta_0$, where $\delta_0 \neq 0$?
2. What are the main factors an experimenter should keep in mind when using an F test to determine the tenability of the assumption $\sigma_1^2 = \sigma_2^2$ prior to testing $H_0: \mu_1 - \mu_2 = \delta_0$?
3. In testing the tenability of the assumption $\sigma_1^2 = \sigma_2^2$ prior to testing $H_0: \mu_1 - \mu_2 = \delta_0$, it is common practice to set $\alpha = 0.20$ or 0.25. What justification for this practice can you offer?
*4. For the data in Chapter 13, Exercise 15 (page 264), test the hypothesis that the population variances are equal versus the alternative that they are not equal. Let $\alpha = 0.10$.
5. For the data in Table 13.4-1 (page 263), test the tenability of the t test assumption that $\sigma_1^2 = \sigma_2^2$. Use $\alpha = 0.20$.
6. It is reasonable to expect 13-year-old boys to exceed 12-year-old boys in strength of grip. In all likelihood the dispersion of strength of grip is greater for 13-year-olds than for 12-year-olds. To test this hypothesis, strength of grip was measured by means of a hand dynamometer for a random sample of 42 boys who had just turned 12. One year later the same boys were remeasured. The variances for the first and second sets of measurements are 196 and 289, respectively. The correlation between the two sets of measurements is 0.83. Compute a t test statistic and make a decision about the scientific hypothesis. Let $\alpha = 0.05$.

14.2 Two-Sample Tests for $p_1 - p_2$

z Test for $p_1 = p_2$ (Independent Samples)

Many variables in the behavioral sciences and education can be partitioned into nonoverlapping and exhaustive classes, and hence are qualitative, for example, sex, year in school, and pass or fail. In such cases, p, the proportion in each class, is a useful descriptive measure. *A statistic for testing the null hypothesis $p_1 = p_2$ versus $p_1 \neq p_2$, where p_1 and p_2 are parameters of two independent binomial distributions, is*

$$z = \frac{\hat{p}_1 - \hat{p}_2}{\sqrt{\hat{p}_{Pd}\hat{q}_{Pd}/n_1 + \hat{p}_{Pd}\hat{q}_{Pd}/n_2}}.$$

Here, $\hat{p}_1$ and $\hat{p}_2$ are the sample estimators of the population proportions, n_1 and n_2 are the sizes of the samples used to estimate the population proportions, $\hat{p}_{Pooled}(\hat{p}_{Pd}) = (n_1\hat{p}_1 + n_2\hat{p}_2)/(n_1 + n_2)$, and $\hat{q}_{Pooled}(\hat{q}_{Pd}) = 1 - \hat{p}_{Pd}$. This z statistic can also be used for testing $H_0: p_1 \leq p_2$ versus $H_1: p_1 > p_2$ and $H_0: p_1 \geq p_2$ versus $H_1: p_1 < p_2$.

It can be shown that the sampling distribution of z approaches a standard normal distribution if the products $n_1\hat{p}_{Pd}$, $n_1\hat{q}_{Pd}$, $n_2\hat{p}_{Pd}$, and $n_2\hat{q}_{Pd}$, are all greater than five. If any one of these is between five and ten, it is desirable to apply a correction for continuity. As discussed in Section 12.4, the correction should be applied when the continuous normal distribution is used to estimate probabilities for a discrete distribution and n is small. The correction consists of subtracting the quantity $(1/2)(1/n_1 + 1/n_2)$ from the absolute value of the z numerator—that is, $|\hat{p}_1 - \hat{p}_2| - (1/2)(1/n_1 + 1/n_2)$.

z Test for $p_1 - p_2 = \delta_0$, where $\delta_0 \neq 0$ (Independent Samples)

In the preceding discussion, the hypothesis that is tested is $p_1 = p_2$. If the hypothesis is true, $\hat{p}_1$ and $\hat{p}_2$ are estimators of the same population parameter p, and the best estimator of p is obtained by computing a weighted mean, $\hat{p}_{Pd}$, of $\hat{p}_1$ and $\hat{p}_2$.[3] If an experimenter wishes to test H_0: $p_1 - p_2 = \delta_0$, where $\delta_0 \neq 0$, *a weighted mean should not be computed*, since $\hat{p}_1$ and $\hat{p}_2$ are estimators for different population parameters. For this hypothesis the appropriate test statistic is

$$z = \frac{(\hat{p}_1 - \hat{p}_2) - \delta_0}{\sqrt{\hat{p}_1 \hat{q}_1/n_1 + \hat{p}_2 \hat{q}_2/n_2}}.$$

z Test for $p_1 = p_2$ (Dependent Samples)

If two samples are dependent, a statistic developed by McNemar (1947) should be used to test H_0: $p_1 = p_2$ versus H_1: $p_1 \neq p_2$, H_0: $p_1 \leq p_2$ versus H_1: $p_1 > p_2$, and H_0: $p_1 \geq p_2$ versus H_1: $p_1 < p_2$. For these tests the data are cast into a 2 × 2 table,

		Sample 2		
		Category 0	Category 1	
Sample 1	Category 1	a	b	$a + b$
	Category 0	c	d	$c + d$
		$a + c$	$b + d$	n

as shown. The cell entry a denotes the number of elements classified in category 1 for sample 1 and in category 0 for sample 2; b denotes the number of elements that are classified in category 1 for both samples 1 and 2; and so on. The number of elements in each sample is n.

An estimator of the population proportion of individuals in category 1 for sample 1 is $\hat{p}_1 = (a + b)/n$. Similarly, the proportion in category 1 for sample 2 is given by $\hat{p}_2 = (b + d)/n$. The difference between the two populations can be expressed either as a proportion, $p_1 - p_2$, or as a frequency, $a - d$. A little algebra is all that is required to show that $n(\hat{p}_1 - \hat{p}_2) = a - d$:

$$\hat{p}_1 - \hat{p}_2 = \frac{a + b}{n} - \frac{b + d}{n} \qquad \text{By definition}$$

$$= \frac{1}{n}(a + b - b - d)$$

$$n(\hat{p}_1 - \hat{p}_2) = a - d.$$

[3] Section 13.4 describes a similar procedure for a t test statistic in which the best estimator of the unknown population variance, σ^2, is a weighted mean of the two sample estimators of σ^2.

If the null hypothesis $p_1 = p_2$ is true, the test statistic

$$z = \frac{a - d}{\sqrt{a + d}}$$

is approximately distributed as the standard normal distribution, provided that $(a + d) \geq 10$. If $10 \leq (a + d) < 20$, a correction for continuity should be applied to the z statistic as follows.

$$z = \frac{|a - d| - 1}{\sqrt{a + d}}$$

Computational Example for z Test with Dependent Samples

Suppose a random sample of 100 students at Thanatos University are polled about whether they approve or disapprove of capital punishment. Following the survey the students are shown a film depicting the effects of crime and acts of violence on the victims and their families. The 100 students are again polled about capital punishment. It is hypothesized that the proportion of students who approve of capital punishment will be higher after seeing the film than before. The statistical hypotheses are $H_0: p_1 \geq p_2$ versus $H_1: p_1 < p_2$. The data, number of students who approve or disapprove of capital punishment before and after seeing the film, are listed in Table 14.2-1. Since $z > z_{0.05}$, $2.02 > 1.64$, and $\hat{p}_1 < \hat{p}_2$, the null hypothesis is rejected. It is concluded that if the population of students was polled, a higher proportion of them would approve of capital punishment after seeing the film than before seeing the film.

Table 14.2-1. Capital Punishment Data

(i) Data:

		Sample 2		
		Disapprove	Approve	
Sample 1	Approve	$a = 2$	$b = 12$	$a + b = 14$
	Disapprove	$c = 76$	$d = 10$	$c + d = 86$
		$a + c = 78$	$b + d = 22$	$n = 100$

(ii) Computation:

$$z = \frac{|a - d| - 1}{\sqrt{a + d}} = \frac{|2 - 10| - 1}{\sqrt{2 + 10}} = \frac{7}{3.46} = 2.02$$

$$z_{0.05} = 1.64$$

$$\hat{p}_1 = \frac{a + b}{n} = \frac{14}{100} = 0.14$$

$$\hat{p}_2 = \frac{b + d}{n} = \frac{22}{100} = 0.22$$

Review Exercises for Section 14.2

*7. In a 1972 Shuffle Poll of $n_2 = 500$ Americans over 18 years old, 11% said they have smoked pot. In 1969 the figure for a sample of $n_1 = 600$ was 4%. Test the hypothesis that $p_1 = p_2$. Let $\alpha = 0.05$.

8. In the 1972 survey cited in Exercise 7, 81% of the interviewees opposed legalization of marijuana. The figure in 1969 was 84%. Test the hypothesis that $p_1 = p_2$. Let $\alpha = 0.05$.

9. According to the 1969 survey cited in Exercise 7, 8% more men than women had tried marijuana. In the 1972 survey, 11% of 300 men and 8% of 200 women reported that they had smoked pot. For these data, test the hypothesis that the difference between the proportion of pot smokers among men and women is 8%.

*10. A national survey of 3000 college and university students conducted by the American Council of Day-Care Centers found that 78% of west coast freshmen return to college for their second year. The comparable figure for freshmen at southern schools is 85%. The percentages are based on $n_1 = 1800$ and $n_2 = 1200$ students, respectively. Test the hypothesis that $p_1 = p_2$. Let $\alpha = 0.001$.

*11. Attitudes of a sample of college students toward taking a required course in music appreciation were measured prior to taking the course and after completing the course. For the data in the table, test the hypothesis that the proportions in favor of the requirement before and after the course are equal. Let $\alpha = 0.05$.

		Postcourse Attitude		
		Unfavorable	Favorable	
Precourse Attitude	Favorable	16	24	40
	Unfavorable	19	27	46
		35	51	86

12. Learning one task often enhances the learning of a similar task; this phenomenon is called *learning to learn*. To investigate this phenomenon, an experimenter asked students to learn 20 lists of nonsense syllables. For the data in the table, test the hypothesis that p_1, the population proportion of students who learned lists 2–6 in less than 25 trials, and p_2, the population proportion who learned lists 16–20 in less than 25 trials, are equal. Let $\alpha = 0.05$.

		Number of Students Who Learned Lists 16–20		
		In 25 Trials or More	In Less than 25 Trials	
Number of Students Who Learned Lists 2–6	In Less than 25 Trials	4	6	10
	In 25 Trials or More	13	13	26
		17	19	36

Table 14.3-1. Summary of Two-Sample Test Statistics

Chapter Section	Statistic	Statistical Hypotheses	Formula	Assumptions
14.1	F	H_0: $\sigma_1^2 = \sigma_2^2$ H_1: $\sigma_1^2 \neq \sigma_2^2$	$F = \dfrac{\hat{\sigma}_1^2}{\hat{\sigma}_2^2}$ $v_1 = n_1 - 1, v_2 = n_2 - 1$	1. Random sampling or random assignment 2. Normality 3. Independence between samples
14.1	t	H_0: $\sigma_1^2 = \sigma_2^2$ H_1: $\sigma_1^2 \neq \sigma_2^2$	$t = \dfrac{\hat{\sigma}_1^2 - \hat{\sigma}_2^2}{\sqrt{[4\hat{\sigma}_1^2 \hat{\sigma}_2^2/(n-2)](1 - r_{12}^2)}}$ $v = n - 2$	1. Random sampling or random assignment 2. Normality 3. Dependence between samples
14.2	z	H_0: $p_1 = p_2$ H_1: $p_1 \neq p_2$	$z = \dfrac{\hat{p}_1 - \hat{p}_2}{\sqrt{\hat{p}_{Pd}\hat{q}_{Pd}/n_1 + \hat{p}_{Pd}\hat{q}_{Pd}/n_2}}$	1. Random sampling or random assignment 2. Binomial distributions 3. Independence between samples 4. $n_1\hat{p}_{Pd} > 5, n_1\hat{q}_{Pd} > 5,$ $n_2\hat{p}_{Pd} > 5, n_2\hat{q}_{Pd} > 5$
14.2	z	H_0: $p_1 - p_2 = \delta_0$ H_1: $p_1 - p_2 \neq \delta_0$ where $\delta_0 \neq 0$	$z = \dfrac{(\hat{p}_1 - \hat{p}_2) - \delta_0}{\sqrt{\hat{p}_1\hat{q}_1/n_1 + \hat{p}_2\hat{q}_2/n_2}}$	1. Random sampling or random assignment 2. Binomial distributions 3. Independence between samples 4. $n_1\hat{p}_1 > 5, n_1\hat{q}_1 > 5,$ $n_2\hat{p}_2 > 5, n_2\hat{q}_2 > 5$
14.2	z	H_0: $p_1 = p_2$ H_1: $p_1 \neq p_2$	$z = \dfrac{a - d}{\sqrt{a + d}}$	1. Random sampling or random assignment 2. Binomial distributions 3. Dependence between samples 4. $a + d > 10$

14.3 Summary

Procedures for testing hypotheses about two variances and two proportions are described in this chapter. And a new F statistic and sampling distribution were introduced.

The z, t, χ^2, and F test statistics presented in Chapters 11–14 have a number of common characteristics that might be overlooked because of differences in their formulas. Each statistic (1) assumes random sampling or random assignment of X's, (2) is used to test hypotheses about a parameter(s) of the sampled population(s), and (3) assumes the sampled population is normal with the exception of the z statistics that are used to test hypotheses about proportions.

The test statistics described in this chapter are presented in Table 14.3-1, along with a summary of the parameters they are used to test, their formulas, and their assumptions.

15

Interval Estimation

15.1 Introduction to Interval Estimation
15.2 Confidence Intervals for μ and $\mu_1 - \mu_2$ when σ^2 Is Unknown
  Confidence Interval for μ
  Computational Example of Confidence Interval for μ when σ^2 Is Unknown
  Interpretation of a Confidence Interval
  One-Sided Confidence Interval
  Confidence Interval for $\mu_1 - \mu_2$ (Independent Samples)
  Confidence Interval for $\mu_1 - \mu_2$ (Dependent Samples)
  Review Exercises for Section 15.2
15.3 Confidence Intervals for σ^2 and σ_1^2/σ_2^2
  Confidence Interval for σ^2
  Computational Example of Confidence Interval for σ^2
  Confidence Interval for σ_1^2/σ_2^2 (Independent Samples)
  Computational Example of Confidence Interval for σ_1^2/σ_2^2
  Review Exercises for Section 15.3
15.4 Confidence Intervals for p and $p_1 - p_2$
  Confidence Interval for p
  Confidence Interval for $p_1 - p_2$ (Independent Samples)
  Review Exercises for Section 15.4
15.5 Confidence Intervals for ρ and $\rho_1 - \rho_2$
  Confidence Interval for ρ
  Computational Example of Confidence Interval for ρ
  Confidence Interval for $\rho_1 - \rho_2$
  Review Exercises for Section 15.5
15.6 Summary

15.1 Introduction to Interval Estimation

The election is only days away and the latest Giddyup poll gives Mr. Jerry Mander 54% of the vote. In the language of the statistician, 54% is a *point estimate*, a single point on the number line that is the best estimate of the population parameter that can be made from sample data. Between periods of euphoria Mr. Jerry Mander ponders the question, should he or should he not cancel the paid political announcement planned for election eve? If he is winning, the expenditure of campaign funds is unnecessary, but what if he is really losing? Is it possible that less than 51% of the voters favor him even though the highly respected poll gives him a majority? Now, Mr. Mander is no statistician, but he does know that polls are subject to sampling error. With anxiety mounting, he decides to forego a vacation to Hawaii and use the campaign funds for their intended purpose. The use of *confidence interval* procedures, the subject of this chapter, might have saved Mr. Mander's vacation.

We noted in Section 11.1 that two complementary topics are categorized under classical statistical inference—hypothesis testing and interval estimation. In many research situations an experimenter isn't interested in testing a hypothesis but rather in obtaining an estimate of some population parameter. *Although we can never know the value of the parameter except by measuring all the elements in the population, a random sample can be used to specify a segment or interval on the number line such that the parameter has a high probability of lying on the segment. The segment is called a* **confidence interval**.

Interval estimation procedures are used much less frequently in the behavioral sciences than hypothesis testing procedures. This is true in spite of the fact that an interval estimate provides more information regarding the outcome of an experiment than simply reporting that the null hypothesis was or wasn't rejected.[1] Behavioral scientists have been slow to adopt the recommendation of statisticians that research reports should include confidence interval information.

15.2 Confidence Intervals for μ and $\mu_1 - \mu_2$ when σ^2 Is Unknown

Confidence Interval for μ

Here we will construct a confidence interval for μ in such a way that the interval has a probability equal to $1 - \alpha$ of including μ. The underlying logic is presented in some detail, since it is applicable to confidence intervals for any parameter, such as σ^2, p, or ρ. Suppose we plan to draw a random sample of size $n = 30$ from a normal population and we want $1 - \alpha$ to equal 0.95. If the population were not normal, we would make n quite large. In either case, we know the sampling distribution of $\bar{X}$ is approximately normal, with mean equal to μ and standard error equal to $\hat{\sigma}/\sqrt{n}$. Therefore, the probability is $1 - 0.05 = 0.95$ that prior to drawing the sample the $\bar{X}$ will lie between $\mu - 2.045\hat{\sigma}/\sqrt{n}$ and $\mu + 2.045\hat{\sigma}/\sqrt{n}$, where -2.045 and 2.045 are the values

---

[1] For an in-depth discussion of the relative merits of the two procedures see Natrella (1960). This article is reproduced in Kirk (1972a).

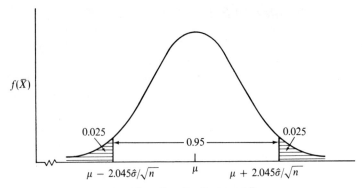

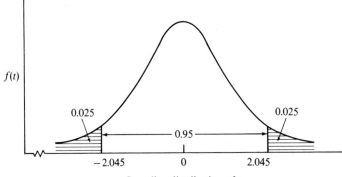

Figure 15.2-1. Sampling distributions of $\bar{X}$ and t. The probability, prior to drawing the sample, that the mean of a random sample will fall between $\mu \pm 2.045\hat{\sigma}/\sqrt{n}$ is equal to 0.95. Similarly, the probability is 0.95 that t will fall between ± 2.045.

of t that cut off the lower and upper 0.025 portions, respectively, of Student's t distribution for $n - 1 = 29$ degrees of freedom. Similarly, the probability is 0.95 that $t = (\bar{X} - \mu)/(\hat{\sigma}/\sqrt{n})$, a transformation of $\bar{X}$, will fall between -2.045 and 2.045. This is illustrated in Figure 15.2-1. Hence, we can write[2]

$$\text{Prob}(-2.045 \leq t \leq 2.045) = 0.95.$$

Substituting $(\bar{X} - \mu)/(\hat{\sigma}/\sqrt{n})$ for t gives

$$\text{Prob}\left(-2.045 \leq \frac{\bar{X} - \mu}{\hat{\sigma}/\sqrt{n}} \leq 2.045\right) = 0.95.$$

Multiplying each term in the inequalities by $\hat{\sigma}/\sqrt{n}$, we have

$$\text{Prob}\left(-\frac{2.045\hat{\sigma}}{\sqrt{n}} \leq \bar{X} - \mu \leq \frac{2.045\hat{\sigma}}{\sqrt{n}}\right) = 0.95.$$

[2] A review of inequalities is given in Appendix Section A.6.

Subtracting $\bar{X}$ from each term, we obtain

$$\text{Prob}\left(-\bar{X} - \frac{2.045\hat{\sigma}}{\sqrt{n}} \leq -\bar{X} + \bar{X} - \mu \leq -\bar{X} + \frac{2.045\hat{\sigma}}{\sqrt{n}}\right) = 0.95,$$

and multiplying by -1, which reverses the direction of the inequalities and the signs of the terms, gives

$$\text{Prob}\left(\bar{X} + \frac{2.045\hat{\sigma}}{\sqrt{n}} \geq \mu \geq \bar{X} - \frac{2.045\hat{\sigma}}{\sqrt{n}}\right) = 0.95.$$

For convenience, this is rearranged to form the confidence statement

$$\text{Prob}\left(\bar{X} - \frac{2.045\hat{\sigma}}{\sqrt{n}} \leq \mu \leq \bar{X} + \frac{2.045\hat{\sigma}}{\sqrt{n}}\right) = 0.95.$$

In words, this says the probability is 0.95 that a random interval from $\bar{X} - 2.045\hat{\sigma}/\sqrt{n}$ to $\bar{X} + 2.045\hat{\sigma}/\sqrt{n}$ contains the parameter μ. This random interval is the confidence interval for μ with *confidence coefficient* equal to $1 - 0.05 = 0.95$. The values of $\bar{X} - 2.045\hat{\sigma}/\sqrt{n}$ and $\bar{X} + 2.045\hat{\sigma}/\sqrt{n}$ are the *lower* and *upper boundaries*, respectively, of the confidence interval; the boundaries are also called *confidence limits* and are designated L_1 and L_2. The general form of a $100(1 - \alpha)\%$ confidence interval on μ is given by

$$\bar{X} - \frac{t_{\alpha/2,v}\hat{\sigma}}{\sqrt{n}} \leq \mu \leq \bar{X} + \frac{t_{\alpha/2,v}\hat{\sigma}}{\sqrt{n}},$$

where $t_{\alpha/2,v}$ is the value that cuts off the upper $\alpha/2$ proportion of Student's t distribution for $n - 1$ degrees of freedom.

To summarize, it is impossible to know the value of a parameter without measuring all the elements in the population. However, for a given parameter it is possible to find two functions L_1 and L_2 of sample values such that before the sample is drawn the probability that the interval from L_1 to L_2 will contain the parameter is equal to $1 - \alpha$. We can be $100(1 - \alpha)\%$ confident that this interval contains the parameter.

Computational Example of Confidence Interval for μ when σ^2 Is Unknown

In Section 12.2 we illustrated a test of the hypothesis that a new class registration procedure at Idle-On-In College would reduce the time required for students to register. A $100(1 - 0.05)\%$ confidence interval for these data in which $\bar{X} = 2.9$, $\hat{\sigma} = 0.158$, $n = 21$, and $t_{0.05/2,20} = 2.086$ is given by

$$2.9 - 2.086\frac{0.158}{\sqrt{21}} \leq \mu \leq 2.9 + 2.086\frac{0.158}{\sqrt{21}}$$

$$2.9 - 0.072 \leq \mu \leq 2.9 + 0.072$$

$$2.828 \leq \mu \leq 2.972.$$

We can feel quite confident that the value of μ is between $L_1 = 2.828$ and $L_2 = 2.972$. The measure of our confidence is 0.95 because, before the sample was drawn, 0.95 was the probability that the interval we were going to construct would cover the true mean. If we want to feel even more confident that we have specified L_1 and L_2 so they capture μ, we can compute a 99% confidence interval. This is accomplished by substituting $t_{0.01/2, 20} = 2.845$ for $t_{0.05/2, 20} = 2.086$. The 99% confidence interval is given by

$$2.9 - 2.845 \frac{0.158}{\sqrt{21}} \le \mu \le 2.9 + 2.845 \frac{0.158}{\sqrt{21}}$$

$$2.9 - 0.098 \le \mu \le 2.9 + 0.098$$

$$2.802 \le \mu \le 2.998.$$

Note that as our confidence that we have captured μ increases, so does the size of the interval $L_1 - L_2$.

Interpretation of a Confidence Interval

We stated that the probability is 0.95 that the random interval from $\bar{X} - 2.086\hat{\sigma}/\sqrt{n}$ to $\bar{X} + 2.086\hat{\sigma}/\sqrt{n}$ contains the parameter μ; in symbols,

$$\text{Prob}\left(\bar{X} - \frac{2.086\hat{\sigma}}{\sqrt{n}} \le \mu \le \bar{X} + \frac{2.086\hat{\sigma}}{\sqrt{n}}\right) = 0.95.$$

This statement is defined on the sample space of all confidence intervals that could be constructed. To clarify, consider obtaining an infinite number of random samples of size n from some population and computing $\bar{X}$ for each sample. This creates a new infinite population with elements that are $\bar{X}$'s. We can conceive of constructing confidence intervals around each $\bar{X}$. Ninety-five percent of these intervals will contain μ between L_1 and L_2 and 5% will not. The probability that a randomly selected interval from this infinite set of intervals will contain μ is 0.95. However, once one of the $\bar{X}$'s has been computed and found to equal, say, 2.9, it is incorrect to state that the probability is 0.95 that the interval from $2.9 - 2.086\hat{\sigma}/\sqrt{n}$ to $2.9 + 2.086\hat{\sigma}/\sqrt{n}$ contains μ; in symbols,

$$\text{Prob}\left(2.9 - \frac{2.086\hat{\sigma}}{\sqrt{n}} \le \mu \le 2.9 + \frac{2.086\hat{\sigma}}{\sqrt{n}}\right) = 0.95$$

is incorrect. Such a probability statement is incorrect because 2.9 is not a random variable. Once a value of $\bar{X}$ has been inserted in the confidence statement, either the interval from $2.9 - 2.086\hat{\sigma}/\sqrt{n}$ to $2.9 + 2.086\hat{\sigma}/\sqrt{n}$ does or doesn't capture μ. Hence, one should not say, for example, "the probability is 0.95 that the confidence interval 2.828–2.972 captures μ." It is correct to say that a 95% confidence interval on μ is 2.828–2.972.

As noted earlier, many statisticians prefer confidence interval procedures to hypothesis testing procedures because the former provide more information than that a particular H_0 was or wasn't rejected at α level of significance. A confidence interval on μ not only specifies the unbiased estimate of the population mean, but it also includes the error variation qualifying that estimate. One can think of a confidence interval as containing all the values of μ_0 that would not lead to rejecting $H_0: \mu = \mu_0$ at α level of significance. Only if μ_0 lies outside the $100(1 - \alpha)\%$ confidence interval can the hypothesis $H_0: \mu = \mu_0$ be rejected. Any null hypothesis can be tested simply by looking at the confidence interval on μ. For example, a 95% confidence interval for the mean time required for students to register using a new procedure at Idle-On-In College was computed previously and found to be 2.828–2.972. A test of the hypothesis $H_0: \mu = \mu_0$ will be rejected only if μ_0 lies outside the interval 2.828–2.972. Thus, $H_0: \mu = 2.85, 2.90, 2.95$, et cetera, would not be rejected, but $H_0: \mu = 2.75, 3.00$, or 3.15 would be rejected.

It is hoped that eventually reporting confidence intervals in research reports will become standard practice. Certainly, a confidence interval communicates far more information than a test of a null hypothesis.

One-Sided Confidence Interval

The confidence intervals computed earlier would be appropriate if we were interested in the possibility that the new registration time is either quicker or slower than the old procedure. In this sense it is analogous to a two-tailed test of the null hypothesis. The statistical hypotheses for the registration example in Section 12.2 are

$$H_0: \mu \geq 3.1$$
$$H_1: \mu < 3.1,$$

where 3.1 is the mean registration time associated with the old procedure. Here the hypotheses are one-tailed. An analogous $100(1 - 0.05)\%$ confidence limit is

$$L_2 = \bar{X} + \frac{t_{\alpha,v}\hat{\sigma}}{\sqrt{n}} = 2.9 + 1.725\frac{0.158}{\sqrt{21}} = 2.9 + 0.059 = 2.959,$$

where $t_{\alpha,v}$ is the value of t that cuts off the upper $\alpha = 0.05$ portion of the sampling distribution of t. The experimenter can be fairly confident that μ for the new registration procedure is not greater than 2.96 hr. Since the mean registration time for the old procedure is 3.1 hr, it can be concluded that the new procedure requires less time. In Section 12.2 we reached the same conclusion using hypothesis testing procedures. However, confidence interval procedures allow us to test all hypotheses of interest, not just $H_0: \mu = 3.1$. For example, since $L_2 = 2.959$, we know that a one-tailed test of significance at the 0.05 level would reject $H_0: \mu \geq 3.06$ and $H_0: \mu \geq 2.98$ but not $H_0: \mu \geq 2.94$ and $H_0: \mu \geq 2.91$.

Confidence Interval for $\mu_1 - \mu_2$ (Independent Samples)

Confidence interval procedures for a single population mean can be generalized to the difference between the means of two independent populations, $\delta_0 = \mu_1 - \mu_2$. This is accomplished by substituting $t = [(\bar{X}_1 - \bar{X}_2) - (\mu_1 - \mu_2)]/\sqrt{\hat{\sigma}^2_{Pooled}(1/n_1 + 1/n_2)}$ for $t = (\bar{X} - \mu_0)/(\hat{\sigma}/\sqrt{n})$ in the probability statement $\text{Prob}(-t_{\alpha/2,v} \leq t \leq t_{\alpha/2,v}) = 1 - \alpha$. A $100(1 - \alpha)\%$ confidence interval for the difference between means of independent

15.2 Confidence Intervals for μ and $\mu_1 - \mu_2$ when σ^2 Is Unknown

samples is given by

$$(\bar{X}_1 - \bar{X}_2) - t_{\alpha/2,\nu}\sqrt{\hat{\sigma}^2_{Pooled}\left(\frac{1}{n_1} + \frac{1}{n_2}\right)}$$

$$\leq \mu_1 - \mu_2 \leq (\bar{X}_1 - \bar{X}_2) + t_{\alpha/2,\nu}\sqrt{\hat{\sigma}^2_{Pooled}\left(\frac{1}{n_1} + \frac{1}{n_2}\right)},$$

where $t_{\alpha/2,\nu}$ is the value of t that cuts off the upper $\alpha/2$ portion of the sampling distribution of t for $\nu = n_1 + n_2 - 2$ and $\hat{\sigma}^2_{Pooled} = [(n_1 - 1)\hat{\sigma}_1^2 + (n_2 - 1)\hat{\sigma}_2^2]/[(n_1 - 1) + (n_2 - 1)]$.

We will use the data in Table 13.4-1 to illustrate the use of this confidence interval. The data were obtained in an experiment designed to compare the effectiveness of distributed practice and massed practice in developing skill on a mirror tracing task. The information from Table 13.4-1 necessary to construct a confidence interval is $\bar{X}_1 = 17$, $\hat{\sigma}_1^2 = 14.421$, $n_1 = 20$, $\bar{X}_2 = 19$, $\hat{\sigma}_2^2 = 23.263$, $n_2 = 20$, and $\hat{\sigma}^2_{Pooled} = 18.842$. A $100(1 - 0.05)\%$ confidence interval for $\mu_1 - \mu_2$ is

$$(17-19) - 2.025\sqrt{18.842\left(\frac{1}{20} + \frac{1}{20}\right)} \leq \mu_1 - \mu_2 \leq (17-19) + 2.025\sqrt{18.842\left(\frac{1}{20} + \frac{1}{20}\right)}$$

$$-2 - 2.780 \leq \mu_1 - \mu_2 \leq -2 + 2.780$$

$$-4.780 \leq \mu_1 - \mu_2 \leq 0.780.$$

Thus, the mean for the distributed practice condition μ_1 could be larger than that for the massed practice condition μ_2, or it could be equal to it or smaller. Because the confidence interval for $\mu_1 - \mu_2$ includes zero, it can't be concluded that performance is better under either condition. Any difference between -4.780 and 0.780 is likely to be a reflection of sampling variation and/or other chance factors.

Confidence Interval for $\mu_1 - \mu_2$ (Dependent Samples)

A $100(1 - \alpha)\%$ confidence interval for the difference between means of dependent samples is given by

$$\bar{X}_D - t_{\alpha/2,\nu}\hat{\sigma}_{\bar{X}_D} \leq \mu_D \leq \bar{X}_D + t_{\alpha/2,\nu}\hat{\sigma}_{\bar{X}_D},$$

where $\bar{X}_D = \sum D_i/n = \bar{X}_1 - \bar{X}_2$, $D_i = X_{i1} - X_{i2}$, $t_{\alpha/2,\nu}$ is the value of t that cuts off the upper $\alpha/2$ portion of the sampling distribution of t for $\nu = n - 1$, $\hat{\sigma}_{\bar{X}_D} = \sqrt{[n\sum D_i^2 - (\sum D_i)^2]/n^2}/\sqrt{n-1}$, and $\mu_D = \mu_1 - \mu_2$.

We will illustrate the use of this confidence interval with the mirror tracing data in Table 13.5-1 (page 268). These data are for matched subjects; one subject of each pair was assigned to the distributed practice condition and the other to the massed practice condition. The information from Table 13.5-1 necessary to construct a confidence interval is $\bar{X}_D = -2$, $\hat{\sigma}_{\bar{X}_D} = 0.657$, and $n = 20$. A $100(1 - 0.05)\%$ confidence interval for $\mu_1 - \mu_2$ is

$$-2 - 2.093(0.657) \leq \mu_D \leq -2 + 2.093(0.657)$$

$$-2 - 1.375 \leq \mu_D \leq -2 + 1.375$$

$$-3.375 \leq \mu_D \leq -0.625.$$

Since the interval -3.375 to -0.625 doesn't include zero, the experimenter can be confident that performance, the time required to trace the star pattern, is superior under the distributed practice condition.

Review Exercises for Section 15.2

1. What assumptions are associated with the statement
$$\text{Prob}\left(\bar{X} - \frac{t_{\alpha/2,\nu}\hat{\sigma}}{\sqrt{n}} \leq \mu \leq \bar{X} + \frac{t_{\alpha/2,\nu}\hat{\sigma}}{\sqrt{n}}\right) = 1 - \alpha?$$

*2. Derive a confidence interval for μ from
$$\text{Prob}(-z_{\alpha/2} \leq z \leq z_{\alpha/2}) = 1 - \alpha.$$

3. What are the advantages of confidence interval procedures over hypothesis testing procedures?

*4. Ordinarily, $L_1 < L_2$. Can you think of two conditions under which L_1 and L_2 would be equal?

*5. A soft-drink machine is designed to dispense a measured amount of a popular drink. Construct a 99% confidence interval on μ if a random sample of 49 drinks has a $\bar{X} = 7.2$ oz. Assume the distribution is approximately normal with $\hat{\sigma} = 0.42$ oz.

*6. If 23–36 is a 95% confidence interval for μ, indicate which of the following statements are correct (C) and which are incorrect (I).
   a. The probability is 0.95 that the interval from 23 to 36 captures the population mean.
   b. The probability that the interval $\bar{X} \pm 1.96\hat{\sigma}/\sqrt{n}$ captures μ is 0.95.
   c. Prob $(23 \leq \mu \leq 36) = 0.95$.
   d. An experimenter can be 95% confident that the interval 23–36 captures μ.
   e. A 95% confidence interval on μ is 23–36.
   f. Prob $(\bar{X} - 1.96\hat{\sigma}/\sqrt{n} \leq \mu \leq \bar{X} + 1.96\hat{\sigma}/\sqrt{n}) = 0.95$, where $\bar{X} = 29.5$.

*7. Students desiring to enter graduate school at Kandykane Technical Institute (KTI) are required to submit Graduate Record Examination (GRE) scores with their applications. The verbal scores for the first 20 applications received this year are given in the table. Construct a 95% confidence interval for these data.

GRE Scores for Verbal Section of Test

402	390	429	391
381	407	410	403
430	413	406	398
376	424	382	410
395	360	410	404

8. (a) If the first 20 applicants in Exercise 7 can be considered a random sample of applicants who will apply, what is the best estimate of the mean for this year's applicants? (b) Last year the mean GRE verbal score of all KTI applicants was 428. Is the mean verbal aptitude for this year's applicants different from that for last year?

*9. The nicotine content of random samples of two brands of cigarettes was measured. The $\bar{X}$'s for brands 1 and 2 were, respectively, 18.6 and 16.1 milligrams, with $\hat{\sigma}$'s 2.4 and 2.6 milligrams. The sample n's were both 36. Construct a 99% confidence interval for $\mu_1 - \mu_2$.

10. An experiment was performed to compare simple and disjunctive reaction times. In the former condition a subject responded to a single light by pressing a button below the light; the disjunctive condition required a subject to press the right button if the right light was illuminated

and the left button if the left light was illuminated. Twenty-four subjects were randomly assigned to the two conditions with the restriction that an equal number participated under each condition. Construct a 95% confidence interval for $\mu_1 - \mu_2$ for the data in the table.

Reaction Time (Hundredths of a Second)

Simple			Disjunctive		
24	24	24	27	34	33
27	26	22	31	32	31
23	21	25	29	30	32
25	23	24	35	28	31

*11. Learning one task enhances the learning of a different but similar task; this phenomenon is called *learning to learn*. To investigate this phenomenon, 18 students learned 15 lists of nonsense syllables with equal association value. The total time required to learn the first three lists and the last three lists is given in the table. Construct a 99% confidence interval for $\mu_1 - \mu_2$.

Time to Learn Lists of Nonsense Syllables (Minutes)

Subject	First Three Lists	Last Three Lists	Subject	First Three Lists	Last Three Lists
1	31	26	10	32	30
2	34	32	11	35	28
3	27	26	12	22	22
4	38	34	13	26	21
5	30	25	14	30	27
6	26	25	15	34	31
7	33	26	16	29	26
8	30	28	17	30	24
9	23	20	18	32	27

*12. How is the size of a confidence interval related to the following?
   a. Size of population standard deviation
   b. Sample size
   c. Confidence coefficient
13. Terms to remember:
   a. Point estimate
   b. Confidence interval
   c. Confidence coefficient
   d. Confidence limits

†15.3 Confidence Intervals for σ^2 and σ_1^2/σ_2^2

Confidence Interval for σ^2

A confidence interval for the unknown variance of a population, σ^2, can be derived from the statement

$$\text{Prob}(\chi^2_{1-\alpha/2,\nu} \leq \chi^2 \leq \chi^2_{\alpha/2,\nu}) = 1 - \alpha,$$

† This section can be omitted without loss of continuity.

where $\chi^2 = [(n-1)\hat{\sigma}^2]/\sigma^2$ and $\chi^2_{1-\alpha/2,v}$ and $\chi^2_{\alpha/2,v}$ are the values of the chi-square distribution for $v = n - 1$ degrees of freedom that cut off the lower and upper $\alpha/2$ regions, respectively. Following the derivation described in Section 15.2, we substitute $[(n-1)\hat{\sigma}^2]/\sigma^2$ for χ^2 in the above probability statement:

$$\text{Prob}\left[\chi^2_{1-\alpha/2,v} \le \frac{(n-1)\hat{\sigma}^2}{\sigma^2} \le \chi^2_{\alpha/2,v}\right] = 1 - \alpha.$$

Dividing each term in the inequality by $(n-1)\hat{\sigma}^2$ and then inverting each term, thereby reversing the direction of the inequalities, we obtain

$$\text{Prob}\left[\frac{(n-1)\hat{\sigma}^2}{\chi^2_{\alpha/2,v}} \le \sigma^2 \le \frac{(n-1)\hat{\sigma}^2}{\chi^2_{1-\alpha/2,v}}\right] = 1 - \alpha.$$

Thus, a $100(1-\alpha)\%$ confidence interval for σ^2 is given by

$$\frac{(n-1)\hat{\sigma}^2}{\chi^2_{\alpha/2,v}} \le \sigma^2 \le \frac{(n-1)\hat{\sigma}^2}{\chi^2_{1-\alpha/2,v}}.$$

The assumptions associated with this interval are (1) the population distribution of X is normal in form and (2) the collection of X's is a random sample from the population of interest.

Computational Example of Confidence Interval for σ^2

To illustrate the construction of a confidence interval for σ^2 we will use the driver license test score data presented in Table 12.3-1 (page 246) in which $\hat{\sigma}^2 = 150.6958$ and $n = 16$. A $100(1 - 0.10)\%$ confidence interval for these data is

$$\frac{(16-1)150.6958}{24.996} \le \sigma^2 \le \frac{(16-1)150.6958}{7.261}$$

$$90.432 \le \sigma^2 \le 311.312.$$

Thus, an experimenter can be confident that σ^2 is on the interval from 90.432 to 311.312.

Confidence Interval for σ_1^2/σ_2^2 (Independent Samples)

A point estimator for the ratio of two population variances, σ_1^2/σ_2^2, is the ratio $\hat{\sigma}_1^2/\hat{\sigma}_2^2$ of sample variances. If σ_1^2 and σ_2^2 are the variances of independent normal populations, we can construct a confidence interval for σ_1^2/σ_2^2 by using the statistic[3] $F = \sigma_2^2\hat{\sigma}_1^2/\sigma_1^2\hat{\sigma}_2^2$. Following the pattern of previous derivations, we begin with the statement

$$\text{Prob}(F_{1-\alpha/2;v_1,v_2} \le F \le F_{\alpha/2;v_1,v_2}) = 1 - \alpha$$

and substitute $\sigma_2^2\hat{\sigma}_1^2/\sigma_1^2\hat{\sigma}_2^2$ for F to obtain

$$\text{Prob}\left(F_{1-\alpha/2;v_1,v_2} \le \frac{\sigma_2^2\hat{\sigma}_1^2}{\sigma_1^2\hat{\sigma}_2^2} \le F_{\alpha/2;v_1,v_2}\right) = 1 - \alpha,$$

where $F_{1-\alpha/2;v_1,v_2}$ and $F_{\alpha/2;v_1,v_2}$ are the values of F that cut off the lower and upper

---

[3] See footnote 2 in Section 14.1 (page 275).

$\alpha/2$ portions of the sampling distribution of F, respectively, with $v_1 = n_1 - 1$ and $v_2 = n_2 - 1$. Multiplying each term in the inequality by $\hat{\sigma}_2^2/\hat{\sigma}_1^2$ and then inverting each term, which changes the direction of the inequality, we obtain

$$\text{Prob}\left(\frac{\hat{\sigma}_1^2}{\hat{\sigma}_2^2}\frac{1}{F_{\alpha/2;v_1,v_2}} \leq \frac{\sigma_1^2}{\sigma_2^2} \leq \frac{\hat{\sigma}_1^2}{\hat{\sigma}_2^2}\frac{1}{F_{1-\alpha/2;v_1,v_2}}\right) = 1 - \alpha.$$

As discussed in Section 14.1, $1/F_{1-\alpha/2;v_1,v_2} = F_{\alpha/2;v_2,v_1}$. Making this substitution, we obtain the following $100(1 - \alpha)\%$ confidence interval for σ_1^2/σ_2^2:

$$\frac{\hat{\sigma}_1^2}{\hat{\sigma}_2^2}\frac{1}{F_{\alpha/2;v_1,v_2}} \leq \frac{\sigma_1^2}{\sigma_2^2} \leq \frac{\hat{\sigma}_1^2}{\hat{\sigma}_2^2}F_{\alpha/2;v_2,v_1}.$$

Computational Example of Confidence Interval for σ_1^2/σ_2^2

In Section 14.1 English-achievement data were presented for 25 disadvantaged children who had been placed in a contingency management classroom and 25 who had been placed in a traditional classroom. At the end of the school year unbiased estimates of the population variances for the two samples were $\hat{\sigma}_1^2 = 64$ and $\hat{\sigma}_2^2 = 196$. A $100(1 - 0.02)\%$ confidence interval for the ratio σ_1^2/σ_2^2 is

$$\frac{64}{196}\frac{1}{2.66} \leq \frac{\sigma_1^2}{\sigma_2^2} \leq \frac{64}{196}2.66$$

$$(0.3265)(0.3759) \leq \frac{\sigma_1^2}{\sigma_2^2} \leq (0.3265)(2.66)$$

$$0.1227 \leq \frac{\sigma_1^2}{\sigma_2^2} \leq 0.8685.$$

Since the interval doesn't include one, we can be confident that the two variances are not equal. The best guess that we can make regarding the ratio σ_1^2/σ_2^2 is $\hat{\sigma}_1^2/\hat{\sigma}_2^2 = 0.3265$; we can be 98% confident that the ratio is not less than 0.1227 or greater than 0.8685.

Review Exercises for Section 15.3

*14. For the data in Exercise 7 (page 291), construct a 98% confidence interval for σ^2.

15. Data for this year's KTI graduate school applicants are presented in Exercise 7 (page 291). Suppose the variance for last year's 20 applicants, sample 1, is 306.1. (a) Construct a 98% confidence interval for the ratio of the variances. (b) Is the variance for this year's applicants different from that for last year's?

*16. A 99% confidence interval for σ_1^2/σ_2^2 is $-0.6–2.7$. Could the population variances be equal? Why?

*17. (a) For the data in Exercise 9 (page 291), construct a 98% confidence interval for σ_1^2/σ_2^2. (b) Could the population variances be equal? (c) If you wanted to test the assumption that the population σ^2's are equal prior to performing a t test, what level of significance would you adopt?

18. (a) For the data in Exercise 10 (page 291), construct a 98% confidence interval for σ_1^2/σ_2^2. (b) Could the population variances be equal?

† 15.4 Confidence Intervals for p and $p_1 - p_2$

Confidence Interval for p

We learned in Section 10.1 that the binomial distribution is adequately approximated by the normal distribution if n is large and if np and $n(1 - p)$ are both greater than five. A $100(1 - \alpha)\%$ confidence interval for p, the parameter of a binomial distribution, is given by

$$\hat{p} - z_{\alpha/2}\sqrt{\frac{\hat{p}\hat{q}}{n}} \leq p \leq \hat{p} + z_{\alpha/2}\sqrt{\frac{\hat{p}\hat{q}}{n}}.$$

If n is small, a correction for continuity should be applied to the confidence interval as follows:

$$\left(\hat{p} - \frac{1}{2n}\right) - z_{\alpha/2}\sqrt{\frac{\hat{p}\hat{q}}{n}} \leq p \leq \left(\hat{p} + \frac{1}{2n}\right) + z_{\alpha/2}\sqrt{\frac{\hat{p}\hat{q}}{n}}.$$

Let us return to our politician, Mr. Jerry Mander, who in Section 15.1 had decided not to cancel a political television announcement scheduled for election eve. Was Mr. Mander's decision a wise one? According to the Giddyup poll, a majority (54%) of the voters said they planned to vote for Mr. Mander. Suppose this percentage is based on a random sample of $n = 1500$ registered voters. A $100(1 - 0.01)\%$ confidence interval for p, the proportion of voters favoring Mr. Mander, is given by

$$0.54 - 2.576\sqrt{\frac{(0.54)(0.46)}{1500}} \leq p \leq 0.54 + 2.576\sqrt{\frac{(0.54)(0.46)}{1500}}$$

$$0.54 - 0.033 \leq p \leq 0.54 + 0.033$$

$$0.507 \leq p \leq 0.573.$$

We can be 99% confident that the true proportion is covered by the interval 0.507–0.573. Had Mr. Jerry Mander known this, he would have undoubtedly decided to listen to election returns while vacationing in Hawaii.

Confidence Interval for $p_1 - p_2$ (Independent Samples)

The most efficient point estimator for $p_1 - p_2$, the difference between the parameters of two binomial distributions, is the statistic $\hat{p}_1 - \hat{p}_2$. If $n_1\hat{p}_1, n_1\hat{q}_1, n_2\hat{p}_2, n_2\hat{q}_2$ are all greater than five and if the samples are statistically independent, a $100(1 - \alpha)\%$ confidence interval for $p_1 - p_2$ is given by

$$(\hat{p}_1 - \hat{p}_2) - z_{\alpha/2}\sqrt{\frac{\hat{p}_1\hat{q}_1}{n_1} + \frac{\hat{p}_2\hat{q}_2}{n_2}} \leq p_1 - p_2 \leq (\hat{p}_1 - \hat{p}_2) + z_{\alpha/2}\sqrt{\frac{\hat{p}_1\hat{q}_1}{n_1} + \frac{\hat{p}_2\hat{q}_2}{n_2}}.$$

This confidence interval, like that for p, is approximate, since the standard error of p depends on the true p—that is, $\sigma_p = \sqrt{pq/n}$. Since p is unknown, a sample estimate of p must be used in the confidence interval. This is satisfactory for all practical purposes if n is large, say at least 100.

† This section can be omitted without loss of continuity.

Review Exercises for Section 15.4

*19. A national survey of unmarried women between the ages of 15 and 19 found that 46% of the 19-year-olds had experienced sexual intercourse. If $n = 200$, construct a 95% confidence interval on p.

20. In the survey cited in Exercise 19, 26.6% of the 17-year-olds reported they had experienced sexual intercourse. If the sample contained $n = 150$ 17-year-olds, construct a 95% confidence interval on p.

*21. One hundred men who had suffered one heart attack participated in a supervised physical fitness program. Only seven of the men had a second attack during the 12 months after beginning the program. Construct a 99% confidence interval on p.

22. According to national statistics, the chances of a man having a second heart attack are one in 20 each year after the first seizure. Was the physical fitness program cited in Exercise 21 effective? Why?

23. In a random sample of 50 homes in Junction City, it was found that 32 have color television sets. Construct a 99% confidence interval on p; use a correction for continuity.

*24. For the data reported in Exercises 19 and 20, construct a 95% confidence interval for the difference between the proportion of 19-year-olds, p_1, and 17-year-olds, p_2, who had experienced sexual intercourse.

25. Twenty-four of 100 men and 13 of 100 women surveyed at a southwest college reported they had a car on campus. Construct a 99% confidence interval on $p_1 - p_2$.

26. A test comparing the detectability of two hues of stoplights under simulated fog conditions found that the relative frequencies of detection for red and yellow lights were 0.56 and 0.62, respectively. The subjects were randomly assigned to view one or the other condition; 321 viewed the red light and 315 viewed the yellow light. Construct a 95% confidence interval on $p_1 - p_2$.

†15.5 Confidence Intervals for ρ and $\rho_1 - \rho_2$

Confidence Interval for ρ

The Fisher r-to-Z' transformation described in Section 12.5 is used to construct a confidence interval for a population correlation coefficient ρ. Recall that the sampling distribution of Z' is approximately normal in form if (1) ρ is not too close to 1 or -1, (2) the population is bivariate normal in form, and (3) the sample size is at least moderately large, say $n > 50$. The r-to-Z' transformation is accomplished by means of Appendix Table D.8.

A $100(1 - \alpha)\%$ confidence interval for Z'_{Pop} is given by

$$Z' - z_{\alpha/2}\sqrt{\frac{1}{n-3}} \leq Z'_{Pop} \leq Z' + z_{\alpha/2}\sqrt{\frac{1}{n-3}},$$

where Z' is the transformed sample r, $z_{\alpha/2}$ is the value of z that cuts off the upper $\alpha/2$ portion of the sampling distribution of z, n is the size of the sample used to estimate ρ, and Z'_{Pop} is the transformed population correlation coefficient. Once the lower and upper limits of Z'_{Pop} have been determined they can be transformed into limits for ρ by means of Table D.8.

† This section can be omitted without loss of continuity.

Computational Example of Confidence Interval for ρ

Suppose an experimenter has correlated college grades and income 10 years after graduation for a random sample of 62 graduates of Lollipop Day-Care College. The sample estimate of the population correlation coefficient is 0.16. A $100(1 - 0.05)\%$ confidence interval is given by

$$0.161 - 1.96\sqrt{\frac{1}{62-3}} \leq Z'_{Pop} \leq 0.161 + 1.96\sqrt{\frac{1}{62-3}}$$

$$0.161 - 0.255 \leq Z'_{Pop} \leq 0.161 + 0.255$$

$$-0.094 \leq Z'_{Pop} \leq 0.416.$$

Transforming the lower and upper limits on Z' into correlations yields the 95% confidence interval on ρ, which is $-0.09 \leq \rho \leq 0.39$. Since the confidence interval includes zero, it is apparent that a test of the null hypothesis $H_0: \rho = 0$ or any other hypothesis in which ρ_0 is on the interval -0.09–0.39 could not be rejected.

Confidence Interval for $\rho_1 - \rho_2$

The Fisher r-to-Z' transformation can be used to construct a confidence interval for $\rho_1 - \rho_2$, the difference between population correlation coefficients. A $100(1 - \alpha)\%$ confidence interval for $Z'_{Pop_1} - Z'_{Pop_2}$ is given by

$$(Z'_1 - Z'_2) - z_{\alpha/2}\sqrt{\frac{1}{n_1-3} + \frac{1}{n_2-3}} \leq Z'_{Pop_1} - Z'_{Pop_2} \leq (Z'_1 - Z'_2) + z_{\alpha/2}\sqrt{\frac{1}{n_1-3} + \frac{1}{n_2-3}},$$

where the terms are as defined on page 296.

Review Exercises for Section 15.5

27. Derive a confidence interval for Z' from $\text{Prob}(-z_{\alpha/2} \leq z \leq z_{\alpha/2}) = 1 - \alpha$, where $z = (Z' - Z'_{Pop})/(1/\sqrt{n-3})$.
*28. The correlation between scores on the TAC (a college entrance test) and grade point averages for a random sample of $n = 84$ freshmen was 0.54. Construct a 95% confidence interval on ρ.
29. Psychological Associates, a consulting firm, has developed a test to select managers for a large hamburger chain. The test was given to a random sample of 52 managers. The correlation between their test scores and a measure of their stores' net incomes was 0.20. (a) Construct a 95% confidence interval on ρ. (b) Should the test be used in selecting future managers for the chain? Why?
30. The correlation between the recreational interests of a random sample of $n = 67$ pairs of husbands and wives who had contacted a large travel agency was 0.52. Construct a 99% confidence interval on ρ.
*31. The correlation between scores on the BEEC (a college entrance test) and grade point averages for a random sample of $n = 52$ freshmen was 0.46. Assume this sample was obtained from the college cited in Exercise 28. Construct a 95% confidence interval for the difference between the TAC and BEEC correlations.
32. The correlations between IQ and school achievement for random samples of male and female high school seniors were 0.37 and 0.46, respectively. Both samples contained 103 students. Construct a 95% confidence interval for $\rho_1 - \rho_2$.

Table 15.6-1. Summary of One- and Two-Sample Confidence Intervals

Chapter Section	Parameter	Confidence Interval	Assumptions
15.2	μ	$\bar{X} - t_{\alpha/2,\nu}\hat{\sigma}_{\bar{X}} \leq \mu \leq \bar{X} + t_{\alpha/2,\nu}\hat{\sigma}_{\bar{X}}$ where $\hat{\sigma}_{\bar{X}} = \hat{\sigma}/\sqrt{n}$	1. Random sampling 2. Normality 3. Variance unknown
15.2	$\mu_1 - \mu_2$	$(\bar{X}_1 - \bar{X}_2) - t_{\alpha/2,\nu}\hat{\sigma}_{\bar{X}_1-\bar{X}_2} \leq \mu_1 - \mu_2 \leq (\bar{X}_1 - \bar{X}_2) + t_{\alpha/2,\nu}\hat{\sigma}_{\bar{X}_1-\bar{X}_2}$ where $\hat{\sigma}_{\bar{X}_1-\bar{X}_2} = \sqrt{\hat{\sigma}^2_{Pooled}\left(\dfrac{1}{n_1} + \dfrac{1}{n_2}\right)}$	1. Random sampling or random assignment 2. Normality 3. Variances unknown but assumed equal 4. Independence between samples
15.2	$\mu_1 - \mu_2$	$(\bar{X}_1 - \bar{X}_2) - t_{\alpha/2,\nu}\hat{\sigma}_{\bar{X}_D} \leq \mu_1 - \mu_2 \leq (\bar{X}_1 - \bar{X}_2) + t_{\alpha/2,\nu}\hat{\sigma}_{\bar{X}_D}$ where $\hat{\sigma}_{\bar{X}_D} = \dfrac{\sqrt{[n\sum D_i^2 - (\sum D_i)^2]/n^2}}{\sqrt{n-1}}$	1. Random sampling or random assignment 2. Normality 3. Variances unknown but assumed equal 4. Dependence between samples
15.3	σ^2	$\dfrac{(n-1)\hat{\sigma}^2}{\chi^2_{\alpha/2,\nu}} \leq \sigma^2 \leq \dfrac{(n-1)\hat{\sigma}^2}{\chi^2_{1-\alpha/2}}$	1. Random sampling 2. Normality

Table 15.6-1 (continued)

Chapter Section	Parameter	Confidence Interval	Assumptions
15.3	$\dfrac{\sigma_1^2}{\sigma_2^2}$	$\dfrac{\hat{\sigma}_1^2}{\hat{\sigma}_2^2}\dfrac{1}{F_{\alpha/2;\nu_1,\nu_2}} \leq \dfrac{\sigma_1^2}{\sigma_2^2} \leq \dfrac{\hat{\sigma}_1^2}{\hat{\sigma}_2^2} F_{\alpha/2;\nu_2,\nu_1}$	1. Random sampling or random assignment 2. Normality 3. Independence between samples
15.4	p	$\hat{p} - z_{\alpha/2}\hat{\sigma}_{\hat{p}} \leq p \leq \hat{p} + z_{\alpha/2}\hat{\sigma}_{\hat{p}}$ where $\hat{\sigma}_{\hat{p}} = \sqrt{\dfrac{\hat{p}\hat{q}}{n}}$	1. Random sampling 2. Binomial distribution 3. $n\hat{p} > 5,\ n(1-\hat{p}) > 5$
15.4	$p_1 - p_2$	$(\hat{p}_1 - \hat{p}_2) - z_{\alpha/2}\hat{\sigma}_{\hat{p}_1-\hat{p}_2} \leq p_1 - p_2 \leq (\hat{p}_1 - \hat{p}_2) + z_{\alpha/2}\hat{\sigma}_{\hat{p}_1-\hat{p}_2}$ where $\hat{\sigma}_{\hat{p}_1-\hat{p}_2} = \sqrt{\dfrac{\hat{p}_1\hat{q}_1}{n_1} + \dfrac{\hat{p}_2\hat{q}_2}{n_2}}$	1. Random sampling or random assignment 2. Binomial distributions 3. Independence between samples 4. $n_1\hat{p}_1 > 5,\ n_1\hat{q}_1 > 5,$ $n_2\hat{p}_2 > 5,\ n_2\hat{q}_2 > 5$
15.5	ρ	$Z' - z_{\alpha/2}\sigma_{Z'} \leq Z'_{Pop} \leq Z' + z_{\alpha/2}\sigma_{Z'}$ where $\sigma_{Z'} = \sqrt{\dfrac{1}{n-3}}$	1. Random sampling 2. Bivariate normality 3. ρ not too close to ± 1 4. n at least moderately large
15.5	$\rho_1 - \rho_2$	$(Z'_1 - Z'_2) - z_{\alpha/2}\hat{\sigma}_{Z'_1-Z'_2} \leq Z'_{Pop_1} - Z'_{Pop_2} \leq (Z'_1 - Z'_2) + z_{\alpha/2}\hat{\sigma}_{Z'_1-Z'_2}$ where $\hat{\sigma}_{Z'_1-Z'_2} = \sqrt{\dfrac{1}{n_1-3} + \dfrac{1}{n_2-3}}$	1. Random sampling or random assignment 2. Bivariate normality 3. ρ_1 and ρ_2 not too close to ± 1 4. n_1 and n_2 at least moderately large

15.6 Summary

Estimation procedures are used when an experimenter is interested in estimating an unknown population parameter rather than in testing a particular hypothesis about the parameter. Point estimation provides us with a single numerical value that is the best guess we can make concerning the unknown parameter. Interval estimation, which is the subject of this chapter, provides us with a range of values that have a specified probability of capturing the parameter. They indicate the reliability of our point estimate.

Given a parameter, say θ, we can find two functions L_1 and L_2 of the sample values such that before the sample is drawn the probability that the interval from L_1 to L_2 will contain θ is equal to $1 - \alpha$. The interval from L_1 to L_2 is called a confidence interval on θ with a confidence coefficient equal to $1 - \alpha$. We can be $100(1 - \alpha)\%$ confident that θ is contained within the confidence limits L_1 and L_2.

A confidence interval indicates the values of the parameter that are consistent with the observed sample statistic. Thus, a $100(1 - \alpha)\%$ confidence interval contains a range of values of μ_0 for which the null hypothesis is nonrejectable at α level of significance.

The size of a confidence interval is determined by (1) the level of the confidence coefficient, $(1 - \alpha)$, (2) the size of the sample, and (3) the size of the population variance.

The confidence intervals we have described in this chapter are presented in Table 15.6-1, along with a summary of the parameters with which they are used, their formulas, and their assumptions.

16

Introduction to the Analysis of Variance

16.1 Purposes of Analysis of Variance
   Testing the Null Hypothesis that $\mu_1 = \mu_2 = \cdots = \mu_k$
   Other Applications
   Review Exercises for Section 16.1
16.2 Basic Concepts of ANOVA
   The Composite Nature of a Score
   The Linear Model Associated with a Score
   Estimating the Parameters of the Linear Model
   Partition of the Total Sum of Squares
   Degrees of Freedom
   Mean Squares and the F Ratio
   The Nature of MS_{BG} and MS_{WG}
   Expectations of MS_{BG} and MS_{WG}
   Review Exercises for Section 16.2
16.3 Completely Randomized Design
   Computational Formulas
   Computational Examples for Type CR-3 Design
   Review Exercises for Section 16.3
16.4 Assumptions Associated with a Type CR-k Design
   Assumption that $X_{ij} = \mu + \beta_j + e_{ij}$ (B1)
   Assumption that the Experiment Contains All Treatment Levels of Interest (B2)
   Assumptions of Independence of Experimental Errors (B3a) and Random Sampling (A4)
   Assumptions of Normally Distributed Errors (B3b) and Populations (A1), and Independence of MS_{BG} and MS_{WG} (A2)
   Assumption that $E(e_{ij}) = 0$ for Each Treatment Population (B3c)
   Assumption of Homogeneity of Variance (B3d)
   Review Exercises for Section 16.4
16.5 Introduction to Multiple Comparisons
   Scheffé's S Multiple Comparison Procedure
   Contrasts among Means
   Computational Example for S Test
   Review Exercises for Section 16.5
16.6 Introduction to Factorial Designs
16.7 Summary

16.1 Purposes of Analysis of Variance

Testing the Null Hypothesis that $\mu_1 = \mu_2 = \cdots = \mu_k$

One of the most frequently used statistical procedures in the behavioral sciences is analysis of variance (see Edgington, 1974), often referred to as ANOVA. The procedure was developed by R. A. Fisher in the early 1920s to test null hypotheses of the form

$$H_0: \mu_1 = \mu_2 = \cdots = \mu_k,$$

where $\mu_1, \mu_2, \ldots, \mu_k$ denote the means of $k \geq 2$ populations. If H_0 is rejected, the alternative hypothesis that at least two of the population means are not equal is tenable—that is, $\mu_j \neq \mu_{j'}$ for some pair of means μ_j and $\mu_{j'}$.

In Chapter 13 we described z and t statistics for testing hypotheses of the form $H_0: \mu_j - \mu_{j'} = 0$. Since these statistics can be used to test hypotheses about the equality of means, you may wonder why we use a procedure like ANOVA that utilizes an F statistic. To see that such a procedure is needed, let's consider an experiment with $k = 3$ experimental conditions. How many hypotheses can be tested using a t statistic? There are $k(k-1)/2 = 3$ hypotheses of the form $\mu_j - \mu_{j'} = \delta_0$, where $\delta_0 = 0$; they are $H_0: \mu_1 - \mu_2 = 0$, $H_0: \mu_1 - \mu_3 = 0$, and $H_0: \mu_2 - \mu_3 = 0$. In addition to these three hypotheses, we might be interested in using a t statistic to test hypotheses involving three means,[1] such as

$$H_0: \mu_1 - \frac{\mu_2 + \mu_3}{2} = 0 \qquad H_0: \mu_2 - \frac{\mu_1 + \mu_3}{2} = 0 \qquad H_0: \mu_3 - \frac{\mu_1 + \mu_2}{2} = 0.$$

These hypotheses state that one population mean is equal to the average of two other population means. Unlike the set of two-mean hypotheses, there is an infinite number of three-mean hypotheses that can be formulated by selecting weights[2] other than 1, $-\frac{1}{2}$, and $-\frac{1}{2}$, as in

$$H_0: \mu_1 - \frac{0.5\mu_2 + 1.5\mu_3}{2} = 0 \qquad \text{or} \qquad H_0: \mu_1 - \frac{0.2\mu_2 + 1.8\mu_3}{2}.$$

If an experimenter tests a large number of these hypotheses, each at α level of significance using a t statistic, the probability that one or more null hypotheses will be falsely rejected is larger than α—and the probability increases as the number of t tests increases. The advantage of the ANOVA approach is that by performing one F test an experimenter tests simultaneously all possible null hypotheses for k means, and the probability of making a type I error is equal to α. If the overall null hypothesis $H_0: \mu_1 = \mu_2 = \cdots = \mu_k$ is rejected, the experimenter can conclude that at least two of the population means are not equal. Then the next step is to use the multiple comparison procedure described in Section 16.5 to determine which means are not equal.

The principal difference between the overall ANOVA F approach and the multiple t approach can be summarized as follows. For ANOVA, the probability of making a

---

[1] The t formula for testing these hypotheses and the one for Scheffé's S statistic (see Section 16.5) are identical.
[2] The only restriction on the weights is that they must sum to zero, for example, $1 + (-\frac{1}{2}) + (-\frac{1}{2}) = 0$.

type I error is equal to α for the overall null hypothesis $\mu_1 = \mu_2 = \cdots = \mu_k$. For t, the probability of a type I error is equal to α for each null hypothesis; this allows the probability of erroneously rejecting one or more true null hypotheses to increase as a function of the number of hypotheses tested. Thus, a persistent experimenter who performs enough t tests, each at α level of significance, will certainly reject too many null hypotheses.

For the special case in which the experiment contains only two experimental conditions and the null hypothesis is $\mu_j - \mu_{j'} = 0$, the F and t tests have the same probability of making a type I error, since in both cases only one hypothesis is tested.

Other Applications

Although ANOVA is most frequently used to test hypotheses about means, this isn't its only application. It can be used, for example, to determine (1) the strength of the association between independent and dependent variables, (2) whether two independent variables interact, and (3) whether the regression of the dependent variable on the independent variable is linear or nonlinear.[3]

Review Exercises for Section 16.1

*1. Suppose five methods of teaching foreign language vocabulary are compared in an experiment. The dependent variable is performance on a 25 item vocabulary test. (a) State the null hypothesis. (b) How many t tests would be required to test just hypotheses of the form $\mu_j - \mu_{j'} = 0$?
*2. If the overall null hypothesis in Exercise 1a is rejected by means of an ANOVA F test, what does this tell the experimenter?
3. Four colors of warning lights on an automobile instrument panel were compared. The dependent measure was reaction time to the onset of a light. (a) State the null hypothesis. (b) If the overall null hypothesis is rejected, what does this tell the experimenter?
4. Under what conditions do the F and t approaches lead to the same probability of making a type I error?
5. For experiments in which the number of experimental conditions is greater than two, what advantage does the ANOVA F approach have over the multiple t approach?
**6. (a) Give two examples of independent variables for which the ANOVA F and multiple t approaches would lead to identical conclusions. *(b) What characteristic do the examples have in common?

16.2 Basic Concepts of ANOVA

The material in this section provides a glimpse of some of the basic concepts associated with a completely randomized ANOVA, the simplest of all the ANOVA designs. An aquaintance with, rather than a mastery of, this section is sufficient for

[3] For a discussion of these topics see Edwards (1972), Kirk (1968), Myers (1972), and Winer (1971).

understanding Section 16.3, which deals with the computational procedures for ANOVA. It should be helpful to review Section 16.2 after working through an ANOVA problem.

The Composite Nature of a Score

The value of a score in an experiment is determined by a variety of variables. Therefore, a score can be thought of as a composite, reflecting, for example, the effects of (1) experimental conditions, (2) individual characteristics of the subject or experimental unit, (3) chance fluctuation in the subject's performance, and (4) environmental conditions. Similarly, the variance of scores is also a composite. ANOVA is used to determine how much of the total variability among scores to attribute to the various sources of variation and to test hypotheses concerning some of them.

The composite nature of a score will be illustrated by an example. An experiment is designed to determine the effectiveness of three diets for obese teen-age girls. Thirty girls who want to lose weight are randomly assigned to the diets, with ten girls using each diet. The independent variable is type of diet; the dependent variable is weight loss in pounds. *For notational convenience the diets are called* **treatment**[4] **B**. *The levels of treatment B, corresponding to the specific diets, are designated by the lowercase letter* **b** *and number subscripts—***b_1***,* **b_2***, and* **b_3***. A particular but unspecified score is denoted by* **X_{ij}***, where the first subscript designates one of the $i = 1, \ldots, n$ subjects in a treatment level and the second subscript designates one of the $j = 1, \ldots, k$ levels of treatment B.* Thus, if Bella Ablipid's score is denoted by X_{72}, we know that she is subject seven and that she used diet b_2. What factors have affected the value of her score? If she stuck to her diet, one major factor is the efficacy of diet b_2. Other factors are degree of obesity, day-to-day fluctuations in eating and exercise habits, time of day that weight loss was measured, and so on. In summary, Bella's weight loss X_{72} reflects (1) the effect of treatment level b_2, (2) effects unique to her, (3) effects attributable to chance fluctuations in her behavior, and (4) effects attributable to environmental conditions.

The Linear Model Associated with a Score

Our conjectures about X_{72} or any other score can be expressed more formally by a *linear model*. We assume that each of the 30 scores is the sum of three components in the linear model

$$X_{ij} = \mu + \beta_j + e_{ij},$$

where
X_{ij} is the score for subject i in treatment level j;
μ is the grand mean of μ_1, μ_2, and μ_3, the means of the three treatment populations (μ is a constant for the 30 scores and reflects their general elevation);

---

[4] Some writers use the term *factor* instead of treatment.

β_j is the treatment effect of population j and is equal to $\mu_j - \mu$, the deviation of the jth population mean from the grand mean (β_j is a constant for the ten scores in population j and reflects the elevation or depression of these scores resulting from the use of the jth diet); and

e_{ij} is the "error" effect associated with X_{ij} and is equal to $X_{ij} - \mu - \beta_j$, the portion of a score that remains after the grand mean and jth treatment effect are subtracted from it (e_{ij} reflects effects unique to subject i, effects attributable to chance fluctuations in subject i, and effects attributable to environmental conditions—in other words, all effects not attributable to treatment level j).

The values of the parameters μ, β_j, and e_{ij} are unknown, but we will see that they can be estimated from sample data and the variability among the X_{ij}'s can be partitioned by ANOVA so as to test the null hypothesis that $\mu_1 = \mu_2 = \cdots = \mu_k$. If the null hypothesis is true, each of the treatment effects, β_j, must equal zero. This follows, since in that case all the population means equal the grand mean, $\mu_1 = \mu_2 = \cdots = \mu_k = \mu$, and $\beta_1 = \mu_1 - \mu = 0$, $\beta_2 = \mu_2 - \mu = 0$, and so on. Thus, a test of $\mu_1 = \mu_2 = \cdots = \mu_k$ is equivalent to a test of the hypothesis $\beta_1 = \beta_2 = \cdots = \beta_k = 0$—that is, that all population treatment effects are equal to zero.

Estimating the Parameters of the Linear Model

Suppose the data in Table 16.2-1 have been obtained in our diet experiment. The treatment means, $\bar{X}_{.1}$, $\bar{X}_{.2}$, and $\bar{X}_{.3}$, and the grand mean, $\bar{X}_{..}$, are shown in the table. The use of a dot in the notation indicates that the mean was obtained by averaging over the subscript replaced by the dot. For example, treatment means are obtained by

$$\bar{X}_{.1} = \sum_{i=1}^{n} \frac{X_{i1}}{n} = \frac{X_{11} + X_{21} + X_{31} + \cdots + X_{10,1}}{n} = \frac{80}{10} = 8$$

$$\bar{X}_{.2} = \sum_{i=1}^{n} \frac{X_{i2}}{n} = \frac{X_{12} + X_{22} + X_{32} + \cdots + X_{10,2}}{n} = \frac{90}{10} = 9$$

$$\bar{X}_{.3} = \sum_{i=1}^{n} \frac{X_{i3}}{n} = \frac{X_{13} + X_{23} + X_{33} + \cdots + X_{10,3}}{n} = \frac{120}{10} = 12.$$

The grand mean is obtained by summing all the kn scores and dividing by kn:

$$\bar{X}_{..} = \sum_{j=1}^{k} \sum_{i=1}^{n} \frac{X_{ij}}{kn} = [(X_{11} + X_{21} + \cdots + X_{10,1}) + \cdots + (X_{13} + X_{23} + \cdots + X_{10,3})]/kn$$

$$= \frac{290}{(3)(10)} = 9.67.$$

Table 16.2-1 contains scores and means, but the values of the parameters μ, β_j, and e_{ij} in the linear model are unknown. However, unbiased estimators of the parameters can be obtained from the sample data, as shown in Table 16.2-2.

We can rewrite the linear model $X_{ij} = \mu + \beta_j + e_{ij}$ using estimators of the parameters as follows:

$$X_{ij} = \bar{X}_{..} + (\bar{X}_{.j} - \bar{X}_{..}) + (X_{ij} - \bar{X}_{.j}).$$

Table 16.2-1. One Month Weight Losses Measured to the Nearest Pound

(i) Data:

	Treatment Levels (Diets)		
	b_1	b_2	b_3
	7	10	12
	9	13	11
	8	9	15
	12	11	7
	8	5	14
	7	9	10
	4	8	12
	10	10	12
	9	8	13
	6	7	14

Sum of $i = 1, \ldots, n$ scores in each treatment level $\rightarrow \sum_{i=1}^{n} X_{i1} = 80 \qquad \sum_{i=1}^{n} X_{i2} = 90 \qquad \sum_{i=1}^{n} X_{i3} = 120 \qquad$ Sum of all scores $\rightarrow \sum_{j=1}^{k}\sum_{i=1}^{n} X_{ij} = 290$

Mean of each treatment level $\rightarrow \bar{X}_{\cdot 1} = 8 \qquad \bar{X}_{\cdot 2} = 9 \qquad \bar{X}_{\cdot 3} = 12 \qquad$ Grand mean $\rightarrow \bar{X}_{\cdot\cdot} = 9.67$

Table 16.2-2. Estimators for Parameters of the Linear Model

Statistic	Parameter Estimated	Interpretation of Parameter
$\bar{X}_{..}$	μ	General elevation of scores
$\bar{X}_{.j} - \bar{X}_{..}$	β_j	Elevation or depression of scores attributable to the jth treatment level
$X_{ij} - \bar{X}_{.j}$	e_{ij}	Error effect unique to subject i in treatment level j

Consider Bella's score, X_{72}, in Table 16.2-1. According to the equation, page 305, this score is a composite reflecting all the factors in the experiment; it can be expressed as

$$X_{72} = \bar{X}_{..} + (\bar{X}_{.2} - \bar{X}_{..}) + (X_{72} - \bar{X}_{.2})$$
$$8 = 9.67 + (9 - 9.67) + (8 - 9)$$
$$8 = 9.67 + (-0.67) + (-1)$$
$$8 = 9.67 - 1.67.$$

By substituting the appropriate values in the $(\bar{X}_{.2} - \bar{X}_{..})$ term of the linear model, we see that the mean weight loss for girls using her diet was 0.67 lb less than the mean loss for all 30 girls (or the mean for all three diets). The last term $(X_{72} - \bar{X}_{.2})$ shows that Bella's weight loss is 1 lb less than the mean loss for the ten girls who used her diet. It is apparent that the score X_{72} is affected by a number of variables, but we can make only three estimates—$\hat{\mu}$, $\hat{\beta}_2$, and $\hat{e}_{72}$. All the variables other than treatment level b_2 are lumped together in the error component $\hat{e}_{72} = (X_{72} - \bar{X}_{.2})$. Thus, $\hat{e}_{72} = -1$ reflects the combination of all the characteristics peculiar to this subject, such as degree of obesity and eating and exercise habits, and those peculiar to the testing conditions, such as time of day. The linear models associated with the more complex ANOVA designs contain more than three components, since certain sources of variation believed to be important are estimated separately from the error component, so the error term is usually smaller in complex ANOVA designs.[5]

Above, we have illustrated the composite nature of a score and procedures for estimating the components of the linear model. Now we will develop procedures for testing the null hypothesis that all β_j's are equal to zero.

Partition of the Total Sum of Squares

We will now show that a measure of the total variability among scores in an experiment, called the *total sum of squares* (SS_{Total}), can be partitioned into two parts—variability between treatment levels and variability within treatment levels. These two parts are called the *between-groups sum of squares* (SS_{BG}) and the *within-groups sum*

---

[5] A discussion of complex ANOVA designs is beyond the scope of this book. For an elementary description, see Kirk (1972b); for an advanced discussion including computational procedures, see Edwards (1972), Keppel (1973), Kirk (1968), Lee (1975), Lindman (1974), Myers (1972), and Winer (1971).

of squares (SS_{WG}). We begin by subtracting $\bar{X}..$ from both sides of the equation $X_{ij} = \bar{X}.. + (\bar{X}._j - \bar{X}..) + (X_{ij} - \bar{X}._j)$, and obtain

$$X_{ij} - \bar{X}.. = \bar{X}.. - \bar{X}.. + (\bar{X}._j - \bar{X}..) + (X_{ij} - \bar{X}._j)$$
$$= (\bar{X}._j - \bar{X}..) + (X_{ij} - \bar{X}._j).$$

This equation expresses the deviation of the grand mean from a score in terms of the deviation of the grand mean from a treatment mean plus the deviation of the treatment mean from a score. Next, we square both sides of the equation:

$$(X_{ij} - \bar{X}..)^2 = [(\bar{X}._j - \bar{X}..) + (X_{ij} - \bar{X}._j)]^2.$$

This equation is for a single score, but we can treat each of the kn scores in the same way and sum the resulting equations to obtain

$$\sum_{j=1}^{k} \sum_{i=1}^{n} (X_{ij} - \bar{X}..)^2 = \sum_{j=1}^{k} \sum_{i=1}^{n} [(\bar{X}._j - \bar{X}..) + (X_{ij} - \bar{X}._j)]^2.$$

The quantity on the left is the total sum of squares. It can be shown using elementary algebra and the summation rules in Technical Note 3.9-1 that the quantity on the right is equal to $SS_{BG} + SS_{WG}$.

$$\sum_{j=1}^{k} \sum_{i=1}^{n} (X_{ij} - \bar{X}..)^2 = \sum_{j=1}^{k} \sum_{i=1}^{n} [(\bar{X}._j - \bar{X}..)^2 + 2(\bar{X}._j - \bar{X}..)(X_{ij} - \bar{X}._j) + (X_{ij} - \bar{X}._j)^2]$$

$$= n \sum_{j=1}^{k} (\bar{X}._j - \bar{X}..)^2 + 2 \sum_{j=1}^{k} (\bar{X}._j - \bar{X}..) \sum_{i=1}^{n} (X_{ij} - \bar{X}._j)$$

$$+ \sum_{j=1}^{k} \sum_{i=1}^{n} (X_{ij} - \bar{X}._j)^2.$$

Since $\sum_{i=1}^{n} (X_{ij} - \bar{X}._j) = 0$ (see Technical Note 3.9-2), the middle term on the right equals zero, and we are left with

$$\sum_{j=1}^{k} \sum_{i=1}^{n} (X_{ij} - \bar{X}..)^2 = n \sum_{j=1}^{k} (\bar{X}._j - \bar{X}..)^2 + \sum_{j=1}^{k} \sum_{i=1}^{n} (X_{ij} - \bar{X}._j)^2$$
$$SS_{Total} = \quad SS_{BG} \quad + \quad SS_{WG}.$$

Now that we have partitioned SS_{Total} into SS_{BG} and SS_{WG}, we need to show what this has to do with testing the hypothesis that $\mu_1 = \mu_2 = \cdots = \mu_k$ or the equivalent hypothesis that $\beta_1 = 0, \beta_2 = 0, \ldots, \beta_k = 0$. But before we can do this we must develop some additional concepts.

Degrees of Freedom

The term *degrees of freedom* refers to the number of observations with values that can be assigned arbitrarily, as we saw in Section 12.2. We will now determine the degrees of freedom associated with SS_{BG}, SS_{WG}, and SS_{Total}.

Consider $SS_{BG} = n \sum_{j=1}^{k} (\bar{X}._j - \bar{X}..)^2$. If we take, say $k = 4$ sample means, they are related to the grand mean by the equation

$$\frac{\bar{X}._1 + \bar{X}._2 + \bar{X}._3 + \bar{X}._4}{4} = \bar{X}...$$

If $\bar{X}_{..} = 6$ and we arbitrarily specify $\bar{X}_{.1} = 6$, $\bar{X}_{.2} = 8$, and $\bar{X}_{.3} = 7$, then $\bar{X}_{.4}$ must equal 3, since $(6 + 8 + 7 + 3)/4 = 6$. Alternatively, if we specify $\bar{X}_{.1} = 5$, $\bar{X}_{.2} = 7$, and $\bar{X}_{.3} = 4$, then $\bar{X}_{.4}$ must equal 8, since $(5 + 7 + 4 + 8)/4 = 6$. Given the value of the grand mean, we are free to assign any values to three of the four treatment means, but having done so the fourth mean is determined. Hence, *the number of degrees of freedom for SS_{BG} is $k - 1$, one less than the number of treatment means.*

The degree of freedom associated with SS_{WG} is $k(n - 1)$. To see why this is true, consider $SS_{WG} = \sum_{j=1}^{k} \sum_{i=1}^{n} (X_{ij} - \bar{X}_{.j})^2$ and let $k = 4$ and $n = 8$. For each of the $j = 1, \ldots, 4$ treatment levels, we can compute $n = 8$ deviations.

$$\overbrace{(X_{1j} - \bar{X}_{.j})^2}^{\text{1st}} + \overbrace{(X_{2j} - \bar{X}_{.j})^2}^{\text{2nd}} + \cdots + \overbrace{(X_{8j} - \bar{X}_{.j})^2}^{\text{8th}}$$

Each score in the jth treatment level is related to the jth mean by

$$\frac{X_{1j} + X_{2j} + \cdots + X_{8j}}{8} = \bar{X}_{.j}.$$

Seven of the scores can take any value, but the eighth is determined, since the sum of the scores divided by eight must equal $\bar{X}_{.j}$. Hence, there are $n - 1 = 8 - 1 = 7$ degrees of freedom associated with the jth treatment level. If $n_1 = n_2 = \cdots = n_k$, each of the k treatment levels has $n - 1$ degrees of freedom; thus, there are $k(n - 1)$ degrees of freedom associated with SS_{WG}. If the n_j's are not equal, the degree of freedom for SS_{WG} is $(n_1 - 1) + (n_2 - 1) + \cdots + (n_k - 1) = N - k$, where N is the total number of scores.

The same line of reasoning can be used to show that when $n_1 = n_2 = \cdots = n_k$, the total sum of squares has $kn - 1$ degrees of freedom. This follows, since each of the $kn = 4(8) = 32$ scores is related to the grand mean by

$$\frac{X_{11} + X_{21} + \cdots + X_{84}}{32} = \bar{X}_{...}$$

Hence, $kn - 1 = 31$ of the scores can take any value, but the 32nd must be assigned so that the mean of the scores equals $\bar{X}_{...}$. If the n_j's are not equal, the number of degrees of freedom is $n_1 + n_2 + \cdots + n_k - 1 = N - 1$.

Mean Squares and the F Ratio

A **mean square** *(MS)* is obtained by dividing a sum of squares *(SS)* by its degrees of freedom *(df)*. Thus, $MS_{Total} = SS_{Total}/(kn - 1)$, $MS_{BG} = SS_{BG}/(k - 1)$, and $MS_{WG} = SS_{WG}/[k(n - 1)]$. The term *MS* is new, but the concept isn't; *MS* is simply another name for variance. The hypothesis $\mu_1 = \mu_2 = \cdots = \mu_k$ is tested by means of an *F* test statistic, which is the ratio of the between-groups variance to the within-groups variance; $F = MS_{BG}/MS_{WG}$. The degrees of freedom for the numerator and denominator of the statistic are, respectively, $v_1 = k - 1$ and $v_2 = k(n - 1)$. The *F* statistic is referred to the sampling distribution of *F*, which is tabled in Appendix Table D.5. If the *F* statistic is equal to or greater than the tabled value, $F_{\alpha;v_1,v_2}$, the null hypothesis is rejected.

The Nature of MS_{BG} and MS_{WG}

It may seem paradoxical to test a hypothesis about means by testing the ratio of two variances, so perhaps we should consider the nature of the population variances estimated by MS_{WG} and MS_{BG}. We will examine MS_{WG} first and show that it is an estimator of the population *error variance* (σ_e^2), the variance of scores in a population. Suppose that random samples of size n are drawn[6] from k normally distributed populations having equal means ($\mu_1 = \mu_2 = \cdots = \mu_k = \mu$) and equal variances ($\sigma_1^2 = \sigma_2^2 = \cdots = \sigma_k^2 = \sigma_e^2$). Unbiased estimators of the k population variances are given by

$$\hat{\sigma}_1^2 = \frac{\sum_{i=1}^{n}(X_{i1} - \bar{X}_{.1})^2}{n-1}, \quad \hat{\sigma}_2^2 = \frac{\sum_{i=1}^{n}(X_{i2} - \bar{X}_{.2})^2}{n-1}, \quad \ldots, \quad \hat{\sigma}_k^2 = \frac{\sum_{i=1}^{n}(X_{ik} - \bar{X}_{.k})^2}{n-1}.$$

The best estimator of the population error variance, σ_e^2, is obtained by combining the k variance estimators according to the formula

$$\frac{\sum_{i=1}^{n}(X_{i1} - \bar{X}_{.1})^2/(n-1) + \sum_{i=1}^{n}(X_{i2} - \bar{X}_{.2})^2/(n-1) + \cdots + \sum_{i=1}^{n}(X_{ik} - \bar{X}_{.k})^2/(n-1)}{k}$$

$$= \frac{\sum_{j=1}^{k}\sum_{i=1}^{n}(X_{ij} - \bar{X}_{.j})^2}{k(n-1)}.$$

The term on the right is MS_{WG}, as previously defined. Thus, MS_{WG} is an estimator of σ_e^2 and is simply the average of k variance estimators—that is, $MS_{WG} = [(\hat{\sigma}_1^2 + \cdots + \hat{\sigma}_k^2)/k] = \hat{\sigma}_e^2$.

We will now show that MS_{BG} also estimates σ_e^2 when k random samples of size n are drawn from k normally distributed populations with equal means and equal variances. These are the same conditions under which we said that MS_{WG} estimates σ_e^2. From Section 10.3 we know that the means of k random samples can be regarded as random variables from a theoretical sampling distribution of means, with mean and standard deviation μ and $\sigma_{\bar{X}}$, respectively. We also know from Section 10.3 that the standard error of a mean can be computed from $\sigma_{\bar{X}} = \sigma_e/\sqrt{n}$; hence, $\sigma_{\bar{X}}^2 = \sigma_e^2/n$. If we could estimate the variance of population means $\sigma_{\bar{X}}^2$, we could then estimate σ_e^2 by $\hat{\sigma}_e^2 = n\hat{\sigma}_{\bar{X}}^2$. In Section 10.3 it was stated that $\sum_{i=1}^{n}(X_{ij} - \bar{X}_{..})^2/(n-1)$, which is computed from a random sample of n scores, is an unbiased estimator of the variance of population scores. Similarly, $\sum_{j=1}^{k}(\bar{X}_{.j} - \bar{X}_{..})^2/(k-1)$, which is computed from a random sample of k means, is an unbiased estimator of the variance of population means. Hence,

$$MS_{BG} = n\sum_{j=1}^{k}\frac{(\bar{X}_{.j} - \bar{X}_{..})^2}{k-1} = n\hat{\sigma}_{\bar{X}}^2 = \hat{\sigma}_e^2.$$

Thus, if k random samples are drawn from normally distributed populations with equal means and equal variances, MS_{BG} is an estimator of the population error variance, σ_e^2.

---

[6] The use of equal n_j's is a convenience, not a requirement.

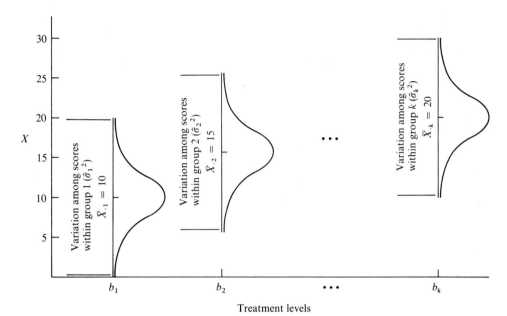

Figure 16.2-1. A within-groups estimator of the population error variance is given by $\hat{\sigma}_e^2 = (\hat{\sigma}_1^2 + \hat{\sigma}_2^2 + \cdots + \hat{\sigma}_k^2)/k$. This estimator is not affected by differences among the sample means. An estimator of the variance among the k population means is given by $\hat{\sigma}_{\bar{X}}^2 = \sum_{j=1}^{k} (\bar{X}_{\cdot j} - \bar{X}_{\cdot \cdot})^2/(k-1)$.

It can be shown that under these conditions, MS_{BG} and MS_{WG} are independent estimators of σ_e^2, and that the ratio $F = MS_{BG}/MS_{WG}$ is distributed as the F distribution with $k-1$ and $k(n-1)$ degrees of freedom. Small values of the F ratio (around one) are likely if the null hypothesis is true—that is, if $\mu_1 = \mu_2 = \cdots = \mu_k$.

But what if the null hypothesis is false? Suppose random samples are drawn from k normally distributed populations with two or more unequal means but equal variances. We would expect more variation among the sample means than can be accounted for by chance sampling variability. However, variation among the sample means is independent of the variation within each of the groups, $\hat{\sigma}_1^2, \hat{\sigma}_2^2, \ldots, \hat{\sigma}_k^2$; this can be seen from Figure 16.2-1. As before, $MS_{WG} = [(\hat{\sigma}_1^2 + \cdots + \hat{\sigma}_k^2)/k] = \hat{\sigma}_e^2$. However, the presence of unequal means will increase the variance of the sample means. Therefore, when the null hypothesis is false, $MS_{BG} = n\hat{\sigma}_{\bar{X}}^2$ tends to be larger than $\hat{\sigma}_e^2$ and the ratio $F = MS_{BG}/MS_{WG}$ tends to be larger than one.

How large must F be in order to conclude that the discrepancy between MS_{BG} and MS_{WG} is more likely due to unequal population means rather than to chance? The most usual practice, according to hypothesis testing conventions, is to reject the null hypothesis of equal means if F falls in the upper 0.05 portion[7] of the sampling distribution of F. Values of F that cut off various portions of the upper tail of this sampling distribution are given in Appendix Table D.5.

[7] The selection of α is discussed in Section 11.4.

Expectations of MS_{BG} and MS_{WG}

We can express the ideas in the previous section in terms of the expectations (expected values) of MS_{BG} and MS_{WG}. An expectation (E) is, according to the empirical relative-frequency view of probability, the long-run average of a random variable. We learned in Section 10.3 that $E(\bar{X}) = \mu$. If we randomly sample from a population an indefinitely large number of times and for each sample compute a mean, $\bar{X}$, the long-run average of the sample means is μ. It can be shown (Kirk, 1968, pp. 50–57) that the expectation of MS_{WG} is the population error variance,

$$E(MS_{WG}) = \sigma_e^2.$$

If the null hypothesis is true, the expectation of MS_{BG} is also the population error variance,

$$E(MS_{BG}) = \sigma_e^2.$$

However, if at least two of the population means are not equal, the expectation of MS_{BG} is

$$E(MS_{BG}) = \sigma_e^2 + \frac{n \sum_{j=1}^{k} (\mu_j - \mu)^2}{k - 1}$$

$$= \sigma_e^2 + \frac{n \sum_{j=1}^{k} \beta_j^2}{k - 1}.$$

The ratio $F = MS_{BG}/MS_{WG}$ is distributed as the F distribution with $v_1 = k - 1$ and $v_2 = k(n - 1)$ if the null hypothesis is true. If the ratio is sufficiently large, it suggests that the null hypothesis is false. The probability of obtaining an F statistic as large or larger than that observed if H_0 is true can be determined from the F table. If the probability is equal to or less than some preselected value, say $\alpha = 0.05$, the null hypothesis that all the means are equal, $\mu_1 = \mu_2 = \cdots = \mu_k$, or that all the treatment effects equal zero, $\beta_j = 0$ for all j, is rejected.

Review Exercises for Section 16.2

*7. Identify the following.
  a. b_2
  b. X_{24}
  c. $X_{16,1}$
  d. $\bar{X}_{.4}$
  e. $\bar{X}_{..}$
  f. $X_{73} = \mu + \beta_3 + e_{73}$
  g. $\mu_2 - \mu$
  h. $X_{13} - \mu - \beta_3$

8. In the derivation of $\sum_{j=1}^{k} \sum_{i=1}^{n} (X_{ij} - \bar{X}_{..})^2 = n \sum_{j=1}^{k} (\bar{X}_{.j} - \bar{X}_{..})^2 + \sum_{j=1}^{k} \sum_{i=1}^{n} (X_{ij} - \bar{X}_{.j})^2$ we assume that $n_1 = n_2 = \cdots = n_k = n$. Assume instead that the n_j's are not all equal and derive $\sum_{j=1}^{k} \sum_{i=1}^{n_j} (X_{ij} - \bar{X}_{..})^2 = \sum_{j=1}^{k} n_j (\bar{X}_{.j} - \bar{X}_{..})^2 + \sum_{j=1}^{k} \sum_{i=1}^{n_j} (X_{ij} - \bar{X}_{.j})^2$.

*9. Which of the following alternative hypotheses are correctly stated? (a) $\mu_j - \mu_{j'} \neq 0$ for some j and j', (b) $\mu_1 \neq \mu_2 \neq \mu_3$, (c) $\sum_{j=1}^{k} (\mu_j - \mu) \neq 0$, (d) $\beta_1 \neq 0, \beta_2 \neq 0, \beta_3 \neq 0$, (e) $\beta_j \neq 0$ for some j.

**10. Express the following scores in terms of estimates of the parameters of the linear model: *(a) X_{83}, (b) X_{52}, (c) X_{24}.

11. Express the hypothesis $\mu_1 = \mu_2 = \mu_3 = \mu_4$ in terms of treatment effects (β_j's).

**12. Calculate the degrees of freedom for MS_{Total}, MS_{BG}, and MS_{WG} with the following conditions. *(a) $k = 4$, $n = 21$, *(b) $k = 5$, $n = 11$, (c) $k = 4$, $n = 8$, *(d) $k = 3$, $n_1 = 6$, $n_2 = 5$, $n_3 = 6$, (e) $k = 4$, $n_1 = 10$, $n_2 = 10$, $n_3 = 9$, $n_4 = 8$.

*13. Under what conditions do MS_{BG} and MS_{WG} both estimate only the population error variance, σ_e^2?

*14. Under what conditions is $F = MS_{BG}/MS_{WG}$ distributed as the F distribution with $k - 1$ and $k(n - 1)$ degrees of freedom?

*15. Under what conditions does $F = MS_{BG}/MS_{WG}$ tend to be larger than one?

16. Terms to remember:
    a. Treatment
    b. Treatment level
    c. Linear model
    d. Treatment effect
    e. Error effect
    f. Sum of squares
    g. Degrees of freedom
    h. Mean square

16.3 Completely Randomized Design

This section presents the computational procedures associated with the simplest of all ANOVA designs—the *completely randomized design*. Here you will see how nicely some of the complex ideas presented previously fit together to produce a decision about some question of interest. In fact, after pondering over the next two tables, you will be amazed at how clearly the logic emerges and how the complex notions we've developed are really simple and logical in the context of what the experimenter is doing.

This design is appropriate for experiments with one treatment (or independent variable) with $k \geq 2$ treatment levels. A total of $N = n_1 + n_2 + \cdots + n_k$ subjects are randomly assigned to the k levels. If the treatment levels are of equal interest to the experimenter, it is desirable to assign the same number of subjects to each level, although this is not necessary. *The completely randomized design is so named because the assignment of subjects to treatment levels is completely random, with each subject assigned to participate under only one level.* For convenience it is referred to as a **type CR-k design**,[8] where k designates the number of levels of treatment B.

Computational Formulas

In Section 16.2 we showed that

$$SS_{Total} = SS_{BG} + SS_{WG}$$

[8] For a comprehensive design classification system and nomenclature see Kirk (1972b).

is equal to

$$\sum_{j=1}^{k} \sum_{i=1}^{n} (X_{ij} - \bar{X}..)^2 = n \sum_{j=1}^{k} (\bar{X}_{.j} - \bar{X}..)^2 + \sum_{j=1}^{k} \sum_{i=1}^{n} (X_{ij} - \bar{X}_{.j})^2.$$

These sums of squares formulas aren't the most convenient for computational purposes. It can be shown[9] that they are equivalent to

$$\sum_{j=1}^{k} \sum_{i=1}^{n} X_{ij}^2 - \frac{\left(\sum_{j=1}^{k} \sum_{i=1}^{n} X_{ij}\right)^2}{kn} = \sum_{j=1}^{k} \frac{\left(\sum_{i=1}^{n} X_{ij}\right)^2}{n} - \frac{\left(\sum_{j=1}^{k} \sum_{i=1}^{n} X_{ij}\right)^2}{kn}$$

$$+ \sum_{j=1}^{k} \sum_{i=1}^{n} X_{ij}^2 - \sum_{j=1}^{k} \frac{\left(\sum_{i=1}^{n} X_{ij}\right)^2}{n}$$

$$SS_{Total} = SS_{BG} + SS_{WG}.$$

The use of these *raw-score formulas* is illustrated below.

Computational Example for Type CR-3 Design

The computational procedures associated with a completely randomized design will be illustrated for the data from the diet experiment. You will recall that 30 girls were randomly assigned to the three diets, with the restriction that an equal number were assigned to each diet. The decision rule and the steps to be followed in testing the null hypothesis are as follows.

Step 1: State the statistical hypotheses: H_0: $\mu_1 = \mu_2 = \mu_3$
H_1: $\mu_j \neq \mu_{j'}$ for some j and j'.

Step 2: Specify the test statistic: $F = \dfrac{MS_{BG}}{MS_{WG}}$.

Step 3: Specify the sample size:[10] $kn = 30$;
and the sampling distribution: F distribution with $v_1 = k - 1$ and $v_2 = k(n - 1)$, because X is assumed to be approximately normal in form for the b_1, b_2, and b_3 population distributions.

Step 4: Specify the significance level: $\alpha = 0.05$.

Step 5: Obtain a random sample of size kn and/or randomly assign n subjects to each of k treatment levels, compute F, and make a decision.

Decision rule:
Reject H_0 if F falls in the upper 5% of the sampling distribution of F; otherwise, don't reject H_0.

---

[9] See Exercise 18 (page 316).
[10] A discussion of procedures for making a rational specification of sample size for a type CR-k design is beyond the scope of this book. The interested reader is referred to Cohen (1969, chap. 8) and Kirk (1968, pp. 9–11, 107–110).

Table 16.3-1. Computational Procedures for a Type CR-3 Design

(i) Data: Notation: $j = 1, \ldots, k$ treatment levels (b_j); $i = 1, \ldots, n$ subjects in each treatment level; X_{ij} denotes a score for subject i in treatment level j.

BS Summary Table[a]

	b_1	b_2	b_3
	7	10	12
	9	13	11
	8	9	15
	12	11	7
	8	5	14
	7	9	10
	4	8	12
	10	10	12
	9	8	13
	6	7	14
$\sum_{i=1}^{n} X_{ij} =$	80	90	120
$\bar{X}_{\cdot j} =$	8	9	12

(ii) Computational sums:[b]

$$\sum_{j=1}^{k}\sum_{i=1}^{n} X_{ij} = 7 + 9 + 8 + \cdots + 14 = 290.00$$

$$\sum_{j=1}^{k}\sum_{i=1}^{n} X_{ij}^2 = [BS] = (7)^2 + (9)^2 + (8)^2 + \cdots + (14)^2 = 3026.00$$

$$\frac{\left(\sum_{j=1}^{k}\sum_{i=1}^{n} X_{ij}\right)^2}{kn} = [X] = \frac{(290)^2}{(3)(10)} = 2803.33$$

$$\sum_{j=1}^{k} \frac{\left(\sum_{i=1}^{n} X_{ij}\right)^2}{n} = [B] = \frac{(80)^2}{10} + \cdots + \frac{(120)^2}{10} = 2890.00$$

(iii) Computational formulas:

$$SS_{Total} = [BS] - [X] = 3026.00 - 2803.33 = 222.67$$
$$SS_{BG} = [B] - [X] = 2890.00 - 2803.33 = 86.67$$
$$SS_{WG} = [BS] - [B] = 3026.00 - 2890.00 = 136.00$$

[a] B denotes treatment B and S denotes subjects; the table is so named because it reflects variation attributable to treatment levels (B) and subjects (S).

[b] The simplified symbols $[BS]$, $[X]$, and $[B]$ are used in the computational formulas in place of the complex summation notation. The simplified symbols can be manipulated like algebraic symbols. The letters in the brackets denote the various sources of variation. For example, $[BS]$ reflects variation due to treatment B and subjects; $[B]$ reflects variation due to treatment B; and $[X]$ is a correction term. The sum of squares attributable to treatment B and subjects (SS_{Total}) is computed from $[BS] - [X]$; that attributable to the treatment (SS_{BG}) from $[B] - [X]$; and that attributable to subjects (SS_{WG}) from $[BS] - [B]$.

Table 16.3-1 presents the data for the diet experiment as well as details of the computational procedures. It is customary to summarize the ANOVA as in Table 16.3-2. The sums of squares values in Table 16.3-2 were obtained from Table 16.3-1, and the MS's were obtained by dividing the SS's by their respective degrees of freedom. The F was obtained by dividing the MS_{BG} in row 1 by the MS_{WG} in row 2; this operation is indicated in the table by the symbol $[\frac{1}{2}]$.

Table 16.3-2. ANOVA Table for Type CR-3 Design

Source	SS	df	MS		F
1. Between groups (BG)	86.67	$k - 1 = 3 - 1 = 2$	43.335	$[\frac{1}{2}]$	8.60[a]
2. Within groups (WG)	136.00	$k(n - 1) = 3(10 - 1) = 27$	5.037		
3. Total	222.67	$kn - 1 = (3)(10) - 1 = 29$			

[a] $p < 0.01$; $[\frac{1}{2}]$ indicates that F was obtained by dividing the value of MS in row 1 by the value of MS in row 2.

According to Appendix Table D.5, an F of 3.36 with 2 and 27 degrees of freedom cuts off the upper 0.05 region of the sampling distribution—that is, $F_{0.05;2,27} = 3.36$.[11] Since $8.60 > 3.36$, the null hypothesis is rejected and it is concluded that at least two of the diets aren't equally effective. Once the overall null hypothesis is rejected, the question becomes "Which means aren't equal?" A procedure for performing multiple comparisons is described in Section 16.5. But before turning to that topic we will examine the assumptions underlying the F test for a type CR-k design.

Review Exercises for Section 16.3

17. Under what conditions is a completely randomized design appropriate?
*18. Show that

$$n \sum_{j=1}^{k} (\bar{X}_{.j} - \bar{X}_{..})^2 = \sum_{j=1}^{k} \frac{\left(\sum_{i=1}^{n} X_{ij}\right)^2}{n} - \frac{\left(\sum_{j=1}^{k}\sum_{i=1}^{n} X_{ij}\right)^2}{kn}$$

and

$$\sum_{j=1}^{k}\sum_{i=1}^{n} (X_{ij} - \bar{X}_{.j})^2 = \sum_{j=1}^{k}\sum_{i=1}^{n} X_{ij}^2 - \sum_{j=1}^{k} \frac{\left(\sum_{i=1}^{n} X_{ij}\right)^2}{n}.$$

19. Fill in the blanks in the ANOVA table.

[11] F values for 2,26 and 2,28 degrees of freedom are given in the table. The value for $F_{0.05;2,27}$ was obtained by interpolation.

Source	SS	df	MS	F
Between groups	36	3	12	( )
Within groups	( )	( )	( )	
Total	164	35		

*20. An experiment was performed to investigate the effects of meaningfulness, or association value, of nonsense syllables on learning. Thirty-two subjects were randomly assigned to four treatment levels, with eight in each level. The nonsense syllables were selected from the list compiled by C. E. Noble. The association values of the lists were 25% for b_1, 50% for b_2, 75% for b_3, and 100% for b_4. The dependent variable was time (in minutes) to learn the list to the criterion of two correct recitations. For the data in the table, test the hypothesis that $\mu_1 = \mu_2 = \mu_3 = \mu_4$ by means of ANOVA. Use $\alpha = 0.05$.

b_1	b_2	b_3	b_4
22	22	18	18
21	20	20	17
20	18	17	16
21	21	16	18
22	20	18	19
24	19	19	15
22	21	18	16
23	19	17	17

21. List the steps used in testing the null hypothesis in Exercise 20, and state the decision rule.
22. The learning of one task enhances the learning of different but similar tasks. To investigate this phenomenon (called *learning to learn*), 30 subjects were randomly assigned to three conditions: subjects in condition b_1 learned two lists of nonsense syllables, those in b_2 learned eight lists, and those in b_3 learned 14 lists. The next day all the subjects learned another list. The dependent variable was the number of trials required to learn this list. Use ANOVA to test the hypothesis that $\mu_1 = \mu_2 = \mu_3$ for the data in the table. Let $\alpha = 0.05$.

b_1	b_2	b_3
7	6	3
9	5	2
5	7	3
7	3	6
8	4	3
7	5	4
6	6	5
8	5	5
7	4	4
6	5	4

*23. Reaction time to red, green, and yellow instrument panel warning lights was investigated. Thirty-one subjects, who were randomly assigned to the conditions, pressed a microswitch as soon as they noticed the warning light. The dependent variable was reaction time in hundredths of a second. Test the hypothesis that $\mu_1 = \mu_2 = \mu_3$ using ANOVA. Let $\alpha = 0.05$.

b_1 (Yellow)	b_2 (Red)	b_3 (Green)
20	23	21
20	20	21
21	21	20
22	21	23
21	23	22
20	22	20
19	22	21
21	21	22
19	22	22
20	22	20
		19

24. Presidents of companies employing between 5000 and 8000 employees were randomly sampled from five geographic areas: b_1 = southeast, b_2 = east, b_3 = midwest, b_4 = southwest, and b_5 = west. Use ANOVA to determine whether mean income for the presidents is the same for the entire country. Let $\alpha = 0.05$. The table gives income in thousands of dollars.

b_1	b_2	b_3	b_4	b_5
40	42	37	36	46
31	40	46	40	40
32	46	45	34	45
35	45	42	34	48
37	37	42	33	46
38	43	43	39	47
35	43	40	38	
33	44	39	37	
35	42		34	
37	39			

16.4 Assumptions Associated with a Type CR-k Design

Like all statistical tests, the ANOVA F involves assumptions. The assumptions fall into two categories—those associated with the derivation of the F sampling distribution and those associated with the linear model. Several assumptions are common to both categories. The assumptions associated with the derivation of the sampling distribution of F were discussed in Section 14.1 and can be summarized as follows.

A1. Observations are drawn from normally distributed populations.
A2. The numerator and denominator of the F ratio are independent.
A3. The numerator and denominator of the F ratio are estimates of the same population variance, σ_e^2.
A4. Observations are random samples from the populations.

The assumptions associated with the linear model for a type CR-k design are as follows.

  B1. The linear model $X_{ij} = \mu + \beta_j + e_{ij}$ reflects all the sources of variation that affect X_{ij}.
  B2. The experiment contains all the treatment levels, b_j's, of interest.
  B3. The error effect, e_{ij}, (a) is independent of all other e's, and (b) is normally distributed within each treatment population, with (c) mean equal to zero and (d) variance equal to σ_e^2.

The assumptions associated with the derivation of the sampling distribution apply to all F ratios. Those associated with the linear model vary somewhat from one ANOVA design to another, since the linear models vary. We will now discuss the two sets of assumptions and the consequences of violating some of them. (Assumption A3 states, in effect, that the null hypothesis is true; it will not be discussed further.)

Assumption that $X_{ij} = \mu + \beta_j + e_{ij}$ (B1)

Assumption B1 states that a score, X_{ij}, is the sum of three components: the overall elevation of scores, μ; the elevation or depression of scores attributable to the jth treatment level, β_j; and all effects not attributable to the jth treatment level, e_{ij}, which includes effects often referred to as *individual differences* and *measurement error*. A type CR-k design deals with one treatment. If an experiment contains two or more treatments, a different ANOVA design must be used—one that permits the additional treatment effects to be estimated. All sources of variation in a type CR-k design that are not estimated separately are lumped with the estimate of e_{ij}.

Assumption that the Experiment Contains All Treatment Levels of Interest (B2)

Assumption B2 states that the experiment contains all the treatment levels about which an experimenter wants to draw conclusions. The model $X_{ij} = \mu + \beta_j + e_{ij}$ for such an experiment is called a **fixed-effects model** or **model I**, and β_j is called a **fixed-treatment effect**. Sometimes an experimenter wants to draw conclusions about more treatment levels than can be included in the experiment. This can be done by performing the experiment using a random sample of k treatment levels from the population of K levels. The results of the experiment apply to the population. The model $X_{ij} = \mu + b_j + e_{ij}$ for this case is called a *random-effects model* or *model II*, and b_j is called a *random-treatment effect*. In defining the two models we have followed the convention of denoting fixed effects by Greek letters and random effects by Roman letters.

The expectation of MS_{BG} is different for models I and II as is the nature of the null hypothesis tested. The expectations for MS_{BG} and MS_{WG} are

Fixed-Effects Model (I)
$$\frac{E(MS_{BG})}{E(MS_{WG})} = \frac{\sigma_e^2 + n \sum_{j=1}^{k} \beta_j^2/(k-1)}{\sigma_e^2}$$

Random-Effects Model (II)
$$\frac{E(MS_{BG})}{E(MS_{WG})} = \frac{\sigma_e^2 + n\sigma_B^2}{\sigma_e^2}$$

The null hypothesis for model I concerns only the k population treatment effects (β_j's) that are represented in the experiment. For model II the null hypothesis concerns all the K population treatment effects (b_j's), not just the k represented in the experiment. If the null hypothesis for model I is rejected, it can be concluded that at least two of the k treatment effects are not equal; if the null hypothesis for model II is rejected, at least two of the treatment effects in the population of K effects are not equal. The computational procedures for both models are identical, as is the form of the F ratio. The difference is in the $E(MS_{BG})$'s and in the nature of the null hypothesis tested.[12]

Assumptions of Independence of Experimental Errors (B3a) and Random Sampling (A4)

The validity of an experiment depends on random sampling and/or random assignment of subjects to k treatment levels. Random sampling and/or assignment ensures that idiosyncratic characteristics of subjects will be randomly distributed over the treatment levels and thus will not selectively bias the outcome of the experiment.[13]

Random sampling and/or assignment, in addition to being an assumption associated with the derivation of the F sampling distribution (A4), ensures that the experimental errors (e_{ij}'s) are independent (assumption B3a). Applying the definition of statistical independence from Section 7.3, the errors for subjects ij and $i'j$ are independent if $p(e_{ij}|e_{i'j}) = p(e_{ij})$. In this case, knowing subject ij's error would tell us nothing about subject $i'j$'s error. Designs that permit nonindependence of e_{ij}'s resulting from the use of repeated measures on the same subjects are discussed in advanced books on ANOVA.

Assumptions of Normally Distributed Errors (B3b) and Populations (A1), and Independence of MS_{BG} and MS_{WG} (A2)

The assumption (B3b) that the e's within each of the populations are normally distributed is equivalent to the assumption that the X_{ij}'s in each of the populations are normally distributed (A1). This follows, since according to the linear model $X_{ij} = \mu + \beta_j + e_{ij}$ the e's are the only source of variation within a particular treatment population. A rough check on the normality assumption can be made by constructing a frequency distribution for the scores in each treatment level and inspecting the distributions for evidence of skewness and kurtosis. Marked departures from normality raise questions concerning normality of the populations. A statistical test of the assumption is ordinarily not performed because the F test, like the t test, is robust with respect to departures from normality. Studies indicate that if the treatment populations have the same shape, for example, all positively skewed or all leptokurtic, the actual probability

---

[12] For a discussion of the random-effects model, the reader is referred to Edwards (1972), Kirk (1968), Myers (1972), and Winer (1971).

[13] Sometimes factors beyond the investigator's control preclude the random assignment of subjects to treatment levels and the control of important extraneous variables. Campbell and Stanley (1963) refer to such experiments as "quasi-experimental designs." Campbell (1957) examines potential sources of bias inherent in these designs. His paper is reprinted in Kirk (1972a).

of making a type I error will be fairly close to the nominal or specified probability.[14]

It can be shown that the numerator and denominator of the F ratio are independent (assumption A2) if the populations are normally distributed or approximately so.[15]

Assumption that $E(e_{ij}) = 0$ for Each Treatment Population (B3c)

Assumption B3c states that the expectation of the error effects within each treatment population is equal to zero. This follows from the fact that e_{ij} is the deviation of a population mean from a random variable ($e_{ij} = X_{ij} - \mu_j$) and the fact that the expectation or long-run average of this deviation is zero, which can be shown as follows. The expectation $E(X_{ij} - \mu_j)$ can be written as $E(X_{ij}) - E(\mu_j)$. By definition, the expectation of the random variable X_{ij} is its population mean and the expectation[16] of the constant μ_j is μ_j. Hence,

$$E(X_{ij} - \mu_j) = E(X_{ij}) - E(\mu_j) = \mu_j - \mu_j = 0.$$

Assumption of Homogeneity of Variance (B3d)

Assumption B3d states that the $j = 1, \ldots, k$ population variances are equal to σ_e^2—that is, that $\sigma_1^2 = \sigma_2^2 = \cdots = \sigma_k^2 = \sigma_e^2$. This is referred to as the **homogeneity of variance assumption**. Fortunately, the ANOVA F test is robust with respect to heterogeneity of population variances provided there are an equal number of observations in the treatment levels.[17]

Review Exercises for Section 16.4

*25. Qualify the statement "the F test in ANOVA is robust with respect to departures from normality."

**26. A rough but adequate check on the tenability of the normality assumption consists of making a frequency distribution of the scores in each treatment level and inspecting them for evidence of skewness and kurtosis. Decide on the tenability of this assumption for the data in exercises *(a) 20 (page 317), (b) 22 (page 317), and (c) 23 (page 318).

27. What is meant by the assumption of homogeneity of variance?

28. Qualify the statement "the F test in ANOVA is robust with respect to heterogeneity of variance."

29. Distinguish between fixed-effects and random-effects models in terms of (a) expectations of the MS's and (b) the null hypothesis tested.

30. Terms to remember:

---

[14] The reader may want to take the robustness of the F test on faith, since the classic studies by Box and Anderson (1955), Norton as cited by Lindquist (1953), and Pearson (1931) were written for a technical audience.

[15] See Lindquist (1953, pp. 31–35).

[16] For a discussion of the algebra of expected values, see Hays (1973, pp. 871–875) and Kirk (1968, pp. 509–512).

[17] The robustness of the test to heterogeneity of population variances has been shown by Cochran (1947), Godard and Lindquist (1940), Horsnell (1953), and Norton as cited by Lindquist (1953).

a. Homogeneity of variance
b. Model I
c. Fixed-treatment effect
d. Model II
e. Random-treatment effect

16.5 Introduction to Multiple Comparisons

The ANOVA F test is used to determine the tenability of the hypothesis $\mu_1 = \mu_2 = \cdots = \mu_k$. If this hypothesis is rejected, interest centers on determining which of the means are not equal. A number of test statistics have been developed for ferreting out significant comparisons among means, or, as it is often called, *data snooping*. The one described here—Scheffé's S test[18]—is referred to as an a posteriori or post hoc procedure, because it tests comparisons suggested by an inspection of one's data.

We observed in Section 16.1 that if an experiment contains more than two treatment levels and if Student's t statistic is used to test the significance of comparisons, the probability of erroneously rejecting one or more true null hypotheses increases as a function of the number of comparisons tested. The persistent experimenter who performs enough t tests, each at α level of significance, will reject too many true H_0's. Scheffé's multiple comparison procedure described below was developed to solve this problem.

Scheffé's S Multiple Comparison Procedure

Scheffé's S test was designed to test all 2, 3, ..., k mean null hypotheses if the overall null hypothesis has been rejected. It sets the probability of falsely rejecting one or more null hypotheses equal to or less than α. The test assumes that the $j = 1, \ldots, k$ populations are normally distributed and that their variances are equal. However, the S test, which is based on the F distribution, is robust with respect to violation of these assumptions.

Contrasts among Means

A comparison between two population means is a special case of **contrast**. A contrast, designated by ψ, among k population means, $\mu_1, \mu_2, \cdots, \mu_k$, is a linear combination

$$\psi = c_1\mu_1 + c_2\mu_2 + \cdots + c_k\mu_k$$

such that the coefficients $c_1, \ldots, c_k$ (1) are real numbers, (2) are not all equal to zero, and (3) sum to zero. For convenience, the coefficients are usually chosen so that the sum of their absolute values equals two—that is, $|c_1| + |c_2| + \cdots + |c_k| = 2$. Examples of contrasts that satisfy the three requirements and for which $\sum_{j=1}^{k} |c_j| = 2$ are the

---

[18] See Kirk (1968, chap. 3) for an in-depth discussion of this and other multiple comparison procedures, including Tukey's test, Dunnett's test, the Newman–Keuls test, and the controversial test proposed by Duncan. An excellent but more technical discussion is given by Games (1971).

following.

$$\psi_1 = \mu_1 - \mu_2 \qquad \text{where} \quad c_1 = 1, c_2 = -1$$

$$\psi_2 = \mu_1 - \frac{1}{2}\mu_2 - \frac{1}{2}\mu_3 \qquad \text{where} \quad c_1 = 1, c_2 = -\frac{1}{2}, c_3 = -\frac{1}{2}$$

$$\psi_3 = \frac{1}{2}\mu_1 + \frac{1}{2}\mu_2 - \frac{1}{2}\mu_3 - \frac{1}{2}\mu_4 \qquad \text{where} \quad c_1 = \frac{1}{2}, c_2 = \frac{1}{2}, c_3 = -\frac{1}{2}, c_4 = -\frac{1}{2}$$

$$\psi_4 = \mu_1 - \frac{1}{3}\mu_2 - \frac{1}{3}\mu_3 - \frac{1}{3}\mu_4 \qquad \text{where} \quad c_1 = 1, c_2 = -\frac{1}{3}, c_3 = -\frac{1}{3}, c_4 = -\frac{1}{3}$$

Contrast ψ_4, for example, is the difference between population mean one and the average of population means two, three, and four; note that $1 - \frac{1}{3} - \frac{1}{3} - \frac{1}{3} = 0$ and that $|1| + |-\frac{1}{3}| + |-\frac{1}{3}| + |-\frac{1}{3}| = 2$. A contrast among population means is estimated by replacing the μ_j's with the sample mean values, $\bar{X}_{.j}$'s.

Scheffé's S statistic is given by

$$S = \frac{c_j \bar{X}_{.j} + c_{j'} \bar{X}_{.j'} + \cdots + c_{j''} \bar{X}_{.j''}}{\sqrt{MS_{WG}\left(\frac{c_j^2}{n_j} + \frac{c_{j'}^2}{n_{j'}} + \cdots + \frac{c_{j''}^2}{n_{j''}}\right)}}.$$

To reject the null hypothesis that $c_j \mu_j + c_{j'} \mu_{j'} + \cdots + c_{j''} \mu_{j''} = 0$, the absolute value of S must equal or exceed $\sqrt{(k-1)F_{\alpha;k-1,k(n-1)}}$, where $F_{\alpha;k-1,k(n-1)}$ is obtained from Appendix Table D.5. The degrees of freedom are $v_1 = k - 1$ and $v_2 = k(n-1)$ if the n_j's are equal; if they are unequal, $v_2 = N - k$; v_1 is always equal to the number of treatment levels minus one, and v_2 to the degrees of freedom associated with the MS error term. If the overall null hypothesis in ANOVA is rejected, we know that there is at least one significant contrast among means that will be detectable using Scheffé's procedure.

Computational Example for S Test

Scheffé's test will be illustrated using the diet data in Table 16.3-1. For these data the overall ANOVA F statistic was significant at the 0.05 level. After examining the data, an experimenter might be interested in the following null hypotheses: $\mu_1 - \mu_2 = 0$, $\mu_1 - \mu_3 = 0$, $\mu_2 - \mu_3 = 0$, and $\mu_3 - (1/2)\mu_1 - (1/2)\mu_2 = 0$. The test statistics are

$$S = \frac{(1)\bar{X}_{.1} + (-1)\bar{X}_{.2}}{\sqrt{MS_{WG}[(1)^2/n_1 + (-1)^2/n_2]}} = \frac{8 - 9}{\sqrt{5.037[1/10 + 1/10]}} = -1.00$$

$$S = \frac{(1)\bar{X}_{.1} + (-1)\bar{X}_{.3}}{\sqrt{MS_{WG}[(1)^2/n_1 + (-1)^2/n_3]}} = \frac{8 - 12}{\sqrt{5.037[1/10 + 1/10]}} = -3.99$$

$$S = \frac{(1)\bar{X}_{.2} + (-1)\bar{X}_{.3}}{\sqrt{MS_{WG}[(1)^2/n_2 + (-1)^2/n_3]}} = \frac{9 - 12}{\sqrt{5.037[1/10 + 1/10]}} = -2.99$$

$$S = \frac{(1)\bar{X}_{.3} + (-1/2)\bar{X}_{.1} + (-1/2)\bar{X}_{.2}}{\sqrt{MS_{WG}[(1)^2/n_3 + (-1/2)^2/n_1 + (-1/2)^2/n_2]}}$$

$$= \frac{12 - (8 + 9)/2}{\sqrt{5.037[1/10 + (1/4)/10 + (1/4)/10]}} = 4.03.$$

The critical value that $|S|$ must equal or exceed to be significant at the 0.05 level is $\sqrt{(3-1)3.36} = 2.59$. The only $|S|$ that doesn't exceed 2.59 is that for $\mu_1 - \mu_2 = 0$; all the other null hypotheses are rejected.

Regardless of the number of tests performed among k means using the S statistic, the probability of falsely rejecting one or more null hypotheses is always equal to or less than α. Scheffé's test is probably the most versatile of the data-snooping procedures. It can be used with unequal n's, can be used to construct confidence intervals, and is robust with respect to violation of its normality and equal variance assumptions. But this versatility is accompanied by a disadvantage. The power of Scheffé's test is lower than that of more specialized multiple comparison procedures. For a discussion of this point, see Games (1971) and Kirk (1968, chap. 3).

Review Exercises for Section 16.5

31. List the requirements for using Scheffé's S test.
32. Distinguish between a comparison and a contrast.
**33. List the coefficients, c_j, for the following population contrasts: *(a) μ_1 versus μ_2, *(b) μ_1 versus mean of μ_2 and μ_3, (c) μ_1 versus mean of μ_2, μ_3, and μ_4, (d) mean of μ_1 and μ_2 versus mean of μ_3 and μ_4, (e) mean of μ_1 and μ_2 versus mean of μ_3, μ_4, and μ_5, *(f) μ_1 versus the weighted mean of μ_2 and μ_3, where μ_2 is weighted twice as much as μ_3, (g) the weighted mean of μ_1 and μ_2 versus the weighted mean of μ_3 and μ_4, where μ_1 and μ_3 are weighted twice as much as μ_2 and μ_4.
**34. Which of the following meet the three requirements for a contrast? *(a) $\mu_1 - \mu_2$, *(b) $2\mu_1 - \mu_2 - \mu_3$, (c) $\mu_1 - (1/3)\mu_2 - (1/3)\mu_3$, (d) $(1\frac{1}{2})\mu_1 - \mu_2 - (1/2)\mu_3$, (e) $(1/2)\mu_1 + (1/2)\mu_2 - (1/3)\mu_3 - (1/3)\mu_4 - (1/3)\mu_5$.
**35. Which contrasts in Exercise 34 satisfy $|c_1| + |c_2| + \cdots + |c_k| = 2$?
**36. Determine the value of $S_{\alpha;v_1,v_2}$ for *(a) $k = 4$, $n = 11$, $\alpha = 0.01$; (b) $k = 5$, $n = 13$, $\alpha = 0.05$; *(c) $k = 3$, $n_1 = 6$, $n_2 = 7$, $n_3 = 8$, $\alpha = 0.05$; (d) $k = 4$, $n_1 = 5$, $n_2 = 5$, $n_3 = 6$, $n_4 = 8$, $\alpha = 0.01$.
*37. The effectiveness of three approaches to drug education in junior high school was investigated. The approaches were scare tactics, treatment level b_1; providing objective scientific information about physiological and psychological effects, b_2; and examining the psychology of drug use, b_3. Forty-one students who didn't use drugs were randomly assigned to each treatment level. At the conclusion of the education program they evaluated its effectiveness; a high score signified effectiveness. The sample means were $\bar{X}_{.1} = 23.1$, $\bar{X}_{.2} = 23.8$, $\bar{X}_{.3} = 26.7$; $MS_{WG} = 16.4$; and $v_2 = 3(41 - 1) = 120$. The hypothesis $\mu_1 = \mu_2 = \mu_3$ was rejected at the 0.01 level of significance using a type CR-3 design. Use Scheffé's S statistic to test $\mu_1 - \mu_2 = 0$, $\mu_1 - \mu_3 = 0$, $\mu_2 - \mu_3 = 0$, $(\mu_1 + \mu_2)/2 - \mu_3 = 0$.
*38. The effects of three dosages of ethylene glycol on the reaction time of chimpanzees was investigated. The animals were randomly assigned to the dosage levels so that five received 2 cc of the drug, treatment level b_1; five received 4 cc, b_2; and five received 6 cc, b_3. The sample means were $\bar{X}_{.1} = 0.29$ sec, $\bar{X}_{.2} = 0.31$ sec, $\bar{X}_{.3} = 0.39$ sec; $MS_{WG} = 0.002$; and $v_2 = 3(5 - 1) = 12$. The hypothesis that $\mu_1 = \mu_2 = \mu_3$ was rejected at the 0.05 level of significance using a type CR-3 design. Perform all two-mean contrasts using Scheffé's S statistic.
39. The religious dogmatism of four church denominations in a large midwestern city was investigated. A random sample of 30 members from each denomination took a paper and pencil test of dogmatism. The sample means were $\bar{X}_{.1} = 64$, $\bar{X}_{.2} = 73$, $\bar{X}_{.3} = 61$, $\bar{X}_{.4} = 49$; $MS_{WG} = 120$; and $v_2 = 4(30 - 1) = 116$. The hypothesis that $\mu_1 = \mu_2 = \mu_3 = \mu_4$ was rejected at the 0.01 level of significance using a type CR-4 design. Use Scheffé's S statistic to test $\mu_2 - \mu_3 = 0$, $\mu_2 - (1/2)\mu_1 - (1/2)\mu_3 = 0$, $\mu_4 - (1/2)\mu_1 - (1/2)\mu_3 = 0$.

40. A hospital questionnaire designed to measure concern about the recovery period was administered to a random sample of 19 coronary patients, b_1; 23 accident victims, b_2; and 21 pulmonary patients, b_3. The sample means were $\bar{X}_{.1} = 82$, $\bar{X}_{.2} = 83$, $\bar{X}_{.3} = 63$; $MS_{WG} = 5.1$; and $v_2 = 63 - 3 = 60$. A low score indicated insecurity about physical well-being and lowered self-esteem. The hypothesis that $\mu_1 = \mu_2 = \mu_3$ was rejected at the 0.05 level of significance using a type CR-3 design. Use Scheffé's S statistic to test $\mu_1 - \mu_2 = 0$, $\mu_1 - \mu_3 = 0$, $\mu_2 - \mu_3 = 0$, and $\mu_3 - (1/2)\mu_1 - (1/2)\mu_2 = 0$.
41. For the data in Exercise 22 (page 317) use Scheffé's S statistic to test the hypothesis that $\mu_1 - \mu_2 = -1$ and $\mu_1 - \mu_3 = -3$. (*Hint*: Examine the numerator of the t formula on page 261.)
42. Terms to remember:
    a. Contrast
    b. Data snooping
    c. A posteriori or post hoc comparisons

16.6 Introduction to Factorial Designs

The term *analysis of variance* is apropos. As we have seen, the total variation among scores for a completely randomized design is partitioned into two sources of variation. Complex ANOVA's partition the total variation into more sources; designs with as many as 16 sources are common in behavioral and educational research. A discussion of the computational procedures associated with such designs is beyond the scope of this book. We will limit our discussion to the main features of one of the more widely used complex ANOVA designs—the completely randomized factorial design.[19]

A factorial design is distinguished from other ANOVA designs in that it has two or more treatments and the levels of each treatment are investigated in combination with those of other treatments. For example, a subject's performance on a learning task might be observed for the combined conditions of large monetary reward for good performance (a level of one treatment) and absence of distractions in the learning environment (a level of a second treatment). A subject in a factorial design is always simultaneously exposed to one level each of two or more treatments. The combination of conditions that the subject is simultaneously exposed to is called a *treatment combination*. By way of contrast, the CR-k design described in Section 16.3 can be used to investigate only one treatment at a time, and a subject is exposed to a single treatment level.

In a factorial design the levels of the treatments can be *completely crossed*, meaning that each level of one treatment occurs once with each level of the other treatments and vice versa, or *partially crossed*, meaning that each level of one treatment occurs with some of the levels of the other treatments but not all of them. A completely randomized factorial design (CRF), the simplest of the factorial designs, is constructed by combining (completely crossing) the treatment levels of two or more completely randomized designs in a single experiment and randomly assigning subjects to the treatment combinations. A design with two treatments, say A and B, is designated by the letters CRF-pq, where p refers to the number of levels of treatment A and q to the number of levels of treatment B.[20]

---

[19] The computational procedures for this design are discussed by Edwards (1972), Keppel (1973), Kirk (1968), Lee (1975), Lindman (1974), Myers (1972), and Winer (1971).

[20] Some writers refer to this design as a *two-way ANOVA*, but this is imprecise, since the name could refer to any one of nine different kinds of factorial designs.

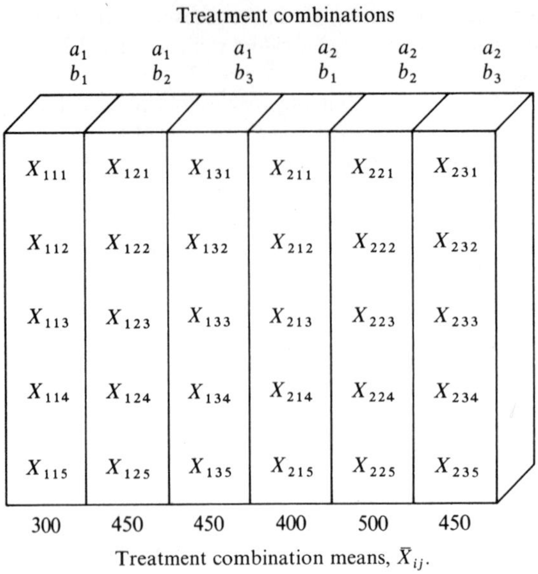

Treatment combinations

	a_1 b_1	a_1 b_2	a_1 b_3	a_2 b_1	a_2 b_2	a_2 b_3	
	X_{111}	X_{121}	X_{131}	X_{211}	X_{221}	X_{231}	$\bar{X}_{1..} = \dfrac{300 + 450 + 450}{3} = 400$
	X_{112}	X_{122}	X_{132}	X_{212}	X_{222}	X_{232}	$\bar{X}_{2..} = \dfrac{400 + 500 + 450}{3} = 450$
	X_{113}	X_{123}	X_{133}	X_{213}	X_{223}	X_{233}	$\bar{X}_{.1.} = \dfrac{300 + 400}{2} = 350$
	X_{114}	X_{124}	X_{134}	X_{214}	X_{224}	X_{234}	$\bar{X}_{.2.} = \dfrac{450 + 500}{2} = 475$
	X_{115}	X_{125}	X_{135}	X_{215}	X_{225}	X_{235}	$\bar{X}_{.3.} = \dfrac{450 + 450}{2} = 450$
	300	450	450	400	500	450	

Treatment combination means, $\bar{X}_{ij.}$

Figure 16.6-1. Layout for a type CRF-23 design with $n = 30$ subjects randomly assigned to the $2 \times 3 = 6$ treatment combinations. Each subject receives one level of treatment A in combination with one level of treatment B. Scores are denoted by X_{ijm}, where the subscripts i and j specify the levels, respectively, of A and B, and m specifies one of the five subjects in cell ij. Treatment means for A are denoted by $\bar{X}_{i..}$, means for B by $\bar{X}_{.j.}$, and treatment combination means by $\bar{X}_{ij.}$.

Consider an experiment to investigate the effects of treatments A and B on reading speed. Suppose treatment A consists of two levels of room illumination: $a_1 = 5$ foot-candles and $a_2 = 30$ foot-candles. Treatment B consists of three levels of type size: $b_1 = 6$ point type, $b_2 = 12$ point type, and $b_3 = 18$ point type. Each level of A is combined with all levels of B to form $2 \times 3 = 6$ treatment combinations: ab_{11}, ab_{12}, $ab_{13}, \ldots, ab_{23}$. The layout for such a CRF-23 design with five subjects randomly assigned to each treatment combination is shown in Figure 16.6-1. This design partitions the total variation among the scores into four sources: treatment A, treatment B, AB interaction, and within-cell (error) variation. It enables the experimenter to test the hypotheses that the p population means for treatment A are equal, the q means for treatment B are equal, and the two treatments do not interact. The third hypothesis concerning interaction is unique to factorial designs. *Two treatments are said to* **interact** *if differences in performance under the levels of one treatment are different at two or more levels of the other treatment.* An interaction between treatments A and B for the data in Figure 16.6-1 is illustrated in Figure 16.6-2. Figure 16.6-2(a) shows that the mean for treatment level a_2 is considerably larger than that for a_1 when $b_1 = 6$ point type, is slightly larger when $b_2 = 12$ point type, and is equal to that for a_1 when $b_3 = 18$ point type. A similar picture of interaction emerges in Figure 16.6-2(b) when the means for treatment B at the two levels of a_i are examined. An interaction always signals that the interpretation of treatment effects must be qualified. For example, although mean

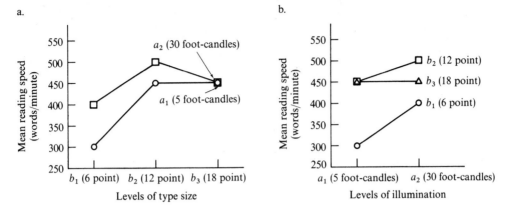

Figure 16.6-2. Two graphs of the interaction between treatments A and B for the data in Figure 16.6-1. Nonparallelism of the lines signifies interaction.

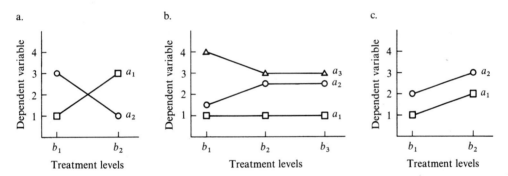

Figure 16.6-3. Interaction is illustrated in parts a and b; part c shows treatments that don't interact.

reading speed is faster under 30 foot-candles of illumination ($\bar{X}_{2..} = 450$) than under 5 foot-candles ($\bar{X}_{1..} = 400$), the former condition is superior only for the 6 and 12 point type conditions. If the experiment had been performed with only treatment A and 18 point type, the experimenter would have concluded that there was no difference in reading speed between the two levels of illumination; if either 6 or 12 point type had been used, the conclusion would have been that there was a difference.

If a test of significance indicates that two treatments interact, a graph like those in Figures 16.2-2 and 16.2-3 is helpful in interpreting the interaction.[21] When the interaction null hypothesis is rejected, we know that a graph will reveal at least two non-parallel lines between at least two levels of a treatment; it doesn't mean that all lines throughout their length are nonparallel. Such a case is shown in Figure 16.6-3(b), where the a_i lines are not parallel between b_1 and b_2 but are parallel between b_2 and b_3. We can

[21] Boik and Kirk (1977) discuss more advanced techniques for interpreting interactions.

think about this in another way. Consider the interaction shown in Figure 16.6-3(b). The value of some contrast among the A means, say $\hat{\psi} = (\bar{X}_{1..} + \bar{X}_{2..})/2 - \bar{X}_{3..}$, at b_1 is different from its value at b_2, but the value of the contrast is the same at b_2 and b_3.

The ability to test interactions is an important feature of factorial designs. But the designs have other features that help account for their wide use. For example, they make efficient use of subjects. The CRF-23 design shown in Figure 16.6-1 uses all 30 subjects simultaneously in evaluating the effects of treatments A and B. If treatment A were evaluated by means of a type CR-2 design and treatment B by means of a separate type CR-3 design, 60 subjects—30 in each experiment—would be required to achieve the precision of the CRF-23 design. In view of these advantages, it is easy to understand the popularity of factorial designs.

The major advantages of factorial designs can be summarized as follows.

1. All subjects are used in evaluating the effects of two or more treatments. Since each treatment is evaluated with the same precision as if the entire experiment had been devoted to that treatment alone, factorial experiments provide efficient use of resources.
2. They enable an experimenter to evaluate interaction effects.

The major disadvantages of factorial designs are the following.

1. If numerous treatments are included in an experiment, the number of subjects required may be prohibitive. For example, a four-treatment type CRF-2433 design has $2 \times 4 \times 3 \times 3 = 72$ treatment combinations. If only two subjects were assigned to each combination, the experiment would require 144 subjects.
2. The interpretation of a factorial design is complex if interaction effects are present. This is not a criticism of the design, but rather an acknowledgment of the fact that interactions always call for some qualification of tests of differences among means for the treatments.
3. The use of a factorial design commits an investigator to a relatively large experiment, but small one-treatment exploratory experiments may indicate much more promising lines of investigation than those originally envisioned.

16.7 Summary

Analysis of variance, ANOVA, is a statistical procedure for (1) determining how much of the total variability among scores to attribute to each source of variation in an experiment and (2) testing hypotheses about some of these sources. The principal application of ANOVA is testing the null hypothesis that two or more population means are equal. A completely randomized design (type CR-k), the simplest ANOVA design, is described in this chapter. It is appropriate for experiments that meet the following conditions.

1. One treatment or independent variable with two or more treatment levels. The levels of the treatment can differ either quantitatively or qualitatively.
2. Subjects randomly assigned to the treatment levels so that each subject is observed under one level.

Although ANOVA appears to be a complicated procedure, the basic notions are relatively simple. A score X_{ij} is assumed to be a composite that estimates the sum of three components, as is stated in the linear model

$$X_{ij} = \mu + \beta_j + e_{ij},$$

where μ is a constant that reflects the general elevation of scores; β_j is the effect of the jth treatment level; and e_{ij} reflects any other effects on X_{ij}, which is apparent when the terms in the model are rearranged to $e_{ij} = X_{ij} - \mu - \beta_j$.

The total variation among the scores, designated by SS_{Total}, is also a composite and can be partitioned into two parts: the sum of squares between groups, SS_{BG}, and the sum of squares within groups, SS_{WG}. A variance or mean square is obtained by dividing a sum of squares by its degrees of freedom, for example, $SS_{BG}/df_{BG} = MS_{BG}$ and $SS_{WG}/df_{WG} = MS_{WG}$. The statistic for testing $\mu_1 = \mu_2 = \cdots = \mu_k$ is $F = MS_{BG}/MS_{WG}$. To use the ratio of two variances to test a hypothesis about means is not strange if you consider the expected values of MS_{BG} and MS_{WG} for the case in which the null hypothesis is true and the case in which it is false. If H_0 is true, all the squared treatment effects equal zero $[\beta_1^2 = (\mu_1 - \mu)^2 = 0, \beta_2^2 = (\mu_2 - \mu)^2 = 0, \ldots, \beta_k^2 = (\mu_k - \mu)^2 = 0]$, in which case

$$\frac{E(MS_{BG})}{E(MS_{WG})} = \frac{\sigma_e^2}{\sigma_e^2}.$$

If H_0 is false, at least two of the squared treatment effects are greater than zero ($\beta_j^2 \neq 0$ and $\beta_{j'}^2 \neq 0$ for some j and j'), in which case

$$\frac{E(MS_{BG})}{E(MS_{WG})} = \frac{\sigma_e^2 + n \sum_{j=1}^{k} \beta_j^2/(k-1)}{\sigma_e^2}.$$

It is apparent that the larger the ratio $F = MS_{BG}/MS_{WG}$, the less tenable the null hypothesis. How large should the ratio be in order to reject H_0? According to hypothesis testing conventions, an F is considered large enough if it falls in the upper 5% of the sampling distribution of F. To determine whether F falls in this region it is referred to the F sampling distribution for degrees of freedom equal to $k - 1$ and $k(n - 1)$. This is why a statistic involving two variances is used to test a hypothesis about means.

If the overall null hypothesis is rejected, the experimenter must still decide which means aren't equal. This can be accomplished by using Scheffé's S data-snooping test statistic. The principal advantage of this multiple comparison procedure over Student's t is that the probability of erroneously rejecting one or more null hypotheses doesn't increase as a function of the number of hypotheses tested. Regardless of the number of tests performed among k means, this probability remains equal to or less than α.

17

Statistical Inference for Frequency Data

17.1 Applications of Pearson's Chi-Square Statistic
17.2 Testing Goodness of Fit
    Computational Example
    Characteristics of Pearson's Statistic
    Degrees of Freedom when E_j's Are Based on a Theoretical Distribution
    Assumptions of the Test
    Review Exercises for Section 17.2
17.3 Testing Independence
    Computational Example
    Degrees of Freedom for a Contingency Table
    Statistical Hypotheses
    Contingency Tables with Three or More Rows or Columns
    Measuring Strength of Association
    Assumptions of the Test
    Review Exercises for Section 17.3
17.4 Testing Equality of $c \geq 2$ Proportions
    Computational Example
    Comparison of Independence Test and Equality of Proportions Test
    Extension of the Test of Equality to More than Two Response Categories
    Review Exercises for Section 17.4
17.5 Summary
17.6 Technical Notes
    Two Ways to Test the Hypothesis that $p = p'$ (or $p = p_0$)
    Outline of an Exact Test of $p_1 = p_1', p_2 = p_2', \ldots, p_k = p_k'$
    Special Computational Procedure for a 2×2 Contingency Table

17.1 Applications of Pearson's Chi-Square Statistic

So far our discussion has been limited to test statistics that are appropriate for measured characteristics such as IQ, time to learn a list of nonsense syllables, and age. Instead of measuring a variable we may choose to count the number of observations in two or more mutually exclusive categories of the variable, in which case the data are a set of frequencies. A statistic for testing hypotheses about frequency data was developed by Karl Pearson in 1900. It is called *Pearson's chi-square statistic* because it is approximately distributed as the chi-square distribution.[1] We will describe three applications of the statistic: testing goodness of fit, testing independence, and testing equality of k proportions.[2] The three tests are often confused, since they all use Pearson's test statistic, but they can be distinguished as follows.

1. Testing goodness of fit. Pearson's chi-square statistic is used to determine whether the population distribution estimated by a single random sample containing n independent observations is identical to some hypothesized or expected population distribution. Depending on the experimenter's interests, expectations may be based on one of the theoretical distributions (such as the normal curve) or on the results of an earlier empirical investigation. Pearson's chi-square statistic is used to test the hypothesis that the observed frequencies $O_1, O_2, \ldots, O_k$ in k mutually exclusive categories represent a population with $E_1, E_2, \ldots, E_k$ expected frequencies.
2. Testing independence. Each of n independent observations for a single random sample is classified in terms of two variables, A and B. Variable A might represent a person's sex and variable B political affiliation (Democrat, Republican, independent, or other). Pearson's chi-square statistic is used to determine whether the variables are statistically independent. If the variables are independent, then $p(\text{Male}|\text{Democrat}) = p(\text{Male})$, $p(\text{Male}|\text{Republican}) = p(\text{Male})$, and so on, which means that a knowledge of political affiliation indicates nothing about the individual's sex and vice versa.
3. Testing equality of $c \geq 2$ population proportions. An experimenter draws c random samples from c populations, and the proportion of elements possessing some characteristic is computed for each sample. Pearson's chi-square statistic is used to test the hypothesis that the population proportions are equal—that is,[3] $p_1 = p_2 = \cdots = p_c$.

Table 17.1-1. Comparison of Three Tests that Use Pearson's Chi-Square Statistic

Purpose	Null Hypothesis	Number of Samples
1. Testing goodness of fit	$O_{Pop_1} = E_{Pop_1}, O_{Pop_2} = E_{Pop_2}, \ldots, O_{Pop_k} = E_{Pop_k}$	One random sample; n observations assigned to k categories
2. Testing independence	$p(A \text{ and } B) = p(A)p(B)$	One random sample; each of n observations is classified in terms of two variables
3. Testing equality of proportions	$p_1 = p_2 = \cdots = p_c$	c random samples, where $c \geq 2$

[1] Pearson's chi-square statistic is not related to $\chi^2 = (n-1)\hat{\sigma}^2/\sigma_0^2$, which is used to test a hypothesis about a population variance, although both are referred to the chi-square distribution.
[2] For an in-depth study of the use and misuse of Pearson's chi-square statistic see Edwards (1950), Lewis and Burke (1949, 1950), Pastore (1950), and Peters (1950).
[3] The letter p is used to denote a proportion as well as a probability. The meaning of the letter will be stated if it isn't clear from the context.

17.2 Testing Goodness of Fit

The goodness-of-fit test was developed to test the hypothesis that a population distribution estimated by a random sample is identical to an hypothesized or expected distribution. If $O_1, O_2, \ldots, O_k$ represent observed frequencies and $E_1, E_2, \ldots, E_k$, expected frequencies, the hypothesis of equality is rejected if Pearson's statistic,

$$\chi^2 = \sum_{j=1}^{k} \frac{(O_j - E_j)^2}{E_j},$$

exceeds the value of $\chi^2_{\alpha,v}$ (given in Table D.4) for $v = k - 1$ degrees of freedom and α level of significance. The test is approximate, since it uses the continuous chi-square distribution to estimate a probability for a discrete sampling distribution.[4] There is an alternative exact test that can be used. When $k = 2$, the exact test is based on the binomial distribution; when $k > 2$, it is based on the multinomial distribution. Exact tests are rarely used because they involve a prohibitive amount of computation.[5]

Computational Example

Suppose we want to know whether the academic potential of this year's graduate school applicants, as measured by the Graduate Record Examination, is the same as that for past years'. Data for previous applicants are given in Table 17.2-1. The statistical hypotheses for Pearson's test are usually stated in terms of proportions, although the test statistic uses frequencies. This is perfectly consistent because frequencies are readily converted into proportions and vice versa. An observed proportion in the jth category is denoted by $\hat{p}_j$ and the expected proportion by p_j'. Proportions can be computed from frequencies by $\hat{p}_j = O_j/n$ and $p_j' = E_j/n$, where n is the sample size. If proportions are known, the frequencies are given by $O_j = n\hat{p}_j$ and $E_j = np_j'$. The null and alternative hypotheses for the data in Table 17.2-1 are as follows.

H_0: $p_1 = 0.08, p_2 = 0.22, p_3 = 0.32, p_4 = 0.20, p_5 = 0.14, p_6 = 0.04$
H_1: $p_j \neq p_j'$ for one or more of the j categories

We want to know whether the proportions in the six categories for the population represented by the sample differ from the proportions for the population of previous applicants. We are faced with a minor interpretation problem, because this year's applicants were not obtained by random sampling. We circumvent the problem by assuming that there is a population for which random sampling could have produced the sample we obtained. Conclusions about equality of the proportions apply only to this population.

The observed frequencies for 200 students who applied to graduate school this year, along with expected frequencies based on data for previous applicants, are given in Table 17.2-2. The computation of Pearson's statistic is illustrated in the table. Degrees of freedom equals one less than the number of categories, $k - 1 = 6 - 1 = 5$. If the 0.05 level of significance is adopted, the tabled value of $\chi^2_{0.05,5}$ is 11.070. Although the null hypothesis is nondirectional, the critical region always lies in the upper tail of the

---

[4] For the case in which $k = 2$, a similar approximate test is based on the normal distribution. A comparison of this test and Pearson's test based on the chi-square distribution is presented in Technical Note 17.6-1.

[5] This point is discussed in Technical Note 17.6-2.

Table 17.2-1. Proportion of Former Applicants with Various Graduate Record Examination (GRE) Scores

GRE Score	p_j'
1400–1499	0.04
1300–1399	0.14
1200–1299	0.20
1100–1199	0.32
1000–1099	0.22
900–999	0.08

Table 17.2-2. Computation of Pearson's Chi-Square Statistic for 200 Graduate School Applicants

GRE Scores	O_j	$np_j' = E_j$	$(O_j - E_j)$	$\dfrac{(O_j - E_j)^2}{E_j}$
1400–1499	13	200(0.04) = 8	5	3.125
1300–1399	35	200(0.14) = 28	7	1.750
1200–1299	49	200(0.20) = 40	9	2.025
1100–1199	57	200(0.32) = 64	−7	0.766
1000–1099	38	200(0.22) = 44	−6	0.818
900–999	8	200(0.08) = 16	−8	4.000
				$\chi^2 = 12.484$

sampling distribution of χ^2. This seems reasonable when we recall that all $O - E$ discrepancies are squared; consequently, the test statistic is insensitive to the direction of the discrepancies. Large discrepancies can only produce large values of the test statistic. An examination of the chi-square table reveals that it is the upper tail that contains the larger values. According to Table 17.2-2, the observed test statistic (12.484) exceeds the value for $\chi^2_{0.05,5}$, so the null hypothesis is rejected and it is concluded that the frequencies in one or more of the categories are not equal to the corresponding expected frequencies.[6]

Characteristics of Pearson's Statistic

An examination of the formula $\sum (O_j - E_j)^2/E_j$ for Pearson's statistic reveals the following.

1. The statistic is never negative, since all $O - E$ discrepancies are squared; and, since the statistic makes no distinction between positive and negative discrepancies, the hypothesis tested is nondirectional. This is true even though the critical region of the sampling distribution of χ^2 is always in the upper tail.
2. The only way the statistic can equal zero is for each observed frequency to equal the corresponding expected frequency.

[6] Procedures for determining which observed and expected frequencies are not equal are described by Marascuilo (1971, pp. 380–382) and Marascuilo and McSweeney (1977).

3. The larger the $O - E$ discrepancy, the larger χ^2. However, the contribution of a discrepancy to χ^2 is affected by the size of E_j, since $(O_j - E_j)^2$ is divided by E_j. This seems reasonable. If we tossed ten coins and observed nine heads where the expected number is five, we would question the fairness of the coins. If the same discrepancy occurred when 100 coins were tossed where the expected frequency is 50, we wouldn't be surprised. A discrepancy of four is viewed one way when $E = 5$ and a different way when $E = 50$. The formula takes this into account by expressing the size of the discrepancy relative to the magnitude of the expected frequency.
4. The larger the number of categories, the larger the degrees of freedom and the observed χ^2. As the number of degrees of freedom increases, the chi-square value required for significance also increases, since degrees of freedom determine which member of the family of chi-square distributions is used in evaluating χ^2. Thus, the test procedure takes the number of categories into account.

Degrees of Freedom when E_j's Are Based on a Theoretical Distribution

A modification of the goodness-of-fit test is required if parameters of a theoretical distribution must be estimated in computing the expected frequencies. In the previous example, the E_j's were computed from data, and no distribution parameters were estimated. For this case $v = k - 1$, where k is the number of categories. Suppose, however, that we compared a frequency distribution with that predicted by, say the normal distribution. We would have to estimate two parameters, μ and σ, to compute E_j. For each parameter estimated, the degrees of freedom are reduced by one. The formula for degrees of freedom is $v = k - 1 - e$, where e is the number of theoretical distribution parameters estimated. Except for this modification, the test procedure is the same.

Assumptions of the Test

The goodness-of-fit test can be used for any hypothetical population, provided the population is discrete or can be grouped into a manageable number of categories. The assumptions of the test are minimal: (1) every sample observation must fall in one and only one category, (2) the observations must be independent, and (3) the sample n must be large. How large is large? This is difficult to specify because the satisfactoriness of the chi-square distribution as an approximation to the true distribution of Pearson's statistic depends on n, the true proportions in the k categories, and the number of degrees of freedom, among other things. As a rule of thumb, when the degrees of freedom equal one, each expected frequency should be at least ten. Each expected frequency should be at least five when the degrees of freedom are greater than one. One remedy if $k > 2$ and expected frequencies are below the minimum is to combine categories until all frequencies are large enough.[7]

When the test has one degree of freedom, *Yates' correction for continuity*[8] can be applied to make the discrete sampling distribution of the test statistic when the null

---

[7] An alternative approach is to use the Kolomogorov–Smirnov goodness-of-fit test, which has greater power than Pearson's test. For a discussion of this test see Korin (1975, pp. 297–298) and Taylor (1972, pp. 141–143). Cochran (1954) discusses other rules of thumb for using Pearson's statistic.

[8] Proposed by Frank Yates (1902), a British statistician.

hypothesis is true more consistent with the theoretical model—the continuous chi-square distribution. The correction can be included in the test statistic as follows.

$$\chi^2 = \sum_{j=1}^{k} \frac{(|O_j - E_j| - 0.5)^2}{E_j}$$

It is good practice to apply the correction when the degrees of freedom equal one and any expected frequency is not appreciably greater than ten.

Review Exercises for Section 17.2

1. Why is the hypothesis tested by Pearson's statistic always nondirectional?
*2. Students in a random sample were asked if they favor a change from the semester system to the quarter system. Thirty-three said yes, 17 said no. (a) List the steps you would follow in testing the null hypothesis that opinion is equally divided on the issue. Let $\alpha = 0.05$. (b) Do the data suggest that opinion is not equally divided on this issue?
*3. According to the most recent public opinion poll in Johnson County, 71 eligible voters were Democrats, 52 were Republicans, and 33 belonged to other parties. Traditionally, the ratio of Democrats to Republicans to others has been 4:3:2. Does the poll suggest a change in party affiliation? Let $\alpha = 0.05$.
4. The dean believes that students in M–W–F classes are more likely to be absent on Monday and Friday than on Wednesday. A random sample of 200 students revealed that 68 were absent on Monday, 48 on Wednesday, and 84 on Friday. (a) List the steps you would follow in testing the hypothesis that the ratio of absences is M:W:F = 3:2:3. Let $\alpha = 0.05$. (b) Perform the test and make a decision.
*5. A student in a statistics class tossed a die 300 times and obtained the results shown in the table. Is the die fair? Let $\alpha = 0.05$.

Outcome	1	2	3	4	5	6
Frequency	53	41	60	47	38	61

6. A random sample of 300 music majors took a test of creativity. (a) Test the hypothesis that their scores are normally distributed. Let $\alpha = 0.01$. The E column in the table presents expected frequencies given that the null hypothesis is true; the O column presents the observed frequencies. (b) How many degrees of freedom does the test have?

Test Score	O_j	E_j
140 and above	0	5.0
130–139	32	20.2
120–129	48	54.2
110–119	76	83.8
100–109	90	78.4
90–99	36	42.4
80–89	18	13.4
Less than 80	0	2.7

7. What is Yates' correction and why is it used?
8. Terms to remember:
   a. Approximate and exact tests
   b. Yates' correction for continuity

17.3 Testing Independence

Pearson's statistic can be used to obtain an approximate test of the hypothesis that two variables are independent. The test is appropriate where n independent observations for a single random sample have been classified in terms of two variables.

Computational Example

Suppose that a random sample of 200 high school students has been classified in terms of sex, variable A, and use or nonuse of marijuana, variable B. A table representing the classification of elements in terms of two or more variables is called a **contingency table**. A partial summary of the data is given in Table 17.3-1. If the two variables in the population are independent, what frequencies should be in the cells of the contingency table? We know from Section 7.3 that two events are independent if $p(A|B) = p(A)$, in which case $p(A \text{ and } B) = p(A)p(B)$. The probabilities can be converted to frequencies by multiplying both sides of the equations by n. The population probabilities are unknown, but we can use the marginal proportions, $p(A_1)$, $p(A_2)$, and so on, in Table 17.3-1 to estimate the expected cell frequencies $E_{A_i \text{ and } B_j}$. Using the relationship $E_{A_i \text{ and } B_j} = np(A_i \text{ and } B_j) = np(A_i)p(B_j)$, estimates of the expected cell values are as follows.

$$E_{A_1 \text{ and } B_1} = np(A_1 \text{ and } B_1) = np(A_1)p(B_1) = 200(0.20)(0.64) = 25.6$$
$$E_{A_2 \text{ and } B_1} = np(A_2 \text{ and } B_1) = np(A_2)p(B_1) = 200(0.80)(0.64) = 102.4$$
$$E_{A_1 \text{ and } B_2} = np(A_1 \text{ and } B_2) = np(A_1)p(B_2) = 200(0.20)(0.36) = 14.4$$
$$E_{A_2 \text{ and } B_2} = np(A_2 \text{ and } B_2) = np(A_2)p(B_2) = 200(0.80)(0.36) = 57.6$$

Table 17.3-1. Partial Summary of Sex and Marijuana Data

	User, B_1	Nonuser, B_2	
Male, A_1			$n_{A_1} = 40; p(A_1) = \dfrac{40}{200} = 0.20$
Female, A_2			$n_{A_2} = 160; p(A_2) = \dfrac{160}{200} = 0.80$
	$n_{B_1} = 128$	$n_{B_2} = 72$	$n = 200$
	$p(B_1) = \dfrac{128}{200} = 0.64$	$p(B_2) = \dfrac{72}{200} = 0.36$	

Alternatively, the expected cell frequencies can be computed directly from the marginal frequencies by the formula

$$E_{A_i \text{ and } B_j} = \frac{n_{A_i} n_{B_j}}{n},$$

since $E_{A_i \text{ and } B_j} = np(A_i)p(B_j) = n(n_{A_i}/n)(n_{B_j}/n) = n_{A_i}n_{B_j}/n$. For example, $E_{A_1 \text{ and } B_1} = (40)(128)/200 = 25.6$. The expected cell frequencies, along with the observed cell frequencies, are given in Table 17.3-2. The computation of the chi-square statistic is illustrated in the table.[9] The advisability of using Yates' correction for a 2 × 2 contingency table is the subject of continuing debate among statisticians (Conover, 1974a, 1974b; Grizzle, 1967; Mantel, 1974; Miettinen, 1974; Plackett, 1964; Starmer, Grizzle, & Sen, 1974) and accordingly is not illustrated in Table 17.3-2. If the correction is desired, it can be incorporated in the chi-square formula as follows.

$$\chi^2 = \sum \frac{(|O_j - E_j| - 0.5)^2}{E_j}$$

Table 17.3-2. Sex and Marijuana Data

(i) Data:

	User, B_1	Nonuser, B_2	
Male, A_1	O = 32 E = 25.6	O = 8 E = 14.4	$n_{A_1} = 40$
Female, A_2	O = 96 E = 102.4	O = 64 E = 57.6	$n_{A_2} = 160$
	$n_{B_1} = 128$	$n_{B_2} = 72$	$n = 200$

(ii) Computation of χ^2:

O_j	E_j	$O_j - E_j$	$\frac{(O_j - E_j)^2}{E_j}$
32	25.6	6.4	1.600
96	102.4	−6.4	0.400
8	14.4	−6.4	2.844
64	57.6	6.4	0.711
		$\chi^2 =$	5.555

Degrees of Freedom for a Contingency Table

How many degrees of freedom are associated with this χ^2? For the goodness-of-fit test we saw that the degrees of freedom are $k - 1 - e$, where e is the number of parameters estimated. A 2 × 2 table is a special case of an $r \times c$ contingency table, where r

---

[9] A simpler computational formula for a 2 × 2 contingency table is given in Technical Note 17.6-3.

and/or c are greater than two. We will develop the formula for degrees of freedom for the more general case. If we denote the number of rows by r and the number of columns by c, an $r \times c$ contingency table has $k = rc$ categories. Since there are r categories for variable A, we must estimate $r - 1$ expected cell frequencies for this variable. Once we have estimated any $r - 1$ of the expected cell frequencies, the remaining one can be obtained by subtraction, because the sum of the r expected cell frequencies is equal to n. Similarly, we must estimate $c - 1$ expected cell frequencies for variable B. In all, we must make $e = (r - 1) + (c - 1)$ estimates. Thus, the number of degrees of freedom for an $r \times c$ contingency table is

$$\begin{aligned} df &= k - 1 - e \\ &= rc - 1 - [(r - 1) + (c - 1)] \\ &= rc - 1 - r + 1 - c + 1 \\ &= rc - r - c + 1 \\ &= (r - 1)(c - 1). \end{aligned}$$

For a 2×2 contingency table the degrees of freedom are $(2 - 1)(2 - 1) = 1$.

Statistical Hypotheses

The statistical hypotheses for the data in Table 17.3-2 are

$$H_0: \quad p(A \text{ and } B) = p(A)p(B)$$
$$H_1: \quad H_0 \text{ is false.}$$

The hypothesis of statistical independence is rejected if the observed χ^2 exceeds the tabled value at α level of significance for $(r - 1)(c - 1)$ degrees of freedom. For the data in Table 17.3-2, the tabled value at the 0.05 level of significance ($\chi^2_{0.05,1}$) is 3.841. Since the value of the test statistic (5.555) exceeds 3.841, the null hypothesis is rejected and we conclude that sex and marijuana usage are not independent.

Knowing that two variables are related is useful, but it would be even more useful to know the strength of the association. Before describing an index of association we will see how to apply a test of independence to contingency tables with more than two rows or columns.

Contingency Tables with Three or More Rows or Columns

The test for independence can be extended to the case in which variable A has $r > 2$ mutually exclusive and exhaustive categories, $A_1, A_2, \ldots, A_i, \ldots, A_r$, and B has $c > 2$ mutually exclusive and exhaustive categories, $B_1, B_2, \ldots, B_j, \ldots, B_c$.

Suppose we want to know if a college graduate's starting salary is independent of the size of the university from which he or she graduated. Data for a random sample of 200 graduates are given in Table 17.3-3. The expected frequencies are computed from $E_{A_i \text{ and } B_j} = n_{A_i} n_{B_j}/n$. The degrees of freedom are $(r - 1)(c - 1) = (2)(2) = 4$. For a test at the 0.05 level of significance, the value of $\chi^2_{0.05,4}$ is 9.488. Since the value of the test statistic (17.978) exceeds 9.488, the null hypothesis is rejected and it is concluded that starting salary and university size are not independent.

Table 17.3-3. University Size and Starting Salaries of Graduates

(i) Data:

	Less than $5000, B_1	$5000–10,000, B_2	Greater than $10,000, B_3	
Small University, A_1	$O = 13$ $E = 10.08$	$O = 28$ $E = 34.44$	$O = 15$ $E = 11.48$	$n_{A_1} = 56$
Medium-Size University, A_2	$O = 11$ $E = 13.14$	$O = 40$ $E = 44.90$	$O = 22$ $E = 14.96$	$n_{A_2} = 73$
Large University, A_3	$O = 12$ $E = 12.78$	$O = 55$ $E = 43.66$	$O = 4$ $E = 14.56$	$n_{A_3} = 71$
	$n_{B_1} = 36$	$n_{B_2} = 123$	$n_{B_3} = 41$	$n = 200$

(ii) Computation of χ^2:

O_j	E_j	$O_j - E_j$	$\dfrac{(O_j - E_j)^2}{E_j}$
13	10.08	2.92	0.846
11	13.14	−2.14	0.349
12	12.78	−0.78	0.048
28	34.44	−6.44	1.204
40	44.90	−4.90	0.535
55	43.66	11.34	2.945
15	11.48	3.52	1.079
22	14.96	7.04	3.313
4	14.56	−10.56	7.659
		$\chi^2 =$	17.978

Finding a significant χ^2 is often just the first step in analyzing data. To better understand the data it is helpful to convert frequencies to proportions, since the marginal row frequencies normally differ, as do the marginal column frequencies. This has been done in Table 17.3-4, where the frequencies in each row of Table 17.3-3 have been divided by their respective row marginal frequencies. If the data had been independent, each of the observed proportions in column B_1 would equal 0.18, those in column B_2 would equal 0.62, and those in column B_3 would equal 0.20. From a visual inspection it appears that graduates of large universities are less likely than expected to have starting salaries over $10,000, while the converse is true for those from medium and small universities. Such hypotheses must be regarded as tentative, since they are not based on the outcome of significance tests.[10] The significant chi-square test statistic applies to the data taken as a whole and provides no clue as to which cells are responsible for significance.

[10] Procedures for determining the contribution to the overall chi-square statistic of a portion of the contingency table are described by Bresnahan and Shapiro (1966), Castellan (1965), Marascuilo (1966), and Marascuilo and McSweeney (1977).

Table 17.3-4. Data from Table 17.3-3 Expressed as Proportions

	Less than $5000, B_1	$5000–10,000, B_2	Greater than $10,000, B_3	
Small University, A_1	$p = \dfrac{O}{n_{A_1}} = 0.23$ $p = \dfrac{E}{n_{A_1}} = 0.18$	$p = \dfrac{O}{n_{A_1}} = 0.50$ $p = \dfrac{E}{n_{A_1}} = 0.62$	$p = \dfrac{O}{n_{A_1}} = 0.27$ $p = \dfrac{E}{n_{A_1}} = 0.20$	$p_{A_1} = 1.00$
Medium-Size University, A_2	$p = \dfrac{O}{n_{A_2}} = 0.15$ $p = \dfrac{E}{n_{A_2}} = 0.18$	$p = \dfrac{O}{n_{A_2}} = 0.55$ $p = \dfrac{E}{n_{A_2}} = 0.62$	$p = \dfrac{O}{n_{A_2}} = 0.30$ $p = \dfrac{E}{n_{A_2}} = 0.20$	$p_{A_2} = 1.00$
Large University, A_3	$p = \dfrac{O}{n_{A_3}} = 0.17$ $p = \dfrac{E}{n_{A_3}} = 0.18$	$p = \dfrac{O}{n_{A_3}} = 0.77$ $p = \dfrac{E}{n_{A_3}} = 0.62$	$p = \dfrac{O}{n_{A_3}} = 0.06$ $p = \dfrac{E}{n_{A_3}} = 0.20$	$p_{A_3} = 1.00$
	$p_{B_1} = \dfrac{n_{B_1}}{n} = 0.18$	$p_{B_2} = \dfrac{n_{B_2}}{n} = 0.62$	$p_{B_3} = \dfrac{n_{B_3}}{n} = 0.20$	

Measuring Strength of Association

If the null hypothesis $p(A \text{ and } B) = p(A)p(B)$ is rejected, we know the variables are correlated. Several correlation coefficients were discussed in Chapter 5 (r, r_s, and η^2), but none of them is appropriate for unordered qualitative variables. One coefficient that is appropriate is *Cramer's measure of association*,[11] denoted by ϕ'. The formula for estimating ϕ' is

$$\hat{\phi}' = \frac{\sqrt{\hat{\phi}^2_{\text{observed}}}}{\sqrt{\hat{\phi}^2_{\text{maximum}}}} = \frac{\sqrt{\chi^2/n}}{\sqrt{s-1}} = \sqrt{\frac{\chi^2}{n(s-1)}},$$

where s is the smaller of the numbers r and c.[12] If $p(A \text{ and } B) = p(A)p(B)$, the parameter ϕ, which is estimated by $\hat{\phi}_{\text{observed}} = \sqrt{\chi^2/n}$, is equal to zero. When there is complete association, $\phi = s - 1$, which is its maximum value. Cramer's statistic is a relative measure, since it is the ratio of an observed statistic to its maximum possible value. The statistic can range from zero (indicating complete independence) to one (indicating complete dependence, or perfect correlation). Cramer's statistic is only computed when the chi-square test is significant. A nonsignificant test suggests that any $\hat{\phi}'$

---

[11] For a discussion of other coefficients see Hays (1973, pp. 742–753) and Loether and McTavish (1974, chaps. 6 and 7).

[12] For a 2 × 2 contingency table, $\hat{\phi}' = \sqrt{\chi^2/n}$, since $s - 1 = 1$. If the categories of variables A and B are scored 0 and 1, the formula for Pearson's product-moment correlation coefficient is algebraically equivalent to that for $\hat{\phi}'$ (Hays, 1973, pp. 743–745).

greater than zero is due to chance. To put it another way, a test of the hypothesis $H_0: p(A \text{ and } B) = p(A)p(B)$ is equivalent to a test of $H_0: \phi' = 0$. If the null hypothesis isn't rejected, it is meaningless to proceed further and compute $\hat{\phi}'$. Unfortunately, Cramer's statistic does not have a simple intuitively useful interpretation as, say, the proportion of explained variance between two variables. It can only be thought of as reflecting magnitude of association on a zero to one scale; the larger the number, the stronger the association. Cramer's statistic for the data in Tables 17.3-2 and 17.3-3 is, respectively,

$$\hat{\phi}' = \sqrt{\frac{5.555}{200(2-1)}} = 0.17 \quad \text{and} \quad \hat{\phi}' = \sqrt{\frac{17.978}{200(3-1)}} = 0.21.$$

The coefficients and the significance tests indicate a low but statistically significant association for both sets of variables. This is another illustration of the maxim that statistical significance doesn't mean practical significance. For example, the chi-square test statistic for the variables of university size and starting salaries of graduates is large enough to be significant beyond the 0.005 level, yet the strength of association is low. The point cannot be made too often that statistical significance only means that in all likelihood a relationship is not due to chance. Other procedures must be used to assess the usefulness or practical significance of a relationship.

Assumptions of the Test

The assumptions associated with both the test for independence and the estimation of strength of association are as follows: (1) Every observation must fall in one and only one cell of the contingency table. (2) The observations must be independent. One situation in which this assumption is likely to be violated occurs when an individual is represented more than once in a cell or in more than one cell. (3) The sample n should be large enough so that every expected frequency is at least ten when there is one degree of freedom and at least five when there is more than one degree of freedom.[13]

Review Exercises for Section 17.3

9. State in your own words the meaning of the hypothesis $H_0: p(A \text{ and } B) = p(A)p(B)$. If H_0 is rejected, what do you know about the variables?
*10. List the similarities and differences between Pearson's product-moment correlation coefficient and Cramer's measure of association.
11. Why is Cramer's measure of association only computed when the hypothesis $p(A \text{ and } B) = p(A)p(B)$ is rejected?
*12. Two hundred women between the ages of 19 and 25 were asked if they favored the use of birth-control pills. The women were classified according to attitude and religious preference. (a) For the data in the table, list the steps you would use in testing the null hypothesis. Let $\alpha = 0.001$. (b) What is your decision? (c) Compute Cramer's statistic.

[13] Fisher's exact test for a 2 × 2 contingency table can be used to test independence when n is small and expected cell frequencies are less than ten. The test is described in Hays (1973, pp. 737–740) and Taylor (1972, pp. 143–145).

	Protestants	Roman Catholics	Non-Christians	
Favor	58	30	28	116
Oppose	8	23	5	36
Undecided	10	15	23	48
	76	68	56	200

13. Three hundred divorced males were classified by age at time of first marriage and duration of first marriage. (a) List the steps you would use in testing the null hypothesis. Let $\alpha = 0.05$. (b) Perform the test and make a decision. (c) Compute Cramer's statistic.

<table>
<tr><td></td><td></td><td colspan="4">Duration of Marriage (Years)</td><td></td></tr>
<tr><td></td><td></td><td><5</td><td>5–9</td><td>10–14</td><td>≥15</td><td></td></tr>
<tr><td rowspan="4">Age at Marriage</td><td><19</td><td>41</td><td>31</td><td>15</td><td>15</td><td>102</td></tr>
<tr><td>19–24</td><td>30</td><td>27</td><td>22</td><td>21</td><td>100</td></tr>
<tr><td>25–34</td><td>11</td><td>7</td><td>16</td><td>16</td><td>50</td></tr>
<tr><td>≥35</td><td>13</td><td>14</td><td>9</td><td>12</td><td>48</td></tr>
<tr><td></td><td></td><td>95</td><td>79</td><td>62</td><td>64</td><td>300</td></tr>
</table>

14. A random sample of 200 college students was classified according to class standing and political conservatism. (a) Are the variables independent? Let $\alpha = 0.05$. (b) Compute Cramer's statistic.

	Conservative	Neutral	Liberal	
Freshman	38	22	6	66
Sophomore	22	24	5	51
Junior	11	12	19	42
Senior	7	13	21	41
	78	71	51	200

15. Term to remember:
    Contingency table

†17.4 Testing Equality of $c \geq 2$ Proportions

The last application of Pearson's chi-square statistic that will be described is testing the equality of $c \geq 2$ population proportions. If a random sample is obtained from each of $c \geq 2$ binomially distributed populations, the sample proportions $\hat{p}_1, \hat{p}_2, \ldots, \hat{p}_c$ can be used to test the null hypothesis

$$H_0: \quad p_1 = p_2 = \cdots = p_c$$

† This section can be omitted without loss of continuity.

versus

$$H_1: \quad H_0 \text{ is false.}$$

The null hypothesis is rejected if $\chi^2 = \sum (O_j - E_j)^2 / E_j$ *exceeds the tabled value of* χ^2 *at* α *level of significance for* $c - 1$ *degrees of freedom*. Rejection of H_0 means that some population proportions are not equal. It doesn't mean that they are all unequal; perhaps only one is discrepant. The test, like the two described earlier, is approximate, since the continuous chi-square distribution is used to estimate a probability for a discrete sampling distribution. The proportions $\hat{p}_1, \hat{p}_2, \ldots, \hat{p}_c$, which may be based on unequal sample sizes, are assumed to represent independent observations for c independent binomially distributed random variables. Like all binomial experiments, each trial must result in one of two outcomes. We will see later that the test can be extended to the multinomial case in which each trial can result in one of three or more outcomes.

Computational Example

Suppose we surveyed older people living in public housing in Borborygme, Texas, to determine whether satisfaction with living conditions is related to age heterogeneity of people in the neighborhood. Neighborhoods were categorized as high, medium, or low in heterogeneity of residents' ages. Random samples of 100 elderly females from each category were interviewed (recall that equal sample n's are not required) and asked, among other things, "Are you satisfied with your living conditions?" The answers were classified as "satisfied" or "not satisfied." It was anticipated that satisfaction would be different for the three neighborhood categories. Responses to the question are given in Table 17.4-1. The computation of Pearson's chi-square statistic is shown in the table; the procedure is identical to that for an $r \times c$ contingency table for testing independence (see Section 17.3). The degrees of freedom are equal to $c - 1 = 2$, the number of age heterogeneity categories minus one. Alternatively, the degrees of freedom can be computed from $(r - 1)(c - 1) = (1)(2) = 2$. For a test at the 0.05 level of significance, the critical value of $\chi^2_{0.05, 2}$ is 5.991 (according to Table D.4). Since the value of the test statistic (9.240) exceeds 5.991, the null hypothesis is rejected and it is concluded that at least two of the population proportions are not equal.[14]

Comparison of Independence Test and Equality of Proportions Test

The tests for equality of $c \geq 2$ proportions and independence use the same Pearson test statistic and the same formula for computing degrees of freedom, and both are approximately distributed as the chi-square distribution. It is not surprising then that the tests are often confused. The key difference is in the sampling method. In testing independence, (1) a single random sample is obtained from a population and (2) the

---

[14] Marascuilo (1971, pp. 380–382) describes a procedure for determining which population proportions are unequal. The procedure is based on Scheffé's method, which is described in Section 16.5.

Table 17.4-1. Satisfaction and Age Heterogeneity

(i) Data:

Age Heterogeneity

	Low, B_1	Medium, B_2	High, B_3	
Satisfied, A_1	$O = 56$ $E = 50.67$	$O = 58$ $E = 50.67$	$O = 38$ $E = 50.67$	$n_{A_1} = 152$
Not Satisfied, A_2	$O = 44$ $E = 49.33$	$O = 42$ $E = 49.33$	$O = 62$ $E = 49.33$	$n_{A_2} = 148$
	$n_{B_1} = 100$	$n_{B_2} = 100$	$n_{B_3} = 100$	$n = 300$

(ii) Computation of χ^2:

O_j	E_j	$O_j - E_j$	$\dfrac{(O_j - E_j)^2}{E_j}$
56	50.67	5.33	0.561
44	49.33	−5.33	0.576
58	50.67	7.33	1.060
42	49.33	−7.33	1.089
38	50.67	−12.67	3.168
62	49.33	12.67	3.254
		$\chi^2 =$	9.708

elements are classified in terms of two variables. On the other hand, in testing equality of proportions, (1) c independent random samples are obtained from c binomially distributed populations and (2) the sample elements are classified in terms of only one variable (such as satisfaction), since the experimenter has predetermined the elements' status with respect to the other variable (such as neighborhoods) by sampling from the c categories. Note that in a test of independence the status of the sample elements is not predetermined for either variable. In a test of equality an experimenter often attempts to sample elements so as to have an equal number from each of the c populations, although this isn't required. It is because of the different sampling procedures that the hypotheses tested are different; H_0: $p(A \text{ and } B) = p(A)p(B)$ for the test of independence and H_0: $p_1 = p_2 = \cdots = p_c$ for the test of equality.

Extension of the Test of Equality to More than Two Response Categories

The test illustrated in Table 17.4-1 can be extended to the case in which variable A has more than two categories, $A_1, A_2, \ldots, A_r$; for example, very satisfied, somewhat satisfied, somewhat dissatisfied, very dissatisfied. In this form the test is referred to as a *test of homogeneity*. The proportions $p_{A_1|B_1}, p_{A_2|B_1}, \ldots, p_{A_r|B_1}; p_{A_1|B_2}, p_{A_2|B_2}, \ldots, p_{A_r|B_2}; \ldots; p_{A_1|B_c}, p_{A_2|B_c}, \ldots, p_{A_r|B_c}$ are assumed to represent observations for c independent multinomially distributed random variables instead of c binomial variables. The null hypothesis states that the proportions in the r categories are equal across the c categories of B. This can be stated symbolically as

$$H_0: \begin{bmatrix} p_{A_1|B_1} \\ p_{A_2|B_1} \\ \vdots \\ p_{A_r|B_1} \end{bmatrix} = \begin{bmatrix} p_{A_1|B_2} \\ p_{A_2|B_2} \\ \vdots \\ p_{A_r|B_2} \end{bmatrix} = \cdots = \begin{bmatrix} p_{A_1|B_c} \\ p_{A_2|B_2} \\ \vdots \\ p_{A_r|B_c} \end{bmatrix}$$

versus H_1: H_0 is false. If A has only two categories (the binomial case), the null hypothesis can be written as

$$H_0: \quad p_{A_1|B_1} = p_{A_1|B_2} = \cdots = p_{A_1|B_c}$$

or simply,

$$H_0: \quad p_1 = p_2 = \cdots = p_c.$$

The computation of Pearson's statistic for a homogeneity test is like that for an $r \times c$ contingency table. The number of degrees of freedom is equal to $(r-1)(c-1)$. The general assumptions discussed in connection with an $r \times c$ contingency table apply as well to testing the equality of c population proportions.

Review Exercises for Section 17.4

16. The equality and independence tests are applicable to an $r \times c$ contingency table and both use the same test statistic. How do they differ?

*17. Company executives were classified as smokers or nonsmokers. Random samples from the two populations were obtained and tested for presence of lung cancer. State the null hypothesis and test it at the 0.01 level of significance.

	Smoker	Nonsmoker	
Cancer Present	16	5	21
Cancer Absent	14	25	39
	30	30	60

18. Students were given a choice of writing or not writing a paper for extra credit. Half of them, selected randomly, were made to feel coerced, the other half received no pressure. The number who chose to write or not to write a paper is given in the table. (a) List the steps you would use to test the null hypothesis. Let $\alpha = 0.05$. (b) Did the two conditions affect paper writing?

	Coerced	Not Coerced	
Wrote Paper	13	9	22
Didn't Write Paper	17	21	38
	30	30	60

*19. State in words the meaning of the hypothesis

$$H_0: \begin{bmatrix} 0.30 \\ 0.60 \\ 0.10 \end{bmatrix} = \begin{bmatrix} 0.30 \\ 0.60 \\ 0.10 \end{bmatrix} = \begin{bmatrix} 0.30 \\ 0.60 \\ 0.10 \end{bmatrix}.$$

*20. New employees on an assembly line were randomly assigned to one of four groups and given different amounts of training. After 2 weeks on the job, their performance was rated by their supervisor. For the data in the table (a) state the statistical hypotheses and (b) test the null hypothesis at the 0.05 level of significance.

		Amount of Training (Days)				
		1	2	5	10	
Rating	Excellent	4	4	6	6	20
	Good	5	3	10	11	29
	Fair	9	9	5	8	31
	Poor	7	9	4	0	20
		25	25	25	25	100

21. Random samples of seventh-, eighth-, and ninth-grade students were interviewed following 1 year of busing. They were asked "Did Black and White students mix more, the same, or less than last year?" State the null hypothesis and test it at the 0.05 level of significance.

		Grade			
		7	8	9	
Response	More	26	12	4	42
	Same	18	25	35	78
	Less	6	13	11	30
		50	50	50	150

17.5 Summary

Pearson's chi-square test statistic is one of the few that is appropriate for frequency data. It provides an approximate test when an exact test based on the binomial or multinomial distributions would require a prohibitive amount of computation. The statistic is versatile; the three applications we have described are testing goodness of fit, independence, and equality of proportions. These are summarized in Table 17.5-1. In each application a set of observed frequencies is compared with a set of expected frequencies. The apparent simplicity of the statistic $\sum (O_j - E_j)^2/E_j$ is deceptive. Its use involves important assumptions that are often overlooked or misunderstood by the novice investigator. The problem is compounded because all three applications use the same statistic for testing different hypotheses. The main points to consider in using Pearson's chi-square statistic are the following.

Table 17.5-1. Three Applications of Pearson's Chi-Square Test Statistic

Purpose	Null Hypothesis	Degrees of Freedom	Requirements
1. Testing goodness of fit	$p_1 = p_1', p_2 = p_2', \ldots, p_k = p_k'$	$k - 1 - e$	1. One random sample. 2. If $v = 1$, every expected frequency should exceed ten. If $v > 1$, every expected frequency should exceed five. 3. Random variable is binomially distributed for $k = 2$ and multinomially distributed for $k > 2$.
2. Testing independence	$p(A \text{ and } B) = p(A)p(B)$	$(r-1)(c-1)$	1. Same as for test of goodness of fit. 2. Same as for test of goodness of fit. 3. Random variable is binomially distributed when A and B have two categories, and multinomially distributed otherwise.
3a. Testing equality of proportions	$p_1 = p_2 = \cdots = p_c$	$c - 1$	1. $c \geq 2$ random samples. 2. If $v = 1$, every expected frequency should exceed ten. If $v > 1$, every expected frequency should exceed five. 3. Random variable is binomially distributed.
3b. Testing homogeneity of proportions	$\begin{bmatrix} p_{A_1\|B_1} \\ p_{A_2\|B_1} \\ \vdots \\ p_{A_r\|B_1} \end{bmatrix} = \begin{bmatrix} p_{A_1\|B_2} \\ p_{A_2\|B_2} \\ \vdots \\ p_{A_r\|B_2} \end{bmatrix} = \cdots = \begin{bmatrix} p_{A_1\|B_c} \\ p_{A_2\|B_c} \\ \vdots \\ p_{A_r\|B_c} \end{bmatrix}$	$(r-1)(c-1)$	1. $c \geq 2$ random samples. 2. Every expected frequency should exceed five. 3. Random variable is multinomially distributed.

1. A single random sample is used in testing goodness of fit and independence; more than one random sample is used in testing equality of population proportions.
2. Each observation must be assigned to one and only one category or one cell of a contingency table.
3. Each subject or observational unit should be represented only once. Multiple observations on the same subject almost always result in violation of the assumption of independence of observations.
4. The chi-square approximation to exact probabilities based on the binomial or multinomial distribution is generally unsatisfactory for small samples. As a precautionary rule of thumb, if $v = 1$, all expected frequencies should exceed ten; if $v > 1$, all expected frequencies should exceed five.
5. If the chi-square test statistic is not significant for an $r \times c$ contingency table, no further tests should be performed on subportions of the table.
6. The hypothesis tested by Pearson's statistic is nondirectional, even though the region for rejection always lies in the upper tail of the sampling distribution.

†17.6 Technical Notes

17.6-1 Two Ways to Test the Hypothesis that $p = p'$ (or $p = p_0$)

A goodness-of-fit test with one degree of freedom is formally equivalent to the large-sample z test for a proportion described in Section 12.4. We will show by an example that

$$z = \frac{\hat{p} - p_0}{\sqrt{p_0 q_0/n}}$$

leads to the same decision about p as Pearson's statistic and in addition, that $|z| = \sqrt{\chi^2}$.

In Section 12.4 we tested the hypothesis that the results of a random sample of 900 dwelling units in which 0.34 were found to be substandard doesn't differ from the results of an earlier survey in which the proportion was 0.30. To test the hypothesis that the number of substandard units had not changed, $H_0: p = 0.30$, the z test statistic was computed:

$$z = \frac{\hat{p} - p_0}{\sqrt{p_0 q_0/n}} = \frac{0.34 - 0.30}{\sqrt{(0.30)(0.70)/900}} = \frac{0.04}{\sqrt{0.0002333}} = 2.62.$$

The statistic exceeded the critical value required to reject the null hypothesis at the 0.05 level; hence, we concluded that the proportion of substandard dwelling units had increased.

The same decision is reached if the data are analyzed by Pearson's chi-square statistic. The computation is shown in Table 17.6-1. The critical value of χ^2 for $v = 1$ degree of freedom at the 0.05 level of significance is $\chi^2_{0.05,1} = 3.841$. Since the observed value (6.86) exceeds 3.841, the null hypothesis is rejected. We note also that the value of the z test statistic and the square root of the chi-square test statistic both equal 2.62—that is, $|z| = \sqrt{\chi^2} = 2.62$.

† These technical notes can be skipped without loss of continuity.

Table 17.6-1. Computation of Pearson's Chi-Square Statistic for $k = 2$ Categories

---

(i) Computation of observed (O_j) and expected (E_j) frequencies:

$$\begin{aligned}
O_1 &= \hat{p}n &= 0.34(900) &= 306 \\
O_2 &= (1 - \hat{p})n &= 0.66(900) &= 594 \\
E_1 &= p'n &= 0.30(900) &= 270 \\
E_2 &= (1 - p')n &= 0.70(900) &= 630
\end{aligned}$$

(ii) Computation of chi-square statistic:

O_j	E_j	$O_j - E_j$	$\dfrac{(O_j - E_j)^2}{E_j}$
306	270	36	$(36)^2/270 = 4.80$
594	630	-36	$(-36)^2/630 = 2.06$
			$\chi^2 = 6.86$

17.6-2 Outline of an Exact Test of $p_1 = p_1', p_2 = p_2', \ldots, p_k = p_k'$

The binomial distribution is the appropriate theoretical model when (1) randomly sampling from a population with elements that belong to one of two categories, (2) the probability of obtaining an element in a category remains constant, and (3) the trials are independent.[15] The probability of observing exactly r successes in n trials is given by the function rule

$$p(X = r) = \frac{n!}{r!(n-r)!} p'^r q'^{n-r}.$$

An exact test of the hypothesis $H_0: p = p'$, where p is the population proportion estimated by $\hat{p}$ and p' is the expected proportion, is obtained by computing the sum

$$p(X \geq r) = p(X = r) + p(X = r + 1) + p(X = r + 2) + \cdots + p(X = n).$$

This tells us the probability of observing r or more successes if the population proportion is equal to p'. For large n's this involves a prohibitive amount of computation. Fortunately, the normal distribution or the chi-square distribution can be used to approximate the exact probabilities when n is large.

Consider next the more general theoretical model in which the population elements are classified as belonging to one of three or more categories. The population is referred to as a *multinomial* population.[16] The probability of observing exactly n_1 elements of kind 1, n_2 elements of kind 2, ..., n_k elements of kind k, where $n_1 + n_2 + \cdots + n_k = n$, is given by the multinomial function rule

$$p(X_1 = n_1 \text{ and } X_2 = n_2, \text{ and } \ldots \text{ and } X_k = n_k)$$
$$= \frac{n!}{n_1! n_2! \cdots n_k!} p_1'^{n_1} p_2'^{n_2} \cdots p_k'^{n_k}.$$

---

[15] The binomial distribution is discussed in Section 9.3.
[16] The multinomial distribution is discussed in Section 9.3.

To test the hypothesis

$$H_0: p_1 = p_1', p_2 = p_2', \ldots, p_k = p_k'$$

versus

$$H_1: H_0 \text{ is false}$$

we would need to compute the probability of this result plus those for all results more deviant from expectation.

When n is small it isn't too difficult to compute an exact test based on the multinomial distribution. Suppose we want to compare students' preferences for three soft drinks, brands A, B, and C. A random sample of seven students taste the drinks and indicate their preference. Five prefer brand A; one prefers brand B; and one prefers brand C. If the expected proportions are $p_A' = p_B' = p_C' = \frac{1}{3}$, what is the probability of a result as extreme or more extreme than that observed? We can observe five or more preferences for brand A as follows. The associated probabilities are computed using the multinomial function rule.

$$p(5 \text{ A and } 1 \text{ B and } 1 \text{ C}) = \frac{7!}{5!\,1!\,1!} \left(\frac{1}{3}\right)^5 \left(\frac{1}{3}\right)^1 \left(\frac{1}{3}\right)^1 = \frac{42}{2187} = 0.0192$$

$$p(5 \text{ A and } 2 \text{ B and } 0 \text{ C}) = \frac{7!}{5!\,2!\,0!} \left(\frac{1}{3}\right)^5 \left(\frac{1}{3}\right)^2 \left(\frac{1}{3}\right)^0 = \frac{21}{2187} = 0.0096$$

$$p(5 \text{ A and } 0 \text{ B and } 2 \text{ C}) = \frac{7!}{5!\,0!\,2!} \left(\frac{1}{3}\right)^5 \left(\frac{1}{3}\right)^0 \left(\frac{1}{3}\right)^2 = \frac{21}{2187} = 0.0096$$

$$p(6 \text{ A and } 1 \text{ B and } 0 \text{ C}) = \frac{7!}{6!\,1!\,0!} \left(\frac{1}{3}\right)^6 \left(\frac{1}{3}\right)^1 \left(\frac{1}{3}\right)^0 = \frac{7}{2187} = 0.0032$$

$$p(6 \text{ A and } 0 \text{ B and } 1 \text{ C}) = \frac{7!}{6!\,0!\,1!} \left(\frac{1}{3}\right)^6 \left(\frac{1}{3}\right)^0 \left(\frac{1}{3}\right)^1 = \frac{7}{2187} = 0.0032$$

$$p(7 \text{ A and } 0 \text{ B and } 0 \text{ C}) = \frac{7!}{7!\,0!\,0!} \left(\frac{1}{3}\right)^7 \left(\frac{1}{3}\right)^0 \left(\frac{1}{3}\right)^0 = \frac{1}{2187} = 0.0005$$

$$\Sigma = \overline{0.0453}$$

The probability of observing five or more preferences for brand A is 0.0453. Similarly, the probability of observing five or more preferences for B is 0.0453, and for C is 0.0453. Thus, the exact probability of observing any outcome as extreme or more extreme than five preferences for one brand is $3(0.0453) = 0.136$.

An approximation to the exact probability is given by Pearson's chi-square statistic, where $O_A = 5$, $O_B = 1$, $O_C = 1$; and $E_A = np_A' = 7(\frac{1}{3}) = 2.333$, $E_B = 2.333$, $E_C = 2.333$. We will illustrate the procedure even though all the expected frequencies are below the minimum recommended, which is five. Substituting the frequencies in the formula $\chi^2 = \sum (O_j - E_j)^2 / E_j$, we obtain

$$\chi^2 = \frac{(5 - 2.333)^2}{2.333} + \frac{(1 - 2.333)^2}{2.333} + \frac{(1 - 2.333)^2}{2.333} = 4.572.$$

The probability associated with this statistic, which has two degrees of freedom, is approximately 0.10. For $n = 7$, the chi-square approximation to the exact multinomial probability isn't very good. For larger n's typically encountered in research, the approximation is quite satisfactory.

17.6-3 Special Computational Procedure for a 2 × 2 Contingency Table

A simpler procedure for computing Pearson's chi-square statistic can be used for a 2 × 2 contingency table. Consider the table shown, in which observed frequencies

a	b	$a + b$
c	d	$c + d$
$a + c$	$b + d$	

are denoted by the letters a, b, c, and d. The formula for Pearson's statistic using only observed cell frequencies is

$$\chi^2 = \frac{n(ad - bc)^2}{(a + b)(c + d)(a + c)(b + d)}.$$

For the data in Table 17.3-2 the formula yields

$$\chi^2 = \frac{200[(32)(64) - (8)(96)]^2}{(32 + 8)(96 + 64)(32 + 96)(8 + 64)} = 5.555,$$

which is identical to that obtained using the conventional formula. If desired, Yates' correction for continuity can be included in the formula:

$$\chi^2 = \frac{n(|ad - bc| - n/2)^2}{(a + b)(c + d)(a + c)(b + d)}.$$

18

Statistical Inference for Rank Data

18.1 Introduction to Assumption-Freer Tests
18.2 Mann-Whitney U Test for Two Independent Samples
  Computational Procedure for Small Samples
  Computational Procedures when One or Both n's Exceed 20
  Two Indexes of Efficiency
  Asymptotic Relative Efficiency of the Mann-Whitney U Test
  Review Exercises for Sections 18.1 and 18.2
18.3 Wilcoxon T Test for Dependent Samples
  Computational Procedures for Small Samples
  Computational Procedures when n Is Greater than 50
  Review Exercises for Section 18.3
18.4 Comparison of Parametric Tests and Assumption-Freer Tests for Rank Data
18.5 Summary

18.1 Introduction to Assumption-Freer Tests[1]

So far we have presented procedures for testing hypotheses about a variety of population parameters. In each case it was necessary to assume that the sampled population had a probability distribution of a particular form—usually normal or binomial. For many variables, such an assumption seems warranted. For others, however, we may not know the form of the underlying distribution and may be unwilling to make an assumption about it. The procedures we have presented also required other assumptions. For example, the t statistic for testing the hypothesis that $\mu_1 - \mu_2 = 0$ assumes the population variances are equal. To avoid having to make such assumptions, nonparametric and distribution-free tests have been developed that are assumption-freer—that is, require less stringent assumptions. Before describing them, let us review the three kinds of distributions about which assumptions are made: (1) the sampled population (for example, the population of the observation statistic X), (2) the sampling distribution of the descriptive statistic used in the test (for example, $\bar{X}$), and (3) the sampling distribution of the test statistic (for example, t). Nonparametric and distribution-free tests are assumption-freer with respect to the distribution of the sampled population. *A statistical test is **nonparametric** if it does not test a hypothesis concerning one of the parameters of the sampled population. It is **distribution-free** if it makes no assumptions about the precise form of the sampled population.* Most distribution-free tests, however, do assume that the sampled population is continuous. Although the distinction between nonparametric and distribution-free tests seems clear enough, in practice the distinction is frequently blurred. Consequently, many statisticians use the terms interchangeably. We will follow Ury's (1967) lead and denote both kinds of tests by the more descriptive label *assumption-freer tests*.

Several tests that have already been described are assumption-freer. The chi-square tests in Chapter 17 are appropriately called assumption-freer tests for frequency data. The two tests to be described in this chapter are assumption-freer tests for rank data.

Assumption-freer tests differ from parametric tests in a number of important respects. We will mention one difference now and defer a complete discussion until later. Parametric tests utilize the magnitude information contained in observation statistics, but assumption-freer tests ignore this information. Instead, they use either the frequency with which observation statistics occur, as in the case of chi-square, or their rank (ordinal position). One consequence of focusing on the categorical or ordinal information contained in observation statistics has already been mentioned: we can avoid having to make assumptions regarding the shape of the sampled population. But, as we will see, this freedom is bought at a price. Assumption-freer tests tend to be less efficient than parametric tests when the assumptions of the parametric tests are fulfilled. A happier consequence is that assumption-freer tests require less sophisticated measurement procedures. One of the simplest measurement procedures is ranking people or objects with respect to a characteristic. We observe, for example, that John is a better quarterback than Fred, who is better than Elmer; this piece of pie looks better than that one; or Julie is more resourceful than Dennis. It is convenient to use numbers to denote

---

[1] For a more detailed introduction see Bradley (in Kirk, 1972a). Sections of this chapter were influenced by Bradley's article.

rank order. For example, John is number one; Fred is two; and Elmer is three. However, the numbers don't reflect the magnitude of the difference in quarterbacking skill between John and Fred or whether the difference is the same as between Fred and Elmer. Presumably, measuring instruments could be devised that would assign numbers that reflect the magnitude of differences in quarterbacking, pie attractiveness, and resourcefulness. This has been done in the area of intellectual assessment; the measuring instrument is an intelligence test. But it is not easy to develop such instruments, which is why experimenters frequently resort to counting or ranking.

Two test statistics that utilize only the ordinal information contained in observation statistics are described in Sections 18.2 and 18.3. They are regarded as assumption-freer alternatives to the independent and dependent two-sample t tests for means. There are many other assumption-freer tests. The interested reader should refer to Bradley (1968) and Marascuilo and McSweeney (1977).

18.2 Mann–Whitney U Test for Two Independent Samples

The Mann–Whitney U test[2] is used to test the hypothesis that two population distributions are identical. It assumes the populations are continuous and random samples have been drawn from each. The test statistic is based on the ranks of observations rather than on their numerical values, and hence is appropriate for most data in the behavioral sciences and education. Because of the U test's modest assumptions, it is widely used as an assumption-freer alternative to the two-sample t test for independent samples.

Computational Procedure for Small Samples

The computational procedures described here can be used when both sample sizes are 20 or less; for larger samples, a normal approximation procedure illustrated later can be used. Suppose an experiment was performed to determine whether observing aggression affects amount of aggressive behavior. A sample of $n = 21$ girls was randomly assigned to one of two conditions: viewing a simulated television program containing numerous aggressive acts, the experimental condition, and viewing a program without aggression, the control condition. Following the television viewing, each girl was observed at play and her aggressive acts were counted.

For purposes of statistical analysis, the data for the girls in the experimental and control groups are treated as one sample. The scores, frequencies of aggressive acts for each girl, are ordered from the smallest to the largest. The first n integers are then substituted for the scores, with the smallest score receiving a rank of 1 and the largest a rank of n. The ranks associated with the experimental and control groups are then added separately. The Mann–Whitney test statistic, U, is based on the smaller of the

[2] The test was originally developed by Frank Wilcoxon in 1945 and called the Wilcoxon rank-sum test. Since then, various forms of the test have appeared—a form by Festinger in 1946, the Mann–Whitney form in 1947, and a form by White in 1952.

two sums of ranks. No assumptions regarding the shape of the population of scores is required, because the test statistic is not based on scores, but instead on their ranks. The null hypothesis tested is that the distribution of aggressive acts for girls in the experimental population is identical to that for girls in the control population. The statistical hypotheses for the experiment can be stated as follows.[3]

H_0: Population distributions for the experimental and control groups are identical.

H_1: Population distributions are not identical.

The 0.05 level of significance is adopted. The experimental data are given in Table 18.2-1, along with the computational procedure for the U test. To be significant at the 0.05 level, the value of U must be less than the critical value, $U_{\alpha/2;n_1,n_2}$, in Appendix Table D.10. For the data in Table 18.2-1, $U = 18.5$ and is less than $U_{0.05/2;10,11} = 26$. Hence, we can conclude that the two populations are not identical. Inspection of the data suggests that the girls who watched a television program containing aggression engaged in more aggressive acts than those who didn't.

The U test can also be used to test directional hypotheses if the population distributions are symmetrical. The test statistic U is computed as before. However, U must be less than the one-tailed critical value from Table D.10. In addition, the relative position of the sample distributions must be consistent with the alternative hypothesis. For example, if the alternative hypothesis states that the population distribution for the experimental group is displaced (shifted) above that for the control group, the sample distributions must exhibit a similar displacement.

Computational Procedures when One or Both n's Exceed 20

When one of the samples contains more than 20 scores, a normal approximation procedure can be used. The test statistic is

$$z = \frac{(U + c) - E(U)}{\sigma_U} = \frac{(U + c) - n_1 n_2/2}{\sqrt{(n_1 n_2/12)(n_1 + n_2 + 1)}},$$

where U is defined in Table 18.2-1 and n_1 and n_2 are the two sample sizes. The term c in the formula is a correction for continuity and is equal to 0.5. The decision rule for the z test is as follows: Reject H_0 if z falls in the critical region of the normal distribution; otherwise, don't reject H_0. Because of the way U is defined, this region is always in the lower tail of the normal distribution regardless of whether the test is one-tailed or two-tailed.

If two or more observations have the same value (tied scores), they are assigned the mean of the ranks they would have occupied. The denominator σ_U of the z statistic can be corrected for ties; the corrected formula is

$$\sigma_U = \sqrt{\frac{n_1 n_2}{12}(n_1 + n_2 + 1)\left[1 - \frac{\sum(t_i^3 - t_i)}{(n_1 + n_2)^3 - (n_1 + n_2)}\right]},$$

---

[3] If it can be assumed that the two population distributions are symmetrical, the Mann–Whitney U test is a test of the hypothesis that the population medians are equal.

Table 18.2-1. Computational Procedure for Mann–Whitney U Test

(i) Data: Samples are treated as one combined sample and the scores are ranked from 1 to n, with the smallest receiving a rank of 1. Two or more scores with the same value (tied scores) are assigned the mean of the ranks they would have received. For example, the two scores of 13 would have received ranks of 12 and 13; instead, they both receive the mean rank of 12.5.

Number of Aggressive Acts for Experimental Group	R_1	Number of Aggressive Acts for Control Group	R_2
2	3	8	8
19	17	1	2
13	12.5	0	1
9	9	10	10
17	15	20	18
18	16	5	6
24	21	11	11
15	14	7	7
22	20	3	4
21	19	13	12.5
		4	5
	$\sum R_1 = 146.5$		$\sum R_2 = 84.5$

Computational check:

$$\sum R_1 + \sum R_2 = \frac{n(n+1)}{2}$$

$$146.5 + 84.5 = \frac{21(21+1)}{2} = 231$$

(ii) Computation of U:

$$U = \text{Smaller of} \begin{bmatrix} n_1 n_2 + \dfrac{n_1(n_1+1)}{2} - \sum R_1 \\ n_1 n_2 + \dfrac{n_2(n_2+1)}{2} - \sum R_2 \end{bmatrix}$$

$$U = \text{Smaller of} \begin{bmatrix} (10)(11) + \dfrac{10(10+1)}{2} - 146.5 = 18.5 \\ (10)(11) + \dfrac{11(11+1)}{2} - 84.5 = 91.5 \end{bmatrix} = 18.5$$

To be significant at α level of significance, U must be less than the critical value $U_{\alpha/2;n_1,n_2}$, in Table D.10. This value is $U_{0.05/2;10,11} = 26$. Since $U < U_{\alpha/2;n_1,n_2}$, the null hypothesis is rejected.

where t_i is the number of tied observations in a particular set. The term $(t_i^3 - t_i)$ is computed for each set and then summed for the sets. If $n_1 + n_2$ is large and the number of ties is small, the correction can be ignored.

The normal distribution approximation will be illustrated using the data in Table 18.2-1. This is done for comparison purposes; the approximate test isn't recommended unless n_1 or n_2 exceeds 20. The test statistic is

$$z = \frac{(U + c) - n_1 n_2/2}{\sqrt{(n_1 n_2/12)(n_1 + n_2 + 1)\{1 - (\sum (t_i^3 - t_i)/[(n_1 + n_2)^3 - (n_1 + n_2)])\}}}$$

$$= \frac{(18.5 + 0.5) - (10)(11)/2}{\sqrt{[(10)(11)/12](10 + 11 + 1)\left\{1 - \frac{[(2)^3 - 2]}{[(10 + 11)^3 - (10 + 11)]}\right\}}}$$

$$= \frac{-36.00}{\sqrt{201.667(0.999)}} = -2.54.$$

The critical value of z for a two-tailed test at $\alpha = 0.05$ is -1.96. Since $z < -z_{0.05/2}$, the null hypothesis is rejected. The absolute value of the z test statistic is large enough to be significant at the 0.02 level. A similar conclusion would be reached using the small-sample exact test procedure. As expected, the correction for ties (0.999) had virtually no effect on the test.

We mentioned earlier that assumption-freer tests tend to be less efficient than parametric tests when the assumptions of the parametric tests are fulfilled. Two statistics for comparing the relative efficiency of different tests will now be described.

Two Indexes of Efficiency

We saw in Section 11.4 that power is determined by four factors: (1) level of significance, (2) sample size, (3) population dispersion, and (4) magnitude of the difference between the true and hypothesized parameters. Any desired power can be achieved for a given significance level and true alternative hypothesis by obtaining a sufficiently large sample. If one test statistic requires a smaller sample size to achieve a desired power than another statistic, it is said to be more efficient. *One index compares the efficiency of two test statistics when both are used to test the same null hypothesis at α significance level against the same alternative hypothesis. It is called **power efficiency** and is given by*

$$\text{Power efficiency} = \frac{100(n_S)}{n_L},$$

where n_L is the sample size required by test L to equal the power of test S based on n_S observations. Suppose, for example, that test S requires 40 subjects to reject H_0 in favor of H_1 at α significance level with power equal to 0.90, and test L requires 80 subjects. The power efficiency, PE, of test L relative to S is

$$\text{PE} = \frac{100(40)}{80} = 50\%.$$

The PE index has one serious disadvantage. Its value is determined by the particular values of α, power, H_0, H_1, and the specified sample size of the more efficient comparison test statistic. Mathematicians prefer an index called *asymptotic relative efficiency*, ARE, which doesn't depend on qualifying conditions that vary from one situation to the next. This index assumes the sampling distributions of the test statistics being compared are normal when their n's are infinitely large and the power of each test approaches one as the sample n's approach infinity. The index is independent of α but is computed for a hypothetical situation, where the qualifying conditions specify that the sample size is infinite, H_1 is essentially identical to H_0, and H_1 is one-tailed. The rationale underlying ARE is beyond the scope of this book.[4] ARE has the advantage of comparing the efficiencies of two test statistics under standard conditions. However, in interpreting a particular value of the index, one should keep in mind its unrealistic qualifying conditions—infinitely large samples and a negligible difference between H_0 and H_1. It turns out that when the two indexes PE and ARE are used to rank order various test statistics, the results are almost identical. The ARE index is preferred for comparing the efficiencies of more than two test statistics, because all comparisons are made under a standard set of conditions. Comparisons among power efficiencies are meaningful only if the qualifying conditions are held constant.

Asymptotic Relative Efficiency of the Mann–Whitney U Test

In general, when the assumptions of parametric tests are met, they are more efficient than assumption-freer tests. This is true for the two-sample t test for independent samples when compared with the Mann–Whitney U test. The two provide tests of the same hypothesis if observations are randomly sampled from a normal population, since in that case $\mu = Mdn_{Pop}$. The asymptotic relative efficiency of the U test in comparison with the t test is 95.5%. Thus, the U test is an excellent alternative to the t test and has two compensating advantages: it is applicable to rank data and assumes only that random samples have been obtained from continuous populations.

Review Exercises for Sections 18.1 and 18.2

*1. Recognizing that in practice the distinction between nonparametric and distribution-free tests is often blurred, indicate the principal differences between them. How do these tests differ from parametric tests?
2. Compare the assumptions of the Mann–Whitney U test with those for the two-sample t test for independent samples. What are the relative merits of the tests?
*3. The effect on exploratory behavior of administering at random intervals a noxious stimulus, electric shock, to gerbils during infancy was investigated. Animals were randomly assigned to the experimental condition, shock, or the control condition, nonshock. The dependent variable, duration in minutes of exploratory behavior during 1 day, was measured when the animals were 6 months old. For the data in the table, test the hypothesis that the experimental and control populations are identical. Let $\alpha = 0.05$.

[4] Bradley (1968, pp. 56–62) provides a good introduction.

Control Group	Experimental Group
40 42 25 30	32 26 30
33 31 34	16 24 20

4. A paired-associates learning task is one in which subjects are presented with stimulus–response paired items and must learn to give the second item in each pair when the first is presented. The effect on learning of having as the stimulus item a dirty word versus a neutral word was investigated. Subjects were randomly assigned to the conditions. For the recall scores listed in the table, test the hypothesis that the populations are identical. Let $\alpha = 0.05$.

Dirty Stimulus Item	Neutral Stimulus Item
12 14 16 13	11 10 7 5
10 18 19	8 13 9

*5. The effect on motor skill development of playing with educational toys for 6 months was investigated. The subjects were 4- and 5-year-olds. Half were randomly assigned to the play group; the remaining half didn't play with the toys. For the data in the table, test the hypothesis that the two populations are identical. Let $\alpha = 0.01$. Perform the test with and without a correction for ties.

Motor Skill Scores for Play Group	Motor Skill Scores for Control Group
28 22 19 15	27 18 17 13
26 21 18 14	25 18 17 10
25 20 18 12	23 18 16 8
24 19 17 11	21 17 16 7
23 19 16 9	20 17 16 6

6. The experiment described in Exercise 5 was repeated with 8- and 9-year-old children. For the data in the table, test the hypothesis that the two populations are identical. Let $\alpha = 0.05$. Perform the test with and without a correction for ties.

Motor Skill Scores for Play Group	Motor Skill Scores for Control Group
47 36 43 32	36 23 26 16
46 35 39 31	34 21 25 15
44 34 37 30	32 20 25 14
45 33 37 29	29 19 24 11
44 32 36 27	27 18 24 12

*7. What are the qualifying conditions associated with PE and ARE?

8. Terms to remember:
   a. Nonparametric tests
   b. Distribution-free tests
   c. Assumption-freer tests
   d. Power efficiency
   e. Asymptotic relative efficiency

18.3 Wilcoxon T Test for Dependent Samples

The Wilcoxon matched-pairs signed-ranks test is used to test the hypothesis that two population distributions are identical. It is appropriate for dependent samples. Such samples can result from (1) obtaining repeated measures on the same subjects, (2) using subjects matched on a variable that is known to be correlated with the dependent variable, (3) using identical twins or litter mates, or (4) obtaining pairs of subjects who are matched by mutual selection.[5] The Wilcoxon test assumes the populations are continuous and a random sample of paired elements has been obtained. Its ARE, in comparison with the two-sample t test for dependent samples, is 95.5%, when the latter test's assumptions are met. Thus, Wilcoxon's test is an excellent alternative to the t test.

The test statistic, denoted by T, is based on the rank of the absolute difference between paired observations rather than on the numerical value of the difference. Consequently, the test is appropriate for observations that reflect at least ordinal information.

Computational Procedures for Small Samples

The computational procedure described here can be used for samples containing 50 or fewer pairs of scores; for larger samples, a normal approximation procedure illustrated later can be used.

Suppose a test of assertiveness was administered to female college students and the scores were used to form 16 pairs of women matched on assertiveness. One woman in each pair was randomly assigned to participate in an assertiveness training group; the other member of the pair participated in a psychology seminar, the control condition. It was hypothesized that women in the assertiveness training group would become more assertive relative to those in the control group. The statistical hypotheses are the following:

H_0: Population distributions of assertiveness scores are identical for the two groups.

H_1: The population distribution of scores for women in the training group is displaced (shifted) above that for the control group.

The alternative hypothesis calls for a one-tailed test in which the sample distribution for the training group is displaced above that for the control group. The 0.05 level of

---

[5] Procedures for obtaining dependent samples are discussed in Section 13.5.

Table 18.3-1. Computational Procedure for Wilcoxon T Test

(i) Data:

Pair	Training Group	Control Group	Difference	Rank	Positive Signed Rank, R_+	Negative Signed Rank, R_-
1	34	32	2	6.5	6.5	
2	36	26	10	16.0	16.0	
3	31	28	3	8.5	8.5	
4	42	41	1	4.0	4.0	
5	47	47	0	1.5		-1.5^a
6	32	33	-1	4.0		-4.0
7	33	29	4	10.0	10.0	
8	39	41	-2	6.5		-6.5
9	31	26	5	11.0	11.0	
10	34	28	6	12.0	12.0	
11	32	29	3	8.5	8.5	
12	41	41	0	1.5	1.5	
13	35	28	7	13.0	13.0	
14	45	44	1	4.0	4.0	
15	31	23	8	14.0	14.0	
16	33	24	9	15.0	15.0	
	$\overline{\sum X_T = 576}$	$\overline{\sum X_C = 520}$			$\overline{\sum R_+ = 124.0}$	$\overline{\lvert \sum R_- \rvert = 12.0}$

Computational check:

$$\sum R_+ + \lvert \sum R_- \rvert = \frac{n(n+1)}{2}$$

$$124 + 12 = \frac{16(16+1)}{2} = 136$$

(ii) Test statistic:

$$T = \text{Smaller of } \sum R_+ \text{ and } \lvert \sum R_- \rvert = 12.0$$

To be significant at α level of significance, T must be less than the one-tailed critical value $T_{\alpha,n}$ in Appendix Table D.11, and the training group must be displaced above the control group. From Table D.11, $T_{0.05,16} = 35$. Since $T < T_{\alpha,n}$ and the training group is displaced above the control group, the null hypothesis is rejected.

[a] See text for explanation of why this zero difference is assigned a negative sign.

significance is adopted. The experimental data are given in Table 18.3-1, along with the computational procedure.

The computations required for Wilcoxon's T test are easy to perform. The magnitude of the difference between each pair of observations is determined. These differences are rank ordered in terms of their absolute size; that is, their sign is ignored. The smallest difference receives a rank of 1; the largest is ranked n. Finally, the sign (positive or negative) of the difference for each pair is attached to each. The test statistic T is the smaller of the sum of the positive ranks and the absolute value of negative ranks.

The presence of zero differences requires a computational adjustment. If the number of zero differences is even, each zero difference is assigned the average rank for the set and then half are arbitrarily given a positive sign and half a negative sign. If an odd number of zero differences occurs, one randomly selected difference is discarded and the procedure for an even number of zero differences is followed. When a score is discarded, the sample size, n, is reduced by one.

To be significant, T must be less than the one-tailed critical value $T_{\alpha,n}$ in Appendix Table D.11 and the training group must be displaced above the control group. Inspection of the data indicates the latter condition is satisfied. The value of the test statistic, T, in Table 18.3-1 is 12 and is less than $T_{0.05,16} = 35$. Hence, we can conclude that the two populations are not identical; the training group is displaced above the control group.

Computational Procedures when n Is Greater than 50

A normal approximation procedure can be used when a sample contains more than 50 pairs of scores. The test statistic is

$$z = \frac{(T + c) - E(T)}{\sigma_T} = \frac{(T + c) - n(n + 1)/4}{\sqrt{n(n + 1)(2n + 1)/24}},$$

where T is defined in Table 18.3-1 and n is the number of pairs of scores. The term c in the formula is a correction for continuity and is equal to 0.5. The decision rule for the z test is as follows: Reject H_0 if z is less than $-z_\alpha$; otherwise, don't reject H_0. Because of the way T is defined, the critical region will always be in the lower tail of the normal distribution whether the test is one-tailed or two-tailed.

If two or more ranks have the same value, they are assigned the mean of the ranks they would have occupied. The denominator σ_T of the z statistic can be corrected for ties; the corrected formula is

$$\sigma_T = \sqrt{\frac{n(n + 1)(2n + 1)}{24} - \frac{1}{48}\sum(t_i^3 - t_i)},$$

where t_i is the number of tied observations in a particular set. The term $(t_i^3 - t_i)$ is computed for each set and then summed for the sets. If n is large and the number of ties is small, the correction can be ignored.

The normal distribution approximation will be illustrated using the data in Table 18.3-1. The procedure provides a satisfactory approximation for $n \geq 8$. The test statistic is

$$z = \frac{(T + c) - n(n + 1)/4}{\sqrt{n(n + 1)(2n + 1)/24 - (1/48)\sum(t_i^3 - t_i)}}$$

$$= \frac{(12 + 0.5) - (16)(16 + 1)/4}{\sqrt{16(16 + 1)[(2)(16) + 1]/24 - (1/48)[(2^3 - 2) + (2^3 - 2) + (2^3 - 2)]}}$$

$$= \frac{-55.5}{\sqrt{374 - 0.375}} = \frac{-55.5}{19.33} = -2.87.$$

The critical value of z for a one-tailed test at $\alpha = 0.05$ is -1.64. Since $z < -z_{0.05}$ and the training group is displaced above the control group, the null hypothesis is rejected. The absolute value of the z test statistic is large enough to be significant at the 0.003 level. The table of critical values for T doesn't have significance levels beyond 0.005, but the observed T is small enough to have been significant at this level.

Review Exercises for Section 18.3

9. Compare the assumptions of the Wilcoxon T test with those for the two-sample t test for dependent samples. What are the relative merits of the tests?

*10. The effect on the sense of well-being of participation in a transactional analysis group was investigated. Subjects completed a questionnaire before and after participation in a group. For the data in the table, test the hypothesis that the two populations are identical. Let $\alpha = 0.05$. The higher the score, the higher the individual's sense of well-being.

Subject	Score before Participation	Score after Participation	Subject	Score before Participation	Score after Participation
1	50	56	8	36	40
2	46	50	9	35	34
3	45	50	10	35	35
4	43	48	11	34	34
5	40	44	12	34	32
6	37	40	13	33	33
7	36	38	14	31	31

11. An experiment was performed to determine the effects of sustained physical activity on hand steadiness. For the data in the table, test the hypothesis that the two populations are identical. Let $\alpha = 0.01$.

Subject	Steadiness before Activity	Steadiness after Activity	Subject	Steadiness before Activity	Steadiness after Activity
1	14	12	9	13	9
2	12	11	10	11	10
3	16	13	11	14	12
4	6	6	12	13	11
5	13	14	13	9	6
6	15	10	14	11	9
7	14	10	15	13	12
8	12	12			

*12. The absolute threshold for a 1000 Hertz tone was investigated under the effects of an hallucinogen, hashish, and of a placebo. The order of administration of the conditions was randomized independently for each subject. For the data in the table, test the hypothesis that the two populations are identical versus the alternative that the placebo population is displaced above the hallucinogen population. Let $\alpha = 0.05$. Scores are dB re. 0.0002 dyne/cm^2.

Subject	Hashish	Placebo	Subject	Hashish	Placebo
1	4	6	9	0	0
2	0	0	10	5	8
3	1	1	11	−1	−2
4	0	−1	12	6	7
5	0	1	13	2	2
6	1	2	14	1	1
7	3	3	15	4	5
8	−1	0	16	2	4

*13. It has been claimed that college students' grades improve following marriage. To test the hypothesis, the variables of college aptitude, sex, and size of high school attended were used to form matched pairs of students; one had been married for at least two semesters and the other was unmarried. For the data in the table, test the hypothesis that the populations are identical versus the alternative that the married population is displaced above the unmarried population. Use the approximate z test to analyze the data. Let $\alpha = 0.05$. Don't use the correction for ties.

Pair	Married	Unmarried	Pair	Married	Unmarried
1	3.7	3.8	19	3.1	3.1
2	3.4	3.2	20	3.8	3.6
3	2.9	3.1	21	3.6	3.3
4	2.9	2.7	22	3.6	3.8
5	3.0	2.8	23	3.5	3.1
6	2.2	2.3	24	3.1	3.0
7	3.1	2.7	25	2.8	2.9
8	3.2	2.7	26	2.6	2.8
9	3.4	3.4	27	2.5	2.3
10	3.5	3.4	28	3.6	3.4
11	3.3	2.6	29	3.5	3.0
12	2.7	3.1	30	3.4	3.5
13	3.2	3.1	31	3.3	3.6
14	1.8	2.3	32	3.2	3.0
15	3.4	3.0	33	3.5	2.2
16	3.9	3.4	34	3.4	2.6
17	3.3	3.2	35	2.0	2.4
18	3.2	2.6			

14. Use the approximate z test to analyze the data in Exercise 11. Use the correction for ties.
*15. Use the approximate z test to analyze the data in Exercise 12. Use the correction for ties.
16. Suppose the two-sample t test required 57 subjects to reject the nondirectional null hypothesis at the 0.05 level of significance with power equal to 0.80, and the Wilcoxon T test required 60 subjects. (a) What is the PE of the Wilcoxon test? (b) What are the qualifying conditions associated with your estimate?

18.4 Comparison of Parametric Tests and Assumption-Freer Tests for Rank Data

When assumption-freer tests first appeared they were regarded as no more than quick and dirty substitutes for parametric tests, since their power efficiency was thought to be inferior. Now we have a clearer understanding of the differences between the two kinds of tests, which mainly involve (1) their assumptions, (2) the level of mathematics necessary to understand their rationale, (3) their computational simplicity, and (4) the nature of the hypotheses they test. These differences are examined below.

Most parametric test statistics assume that (1) population elements are randomly sampled or randomly assigned, (2) the population is normally distributed, and (3) the null hypothesis is true. Of course, the null hypothesis is advanced provisionally in the hope that it can be rejected. A fourth assumption is required by some test statistics if the null hypothesis concerns two or more populations—that the population variances are equal.

Assumption-freer test statistics are much less demanding and hence can be used in situations for which parametric methods are not appropriate. Most assumption-freer procedures assume (1) random sampling or random assignment of population elements, (2) continuity of the sampled population, which implies that no two sample observations will have the same value (that is, no tied scores),[6] and (3) the null hypothesis is true. It is relatively easy to determine whether the assumptions of assumption-freer tests are satisfied. For example, a sampling procedure is under an experimenter's control; the experimenter knows whether it is random. And data can easily be checked for tied scores.

On the other hand, the parametric assumptions of normality and equal variances are more difficult to check, since in any practical situation the population is not available for examination. There are statistical tests that can be applied to sample data to test the assumptions. However, for the small samples typically used in the behavioral sciences and education, the tests may fail to detect departures from normality and equal variances.

If all the assumptions for parametric tests are met, these tests are more efficient than, or equally efficient as, their assumption-freer counterparts. If, however, the assumptions aren't met, parametric tests are only approximate. This means the probability of making a type I error is not α as specified by the experimenter. Fortunately, parametric tests are relatively insensitive to violation of the normality and equal variances assumptions.[7] Nevertheless, one may prefer to use an assumption-freer test for which the probability of making a type I error is known rather than to rely on an inexact parametric test.

The second major way in which parametric and assumption-freer tests differ is in the level of mathematics necessary to understand their rationale. The derivation of parametric tests involves mathematics beyond the training of most researchers in the behavioral sciences and education. Many assumption-freer tests, however, can be

---

[6] This follows, since the probability of randomly drawing the same value twice in a finite sample from a continuous population is zero. However, even if the population is continuous, the same value may occur twice in one's sample, because the measuring instrument is calibrated in discrete units.

[7] This point is developed in Sections 10.3 and 12.2.

derived using high school algebra and elementary probability and counting rules. This is a real plus, since most researchers want to understand the rationale for the procedures they use rather than having to accept their validity and appropriateness on faith.

Third, assumption-freer tests differ from parametric tests in being easier to apply. For example, the chi-square tests discussed in Chapter 17 used the simplest kind of measurement—counting the number of observations in categories—and a test statistic that was easy to compute. The tests in the present chapter also use a simple measuring procedure—ranking—and statistics that are easy to compute.

The fourth difference is in the nature of the hypothesis tested. Two-sample parametric methods, for example, test hypotheses about particular population parameters; most assumption-freer methods test hypotheses about equality of population distributions. As we have seen, populations can differ in a number of ways, such as central tendency, dispersion, skewness, and kurtosis. To test the hypothesis $\mu_1 - \mu_2 = 0$ using a t statistic, one must assume the populations have equal variances and are symmetrical and mesokurtic. Thus, to test a hypothesis about one population parameter, we must be willing to make assumptions about other parameters. Assumption-freer tests don't require such assumptions and, accordingly, are less specific in what they tell us.

18.5 Summary

Test statistics are often classified according to whether they are parametric, nonparametric, or distribution-free. The classification scheme is not entirely satisfactory because some tests fall into more than one category. A test is parametric if it tests a hypothesis concerning one of the parameters of the sampled population and if it requires stringent assumptions regarding the precise form of the sampled population; if not, it is nonparametric. A test is distribution-free if it makes no assumptions about the precise form of the sampled population. The Mann–Whitney U test and the Wilcoxon T test fit into both categories, since they don't require assumptions about the precise form of the sampled population nor do they test hypotheses about parameters of the sampled population. Certainly, a classification scheme is less useful if its categories are not mutually exclusive. Even the parametric–distribution-free distinction is not always clear. For example, many parametric tests that assume the sampled population is normally distributed are approximately distribution-free for very large samples. Little is gained by trying to distinguish between nonparametric and distribution-free tests; it is more accurate to label the two categories collectively as assumption-freer.

The Mann–Whitney U test is used to test the hypothesis that two population distributions are identical. It is an excellent alternative to the two-sample t test for independent samples, since its ARE is 95.5%. The Wilcoxon T test is often used in place of the two-sample t test for dependent samples when the assumptions of the latter aren't tenable. Although it involves less stringent assumptions, Wilcoxon's T is nearly as efficient as the two-sample t—its ARE is 95.5%. Like the U test, it tests the hypothesis that two population distributions are identical.

The major differences between assumption-freer and parametric tests can be summarized as follows: The assumption-freer methods based on ranks (1) require less stringent assumptions, (2) involve assumptions that are easier to verify, (3) are usually less powerful when the assumptions of corresponding parametric tests are satisfied, (4) are less complicated to compute, (5) require simpler mathematical procedures for their derivation and understanding, (6) usually test hypotheses about population distributions instead of parameters, and (7) utilize information regarding rank order instead of the numerical value of individual observations.

Appendix A
Review of Basic Mathematics

This appendix provides a brief review of selected arithmetic and algebraic concepts.[1] You have no doubt been exposed to this material in the past, but chances are you have forgotten some of it. If so, this review should help refresh your memory.

The following test is designed to appraise your knowledge of basic mathematics and help you pinpoint concepts that should be reviewed. Answers are given at the end of the test, along with references to relevant review sections.

A.1 Test of Mathematical Skills

Round the following numbers to three digits.

1. 2.576 _____
2. 100.4 _____
3. 1.645 _____
4. 2.328 _____
5. 15.35 _____
6. 16.25 _____

Perform the following basic operations.

7. $|-3| + |3|$ = _____
8. $-5 + 2$ = _____
9. $3 - 2 + 4 - 8$ = _____
10. $-6 - 3$ = _____
11. $5 - (-1)$ = _____
12. $-9 - (-4)$ = _____
13. $(-2)(-6)$ = _____
14. $10/(-2)$ = _____

---

[1] For a more detailed review see Baggaley (1969), Bradshaw (1969), Clark and Tarter (1968), Kearney (1970), and Walker (1951).

15. $0/6$ = _____ 16. $9/0$ = _____
17. $(a/b)(n/n)$ = _____ 18. $(a/b)^2$ = _____
19. $(2/5)(3/6)$ = _____ 20. $(\frac{3}{4})/2$ = _____
21. $3/(\frac{4}{2})$ = _____ 22. 2^0 = _____
23. $(X)(X^2)$ = _____ 24. $(X^2)^3$ = _____
25. 2^{-1} = $()/()$ 26. 3^{-2} = $()/()^{()}$
27. $3^2/3^4$ = $()^{()}$ 28. $3\sqrt{15}$ = $\sqrt{()}15$
29. Factor $X^2 - 2XY + Y^2$ = _____ 30. Factor $pn - p$ = _____
31. $3!$ = _____ 32. $0!$ = _____

Remove the parentheses and brackets.
33. $X_i - (\bar{X} + C)$ _____ 34. $(Y_i - Y_i') + (Y_i' - \bar{Y})$ _____
35. $nS_Y^2(1 - r^2)$ _____ 36. $[(X - \bar{X})/S]S' + \bar{X}'$ _____

Solve the equations and inequalities.
37. $3X - 6 = 12$ $X =$ _____ 38. $2X/3 = 6$ $X =$ _____
39. $\bar{X} = \sum X/n$ $\sum X =$ _____ 40. $z = (X - \bar{X})/S$ $X =$ _____
41. $y = a + bX$ $X =$ _____ 42. $S_{min} = R/\sqrt{2n}$ $R =$ _____
43. $S_{Y\cdot X} = S_Y\sqrt{1 - r^2}$ $r =$ _____ 44. $\chi^2 = [(n-1)\hat{\sigma}^2]/\sigma^2$ $\hat{\sigma}^2 =$ _____
45. $2X - 1 < 3$ $X <$ _____ 46. $-3 \le \dfrac{16-\mu}{7} \le 5$ _____ $\le \mu \le$ _____
47. $-z_{\alpha/2} \le \dfrac{\bar{X} - \mu}{\sigma_{\bar{X}}} \le z_{\alpha/2}$ _____ $\le \mu \le$ _____ 48. $[(n-1)\hat{\sigma}^2]/\sigma^2 \le \chi^2_{\alpha/2, v}$ $\sigma^2 \ge$ _____

The answers to the skills test follow. The numbers in parentheses refer to the review sections that discuss the principle involved.

Principles discussed in Section A.2.
1. 2.58 (1) 2. 100 (2) 3. 1.64 (3)
4. 2.33 (1) 5. 15.4 (3) 6. 16.2 (3)

Principles discussed in Section A.3.
7. 6 (5b, 6a) 8. -3 (6b) 9. -3 (6c)
10. -9 (7) 11. 6 (7) 12. -5 (7)
13. 12 (8) 14. -5 (8) 15. 0 (9c)
16. Undefined (9d) 17. a/b (10c-iii) 18. a^2/b^2 (10c-iv)
19. $\frac{6}{30}$ (10c-i) 20. $\frac{3}{8}$ (10d-i) 21. $\frac{6}{4}$ (10d-ii)
22. 1 (11b) 23. X^3 (11c-i) 24. X^6 (11c-ii)
25. $\frac{1}{2}$ or 0.5 (11d-i) 26. $1/3^2$ (11d-i) 27. 3^{-2} (11d-iv)
28. $\sqrt{9(15)}$ (12c) 29. $(X - Y)^2$ (13) 30. $p(n - 1)$ (13)
31. 6 (14a) 32. 1 (14b)

Principles discussed in Section A.4.
33. $X_i - \bar{X} - C$ (17b) 34. $Y_i - Y_i' + Y_i' - \bar{Y}$ (17b)
35. $nS_Y^2 - nS_Y^2 r^2$ (17c) 36. $XS'/S - \bar{X}S'/S + \bar{X}'$ (17c)

Principles discussed in Section A.5.
37. 6 (20, 21, 22) 38. 9 (20, 22)
39. $n\bar{X}$ (20, 22) 40. $zS + \bar{X}$ (20, 21, 22)
41. $(y - a)/b$ (20, 21, 22) 42. $S_{min}\sqrt{2n}$ (20, 22)
43. $\sqrt{1 - S_{Y\cdot X}^2/S_Y^2}$ (20, 21, 22) 44. $\chi^2\sigma^2/(n-1)$ (20, 22)

Review of Basic Mathematics

Principles discussed in Section A.6.
45. 2 (25)
46. $-19 \leq \mu \leq 37$ (25, 26)
47. $\bar{X} - z_{\alpha/2}\sigma_{\bar{X}} \leq \mu \leq \bar{X} + z_{\alpha/2}\sigma_{\bar{X}}$ (25, 26)
48. $[(n-1)\hat{\sigma}^2]/\chi^2_{\alpha/2,v}$ (25, 27)

A.2 Rounding Numbers

The number of significant digits in a number (all digits except zero when it is used only to position the decimal point) should reflect the precision of a measurement. Therefore, numbers should be rounded to give a correct impression of the measurement precision actually achieved. Rounding involves dropping digits if they are to the right of the decimal or replacing them by zero if they are to the left of the decimal.

1. When the digit to be dropped is greater than five or is a five with nonzero digits to the right, the digit to the left of it is increased by one.
   Examples 246.36 rounded to four significant digits becomes 246.4
   386 rounded to two significant digits becomes 390
   0.0068 rounded to one significant digit becomes 0.007
   6.51 rounded to one significant digit becomes 7
2. When the digit to be dropped is less than five, no change is made in the digit to the left of it.
   Examples 246.31 rounded to four significant digits becomes 246.3
   384 rounded to two significant digits becomes 380
   0.0063 rounded to one significant digit becomes 0.006
3. When the digit to be dropped is five or is five with only zeros to the right, the digit to the left of five is increased by one if it is odd and is not changed if it is even.
   Examples 75 rounded to one significant digit becomes 80
   935.35 rounded to four significant digits becomes 935.4
   674.5 rounded to three significant digits becomes 674
   912.5 rounded to three significant digits becomes 912
4. If the final result of a computation is to be rounded to s digits to the right of the decimal, at least $s + 1$ digits to the right should be retained in intermediate computational steps.
   Example $(115 - 100)/(14/\sqrt{26}) = 5/(14/5.099) = 5/2.746 = 1.82$

A.3 Basic Operations

5. Numbers
   a. A signed number, for example, 4, -2, 9, -11, indicates (1) direction and (2) size. The sign, $+$ or $-$, indicates the direction of movement away from a starting point, zero. Zero has no direction. The number part of the signed number indicates the extent or size of movement away from zero.

Example The size and direction of movement for -2 and 4 are shown here.

b. The absolute value of a number, denoted by $|\ |$, indicates size but not direction. The absolute value is always positive for nonzero numbers. The general rule is

$$|a| = \begin{cases} a & \text{if } a \geq 0 \\ -a & \text{if } a < 0. \end{cases}$$

The second part of the rule appears contradictory but isn't. If a is less than zero, it is a negative number and a minus sign in front of a negative number makes it positive.

Examples $|3| = 3; |-3| = 3; |0| = 0; |-1.96| = 1.96;$
$|X| = X$ if X is a positive number, and $|X| = -X$ if X is a negative number.

6. Addition
   a. Two numbers of like sign: add the absolute values of the numbers and attach the common sign to the sum.
   Examples $3 + 2 = 5; -3 + (-2) = -5$
   b. Two numbers of unlike sign: determine the difference between their absolute values and attach the sign of the larger number.
   Examples $3 + (-2) = 1; -3 + 2 = -1$
   c. More than two numbers with unlike sign: add the absolute values of the positive numbers as in Rule 6a and do the same for the negative numbers; then determine the difference between their absolute values and attach the sign of the larger of the two as in Rule 6b.
   Example $3 + 2 + (-5) + (-3) + (-1) = 5 + (-9) = -4$
7. Subtraction
   To subtract one number from another, change the sign of the number to be subtracted and proceed as in addition.
   Examples $\quad 3 - (2) \quad = \quad 3 + (-2) = \quad 1$
   $\qquad\qquad 3 - (-2) = \quad 3 + 2 \quad = \quad 5$
   $\qquad\quad -3 - (2) \quad = -3 + (-2) = -5$
   $\qquad\quad -3 - (-2) = -3 + 2 \quad = -1$
8. Multiplication and division
   Multiplying two numbers together results in another number called a *product*; dividing one number by another results in another number called a *quotient*. When two numbers have like signs, their product and quotient are positive; when they have unlike signs, their product and quotient are negative.

Examples $(6)(3) = 18$ $\quad\quad\quad 6/3 = 2$
$(6)(-3) = -18$ $\quad\quad 6/-3 = -2$
$(-6)(3) = -18$ $\quad\quad -6/3 = -2$
$(-6)(-3) = 18$ $\quad\quad -6/-3 = 2$

9. Operations with zero
   a. If zero is added to or subtracted from any number, the result is the number itself.
      Examples $3 + 0 = 3; 9 - 0 = 9$
   b. The product of zero and any other number is equal to zero.
      Examples $(3)(0) = 0; (2)(3)(0) = 0$
   c. $0/a = 0$ for all nonzero values of a.
      Examples $\frac{0}{3} = 0; \frac{0}{7} = 0$
   d. The use of zero as a divisor results in a fraction that cannot be evaluated.
      Example $\frac{5}{0}$ is undefined

10. Fractions
    a. A fraction, for example, 6/2 or a/b, is the result of dividing one number or expression by another. The upper part of the fraction is called the *numerator*; the lower part (the divisor) is called the *denominator*.
    b. Addition and subtraction
       i. To add or subtract fractions with the same denominator, perform the indicated operation on the numerator and leave the denominator unchanged.

       Examples $\dfrac{a}{c} + \dfrac{b}{c} = \dfrac{a+b}{c}; \dfrac{a}{c} - \dfrac{b}{c} = \dfrac{a-b}{c}; \dfrac{2}{3} + \dfrac{4}{3} = \dfrac{6}{3}; \dfrac{2}{3} - \dfrac{4}{3} = \dfrac{-2}{3}$

       ii. To add or subtract fractions with different denominators, find a common denominator, change all fractions accordingly, and proceed as above. Some multiple of all the original denominators is selected as the common denominator; each new numerator is formed by multiplying the original numerator by the number of times the original denominator divides into the common denominator.

       Examples $\dfrac{a}{b} + \dfrac{c}{d} = \dfrac{ad}{bd} + \dfrac{bc}{bd} = \dfrac{ad + bc}{bd}; \dfrac{1}{2} + \dfrac{2}{3} = \dfrac{3}{6} + \dfrac{4}{6} = \dfrac{7}{6}$

       iii. In general, if the same quantity is added to or subtracted from both the numerator and denominator, the value of the fraction is changed.

       Examples $\dfrac{\chi_1^2 + n}{\chi_2^2 + n} \neq \dfrac{\chi_1^2}{\chi_2^2}$ unless $\chi_1^2 = \chi_2^2$ or $n = 0$;

       $\dfrac{3}{4} \neq \dfrac{3 + 2}{4 + 2} = \dfrac{5}{6}$

    c. Multiplication
       i. To multiply two or more fractions, multiply their numerators together and their denominators together to obtain, respectively, the numerator and the denominator of the product.

       Examples $\left(\dfrac{a}{b}\right)\left(\dfrac{c}{d}\right) = \dfrac{ac}{bd}; \left(\dfrac{2}{3}\right)\left(\dfrac{3}{4}\right) = \dfrac{(2)(3)}{(3)(4)} = \dfrac{6}{12}$

ii. Multiplying anything by one leaves its value unchanged.

Examples $\dfrac{a}{b}(1) = \dfrac{a}{b}; \dfrac{3}{2}(1) = \dfrac{3}{2}$

iii. Multiplying both the numerator and denominator of a fraction by the same quantity other than zero does not change its value.

Examples $\left(\dfrac{\chi_1^2}{\chi_2^2}\right)\left(\dfrac{n}{n}\right) = \dfrac{(\chi_1^2)(n)}{(\chi_2^2)(n)} = \dfrac{\chi_1^2}{\chi_2^2}; \left(\dfrac{3}{4}\right)\left(\dfrac{2}{2}\right) = \dfrac{(3)(2)}{(4)(2)} = \dfrac{3}{4}$

iv. In general, squaring a fraction or taking its square root changes its value.

Examples $\dfrac{a}{b} \neq \left(\dfrac{a}{b}\right)^2$ and $\dfrac{a}{b} \neq \dfrac{\sqrt{a}}{\sqrt{b}}$ unless $a = b$;

$\dfrac{4}{9} \neq \left(\dfrac{4}{9}\right)^2 = \dfrac{16}{81}$ and $\dfrac{4}{9} \neq \dfrac{\sqrt{4}}{\sqrt{9}} = \dfrac{2}{3}$

d. Division
i. To divide a fraction by a quantity, multiply the denominator of the fraction by that quantity.

Examples $\dfrac{a/b}{c} = \dfrac{a}{bc}; \dfrac{2/3}{4} = \dfrac{2}{(3)(4)} = \dfrac{2}{12}$

ii. To divide a quantity by a fraction, invert the fraction and multiply.

Examples $\dfrac{a}{b/c} = a\dfrac{c}{b} = \dfrac{ac}{b}; \dfrac{2}{3/4} = 2\left(\dfrac{4}{3}\right) = \dfrac{8}{3}$

iii. To divide a fraction by another fraction, invert the second fraction and multiply.

Examples $\dfrac{a/b}{c/d} = \left(\dfrac{a}{b}\right)\left(\dfrac{d}{c}\right) = \dfrac{ad}{bc}; \dfrac{2/3}{4/5} = \left(\dfrac{2}{3}\right)\left(\dfrac{5}{4}\right) = \dfrac{10}{12}$

iv. Any quantity (other than zero) divided by itself equals one.
Examples $a/a = 1$ if $a \neq 0; 3/3 = 1$

v. Dividing any quantity by one leaves its value unchanged.

Examples $\dfrac{a/b}{1} = \dfrac{a}{b}; \dfrac{3/2}{1} = \dfrac{3}{2}$

11. Exponents
a. The number of times a number, the *base*, is multiplied by itself is denoted by a superscript, the *exponent*.
Examples $a^1 = a; a^2 = (a)(a), a^3 = (a)(a)(a)$
b. A number not equal to zero with a zero exponent, a^0, is defined as one.
Examples $a^0 = 1$ for $a \neq 0; 2^0 = 1; 5^0 = 1$

c. Laws of positive exponents
   i. $a^n a^m = a^{n+m}$ Example $(2)^2(2)^3 = 2^{2+3} = 2^5$
   ii. $(a^n)^m = a^{nm}$ Example $(2^2)^3 = 2^{(2)(3)} = 2^6$
   iii. $a^n b^n = (ab)^n$ Example $(2)^2(4)^2 = [(2)(4)]^2 = 8^2$
d. Laws of negative exponents and mixed exponents
   i. $a^{-1} = \dfrac{1}{a}$ Example $2^{-1} = \dfrac{1}{2}$

   $a^{-2} = \dfrac{1}{a^2}$ Example $2^{-2} = \dfrac{1}{2^2}$

   $a^{-n} = \dfrac{1}{a^n}$

   ii. $(a^n)^{-1} = (a^{-1})^n = a^{(-1)(n)} = a^{-n}$
   Example $(2^3)^{-1} = 2^{(3)(-1)} = 2^{-3}$

   iii. $\left(\dfrac{a}{b}\right)^n = a^n \left(\dfrac{1}{b}\right)^n = a^n(b^{-1})^n = a^n b^{-n} = \dfrac{a^n}{b^n}$

   Example $\left(\dfrac{2}{4}\right)^2 = \dfrac{(2)^2}{(4)^2}$

   iv. $\dfrac{a^n}{a^m} = a^n a^{-m} = a^{n-m}$

   Example $\dfrac{(2)^2}{(2)^3} = 2^{2-3} = 2^{-1}$

e. Do not confuse exponents and coefficients. When multiplying terms with coefficients and exponents, add exponents and multiply coefficients.
   Examples $(3X^2)(2X^3) = 6X^5$; $(3X)(5X^2) = 15X^3$

12. Radicals
    a. A radical is used to indicate a specific root of a quantity as, for example, in the expression, $b = \sqrt[n]{a}$; b is said to be the nth root of a, $\sqrt{}$ is a radical sign, and a is called the radicand. If n is not specified, it is understood to equal 2, in which case $b = \sqrt{a}$ is called the square root of a.
    b. To multiply two radicals if both radicands are positive, multiply their radicands under one radical. Similarly, to divide one radical by another, divide (under one radical) the radicand of the first by the radicand of the second.
       Examples $\sqrt{a}\sqrt{b} = \sqrt{ab}$; $\sqrt{2}\sqrt{3} = \sqrt{6}$
       $\sqrt{a}/\sqrt{b} = \sqrt{a/b}$; $\sqrt{2}/\sqrt{3} = \sqrt{2/3}$
    c. To multiply or divide a radical of the nth order by a number a not equal to zero, place the number raised to the nth power under the radical and multiply or divide the radicand by it.
       Examples $a\sqrt{b} = \sqrt{a^2 b}$; $2\sqrt{3} = \sqrt{4(3)}$; $2\sqrt[3]{3} = \sqrt[3]{(2)^3(3)}$
       $\sqrt{a}/c = \sqrt{a/c^2}$; $\sqrt{3}/2 = \sqrt{3/4}$

13. Factoring
    Factoring an expression consists of dividing it into smaller terms or expressions that, when multiplied, will yield the original expression.

Examples $kn - k = k(n - 1)$
$$\sqrt{S_Y^2 - r^2 S_Y^2} = S_Y\sqrt{1 - r^2}$$
$$X^2 - 2\bar{X}X + \bar{X}^2 = (X - \bar{X})(X - \bar{X}) = (X - \bar{X})^2$$

14. Factorials
    a. The product of the first n natural numbers (positive integers) is called n *factorial* and is denoted by $n!$, which equals $n(n - 1)(n - 2) \cdots (3)(2)(1)$.
       Example $4! = 4(4 - 1)(4 - 2)(4 - 3) = (4)(3)(2)(1) = 24$
    b. For $n = 0$, $0!$ is defined as one.

A.4 Order of Performing Operations

15. The order in which numbers are added does not affect the result.
    Examples $a + b = b + a$; $2 + 3 + 5 = 3 + 2 + 5 = 10$;
    $X_1 + X_2 = X_2 + X_1$; $(2 + 3) + 5 = 2 + (3 + 5) = 10$
    $a + b = b + a$ illustrates the *commutative law of addition*.
    $(a + b) + c = a + (b + c)$ illustrates the *associative law of addition*.
16. The order in which numbers are multiplied does not affect the result.
    Examples $ab = ba$; $(2)(3)(5) = (3)(2)(5) = 30$; $X_1 X_2 = X_2 X_1$;
    $[(2)(3)]5 = 2[(3)(5)] = 30$
    $ab = ba$ illustrates the *commutative law of multiplication*.
    $(ab)c = a(bc)$ illustrates the *associative law of multiplication*.
17. Parentheses ( ), braces { }, brackets [ ], and the radical $\sqrt{}$ indicate that the enclosed expression is to be treated as a single number. The bar of a fraction has a similar effect; the numerator and denominator are treated as single numbers.
    Examples $10(16 - 14) = 10(2) = 20$; $(3 - 1)4.21 = (2)4.21 = 8.42$;
    $$\frac{2 + 4}{3 - 1} = \frac{6}{2} = 3$$

    a. When a plus sign precedes parentheses, the parentheses may be removed without changing the signs of terms within the parentheses.
       Examples $a + (b + c) = a + b + c$; $2 + (3 + 5) = 2 + 3 + 5 = 10$
    b. If a minus sign precedes parentheses and the parentheses are removed, the sign of every term within the parentheses must be changed.
       Examples $(a + b) - (c + d) = a + b - c - d$; $2 - (3 + 5) = 2 - 3 - 5 = -6$
    c. When a quantity within parentheses is to be multiplied by a number, each term within the parentheses must be so multiplied.
       Examples $a(b + c) = ab + ac$;
       $10(16 - 14) = 10(16) - 10(14) = 160 - 140 = 20$;
       $$10\left(\frac{1}{20} + \frac{6}{24}\right) = \frac{10}{20} + \frac{60}{24} = \frac{1}{2} + \frac{5}{2} = 3$$

    $a(b + c) = ab + ac$ illustrates the *distributive law*.
18. Unless specifically altered, for example by parentheses, the order for performing operations is as follows: first, exponentiation (raising a number to a power); next, multiplication and division; and last, addition and subtraction.

$$\text{Examples} \quad Mdn = 44.5 + 10\left(\frac{30/2 - 14}{3}\right) = 44.5 + 10\left(\frac{15 - 14}{3}\right)$$

$$= 44.5 + 10\left(\frac{1}{3}\right) = 44.5 + 3.33 = 47.8;$$

$$S = \sqrt{\frac{10(32) - (16)^2}{(10)^2}} = \sqrt{\frac{320 - 256}{100}} = \sqrt{\frac{64}{100}} = 0.8;$$

$$F = \frac{320/(3-1)}{1350/[3(10)-3]} = \frac{320/2}{1350/27} = \frac{160}{50} = 3.2$$

A.5 Equations

19. An equation is a statement asserting that what is on the left side of the equal sign is equal to what is on the right.
    Examples $2 + 4 = 6$ is an example of an *arithmetic equation*, since it contains only numbers.
    $2X - 5 = 7$ is an example of an *algebraic equation*, since it contains a symbol.
    a. An equation in which both sides have the same numerical value or one that is true for all values of the variables employed is called an *identity*.
        Examples $2 + 4 = 6$; $3a + 4a = 7a$
        One that is true only when certain values are substituted for variables is called a *conditional equation*.
        Examples $2X = 6$; $3X - 4 = X + 6$
    b. To solve for an unknown in an algebraic equation find the set of values (called *roots*) that, when substituted for the unknown, makes the two sides of the equation numerically equal. For $2X - 5 = 7$, the root is 6, since $2(6) - 5 = 7$—that is, $7 = 7$.
    c. To find the roots of an equation, perform a series of manipulations that place the unknown alone on the left side (see Rules 20–23).
        Example $2X - 5 = 7$
        $\qquad 2X - 5 + 5 = 7 + 5$ Adding 5 to both sides (see Rule 21)
        $\qquad \dfrac{2X}{2} = \dfrac{7+5}{2}$ Dividing both sides by 2 (see Rule 22)
        $\qquad X = \dfrac{12}{2}$
        $\qquad X = 6$
20. An operation performed on one side of an equation must also be performed on the other. The condition of equality is not affected by
    a. adding the same quantity to both sides,
    b. subtracting the same quantity from both sides,
    c. multiplying both sides by the same quantity,
    d. dividing both sides by the same nonzero quantity,

e. raising both sides to the same power if both sides have the same sign,
f. taking the same root of both sides if both sides have the same sign.
21. Any term on one side of an equation may be transposed to the other side by changing its sign. In essence, the term to be transposed is either added or subtracted from both sides of the equation.

Example Solving for a:
$$a + b = c$$
$$a + b - b = c - b$$
$$a = c - b$$

22. A quantity that multiplies one side of an equation may be transposed to divide the other side or vice versa. In essence, both sides of the equation are subjected to the same operation—either multiplication or division.

Examples Solving for a:

$$ab = c$$
$$\frac{ab}{b} = \frac{c}{b}$$
$$a = \frac{b}{c}$$

Solving for X:

$$\left(\frac{X - \bar{X}}{S}\right) S' + \bar{X}' = z$$
$$\left(\frac{X - \bar{X}}{S}\right) S' = z - \bar{X}'$$
$$\frac{X - \bar{X}}{S} = \frac{z - \bar{X}'}{S'}$$
$$X - \bar{X} = \left(\frac{z - \bar{X}'}{S'}\right) S$$
$$X = \left(\frac{z - \bar{X}'}{S'}\right) S + \bar{X}$$

23. When each side of an equation consists of a fraction, the fractions can be removed by cross-multiplying as follows.

Example
$$\frac{a}{b} = \frac{c}{d}$$
$$ad = bc$$

A.6 Inequalities

24. Two or more expressions connected by one of the ordering symbols $<$, $>$, $\leq$, or $\geq$ is an *inequality*.

Examples $a < b$; $-1.96 \leq t \leq 1.96$; $|z| > z_{0.05}$

25. The solutions to an inequality are not affected by the following operations:
    a. Adding the same quantity to both sides.
    b. Subtracting the same quantity from both sides.
    c. Multiplying both sides by a positive quantity.
    d. Dividing both sides by a positive quantity.
    Example $4X + 2 \geq 10$
    $$4X \geq 10 - 2$$
    $$X \geq 2$$

26. If both sides of an inequality are multiplied or divided by the same negative number, a new inequality is formed, with direction opposite to that of the original.

   Examples
   $-a < b$ $\qquad$ $-3 < 2$
   $-1(-a < b)$ $\qquad$ $-1(-3 < 2)$
   $a > -b$ $\qquad\qquad$ $3 > -2$

   Find μ if $-3 \leq \dfrac{2-\mu}{5} \leq 8$

   $-3(5) \leq 2 - \mu \leq 8(5)$ $\qquad$ Multiply each member by 5
   $-15 - 2 \leq -\mu \leq 40 - 2$ $\qquad$ Subtract 2 from each member
   $-1(-17 \leq -\mu \leq 38)$ $\qquad$ Multiply by -1
   $17 \geq \mu \geq -38$
   $-38 \leq \mu \leq 17$ $\qquad$ Rearrange terms

27. Taking the reciprocal of, or inverting, all expressions in an inequality results in a new inequality with direction opposite to that of the original.

   Examples $a > b$ implies that $\dfrac{1}{a} < \dfrac{1}{b}$ if $ab > 0$

   $$\chi^2_{1-\alpha/2,v} \leq \frac{(n-1)\hat{\sigma}^2}{\sigma^2} \leq \chi^2_{\alpha/2,v}$$

   $$\frac{1}{\chi^2_{1-\alpha/2,v}} \geq \frac{\sigma^2}{(n-1)\hat{\sigma}^2} \geq \frac{1}{\chi^2_{\alpha/2,v}}$$

Appendix A

Appendix B
Glossary of Symbols

Mathematical Symbols

Symbol	Example	Meaning[1]
$+$	$X + Y$	X and Y are added (A.3)
$-$	$X - Y$	Y is subtracted from X (A.3)
$(\;)(\;)$, $\times$	$(X)(Y)$, $X \times Y$, or XY	X and Y are multiplied (A.3)
$/$, $\div$	X/Y or $X \div Y$	X is divided by Y (A.3)
$=$	$X = Y$	X is equal to Y (A.5)
$\neq$	$X \neq Y$	X is not equal to Y
$\simeq$, $\cong$	$X \simeq Y$, $X \cong Y$	X is approximately equal to Y
$>$	$X > Y$	X is greater than Y (A.6)
$\geq$	$X \geq Y$	X is greater than or equal to Y (A.6)
$<$	$X < Y$	X is less than Y (A.6)
$\leq$	$X \leq Y$	X is less than or equal to Y (A.6)
$\leq \leq$	$W \leq X \leq Y$	X is greater than or equal to W and less than or equal to Y (A.6)
$\sum_{i=1}^{n}$	$\sum_{i=1}^{n} X_i$	Sum of X_i, letting i equal $1-n$ (3.3, 3.9)
$\ldots$	$1, 2, 3, \ldots, 6$	Continue the pattern—that is, 1, 2, 3, 4, 5, 6 in this case
$|\;|$	$|X|$	Absolute value of X; for $X = 0$, $|X| = 0$, and for $X \neq 0$, $|X|$ is equal to the positive member of the couple X, $-X$ (A.3)
$\sqrt{\;}$	$\sqrt{X}$	Square root of X (A.3)
$!$	$n!$	n factorial, $n(n-1)(n-2) \cdots (3)(2)(1)$ (A.3)

[1] The letter/number in parentheses refers to the section in which the symbol is discussed.

Greek Letters

Symbol	Meaning
α (alpha)	Probability of a type I error (11.4); significance level (11.2)
β (beta)	Probability of a type II error (11.4)
β_j	Treatment effect in the jth population (16.2)
δ_0 (delta)	Value of the difference between two population parameters specified by the null hypothesis (13.2)
η^2 (eta)	Correlation ratio (5.6)
θ (theta)	Population parameter (10.3)
μ (mu)	Population mean (3.3)
μ_0	Value of the population mean specified by the null hypothesis (11.2)
$\mu_{\bar{X}}$	Mean of means (10.3)
ν (nu)	Degrees of freedom (12.2)
π (pi)	Ratio of the circumference of a circle to the diameter; approximately 3.1416 (10.1)
ρ (rho)	Pearson product-moment population correlation parameter (5.2)
$\sum$ (sigma)	Summation (3.3)
$\sum R_1$ and $\sum R_2$	Sum of ranks, respectively, for variables 1 and 2 (18.2)
$\sum R_+$ and $\sum R_-$	Sum of ranks, respectively, for positive and negative ranks (18.3)
σ (sigma)	Population standard deviation (4.2); a caret over the symbol, $\hat{\sigma}$, denotes an estimate of σ (4.2)
σ_{Mdn}	Standard error of a median (10.3)
σ_S	Standard error of a standard deviation (10.3)
$\sigma_{\bar{X}}$	Standard error of a mean (10.3); a caret over the symbol, $\hat{\sigma}_{\bar{X}}$, denotes an estimate of $\sigma_{\bar{X}}$ (12.2)
$\hat{\sigma}_{\bar{X}_D}$	Estimate of the standard error of the mean of difference scores (13.5)
$\sigma_{\bar{X}_1 - \bar{X}_2}$	Standard error of the difference between means (13.2)
$\hat{\sigma}_{Y \cdot X}$	Estimate of the population standard error of estimate (6.3)
σ^2	Population variance (4.2); a caret over the symbol, $\hat{\sigma}^2$, denotes an estimate of σ^2 (4.2)
σ_0^2	Value of the population variance specified by the null hypothesis (12.3)
σ_e^2	Population error variance (16.2)
σ_{est}^2	Estimate of the population variance (10.5)
σ_U	Standard error of U (18.2)
σ_T	Standard error of T (18.3)
$\hat{\sigma}_{Pooled}^2$, $\hat{\sigma}_{Pd}^2$	Weighted estimate of the population variance (13.4)
ϕ' (phi)	Cramer's measure of association (17.3); a caret over the symbol, $\hat{\phi}'$, denotes an estimate of ϕ' (17.3)
χ^2 (chi)	Chi-square random variable (12.3); Pearson's chi-square random variable (17.2)

Symbol	Meaning
$\chi^2_{\alpha,v}$ and $\chi^2_{1-\alpha,v}$	Values that cut off, respectively, the upper and lower α regions of the sampling distribution of χ^2 for v degrees of freedom (12.3)
ψ (psi)	Contrast among population means (16.5); a caret over the symbol, $\hat{\psi}$, denotes an estimate of ψ (16.5)

English Letters

Symbol	Meaning
ARE	Asymptotic relative efficiency (18.2)
$a_{Y \cdot X}$	Y intercept of a line (6.2)
$b_{Y \cdot X}$	Sample coefficient of linear regression of Y on X (6.2)
c_j	Coefficient of a linear contrast (16.5)
$_nC_r$	Combination of n objects taken r at a time (8.1)
c	A constant (3.9); number of qualitative categories (4.2); correction for continuity (18.2)
Cum f	Cumulative frequency (2.2)
Cum % f	Cumulative percentage frequency (2.2)
Cum prop. f	Cumulative proportionate frequency (2.2)
D	Index of dispersion (4.2)
D_i	Difference between scores for the ith pair of elements (13.5)
DP	Number of distinguishable pairs (4.2)
DP'	Number of distinguishable pairs when observations are equally divided among categories (4.2)
d	Effect size (11.4)
df	Degrees of freedom (12.2); also denoted by v
E	Index of forecasting efficiency (6.4)
E_i	ith event (7.2); sample point for the ith event (7.2)
E_j	Expected frequency in the jth category (17.1)
$E(\chi_v^2)$	Expected value of a chi-square random variable (12.3)
$E(\theta)$	Expected value of an estimator (10.3)
$E(\hat{\sigma}^2)$	Expected value of $\hat{\sigma}^2$ (10.5)
$E(S^2)$	Expected value of S^2 (10.5)
$E(T)$	Expected value of T in the Wilcoxon T statistic (18.3)
$E(MS_{BG})$ and $E(MS_{WG})$	Expected value of MS_{BG} and MS_{WG} (16.2)
$E(U)$	Expected value of U in the Mann–Whitney U statistic (18.2)
$E(X)$	Expected value of X (9.2)
e	Base of the system of natural logarithms; approximately 2.7183 (10.1); number of distribution parameters estimated (16.2)
e_i	Prediction error for the ith element; difference between the observed and predicted scores (6.2)
F	F random variable (14.1)
$F_{\alpha;v_1,v_2}$ and $F_{1-\alpha;v_1,v_2}$	Values that cut off, respectively, the upper and lower α regions of the sampling distribution of F (14.1)

Symbol	Meaning
f_a and f_b	Number of scores, respectively, above the real upper limit of a class interval and below the real lower limit of a class interval (3.4)
f, f_i, f_j	Frequency of a measurement or event class (2.2)
f_i	Number of scores in the class interval containing a particular statistic (3.4)
$\%f$	Percentage frequency (2.2)
H_0	Null hypothesis (11.1)
H_1	Alternative hypothesis (11.1)
i	Class interval size (2.2); index of summation (3.3)
k	Number of class intervals in a frequency distribution (3.3); number of standard deviation units in Tchebycheff's theorem (4.5); coefficient of alienation (6.4); number of levels of a treatment (16.2)
k^2	Coefficient of nondetermination (5.4)
Kur	Kurtosis index (4.6)
Mdn	Sample median (3.4)
Mdn_{Pop}	Population median (18.2)
Mo	Sample mode (3.2)
MS_{BG}	Between-groups mean square (16.2)
MS_{WG}	Within-groups mean square (16.2)
n	Number of observations in a sample (2.2); number of trials in a binomial experiment (9.3)
$n!$	n factorial (8.1)
n_A	Number of equally likely events favoring A (7.1)
n_L and n_S	Sample size of tests L and S (18.2)
n_S	Total number of equally likely events (7.1)
O_j	Number of observations in the jth category (17.1)
$_nP_n$	Permutation of n objects taken n at a time (8.1)
$_nP_r$	Permutation of n objects taken r at a time (8.1)
$_nP_{r_1, r_2, \ldots, r_k}$	Permutation of n objects in which $r_1, r_2, \ldots, r_k$ are alike (8.1)
PE	Power efficiency (18.2)
p	Probability of a success (9.3); population proportion (12.4); a caret over the symbol, $\hat{p}$, denotes an estimate of p (12.4); number of levels of treatment A (16.6)
p_j	Population proportion of observations in the jth category (17.1); a caret over the symbol, $\hat{p}_j$, denotes an estimate of p_j (17.2)
p_j'	Value of the population proportion specified by the null hypothesis (17.2)
p_0	Value of the population proportion specified by the null hypothesis (12.4)
$\hat{p}_{Pooled}, \hat{p}_{Pd}$	Weighted mean of two population proportion estimates (14.2)

Symbol	Meaning
$p(A)$	Probability of event A (7.1)
$p(A\|B)$	Conditional probability of A given B (7.3)
$p(A \text{ and } B)$	Probability of the intersection of events A and B (7.3)
$p(A \text{ or } B)$	Probability of the union of events A and B (7.3)
$p(D_j\|E_i)$ and $p(E_i\|D_j)$	Conditional probability of D_j given E_i and E_i given D_j (8.2)
$p(E_i)$	Probability of event E_i (7.2)
$p(X = r)$	Probability that X is equal to r (9.2)
Prop f	Proportionate frequency (2.2)
Q	Semi-interquartile range (4.2)
Q_1, Q_2, Q_3	First, second, and third quartile points, respectively (4.2)
q	Probability of a failure (9.3); number of levels of treatment B (16.6)
q_0	Value of the population proportion specified by the null hypothesis (12.4)
R	Sample range (4.2)
R_X and R_Y	Ranks, respectively, on variables X and Y (5.7)
r	Sample product-moment correlation coefficient (5.2); number of successes in a binomial experiment (9.3)
r_s	Spearman rank correlation coefficient for a sample (5.7)
r^2	Sample coefficient of determination (5.4)
S	Sample standard deviation (4.2); value of Scheffé's test statistic (16.5)
S^2	Sample variance (4.2)
S'	Desired sample standard deviation (10.2)
S_c	Standard deviation corrected for the grouping error (4.2)
S_{cX}	Standard deviation that has been altered by multiplying each score by a constant (4.8)
Sk	Skewness index (4.6)
S_X and S_Y	Sample standard deviations, respectively, of X and Y (6.2)
S_X^2 and S_Y^2	Sample variances, respectively, of X and Y (5.4)
S_{X+c}	Standard deviation that has been altered by adding a constant to each score (4.8)
S_{XY}	Sample covariance (5.3)
$S_{Y \cdot X}$	Standard error of estimate for predicting Y from X (6.3)
SS_{BG}	Between-groups sum of squares (16.2)
SS_{Total}	Total sum of squares (16.2)
SS_{WG}	Within-groups sum of squares (16.2)
$S_{\alpha; v_1, v_2}$	Critical value of Scheffé's S statistic (16.5)
T	Wilcoxon's T random variable (18.3)
$T_{\alpha,n}$ and $T_{\alpha/2,n}$	Values that cut off, respectively, the α and $\alpha/2$ regions of the sampling distributions of Wilcoxon's T for n pairs of observations (18.3)
t	Student's t random variable (12.1)
t_i	Number of tied observations in a set (18.2)

Symbol	Meaning
$t_{\alpha,v}$, and $t_{\alpha/2,v}$	Values that cut off, respectively, the upper α and $\alpha/2$ regions of the sampling distribution of t for v degrees of freedom (12.2)
U	Mann-Whitney U random variable (18.2)
$U_{\alpha;n_1,n_2}$ and $U_{\alpha/2;n_1,n_2}$	Values that cut off, respectively, the α and $\alpha/2$ regions of the sampling distribution of U for n_1 and n_2 observations in samples 1 and 2 (18.2)
V	A variable (3.9)
$\text{Var}(t)$	Variance of Student's t random variable (12.2)
$\text{Var}(\bar{X})$	Variance of sample means (10.3)
$\text{Var}(\chi_v^2)$	Variance of a chi-square random variable (12.3)
$\text{Var}(Mdn)$	Variance of sample medians (10.3)
$\text{Var}(\hat{\sigma}^2)$	Variance of $\hat{\sigma}^2$ (10.5)
$\text{Var}(\sigma_{est}^2)$	Variance of σ_{est}^2 (10.5)
W	A variable (3.9)
X	A score (1.4); the independent variable in an experiment (5.1)
X_i	A score for the ith measurement or event class (3.3)
X_j	Midpoint of the jth class interval (3.3)
X_{ij}	A score for the ith subject in the jth treatment condition (16.2)
X_{ll}	Real lower limit of a score (4.2) or class interval (3.4)
X_{ul}	Real upper limit of a score (4.2) or class interval (3.4)
$\bar{X}$	Sample arithmetic mean (3.3)
$\bar{X}_D$	Mean of difference scores (13.5)
$\bar{X}'$	Desired sample mean (10.2)
$\bar{X}_{..}$	Arithmetic mean of all scores (16.2)
$\bar{X}_{i..}$	Arithmetic mean of the ith level of treatment A (16.6)
$\bar{X}_{.j}$	Arithmetic mean of scores in the jth treatment condition (16.2)
$\bar{X}_{.j.}$	Arithmetic mean of the jth level of treatment B (16.6)
$\bar{X}_{ij.}$	Arithmetic mean of the ith and jth treatment combination (16.6)
$\bar{X}_W$	Weighted mean of two or more sample means (3.7)
Y	A score (1.4); the dependent variable in an experiment (5.1)
Y_i'	A predicted Y score (6.2)
$\bar{Y}$	Sample arithmetic mean (3.3)
Z'	Fisher's transformation of r (12.5)
Z_0'	Value of population Z' specified by the null hypothesis (12.5)
Z_{Pop}'	Fisher's transformation of ρ (14.6)
z	Standard score (10.1); z random variable (11.2)
z_α and $z_{\alpha/2}$	Values that cut off, respectively, the upper α and $\alpha/2$ regions of the sampling distribution of z (11.4)

Appendix C
Answers to Starred Exercises

Chapter 1

3. (a) white female students in this university, a female student, a measure of career ambivalence
4. one and all but one of the population elements
7. (a) R (b) NR (c) R (d) NR
11. (a) D (b) U (c) C (d) O
13. (a) D (b) U (c) D (d) O
15. (a) ratio (b) nominal (c) between ordinal and interval (d) ordinal
17. nominal: one-to-one substitution, ordinal: monotonic, interval: positive linear, ratio: multiplication by a positive constant
18. (a) interval (b) between ordinal and interval
21. A change of three points on the measurement scale represents the same empirical change from 62 to 65 as from 68 to 71 and from 71 to 74.
23. (a) national statistics, probability theory, and experimental statistics
25. The modern era uses exact inductive procedures appropriate for both large and small samples; the previous period relied on large-sample procedures.

Chapter 2

1.

X	f	X	f
13	1	6	2
12	0	5	4
11	0	4	6
10	1	3	3
9	1	2	2
8	0	1	1
7	1	0	1

$n = 23$

3.

X	f	X	f
25	1	13	1
24	0	12	2
23	0	11	4
22	1	10	4
21	1	9	5
20	0	8	8
19	0	7	10
18	0	6	9
17	1	5	7
16	0	4	3
15	2	3	1
14	0		

$n = 60$

4. see list in Section 2.2
5. (a) 49.5–54.5, 5 (b) 73.5–74.5, 1 (d) 17.95–19.95, 2
6. (a) 16, 3, 21–23 (b) 11, 15, 105–119 (c) 17, 10, 90–99

7.

X	f	X	f
95–99	1	60–64	2
90–94	1	55–59	4
85–89	4	50–54	5
80–84	5	45–49	3
75–79	4	40–44	2
70–74	2	35–39	1
65–69	1		

$n = 35$

9. (a)

X	f	X	f
210–212	1	192–194	3
207–209	0	189–191	5
204–206	1	186–188	6
201–203	2	183–185	4
198–200	2	180–182	1
195–197	2		

$n = 27$

X	f	X	f
210–211	1	194–195	1
208–209	0	192–193	3
206–207	0	190–191	3
204–205	1	188–189	4
202–203	2	186–187	4
200–201	0	184–185	4
198–199	2	182–183	0
196–197	1	180–181	1

$n = 27$

(b) The distribution with $i = 3$ is better than the one with $i = 2$ because the number of scores is relatively small and because it provides a clearer picture of the distribution of scores.

11.

X	Prop. f	X	Prop. f
95–99	0.03	60–64	0.06
90–94	0.03	55–59	0.11
85–89	0.11	50–54	0.14
80–84	0.14	45–49	0.09
75–79	0.11	40–44	0.06
70–74	0.06	35–39	0.03
65–69	0.03		

15.

X	f	Cum f	X	f	Cum f
16	1	32	10	7	17
15	0	31	9	4	10
14	1	31	8	3	6
13	2	30	7	2	3
12	5	28	6	1	1
11	6	23			

$n = 32$

16.

X	f	Prop. f	Cum prop. f	X	f	Prop. f	Cum prop. f
13	1	0.04	0.98	6	2	0.09	0.82
12	0	0.00	0.94	5	4	0.17	0.73
11	0	0.00	0.94	4	6	0.26	0.56
10	1	0.04	0.94	3	3	0.13	0.30
9	1	0.04	0.90	2	2	0.09	0.17
8	0	0.00	0.86	1	1	0.04	0.08
7	1	0.04	0.86	0	1	0.04	0.04

$n = 23$

19.

X	f
Graduate	1
Senior	8
Junior	10
Sophomore	6
Freshman	4

$n = 29$

26.

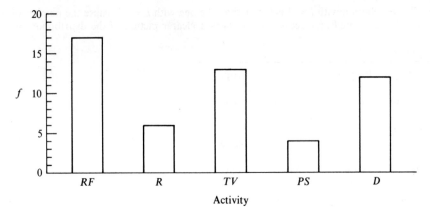

Favorite leisure time activity of college students

31.

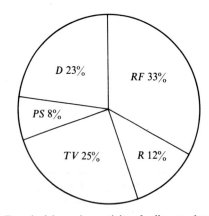

Favorite leisure time activity of college students

36.

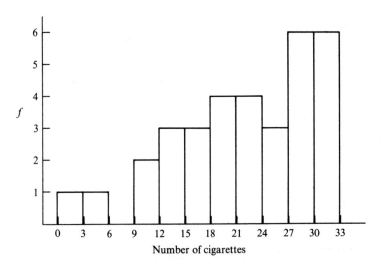

Histogram for number of cigarettes smoked per day by mothers whose first babies were stillborn

41. (a) 22 (c) 132.5 (e) 22

42.

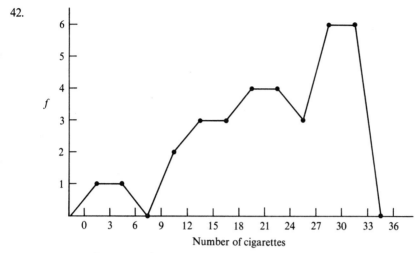

Frequency polygon for number of cigarettes smoked by mothers whose first babies were stillborn

46. (a)

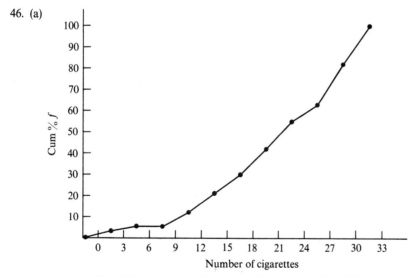

Cumulative percentage frequency polygon for number of cigarettes smoked per day by mothers whose first babies were stillborn

(b) 22.5
49. A cumulative polygon will have an "S" shape if there are more scores in the middle of the corresponding frequency distribution than at the extremes.
51. (a) false (b) true

52. (a)

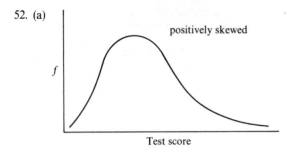

(b)

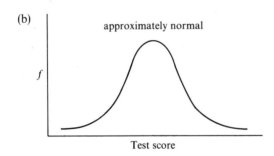

54.

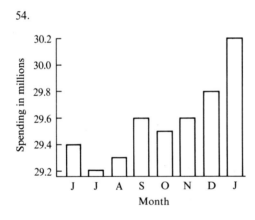
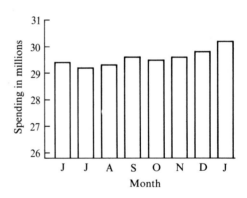

Chapter 3

1. (a) $Mo = S$ (b) ordered qualitative variable
4. (a) The distribution is bimodal; the two maximum values occur at 1 and 3.
   (b) unordered qualitative variable
6. (a) X score for element one (c) mean of population one (h) X score for element i
7. (a) $X_1 + X_2 + \cdots + X_n$ (c) $(Z_1 + Z_2 + Z_4)/3$ (d) $(f_1 X_1 + f_2 X_2 + \cdots + f_k X_k)/n$
8. $\bar{X} = 4.59$
12. (a) 12 (e) 5
13. $\bar{X} = 9.5$
18. (a) 9 (c) 18 (e) 3.25 (f) 3.75
19. $Mdn = 7.5 + 2[(18 - 8)/10] = 9.5$

21. X_{ul} = real upper limit of class interval containing the median, i = class interval size, n = number of scores, $\sum f_a$ = number of scores above X_{ul}, f_i = number of scores in the class interval containing the median
24. (a) mean, quantitative data and the distribution is relatively symmetrical
    (c) mean, quantitative data and the distribution is relatively symmetrical
25. (a) $\bar{X} = 1, Mdn = 2, Mo = 3$ (e) only Mo is appropriate
29. (a) positively skewed (e) positively skewed
30. (a) 43.33

Chapter 4

1. (a) 16 (b) 16
5. (a) $Mdn = 27.33, Q = 1.33$ (b) $P_{10} = 24.70, P_{90} = 30.3$
6. (a) 49.9% (b) The distribution is asymmetrical.
8. (b) 87.50
10. $\bar{X} = 2.467, S = 1.258$
11. $\sqrt{[15(115) - (37)^2]/(15)^2} = 1.258$
14. $S_c = 5.737$
15. (a) Mo = moderately desire career, $D = 0.97$
17. (a) Approximately 68% of the scores fall between 85 and 115.
19. (a) Compute the mean and standard deviation because the variable is quantitative and the distribution is relatively symmetrical.
    (c) Compute the median and semi-interquartile range because the distribution is skewed.
    (d) Compute the mode and index of dispersion because the variable is qualitative.
20. (a) 84.14%
21. (a) 3.37 (b) 20
23. $S_{min} = 3, S_{max} = 15$
25. (a) correct
27. (a) 67.35%
28. (a) Minimum is 0%; probable percent is 68.27%.
    (b) Minimum is 75%; probable percent is 95.45%.
29. (a) $Sk = -1.02$; distribution is negatively skewed.

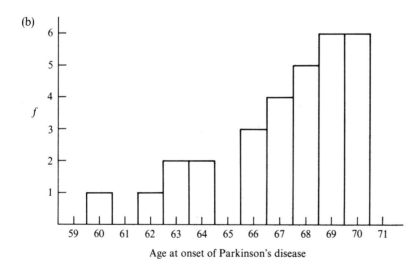

(c) yes
31. Since $Kur = 2.96$, the data do not support the prediction.
32. $Kur = 3.23$; distribution is leptokurtic.

Chapter 5

1. (a)

Test A	Test B										(b) linear
	30	31	32	33	34	35	36	37	38	39	
33									1		
32									1		
31							1				
30								1			
29						1	1				
28					1	1					
27				1	1	1					
26			1		2	1	1				
25				1	2		1				
24			1		1						
23			1								
22					1						
21			1								
20	1										

3. The term means that extreme scores for one variable, that is scores that differ considerably from their mean, are likely to be paired with less extreme scores for the other variable.
6. (a) 1   (e) -0.9 7. (a) positive (b) positive
9. 0.86   10. 0.90
12. (a) 18, quadrants 2 and 4, variables are positively related
    (b) 0, quadrants 1, 2, 3 and 4, variables are not related
15. $1, -1, 0$

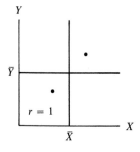 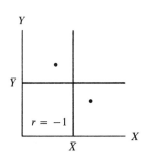 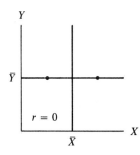

16. $S^2_{X+Y} = \dfrac{\sum[(X+Y) - (\bar{X}+\bar{Y})]^2}{n} = \dfrac{\sum[(X-\bar{X}) + (Y-\bar{Y})]^2}{n}$

$= \dfrac{\sum[(X-\bar{X})^2 + (Y-\bar{Y})^2 + 2\sum(X-\bar{X})(Y-\bar{Y})]}{n}$

$= \dfrac{\sum(X-\bar{X})^2}{n} + \dfrac{\sum(Y-\bar{Y})^2}{n} + \dfrac{2\sum(X-\bar{X})(Y-\bar{Y})}{n},$

but $\sum (X - \bar{X})^2/n = S_X^2$, $\sum (Y - \bar{Y})^2/n = S_Y^2$, and $2rS_XS_Y = 2\sum (X - \bar{X})(Y - \bar{Y})/n$. Hence $S_{X+Y}^2 = S_X^2 + S_Y^2 + 2rS_XS_Y$.

20. (a) The proportion of variance in English grades explained by variation in grades in bowling is 0.048; the proportion that is not explained is 0.952.

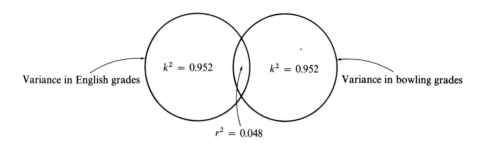

22. (a) Correct interpretation
    (c) This interpretation is incorrect since it uses an arbitrary descriptive label, medium, to denote r's between 0.30 and 0.69.
24. The mean IQ should increase because of regression toward the mean.
27. (a) r underestimates the magnitude of the relationship.

(d) r overestimates the magnitude of the relationship.

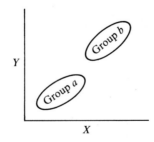

34. 0.76
38. (a) strictly monotonic (b) nonmonotonic

Chapter 6

1. 0.5

3. (a)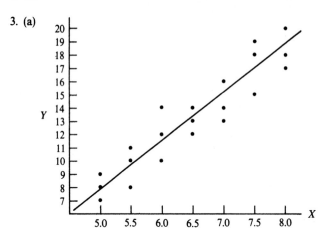

   Data appear to be linearly related.
   (b) $a_{Y \cdot X} = -9.356$
   $b_{Y \cdot X} = 3.476$,
   $r = 0.93$
   (c) Estimate based on the regression equation is 11.5; estimate based on the line of best fit is 11.3.

5. They are best fitting lines in the sense that they minimize the sum of the squared prediction errors.

8. $b_{Y \cdot X} = \dfrac{\sum(X - \bar{X})(Y - \bar{Y})}{\sum(X - \bar{X})^2} = \dfrac{\sum(XY - X\bar{Y} - Y\bar{X} + \bar{X}\bar{Y})}{\sum(X^2 - 2X\bar{X} + \bar{X}^2)}$

$= \dfrac{\sum XY - \bar{Y}\sum X - \bar{X}\sum Y + n\bar{X}\bar{Y}}{\sum X^2 - 2\bar{X}\sum X + n\bar{X}^2} = \dfrac{\sum XY - \dfrac{\sum Y}{n}\sum X - \dfrac{\sum X}{n}\sum Y + n\dfrac{\sum X \sum Y}{n \cdot n}}{\sum X^2 - 2\dfrac{\sum X}{n}\sum X + n\dfrac{(\sum X)^2}{n^2}}$

$= \dfrac{\sum XY - 2\dfrac{\sum X \sum Y}{n} + \dfrac{\sum X \sum Y}{n}}{\sum X^2 - 2\dfrac{(\sum X)^2}{n} + \dfrac{(\sum X)^2}{n}} = \dfrac{\sum XY - \dfrac{\sum X \sum Y}{n}}{\sum X^2 - \dfrac{(\sum X)^2}{n}}$

$= \dfrac{n\sum XY - \sum X \sum Y}{n\sum X^2 - (\sum X)^2}$

10. (a) $Y'_i = \bar{Y} - b\bar{X} + bX_i$
    $\sum Y'_i = \sum(\bar{Y} - b\bar{X} + bX_i)$
    $= n\bar{Y} - nb\bar{X} + b\sum X_i$
    $= n\dfrac{\sum Y_i}{n} - nb\dfrac{\sum X_i}{n} + b\sum X_i$
    $= \sum Y_i$

    (b) r is equal to either -1 or 1.

12. (a) 1.42 (b) $13.2 \pm 1.42 = 11.78$ and 14.62 (c) For $r = 0$, $S_{Y \cdot X} = 2.97$; for $r = 1$, $S_{Y \cdot X} = 0$
   (d) $k = 0.48$ and $E = 0.52$. The standard error of estimate is 48% of the size it would be if the correlation were zero. By taking the value of X_i into account in predicting Y_i, we have reduced the magnitude of the prediction error by 52%.

Chapter 7

3. (a) $\frac{1}{2}$
   (b) It is assumed that there are six possible outcomes, that they are equally likely, and that the number of outcomes favoring an odd number is three.
5. (a) 0.52 (b) 0.5.

6. (a)

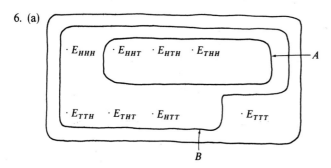

   (b) $p(A) = \frac{3}{8}$ (c) $p(B) = \frac{7}{8}$
9. (a) 3/10 (b) 6/10,000
13. (a) $\frac{4}{52} = \frac{1}{13}$ (b) $\frac{13}{52}$ (c) $\frac{16}{52} = \frac{4}{13}$ (d) $\frac{26}{52} = \frac{1}{2}$ (e) $\frac{12}{52} = \frac{3}{13}$ (f) $\frac{16}{52} = \frac{4}{13}$
14. (c) $\frac{2}{3}$
15. (a) 0.48 (b) $(0.4)(0.2) = 0.08$ (c) $0.80 + 0.60 - 0.48 = 0.92$

16. (a)

	F	Not F	
D	p(D and F) = 0.002	p(D and Not F) = 0.098	p(D) = 0.10
O	p(O and F) = 0.002	p(O and Not F) = 0.898	p(O) = 0.90

   $p(F) = 0.004$ $p(\text{Not } F) = 0.996$

   (b) 0.002

17. $p(\text{Not } D \text{ and } M) = 0.147$ 21. (a) 0.4

Chapter 8

1. (a) 8 (b) 1296 (c) 12   4. 24   6. 5040
10. (a) $(9)(8) = 72$ (b) $9[8!/(8 - 2)!] = 504$ (c) $9\{8!/[2!(8 - 2)!]\} = 252$
13. (a) 0.02274 (b) 0.8707

Answers to Starred Exercises

395

15. (a)

	E_1	E_2	
D_1	$p(D_1 \text{ and } E_1)$ = 0.45	$p(D_1 \text{ and } E_2)$ = 0.15	$p(D_1) = 0.60$
D_2	$p(D_2 \text{ and } E_1)$ = 0.05	$p(D_2 \text{ and } E_2)$ = 0.35	$p(D_2) = 0.40$
	$p(E_1) = 0.50$	$p(E_2) = 0.50$	

(b) $p(E_1|D_1) = 0.75$ (c) $p(E_1|D_2) = 0.125$

Chapter 9

1. Identify the population, decide whether to sample with or without replacement, and select elements using a random sampling procedure.
3. (a) 2118760
4. (a) $8!/[4!(8 - 4)!] = 70$ (b) 0.0143 (c) $(8)(8)(8)(8) = 4096$

9. (a)

r	$p(X = r)$
0	0.0625
1	0.2500
2	0.3750
3	0.2500
4	0.0625

10. (a) 0.93 (b) 0.07 (c) 0.77 (d) 1.24 (e) 0.88
13. $22
19. (a) $p(X = 0) = 0.16, p(X = 1) = 0.48, p(X = 2) = 0.36$

(b)
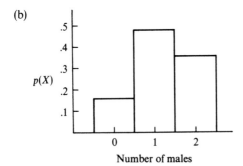

(c) $E(X) = 1.2, \sigma = 0.693$
21. (a) 0.149 (b) $E(X) = 3, \sigma = 1.449$
23. (a) 250 (b) 25 (c) 500
25. 0.071
26. 0.091

Chapter 10

2. (a) normal (b) normal (c)  (d) normal

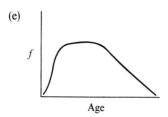 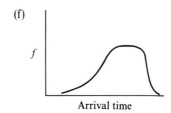

(e) Age   (f) Arrival time

3. (a) 2 (b) -1.6 (c) -1
4. (a) 0.0668 (b) 0.0228 (c) 0.4987 (e) 0.1359
5. (a) 0.6826 (b) 0.8990 (c) 0.9500   6. (a) 190 (b) 120 (c) 212
7. (a) 0 (b) 1.645 (c) 0.25
9. The number eligible will decrease by 9.94%.
12. (a) 0.25 (b) 0.13   13. (a) 0.06 (b) 0.92 (c) 0.04
17. Performance on test 1 was best and performance on test 2 was worst.
19. Yes.   20. (a) 88 (b) 42 (c) 8.4

22. (a)

Sample No.	Sample Values	$\bar{X}$	Sample No.	Sample Values	$\bar{X}$
1	0, 0	0.0	9	2, 0	1.0
2	0, 1	0.5	10	2, 1	1.5
3	0, 2	1.0	11	2, 2	2.0
4	0, 3	1.5	12	2, 3	2.5
5	1, 0	0.5	13	3, 0	1.5
6	1, 1	1.0	14	3, 1	2.0
7	1, 2	1.5	15	3, 2	2.5
8	1, 3	2.0	16	3, 3	3.0

(b) $\mu = 48/32 = 1.5$, $\sigma_{\bar{X}} = 1.11803/\sqrt{2} = 0.7906$ (c) $\mu_{\bar{X}} = 24/16 = 1.5$, $\sigma_{\bar{X}} = 0.7906$
26. (a) 7.07 (b) 5.00
27. $(115 - 120)/(10/\sqrt{25}) = -2.5$, $p = 0.0062$
28. $(182 - 165)/(15/\sqrt{9}) = 3.4$, $p = 0.0003$

Chapter 11

1. (a) scientific hypothesis (b) scientific hypothesis (c) not a scientific hypothesis
   (d) scientific hypothesis
3. (a) yes (b) no (c) yes (d) no (e) yes (f) no (g) yes (h) no (i) yes (j) yes

7. $H_0: \mu \le 8, H_1: \mu > 8$ 11. $H_0: \mu \le 15$
15. When σ^2 is known and the population distribution of X is normal or n is very large
17. (a) $z = (55.1667 - 50)/(10/\sqrt{30}) = 2.83$
    (b) The H_0 is rejected, and this supports the truth of the scientific hypothesis.

20. (a)

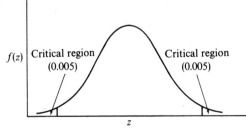

22. (a) H_0 is exact; H_1 is inexact.
    (b) H_0 is inexact; H_1 is inexact.
    (c) H_0 is inexact; H_1 is inexact.
24. (a) 0.67 (b) $[1.645 - (-1.28)]^2/[(118 - 115)^2/(15)^2] = 214$

Chapter 12

3. (a) $t = (88.75 - 100)/(10.1620/\sqrt{16}) = -4.428$. Reject H_0; this year's classes are below average in intelligence.
   (b) $n = 10$
8. $\chi^2 = (16 - 1)(10.1620)^2/(15)^2 = 6.884$. Reject H_0 since $\chi^2 < \chi^2_{1-.10/2,15}$.
11. n should be at least 26 since in that case both $n(0.2)$ and $n(1 - 0.2)$ are greater than 5.
12. $z = [(12 - \frac{1}{2}) - 10]/\sqrt{20(0.5)(0.5)} = 0.67$. The hypothesis that $p = 0.50$ cannot be rejected.
17. $t = (0.24\sqrt{26 - 2})/\sqrt{1 - (0.24)^2} = 1.21$. Do not reject the null hypothesis.
18. (a) 0.497 (b) -0.234 19. (a) 0.50 (b) -0.20
21. (a) No
    (b) The value of r is less than 0.50, which means that it is in the wrong tail of the sampling distribution.

Chapter 13

4. $z = (112 - 116)/\sqrt{(15)^2/50 + (15)^2/50} = -1.333$; do not reject the null hypothesis.
8. The significant result indicates only that chance is an unlikely explanation for the observed difference.
11. (a) random assignment
14. Pooling is appropriate when the variances of the populations are equal.
15. $t = (24.731 - 18.000)/0.712 = 9.45$; the data support the experimenter's scientific hypothesis.
19. The order of presentation of the conditions should be randomized independently for each subject.
21. (a) The larger the positive correlation between samples, the smaller is the standard error of the difference between means.
    (b) The larger the positive correlation, the higher is the probability of rejecting a false null hypothesis.
22. (a) $t = -2.1875/0.2772 = -7.891$. Reject the null hypothesis; viewing the film does result in more favorable attitudes toward legalization of marijuana.
    (b) 11

Chapter 14

1. No
4. $F = 1.24$ or 0.81, depending on which sample variance is made the numerator; do not reject the null hypothesis.
7. $z = (0.04 - 0.11)/0.01563 = -4.48$; reject the null hypothesis.
10. $z = (0.78 - 0.85)/0.01468 = -4.77$; reject the null hypothesis.
11. $z = (16 - 27)/\sqrt{16 + 27} = -1.68$. Do not reject the null hypothesis; the data do not provide evidence that the population proportions differ.

Chapter 15

2. $z = (\bar{X} - \mu)/(\sigma/\sqrt{n})$; hence $-z_{\alpha/2} \leq (\bar{X} - \mu)/(\sigma/\sqrt{n}) \leq z_{\alpha/2} = -\bar{X} - z_{\alpha/2}(\sigma/\sqrt{n}) \leq -\mu \leq -\bar{X} + z_{\alpha/2}(\sigma/\sqrt{n}) = \bar{X} - z_{\alpha/2}(\sigma/\sqrt{n}) \leq \mu \leq \bar{X} + z_{\alpha/2}(\sigma/\sqrt{n})$.
4. L_1 and L_2 would be equal if all of the scores were equal or if the sample exhausted all of the elements in the population in which case $\bar{X} = \mu$.
5. $7.05 \leq \mu \leq 7.35$
6. (a) I (b) C (c) I (d) C (e) C (f) I
7. $392.77 \leq \mu \leq 409.33$
9. $0.94 \leq \mu_1 - \mu_2 \leq 4.06$
11. $2.15 \leq \mu_D \leq 4.97$
12. (a) The larger σ, the larger the interval.
    (b) The larger n, the smaller the interval.
    (c) The larger $1 - \alpha$, the larger the interval.
14. $164.38 \leq \sigma^2 \leq 779.37$
16. The population variances could be equal because the confidence interval includes one.
17. (a) $(5.76/6.76)(1/2.25) \leq \sigma_1^2/\sigma_2^2 \leq (5.76/6.76)(2.25) = 0.38 \leq \sigma_1^2/\sigma_2^2 \leq 1.92$
    (b) yes (c) 0.25
19. $0.391 \leq p \leq 0.529$
21. $0.004 \leq p \leq 0.136$
24. $0.095 \leq p_1 - p_2 \leq 0.293$
28. $0.37 \leq \rho \leq 0.68$
31. $-0.24 \leq \rho_1 - \rho_2 \leq 0.43$

Chapter 16

1. (a) $H_0: \mu_1 = \mu_2 = \cdots = \mu_5$ (b) 10
2. It means that at least two of the population means are not equal.
6. (b) The number of levels of the independent variables is equal to two.
7. (a) level 2 of treatment B
   (b) score for subject 2 in treatment level 4
   (c) score for subject 16 in treatment level 1
   (d) mean of treatment level 4
   (e) grand mean
   (f) linear model for subject 7 in treatment level 3
   (g) treatment effect of population 2
   (h) error effect for subject 1 in treatment level 3
9. (a) correct (b) incorrect (c) incorrect (d) incorrect (e) correct
10. (a) $X_{83} = \bar{X}_{..} + (\bar{X}_{.3} - \bar{X}_{..}) + (X_{83} - \bar{X}_{.3})$
12. (a) Total = 83, $BG = 3$, $WG = 80$
    (b) Total = 54, $BG = 4$, $WG = 50$
    (d) Total = 16, $BG = 2$, $WG = 14$
13. MS_{BG} and MS_{WG} both estimate σ_e^2 when random samples are drawn from normally distributed populations having equal means and equal variances.
14. MS_{BG}/MS_{WG} is distributed as the F distribution when k random samples of size n are drawn from normally distributed populations having equal means and equal variances.

15. MS_{BG}/MS_{WG} tends to be larger than one when random samples are drawn from normally distributed populations having two or more unequal means but equal variances.

18. $n \sum_{j=1}^{k} (\bar{X}_{.j} - \bar{X}_{..})^2 = n \sum_{j=1}^{k} (\bar{X}_{.j}^2 - 2\bar{X}_{..}\bar{X}_{.j} + \bar{X}_{..}^2)$

$$= n \sum_{j=1}^{k} \frac{\left(\sum_{i=1}^{n} X_{ij}\right)^2}{n^2} - 2n \frac{\left(\sum_{j=1}^{k}\sum_{i=1}^{n} X_{ij}\right)\sum_{j=1}^{k}\left(\sum_{i=1}^{n} X_{ij}\right)}{kn \cdot n} + nk \frac{\left(\sum_{j=1}^{k}\sum_{i=1}^{n} X_{ij}\right)^2}{k^2 n^2}$$

$$= \sum_{j=1}^{k} \frac{\left(\sum_{i=1}^{n} X_{ij}\right)^2}{n} - 2\frac{\left(\sum_{j=1}^{k}\sum_{i=1}^{n} X_{ij}\right)^2}{kn} + \frac{\left(\sum_{j=1}^{k}\sum_{i=1}^{n} X_{ij}\right)^2}{kn}$$

$$= \sum_{j=1}^{k} \frac{\left(\sum_{i=1}^{n} X_{ij}\right)^2}{n} - \frac{\left(\sum_{j=1}^{k}\sum_{i=1}^{n} X_{ij}\right)^2}{kn}$$

$\sum_{j=1}^{k}\sum_{i=1}^{n} (X_{ij} - \bar{X}_{.j})^2 = \sum_{j=1}^{k}\sum_{i=1}^{n} (X_{ij}^2 - 2\bar{X}_{.j} X_{ij} + \bar{X}_{.j}^2)$

$$= \sum_{j=1}^{k}\sum_{i=1}^{n} X_{ij}^2 - 2 \sum_{j=1}^{k} \frac{\left(\sum_{i=1}^{n} X_{ij}\right)}{n} \sum_{i=1}^{n} (X_{ij}) + n \sum_{j=1}^{k} \frac{\left(\sum_{i=1}^{n} X_{ij}\right)^2}{n^2}$$

$$= \sum_{j=1}^{k}\sum_{i=1}^{n} X_{ij}^2 - \sum_{j=1}^{k} \frac{\left(\sum_{i=1}^{n} X_{ij}\right)^2}{n}$$

20. $F = 23.49$; reject the null hypothesis   23. $F = 4.54$; reject the null hypothesis
25. The test is robust if the treatment populations have the same shape.
26. (a) Assumption is tenable.   33. (a) $1 - 1$ (b) $1 - \frac{1}{2} - \frac{1}{2}$ (f) $1 - \frac{2}{3} - \frac{1}{3}$
34. (a) meets requirements (b) meets requirements
35. (a) satisfies (b) does not satisfy   36. (a) 3.596 (c) 2.665
37. Do not reject $\mu_1 - \mu_2$; $S = -0.783$. Reject $\mu_1 - \mu_3$; $S = -4.025$. Reject $\mu_2 - \mu_3$; $S = -3.242$. Reject $(\mu_1 + \mu_2)/2 - \mu_3$; $S = -4.196$. $S_{0.01;2,20} = 3.095$.
38. $-0.02/0.028284 = -0.707$, $-0.10/0.028284 = -3.536$, $-0.08/0.028284 = -2.828$; $S_{0.05;2,12} = 2.789$.

Chapter 17

2. (a) State the statistical hypotheses—$H_0: p_1 = 0.50$, $H_1: p_1 \neq 0.50$. Specify the test statistic—$\chi^2 = \sum (O - E)^2/E$. Specify the sample size—$n = 50$, and the sampling distribution—chi-square distribution. Specify the level of significance—$\alpha = 0.05$. Obtain a random sample of size 50, compute χ^2 and make a decision.
   (b) $\chi^2 = 5.12$. Reject H_0; the data suggest that opinion is not equally divided on the issue.
3. $\chi^2 = 0.120$; the data do not suggest that there has been a change in party affiliation.
5. $\chi^2 = 9.280$; the data do not suggest that the die is not fair.
10. Both r and $\hat{\phi}'$ are measures of association, but, unlike r, $\hat{\phi}'$ is appropriate for unordered qualitative variables and ranges over values of 0 to 1. Further, $\hat{\phi}'$ does not have a simple interpretation in terms of proportion of explained variance.

12. (a) State the statistical hypotheses—H_0: $p(A \text{ and } B) = p(A)p(B)$, H_1: H_0 is false. Specify the test statistic—$\chi^2 = \sum (O - E)^2/E$. Specify the sample size—$n = 200$, and the sampling distribution—chi-square distribution. Specify the level of significance—$\alpha = 0.001$. Obtain a random sample of size 200, compute χ^2 and make a decision.
(b) $\chi^2 = 32.28$. Reject the null hypothesis; the variables are not independent.
(c) $\hat{\phi}' = 0.28$
17. H_0: $p_1 = p_2$. $\chi^2 = 8.864$. Reject the null hypothesis.
19. The population proportions in the three categories of variable A, 0.30, 0.60, 0.10, are equal across the three categories of variable B.

20. (a) H_0:
$$\begin{bmatrix} p_{A_1|B_1} \\ p_{A_2|B_1} \\ p_{A_3|B_1} \\ p_{A_4|B_1} \end{bmatrix} = \begin{bmatrix} p_{A_1|B_2} \\ p_{A_2|B_2} \\ p_{A_3|B_2} \\ p_{A_4|B_2} \end{bmatrix} = \begin{bmatrix} p_{A_1|B_3} \\ p_{A_2|B_3} \\ p_{A_3|B_3} \\ p_{A_4|B_3} \end{bmatrix} = \begin{bmatrix} p_{A_1|B_4} \\ p_{A_2|B_4} \\ p_{A_3|B_4} \\ p_{A_4|B_4} \end{bmatrix}$$

H_1: H_0 is false
(b) $\chi^2 = 17.560$; reject the null hypothesis.

Chapter 18

1. A test is nonparametric if it does not test a hypothesis concerning one of the parameters of the sampled population. It is distribution free if it makes no assumptions about the precise form of the sampled population. Parametric tests test hypotheses about one of the parameters of the sampled population, and they make assumptions about the precise form of the sampled population.
3. $U = 5.5$; reject the null hypothesis that the populations are identical.
5. Without the correction, $z = -1.38$; with the correction, -1.38. Do not reject the null hypothesis.
7. PE is dependent on α, $1 - \beta$, H_0, and H_1, and the sample size of the more efficient comparison test statistic; ARE is a theoretical value that applies when the sample size is infinite, H_1 is essentially identical to H_0, and H_1 is directional.
10. $T = 16.5$; reject the null hypothesis.
13. $z = -2.28$; reject the null hypothesis.
12. $T = 30.5$; reject the null hypothesis.
15. $z = -1.94$; reject the null hypothesis.

Appendix D
Tables

D.1 Random Numbers
D.2 Areas under the Standard Normal Distribution
D.3 Percentage Points of Student's t Distribution
D.4 Upper Percentage Points of the Chi-Square Distribution
D.5 Upper Percentage Points of the F Distribution
D.6 Critical Values of the Pearson r
D.7 Critical Values of r_s (Spearman Rank Correlation Coefficient)
D.8 Transformation of r to Z'
D.9 Approximate n Required for Testing Hypotheses about Means
D.10 Critical Values of the Mann–Whitney U
D.11 Critical Values of the Wilcoxon T

Table D.1. Random Numbers[a]

	1 2 3 4 5	6 7 8 9 10	11 12 13 14 15	16 17 18 19 20	21 22 23 24 25
1	10 27 53 96 23	71 50 54 36 23	54 31 04 82 98	04 14 12 15 09	26 78 25 47 47
2	28 41 50 61 88	64 85 27 20 18	83 36 36 05 56	39 71 65 09 62	94 76 62 11 89
3	34 21 42 57 02	59 19 18 97 48	80 30 03 30 98	05 24 67 70 07	84 97 50 87 46
4	61 81 77 23 23	82 82 11 54 08	53 28 70 58 96	44 07 39 55 43	42 34 43 39 28
5	61 15 18 13 54	16 86 20 26 88	90 74 80 55 09	14 53 90 51 17	52 01 63 01 59
6	91 76 21 64 64	44 91 13 32 97	75 31 62 66 54	84 80 32 75 77	56 08 25 70 29
7	00 97 79 08 06	37 30 28 59 85	53 56 68 53 40	01 74 39 59 73	30 19 99 85 48
8	36 46 18 34 94	75 20 80 27 77	78 91 69 16 00	08 43 18 73 68	67 69 61 34 25
9	88 98 99 60 50	65 95 79 42 94	93 62 40 89 96	43 56 47 71 66	46 76 29 67 02
10	04 37 59 87 21	05 02 03 24 17	47 97 81 56 51	92 34 86 01 82	55 51 33 12 91
11	63 62 06 34 41	94 21 78 55 09	72 76 45 16 94	29 95 81 83 83	79 88 01 97 30
12	78 47 23 53 90	34 41 92 45 71	09 23 70 70 07	12 38 92 79 43	14 85 11 47 23
13	87 68 62 15 43	53 14 36 59 25	54 47 33 70 15	59 24 48 40 35	50 03 42 99 36
14	47 60 92 10 77	88 59 53 11 52	66 25 69 07 04	48 68 64 71 06	61 65 70 22 12
15	56 88 87 59 41	65 28 04 67 53	95 79 88 37 31	50 41 06 94 76	81 83 17 16 33
16	02 57 45 86 67	73 43 07 34 48	44 26 87 93 29	77 09 61 67 84	06 69 44 77 75
17	31 54 14 13 17	48 62 11 90 60	68 12 93 64 28	46 24 79 16 76	14 60 25 51 01
18	28 50 16 43 36	28 97 85 58 99	67 22 52 76 23	24 70 36 54 54	59 28 61 71 96
19	63 29 62 66 50	02 63 45 52 38	67 63 47 54 75	83 24 78 43 20	92 63 13 47 48
20	45 65 58 26 51	76 96 59 38 72	86 57 45 71 46	44 67 76 14 55	44 88 01 62 12
21	39 65 36 63 70	77 45 85 50 51	74 13 39 35 22	30 53 36 02 95	49 34 88 73 61
22	73 71 98 16 04	29 18 94 51 23	76 51 94 84 86	79 93 96 38 63	08 58 25 58 94
23	72 20 56 20 11	72 65 71 08 86	79 57 95 13 91	97 48 72 66 48	09 71 17 24 89
24	75 17 26 99 76	89 37 20 70 01	77 31 61 95 46	26 97 05 73 51	53 33 18 72 87
25	37 48 60 82 29	81 30 15 39 14	48 38 75 93 29	06 87 37 78 48	45 56 00 84 47
26	68 08 02 80 72	83 71 46 30 49	89 17 95 88 29	02 39 56 03 46	97 74 06 56 17
27	14 23 98 61 67	70 52 85 01 50	01 84 02 78 43	10 62 98 19 41	18 83 99 47 99
28	49 08 96 21 44	25 27 99 41 28	07 41 08 34 66	19 42 74 39 91	41 96 53 78 72
29	78 37 06 08 43	63 61 62 42 29	39 68 95 10 96	09 24 23 00 62	56 12 80 73 16
30	37 21 34 17 68	68 96 83 23 56	32 84 60 15 31	44 73 67 34 77	91 15 79 74 58
31	14 29 09 34 04	87 83 07 55 07	76 58 30 83 64	87 29 25 58 84	86 50 60 00 25
32	58 43 28 06 36	49 52 83 51 14	47 56 91 29 34	05 87 31 06 95	12 45 57 09 09
33	10 43 67 29 70	80 62 80 03 42	10 80 21 38 84	90 56 35 03 09	43 12 74 49 14
34	44 38 88 39 54	86 97 37 44 22	00 95 01 31 76	17 16 29 56 63	38 78 94 49 81
35	90 69 59 19 51	85 39 52 85 13	07 28 37 07 61	11 16 36 27 03	78 86 72 04 95
36	41 47 10 25 62	97 05 31 03 61	20 26 36 31 62	68 69 86 95 44	84 95 48 46 45
37	91 94 14 63 19	75 89 11 47 11	31 56 34 19 09	79 57 92 36 59	14 93 87 81 40
38	80 06 54 18 66	09 18 94 06 19	98 40 07 17 81	22 45 44 84 11	24 62 20 42 31
39	67 72 77 63 48	84 08 31 55 58	24 33 45 77 58	80 45 67 93 82	75 70 16 08 24
40	59 40 24 13 27	79 26 88 86 30	01 31 60 10 39	53 58 47 70 93	85 81 56 39 38
41	05 90 35 89 95	01 61 16 96 94	50 78 13 69 36	37 68 53 37 31	71 26 35 03 71
42	44 43 80 69 98	46 68 05 14 82	90 78 50 05 62	77 79 13 57 44	59 60 10 39 66
43	61 81 31 96 82	00 57 25 60 59	46 72 60 18 77	55 66 12 62 11	08 99 55 64 57
44	42 88 07 10 05	24 98 65 63 21	47 21 61 88 32	27 80 30 21 60	10 92 35 36 12
45	77 94 30 05 39	28 10 99 00 27	12 73 73 99 12	49 99 57 94 82	96 88 57 17 91
46	78 83 19 76 16	94 11 68 84 26	23 54 20 86 85	23 86 66 99 07	36 37 34 92 09
47	87 76 59 61 81	43 63 64 61 61	65 76 36 95 90	18 48 27 45 68	27 23 65 30 72
48	91 43 05 96 47	55 78 99 95 24	37 55 85 78 78	01 48 41 19 10	35 19 54 07 73
49	84 97 77 72 73	09 62 06 65 72	87 12 49 03 60	41 15 20 76 27	50 47 02 29 16
50	87 41 60 76 83	44 88 96 07 80	83 05 83 38 96	73 70 66 81 90	30 56 10 48 59

Table D.1 is taken from Table XXXIII of Fisher and Yates: *Statistical Tables for Biological, Agricultural and Medical Research*, published by Longman Group Ltd., London (previously published by Oliver & Boyd, Edinburgh), and reprinted by permission of the authors and publishers.

[a] Discussed in Section 9.1.

Table D.1. (Continued)

	1	2	3	4	5	6	7	8	9	10	11	12	13	14	15	16	17	18	19	20	21	22	23	24	25
1	22	17	68	65	84	68	95	23	92	35	87	02	22	57	51	61	09	43	95	06	58	24	82	03	47
2	19	36	27	59	46	13	79	93	37	55	39	77	32	77	09	85	52	05	30	62	47	83	51	62	74
3	16	77	23	02	77	09	61	87	25	21	28	06	24	25	93	16	71	13	59	78	23	05	47	47	25
4	78	43	76	71	61	20	44	90	32	64	97	67	63	99	61	46	38	03	93	22	69	81	21	99	21
5	03	28	28	26	08	73	37	32	04	05	69	30	16	09	05	88	69	58	28	99	35	07	44	75	47
6	93	22	53	64	39	07	10	63	76	35	87	03	04	79	88	08	13	13	85	51	55	34	57	72	69
7	78	76	58	54	74	92	38	70	96	92	52	06	79	79	45	82	63	18	27	44	69	66	92	19	09
8	23	68	35	26	00	99	53	93	61	28	52	70	05	48	34	56	65	05	61	86	90	92	10	70	80
9	15	39	25	70	99	93	86	52	77	65	15	33	59	05	28	22	87	26	07	47	86	96	98	29	06
10	58	71	96	30	24	18	46	23	34	27	85	13	99	24	44	49	18	09	79	49	74	16	32	23	02
11	57	35	27	33	72	24	53	63	94	09	41	10	76	47	91	44	04	95	49	66	39	60	04	59	81
12	48	50	86	54	48	22	06	34	72	52	82	21	15	65	20	33	29	94	71	11	15	91	29	12	03
13	61	96	48	95	03	07	16	39	33	66	98	56	10	56	79	77	21	30	27	12	90	49	22	23	62
14	36	93	89	41	26	29	70	83	63	51	99	74	20	52	36	87	09	41	15	09	98	60	16	03	03
15	18	87	00	42	31	57	90	12	02	07	23	47	37	17	31	54	08	01	88	63	39	41	88	92	10
16	88	56	53	27	59	33	35	72	67	47	77	34	55	45	70	08	18	27	38	90	16	95	86	70	75
17	09	72	95	84	29	49	41	31	06	70	42	38	06	45	18	64	84	73	31	65	52	53	37	97	15
18	12	96	88	17	31	65	19	69	02	83	60	75	86	90	68	24	64	19	35	51	56	61	87	39	12
19	85	94	57	24	16	92	09	84	38	76	22	00	27	69	85	29	81	94	78	70	21	94	47	90	12
20	38	64	43	59	98	98	77	87	68	07	91	51	67	62	44	40	98	05	93	78	23	32	65	41	18
21	53	44	09	42	72	00	41	86	79	79	68	47	22	00	20	35	55	31	51	51	00	83	63	22	55
22	40	76	66	26	84	57	99	99	90	37	36	63	32	08	58	37	40	13	68	97	87	64	81	07	83
23	02	17	79	18	05	12	59	52	57	02	22	07	90	47	03	28	14	11	30	79	20	69	22	40	98
24	95	17	82	06	53	31	51	10	96	46	92	06	88	07	77	56	11	50	81	69	40	23	72	51	39
25	35	76	22	42	92	96	11	83	44	80	34	68	35	48	77	33	42	40	90	60	73	96	53	97	86
26	26	29	13	56	41	85	47	04	66	08	34	72	57	59	13	82	43	80	46	15	38	26	61	70	04
27	77	80	20	75	82	72	82	32	99	90	63	95	73	76	63	89	73	44	99	05	48	67	26	43	18
28	46	40	66	44	52	91	36	74	43	53	30	82	13	54	00	78	45	63	98	35	55	03	36	67	68
29	37	56	08	18	09	77	53	84	46	47	31	91	18	95	58	24	16	74	11	53	44	10	13	85	57
30	61	65	61	68	66	37	27	47	39	19	84	83	70	07	48	53	21	40	06	71	95	06	79	88	54
31	93	43	69	64	07	34	18	04	52	35	56	27	09	24	86	61	85	53	83	45	19	90	70	99	00
32	21	96	60	12	99	11	20	99	45	18	48	13	93	55	34	18	37	79	49	90	65	97	38	20	46
33	95	20	47	97	97	27	37	83	28	71	00	06	41	41	74	45	89	09	39	84	51	67	11	52	49
34	97	86	21	78	73	10	65	81	92	59	58	76	17	14	97	04	76	62	16	17	17	95	70	45	80
35	69	92	06	34	13	59	71	74	17	32	27	55	10	24	19	23	71	82	13	74	63	52	52	01	41
36	04	31	17	21	56	33	73	99	19	87	26	72	39	27	67	53	77	57	68	93	60	61	97	22	61
37	61	06	98	03	91	87	14	77	43	96	43	00	65	98	50	45	60	33	01	07	98	99	46	50	47
38	85	93	85	86	88	72	87	08	62	40	16	06	10	89	20	23	21	34	74	97	76	38	03	29	63
39	21	74	32	47	45	73	96	07	94	52	09	65	90	77	47	25	76	16	19	33	53	05	70	53	30
40	15	69	53	82	80	79	96	23	53	10	65	39	07	16	29	45	33	02	43	70	02	87	40	41	45
41	02	89	08	04	49	20	21	14	68	86	87	63	93	95	17	11	29	01	95	80	35	14	97	35	33
42	87	18	15	89	79	85	43	01	72	73	08	61	74	51	69	89	74	39	82	15	94	51	33	41	67
43	98	83	71	94	22	59	97	50	99	52	08	52	85	08	40	87	80	61	65	31	91	51	80	32	44
44	10	08	58	21	66	72	68	49	29	31	89	85	84	46	06	59	73	19	85	23	65	09	29	75	63
45	47	90	56	10	08	88	02	84	27	83	42	29	72	23	19	66	56	45	65	79	20	71	53	20	25
46	22	85	61	68	90	49	64	92	85	44	16	40	12	89	88	50	14	49	81	06	01	82	77	45	12
47	67	80	43	79	33	12	83	11	41	16	25	58	19	68	70	77	02	54	00	52	53	43	37	15	26
48	27	62	50	96	72	79	44	61	40	15	14	53	40	65	39	27	31	58	50	28	11	39	03	34	25
49	33	78	80	87	15	38	30	06	38	21	14	47	47	07	26	54	96	87	53	32	40	36	40	96	76
50	13	13	92	66	99	47	24	49	57	74	32	25	43	62	17	10	97	11	69	84	99	63	22	32	98

Table D.2. Areas under the Standard Normal Distribution[a]

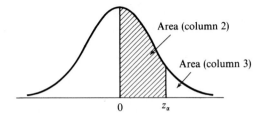

(1) z_α	(2) Area between Mean and z_α	(3) Area beyond z_α	(1) z_α	(2) Area between Mean and z_α	(3) Area beyond z_α	(1) z_α	(2) Area between Mean and z_α	(3) Area beyond z_α
0.00	0.0000	0.5000	0.30	0.1179	0.3821	0.60	0.2257	0.2743
0.01	0.0040	0.4960	0.31	0.1217	0.3783	0.61	0.2291	0.2709
0.02	0.0080	0.4920	0.32	0.1255	0.3745	0.62	0.2324	0.2676
0.03	0.0120	0.4880	0.33	0.1293	0.3707	0.63	0.2357	0.2643
0.04	0.0160	0.4840	0.34	0.1331	0.3669	0.64	0.2389	0.2611
0.05	0.0199	0.4801	0.35	0.1368	0.3632	0.65	0.2422	0.2578
0.06	0.0239	0.4761	0.36	0.1406	0.3594	0.66	0.2454	0.2546
0.07	0.0279	0.4721	0.37	0.1443	0.3557	0.67	0.2486	0.2514
0.08	0.0319	0.4681	0.38	0.1480	0.3520	0.68	0.2517	0.2483
0.09	0.0359	0.4641	0.39	0.1517	0.3483	0.69	0.2549	0.2451
0.10	0.0398	0.4602	0.40	0.1554	0.3446	0.70	0.2580	0.2420
0.11	0.0438	0.4562	0.41	0.1591	0.3409	0.71	0.2611	0.2389
0.12	0.0478	0.4522	0.42	0.1628	0.3372	0.72	0.2642	0.2358
0.13	0.0517	0.4483	0.43	0.1664	0.3336	0.73	0.2673	0.2327
0.14	0.0557	0.4443	0.44	0.1700	0.3300	0.74	0.2704	0.2296
0.15	0.0596	0.4404	0.45	0.1736	0.3264	0.75	0.2734	0.2266
0.16	0.0636	0.4364	0.46	0.1772	0.3228	0.76	0.2764	0.2236
0.17	0.0675	0.4325	0.47	0.1808	0.3192	0.77	0.2794	0.2206
0.18	0.0714	0.4286	0.48	0.1844	0.3156	0.78	0.2823	0.2177
0.19	0.0753	0.4247	0.49	0.1879	0.3121	0.79	0.2852	0.2148
0.20	0.0793	0.4207	0.50	0.1915	0.3085	0.80	0.2881	0.2119
0.21	0.0832	0.4168	0.51	0.1950	0.3050	0.81	0.2910	0.2090
0.22	0.0871	0.4129	0.52	0.1985	0.3015	0.82	0.2939	0.2061
0.23	0.0910	0.4090	0.53	0.2019	0.2981	0.83	0.2967	0.2033
0.24	0.0948	0.4052	0.54	0.2054	0.2946	0.84	0.2995	0.2005
0.25	0.0987	0.4013	0.55	0.2088	0.2912	0.85	0.3023	0.1977
0.26	0.1026	0.3974	0.56	0.2123	0.2877	0.86	0.3051	0.1949
0.27	0.1064	0.3936	0.57	0.2157	0.2843	0.87	0.3078	0.1922
0.28	0.1103	0.3897	0.58	0.2190	0.2810	0.88	0.3106	0.1894
0.29	0.1141	0.3859	0.59	0.2224	0.2776	0.89	0.3133	0.1867

Table D.2 is abridged from Table IIi of Fisher and Yates: *Statistical Tables for Biological, Agricultural and Medical Research*, published by Longman Group Ltd., London (previously published by Oliver & Boyd, Edinburgh), and reprinted by permission of the authors and publishers.

[a] Discussed in Section 10.1.

Table D.2. (Continued)

(1) z_α	(2) Area between Mean and z_α	(3) Area beyond z_α	(1) z_α	(2) Area between Mean and z_α	(3) Area beyond z_α	(1) z_α	(2) Area between Mean and z_α	(3) Area beyond z_α
0.90	0.3159	0.1841	1.35	0.4115	0.0885	1.80	0.4641	0.0359
0.91	0.3186	0.1814	1.36	0.4131	0.0869	1.81	0.4649	0.0351
0.92	0.3212	0.1788	1.37	0.4147	0.0853	1.82	0.4656	0.0344
0.93	0.3238	0.1762	1.38	0.4162	0.0838	1.83	0.4664	0.0336
0.94	0.3264	0.1736	1.39	0.4177	0.0823	1.84	0.4671	0.0329
0.95	0.3289	0.1711	1.40	0.4192	0.0808	1.85	0.4678	0.0322
0.96	0.3315	0.1685	1.41	0.4207	0.0793	1.86	0.4686	0.0314
0.97	0.3340	0.1660	1.42	0.4222	0.0778	1.87	0.4693	0.0307
0.98	0.3365	0.1635	1.43	0.4236	0.0764	1.88	0.4699	0.0301
0.99	0.3389	0.1611	1.44	0.4251	0.0749	1.89	0.4706	0.0294
1.00	0.3413	0.1587	1.45	0.4265	0.0735	1.90	0.4713	0.0287
1.01	0.3438	0.1562	1.46	0.4279	0.0721	1.91	0.4719	0.0281
1.02	0.3461	0.1539	1.47	0.4292	0.0708	1.92	0.4726	0.0274
1.03	0.3485	0.1515	1.48	0.4306	0.0694	1.93	0.4732	0.0268
1.04	0.3508	0.1492	1.49	0.4319	0.0681	1.94	0.4738	0.0262
1.05	0.3531	0.1469	1.50	0.4332	0.0668	1.95	0.4744	0.0256
1.06	0.3554	0.1446	1.51	0.4345	0.0655	1.96	0.4750	0.0250
1.07	0.3577	0.1423	1.52	0.4357	0.0643	1.97	0.4756	0.0244
1.08	0.3599	0.1401	1.53	0.4370	0.0630	1.98	0.4761	0.0239
1.09	0.3621	0.1379	1.54	0.4382	0.0618	1.99	0.4767	0.0233
1.10	0.3643	0.1357	1.55	0.4394	0.0606	2.00	0.4772	0.0228
1.11	0.3665	0.1335	1.56	0.4406	0.0594	2.01	0.4778	0.0222
1.12	0.3686	0.1314	1.57	0.4418	0.0582	2.02	0.4783	0.0217
1.13	0.3708	0.1292	1.58	0.4429	0.0571	2.03	0.4788	0.0212
1.14	0.3729	0.1271	1.59	0.4441	0.0559	2.04	0.4793	0.0207
1.15	0.3749	0.1251	1.60	0.4452	0.0548	2.05	0.4798	0.0202
1.16	0.3770	0.1230	1.61	0.4463	0.0537	2.06	0.4803	0.0197
1.17	0.3790	0.1210	1.62	0.4474	0.0526	2.07	0.4808	0.0192
1.18	0.3810	0.1190	1.63	0.4484	0.0516	2.08	0.4812	0.0188
1.19	0.3830	0.1170	1.64	0.4495	0.0505	2.09	0.4817	0.0183
			1.645	0.4500	0.0500			
1.20	0.3849	0.1151	1.65	0.4505	0.0495	2.10	0.4821	0.0179
1.21	0.3869	0.1131	1.66	0.4515	0.0485	2.11	0.4826	0.0174
1.22	0.3888	0.1112	1.67	0.4525	0.0475	2.12	0.4830	0.0170
1.23	0.3907	0.1093	1.68	0.4535	0.0465	2.13	0.4834	0.0166
1.24	0.3925	0.1075	1.69	0.4545	0.0455	2.14	0.4838	0.0162
1.25	0.3944	0.1056	1.70	0.4554	0.0446	2.15	0.4842	0.0158
1.26	0.3962	0.1038	1.71	0.4564	0.0436	2.16	0.4846	0.0154
1.27	0.3980	0.1020	1.72	0.4573	0.0427	2.17	0.4850	0.0150
1.28	0.3997	0.1003	1.73	0.4582	0.0418	2.18	0.4854	0.0146
1.29	0.4015	0.0985	1.74	0.4591	0.0409	2.19	0.4857	0.0143
1.30	0.4032	0.0968	1.75	0.4599	0.0401	2.20	0.4861	0.0139
1.31	0.4049	0.0951	1.76	0.4608	0.0392	2.21	0.4864	0.0136
1.32	0.4066	0.0934	1.77	0.4616	0.0384	2.22	0.4868	0.0132
1.33	0.4082	0.0918	1.78	0.4625	0.0375	2.23	0.4871	0.0129
1.34	0.4099	0.0901	1.79	0.4633	0.0367	2.24	0.4875	0.0125

Table D.2. (Continued)

(1) z_α	(2) Area between Mean and z_α	(3) Area beyond z_α	(1) z_α	(2) Area between Mean and z_α	(3) Area beyond z_α	(1) z_α	(2) Area between Mean and z_α	(3) Area beyond z_α
2.25	0.4878	0.0122	2.64	0.4959	0.0041	3.00	0.4987	0.0013
2.26	0.4881	0.0119	2.65	0.4960	0.0040	3.01	0.4987	0.0013
2.27	0.4884	0.0116	2.66	0.4961	0.0039	3.02	0.4987	0.0013
2.28	0.4887	0.0113	2.67	0.4962	0.0038	3.03	0.4988	0.0012
2.29	0.4890	0.0110	2.68	0.4963	0.0037	3.04	0.4988	0.0012
2.30	0.4893	0.0107	2.69	0.4964	0.0036	3.05	0.4989	0.0011
2.31	0.4896	0.0104	2.70	0.4965	0.0035	3.06	0.4989	0.0011
2.32	0.4898	0.0102	2.71	0.4966	0.0034	3.07	0.4989	0.0011
2.33	0.4901	0.0099	2.72	0.4967	0.0033	3.08	0.4990	0.0010
2.34	0.4904	0.0096	2.73	0.4968	0.0032	3.09	0.4990	0.0010
2.35	0.4906	0.0094	2.74	0.4969	0.0031	3.10	0.4990	0.0010
2.36	0.4909	0.0091	2.75	0.4970	0.0030	3.11	0.4991	0.0009
2.37	0.4911	0.0089	2.76	0.4971	0.0029	3.12	0.4991	0.0009
2.38	0.4913	0.0087	2.77	0.4972	0.0028	3.13	0.4991	0.0009
2.39	0.4916	0.0084	2.78	0.4973	0.0027	3.14	0.4992	0.0008
2.40	0.4918	0.0082	2.79	0.4974	0.0026	3.15	0.4992	0.0008
2.41	0.4920	0.0080	2.80	0.4974	0.0026	3.16	0.4992	0.0008
2.42	0.4922	0.0078	2.81	0.4975	0.0025	3.17	0.4992	0.0008
2.43	0.4925	0.0075	2.82	0.4976	0.0024	3.18	0.4993	0.0007
2.44	0.4927	0.0073	2.83	0.4977	0.0023	3.19	0.4993	0.0007
2.45	0.4929	0.0071	2.84	0.4977	0.0023	3.20	0.4993	0.0007
2.46	0.4931	0.0069	2.85	0.4978	0.0022	3.21	0.4993	0.0007
2.47	0.4932	0.0068	2.86	0.4979	0.0021	3.22	0.4994	0.0006
2.48	0.4934	0.0066	2.87	0.4979	0.0021	3.23	0.4994	0.0006
2.49	0.4936	0.0064	2.88	0.4980	0.0020	3.24	0.4994	0.0006
2.50	0.4938	0.0062	2.89	0.4981	0.0019	3.25	0.4994	0.0006
2.51	0.4940	0.0060	2.90	0.4981	0.0019	3.30	0.4995	0.0005
2.52	0.4941	0.0059	2.91	0.4982	0.0018	3.35	0.4996	0.0004
2.53	0.4943	0.0057	2.92	0.4982	0.0018	3.40	0.4997	0.0003
2.54	0.4945	0.0055	2.93	0.4983	0.0017	3.45	0.4997	0.0003
2.55	0.4946	0.0054	2.94	0.4984	0.0016	3.50	0.4998	0.0002
2.56	0.4948	0.0052	2.95	0.4984	0.0016	3.60	0.4998	0.0002
2.57	0.4949	0.0051	2.96	0.4985	0.0015	3.70	0.4999	0.0001
2.576	0.4950	0.0050	2.97	0.4985	0.0015	3.80	0.4999	0.0001
2.58	0.4951	0.0049	2.98	0.4986	0.0014	3.90	0.49995	0.00005
2.59	0.4952	0.0048	2.99	0.4986	0.0014	4.00	0.49997	0.00003
2.60	0.4953	0.0047						
2.61	0.4955	0.0045						
2.62	0.4956	0.0044						
2.63	0.4957	0.0043						

Table D.3. Percentage Points of Student's t Distribution[a]

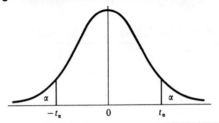

Degrees of Freedom, v	Level of Significance for a One-Tailed Test								
	0.25	0.20	0.15	0.10	0.05	0.025	0.01	0.005	0.0005
	Level of Significance for a Two-Tailed Test								
	0.50	0.40	0.30	0.20	0.10	0.05	0.02	0.01	0.001
1	1.000	1.376	1.963	3.078	6.314	12.706	31.821	63.657	636.619
2	.816	1.061	1.386	1.886	2.920	4.303	6.965	9.925	31.598
3	.765	.978	1.250	1.638	2.353	3.182	4.541	5.841	12.924
4	.741	.941	1.190	1.533	2.132	2.776	3.747	4.604	8.610
5	.727	.920	1.156	1.476	2.015	2.571	3.365	4.032	6.869
6	.718	.906	1.134	1.440	1.943	2.447	3.143	3.707	5.959
7	.711	.896	1.119	1.415	1.895	2.365	2.998	3.499	5.408
8	.706	.889	1.108	1.397	1.860	2.306	2.896	3.355	5.041
9	.703	.883	1.100	1.383	1.833	2.262	2.821	3.250	4.781
10	.700	.879	1.093	1.372	1.812	2.228	2.764	3.169	4.587
11	.697	.876	1.088	1.363	1.796	2.201	2.718	3.106	4.437
12	.695	.873	1.083	1.356	1.782	2.179	2.681	3.055	4.318
13	.694	.870	1.079	1.350	1.771	2.160	2.650	3.012	4.221
14	.692	.868	1.076	1.345	1.761	2.145	2.624	2.977	4.140
15	.691	.866	1.074	1.341	1.753	2.131	2.602	2.947	4.073
16	.690	.865	1.071	1.337	1.746	2.120	2.583	2.921	4.015
17	.689	.863	1.069	1.333	1.740	2.110	2.567	2.898	3.965
18	.688	.862	1.067	1.330	1.734	2.101	2.552	2.878	3.922
19	.688	.861	1.066	1.328	1.729	2.093	2.539	2.861	3.883
20	.687	.860	1.064	1.325	1.725	2.086	2.528	2.845	3.850
21	.686	.859	1.063	1.323	1.721	2.080	2.518	2.831	3.819
22	.686	.858	1.061	1.321	1.717	2.074	2.508	2.819	3.792
23	.685	.858	1.060	1.319	1.714	2.069	2.500	2.807	3.767
24	.685	.857	1.059	1.318	1.711	2.064	2.492	2.797	3.745
25	.684	.856	1.058	1.316	1.708	2.060	2.485	2.787	3.725
26	.684	.856	1.058	1.315	1.706	2.056	2.479	2.779	3.707
27	.684	.855	1.057	1.314	1.703	2.052	2.473	2.771	3.690
28	.683	.855	1.056	1.313	1.701	2.048	2.467	2.763	3.674
29	.683	.854	1.055	1.311	1.699	2.045	2.462	2.756	3.659
30	.683	.854	1.055	1.310	1.697	2.042	2.457	2.750	3.646
40	.681	.851	1.050	1.303	1.684	2.021	2.423	2.704	3.551
60	.679	.848	1.046	1.296	1.671	2.000	2.390	2.660	3.460
120	.677	.845	1.041	1.289	1.658	1.980	2.358	2.617	3.373
∞	.674	.842	1.036	1.282	1.645	1.960	2.326	2.576	3.291

Table D.3 is taken from Table III of Fisher and Yates: *Statistical Tables for Biological, Agricultural and Medical Research*, published by Longman Group Ltd., London (previously published by Oliver & Boyd, Edinburgh), and reprinted by permission of the authors and publishers.

[a] Discussed in Section 12.2.

Table D.4. Upper Percentage Points of the Chi-Square Distribution[a]

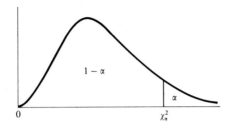

v[b]	0.99	0.98	0.95	0.90	0.80	0.70	0.50	0.30	0.20	0.10	0.05	0.02	0.01	0.001
1	.0³157	.0³628	.00393	.0158	.0642	.148	.455	1.074	1.642	2.706	3.841	5.412	6.635	10.827
2	.0201	.0404	.103	.211	.446	.713	1.386	2.408	3.219	4.605	5.991	7.824	9.210	13.815
3	.115	.185	.352	.584	1.005	1.424	2.366	3.665	4.642	6.251	7.815	9.837	11.345	16.266
4	.297	.429	.711	1.064	1.649	2.195	3.357	4.878	5.989	7.779	9.488	11.668	13.277	18.467
5	.554	.752	1.145	1.610	2.343	3.000	4.351	6.064	7.289	9.236	11.070	13.388	15.086	20.515
6	.872	1.134	1.635	2.204	3.070	3.828	5.348	7.231	8.558	10.645	12.592	15.033	16.812	22.457
7	1.239	1.564	2.167	2.833	3.822	4.671	6.346	8.383	9.803	12.017	14.067	16.622	18.475	24.322
8	1.646	2.032	2.733	3.490	4.594	5.527	7.344	9.524	11.030	13.362	15.507	18.168	20.090	26.125
9	2.088	2.532	3.325	4.168	5.380	6.393	8.343	10.656	12.242	14.684	16.919	19.679	21.666	27.877
10	2.558	3.059	3.940	4.865	6.179	7.267	9.342	11.781	13.442	15.987	18.307	21.161	23.209	29.588
11	3.053	3.609	4.575	5.578	6.989	8.148	10.341	12.899	14.631	17.275	19.675	22.618	24.725	31.264
12	3.571	4.178	5.226	6.304	7.807	9.034	11.340	14.011	15.812	18.549	21.026	24.054	26.217	32.909
13	4.107	4.765	5.892	7.042	8.634	9.926	12.340	15.119	16.985	19.812	22.362	25.472	27.688	34.528
14	4.660	5.368	6.571	7.790	9.467	10.821	13.339	16.222	18.151	21.064	23.685	26.873	29.141	36.123
15	5.229	5.985	7.261	8.547	10.307	11.721	14.339	17.322	19.311	22.307	24.996	28.259	30.578	37.697
16	5.812	6.614	7.962	9.312	11.152	12.624	15.338	18.418	20.465	23.542	26.296	29.633	32.000	39.252
17	6.408	7.255	8.672	10.085	12.002	13.531	16.338	19.511	21.615	24.769	27.587	30.995	33.409	40.790
18	7.015	7.906	9.390	10.865	12.857	14.440	17.338	20.601	22.760	25.989	28.869	32.346	34.805	42.312
19	7.633	8.567	10.117	11.651	13.716	15.352	18.338	21.689	23.900	27.204	30.144	33.687	36.191	43.820
20	8.260	9.237	10.851	12.443	14.578	16.266	19.337	22.775	25.038	28.412	31.410	35.020	37.566	45.315
21	8.897	9.915	11.591	13.240	15.445	17.182	20.337	23.858	26.171	29.615	32.671	36.343	38.932	46.797
22	9.542	10.600	12.338	14.041	16.314	18.101	21.337	24.939	27.301	30.813	33.924	37.659	40.289	48.268
23	10.196	11.293	13.091	14.848	17.187	19.021	22.337	26.018	28.429	32.007	35.172	38.968	41.638	49.728
24	10.856	11.992	13.848	15.659	18.062	19.943	23.337	27.096	29.553	33.196	36.415	40.270	42.980	51.179
25	11.524	12.697	14.611	16.473	18.940	20.867	24.337	28.172	30.675	34.382	37.652	41.566	44.314	52.620
26	12.198	13.409	15.379	17.292	19.820	21.792	25.336	29.246	31.795	35.563	38.885	42.856	45.642	54.052
27	12.879	14.125	16.151	18.114	20.703	22.719	26.336	30.319	32.912	36.741	40.113	44.140	46.963	55.476
28	13.565	14.847	16.928	18.939	21.588	23.647	27.336	31.391	34.027	37.916	41.337	45.419	48.278	56.893
29	14.256	15.574	17.708	19.768	22.475	24.577	28.336	32.461	35.139	39.087	42.557	46.693	49.588	58.302
30	14.953	16.306	18.493	20.599	23.364	25.508	29.336	33.530	36.250	40.256	43.773	47.962	50.892	59.703

Table D.4 is taken from Table IV of Fisher and Yates: *Statistical Tables for Biological, Agricultural and Medical Research*, published by Longman Group Ltd., London (previously published by Oliver & Boyd, Edinburgh), and reprinted by permission of the authors and publishers.

[a] Discussed in Section 12.3.

[b] For $v > 30$, the expression $\sqrt{2\chi^2} - \sqrt{2v - 1}$ may be referred to the standard normal distribution, Table D.2.

Table D.5. Upper Percentage Points of the F Distribution[a]

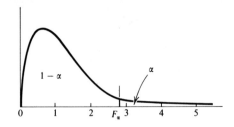

Degrees of Freedom for Denominator, v_2	α	Degrees of Freedom for Numerator, v_1											
		1	2	3	4	5	6	7	8	9	10	11	12
1	.25	5.83	7.50	8.20	8.58	8.82	8.98	9.10	9.19	9.26	9.32	9.36	9.41
	.10	39.9	49.5	53.6	55.8	57.2	58.2	58.9	59.4	59.9	60.2	60.5	60.7
	.05	161	200	216	225	230	234	237	239	241	242	243	244
2	.25	2.57	3.00	3.15	3.23	3.28	3.31	3.34	3.35	3.37	3.38	3.39	3.39
	.10	8.53	9.00	9.16	9.24	9.29	9.33	9.35	9.37	9.38	9.39	9.40	9.41
	.05	18.5	19.0	19.2	19.2	19.3	19.3	19.4	19.4	19.4	19.4	19.4	19.4
	.01	98.5	99.0	99.2	99.2	99.3	99.3	99.4	99.4	99.4	99.4	99.4	99.4
3	.25	2.02	2.28	2.36	2.39	2.41	2.42	2.43	2.44	2.44	2.44	2.45	2.45
	.10	5.54	5.46	5.39	5.34	5.31	5.28	5.27	5.25	5.24	5.23	5.22	5.22
	.05	10.1	9.55	9.28	9.12	9.01	8.94	8.89	8.85	8.81	8.79	8.76	8.74
	.01	34.1	30.8	29.5	28.7	28.2	27.9	27.7	27.5	27.3	27.2	27.1	27.1
4	.25	1.81	2.00	2.05	2.06	2.07	2.08	2.08	2.08	2.08	2.08	2.08	2.08
	.10	4.54	4.32	4.19	4.11	4.05	4.01	3.98	3.95	3.94	3.92	3.91	3.90
	.05	7.71	6.94	6.59	6.39	6.26	6.16	6.09	6.04	6.00	5.96	5.94	5.91
	.01	21.2	18.0	16.7	16.0	15.5	15.2	15.0	14.8	14.7	14.5	14.4	14.4
5	.25	1.69	1.85	1.88	1.89	1.89	1.89	1.89	1.89	1.89	1.89	1.89	1.89
	.10	4.06	3.78	3.62	3.52	3.45	3.40	3.37	3.34	3.32	3.30	3.28	3.27
	.05	6.61	5.79	5.41	5.19	5.05	4.95	4.88	4.82	4.77	4.74	4.71	4.68
	.01	16.3	13.3	12.1	11.4	11.0	10.7	10.5	10.3	10.2	10.1	9.96	9.89
6	.25	1.62	1.76	1.78	1.79	1.79	1.78	1.78	1.78	1.77	1.77	1.77	1.77
	.10	3.78	3.46	3.29	3.18	3.11	3.05	3.01	2.98	2.96	2.94	2.92	2.90
	.05	5.99	5.14	4.76	4.53	4.39	4.28	4.21	4.15	4.10	4.06	4.03	4.00
	.01	13.7	10.9	9.78	9.15	8.75	8.47	8.26	8.10	7.98	7.87	7.79	7.72
7	.25	1.57	1.70	1.72	1.72	1.71	1.71	1.70	1.70	1.69	1.69	1.69	1.68
	.10	3.59	3.26	3.07	2.96	2.88	2.83	2.78	2.75	2.72	2.70	2.68	2.67
	.05	5.59	4.74	4.35	4.12	3.97	3.87	3.79	3.73	3.68	3.64	3.60	3.57
	.01	12.2	9.55	8.45	7.85	7.46	7.19	6.99	6.84	6.72	6.62	6.54	6.47
8	.25	1.54	1.66	1.67	1.66	1.66	1.65	1.64	1.64	1.63	1.63	1.63	1.62
	.10	3.46	3.11	2.92	2.81	2.73	2.67	2.62	2.59	2.56	2.54	2.52	2.50
	.05	5.32	4.46	4.07	3.84	3.69	3.58	3.50	3.44	3.39	3.35	3.31	3.28
	.01	11.3	8.65	7.59	7.01	6.63	6.37	6.18	6.03	5.91	5.81	5.73	5.67
9	.25	1.51	1.62	1.63	1.63	1.62	1.61	1.60	1.60	1.59	1.59	1.58	1.58
	.10	3.36	3.01	2.81	2.69	2.61	2.55	2.51	2.47	2.44	2.42	2.40	2.38
	.05	5.12	4.26	3.86	3.63	3.48	3.37	3.29	3.23	3.18	3.14	3.10	3.07
	.01	10.6	8.02	6.99	6.42	6.06	5.80	5.61	5.47	5.35	5.26	5.18	5.11

Abridged from Table 18 in *Biometrika Tables for Statisticians*, Vol. 1 (3rd Ed.), E. S. Pearson and H. O. Hartley (Eds.). Reprinted by permission of the Biometrika Trustees.

[a] Discussed in Section 14.1.

Table D.5. (Continued)

	Degrees of Freedom for Numerator, v_1												α	Degrees of Freedom for Denominator, v_2
15	20	24	30	40	50	60	100	120	200	500	∞		
9.49	9.58	9.63	9.67	9.71	9.74	9.76	9.78	9.80	9.82	9.84	9.85	.25	
61.2	61.7	62.0	62.3	62.5	62.7	62.8	63.0	63.1	63.2	63.3	63.3	.10	1
246	248	249	250	251	252	252	253	253	254	254	254	.05	
3.41	3.43	3.43	3.44	3.45	3.45	3.46	3.47	3.47	3.48	3.48	3.48	.25	
9.42	9.44	9.45	9.46	9.47	9.47	9.47	9.48	9.48	9.49	9.49	9.49	.10	2
19.4	19.4	19.5	19.5	19.5	19.5	19.5	19.5	19.5	19.5	19.5	19.5	.05	
99.4	99.4	99.5	99.5	99.5	99.5	99.5	99.5	99.5	99.5	99.5	99.5	.01	
2.46	2.46	2.46	2.47	2.47	2.47	2.47	2.47	2.47	2.47	2.47	2.47	.25	
5.20	5.18	5.18	5.17	5.16	5.15	5.15	5.14	5.14	5.14	5.14	5.13	.10	3
8.70	8.66	8.64	8.62	8.59	8.58	8.57	8.55	8.55	8.54	8.53	8.53	.05	
26.9	26.7	26.6	26.5	26.4	26.4	26.3	26.2	26.2	26.2	26.1	26.1	.01	
2.08	2.08	2.08	2.08	2.08	2.08	2.08	2.08	2.08	2.08	2.08	2.08	.25	
3.87	3.84	3.83	3.82	3.80	3.80	3.79	3.78	3.78	3.77	3.76	3.76	.10	4
5.86	5.80	5.77	5.75	5.72	5.70	5.69	5.66	5.66	5.65	5.64	5.63	.05	
14.2	14.0	13.9	13.8	13.7	13.7	13.7	13.6	13.6	13.5	13.5	13.5	.01	
1.89	1.88	1.88	1.88	1.88	1.88	1.87	1.87	1.87	1.87	1.87	1.87	.25	
3.24	3.21	3.19	3.17	3.16	3.15	3.14	3.13	3.12	3.12	3.11	3.10	.10	5
4.62	4.56	4.53	4.50	4.46	4.44	4.43	4.41	4.40	4.39	4.37	4.36	.05	
9.72	9.55	9.47	9.38	9.29	9.24	9.20	9.13	9.11	9.08	9.04	9.02	.01	
1.76	1.76	1.75	1.75	1.75	1.75	1.74	1.74	1.74	1.74	1.74	1.74	.25	
2.87	2.84	2.82	2.80	2.78	2.77	2.76	2.75	2.74	2.73	2.73	2.72	.10	6
3.94	3.87	3.84	3.81	3.77	3.75	3.74	3.71	3.70	3.69	3.68	3.67	.05	
7.56	7.40	7.31	7.23	7.14	7.09	7.06	6.99	6.97	6.93	6.90	6.88	.01	
1.68	1.67	1.67	1.66	1.66	1.66	1.65	1.65	1.65	1.65	1.65	1.65	.25	
2.63	2.59	2.58	2.56	2.54	2.52	2.51	2.50	2.49	2.48	2.48	2.47	.10	7
3.51	3.44	3.41	3.38	3.34	3.32	3.30	3.27	3.27	3.25	3.24	3.23	.05	
6.31	6.16	6.07	5.99	5.91	5.86	5.82	5.75	5.74	5.70	5.67	5.65	.10	
1.62	1.61	1.60	1.60	1.59	1.59	1.59	1.58	1.58	1.58	1.58	1.58	.25	
2.46	2.42	2.40	2.38	2.36	2.35	2.34	2.32	2.32	2.31	2.30	2.29	.10	8
3.22	3.15	3.12	3.08	3.04	3.02	3.01	2.97	2.97	2.95	2.94	2.93	.05	
5.52	5.36	5.28	5.20	5.12	5.07	5.03	4.96	4.95	4.91	4.88	4.86	.01	
1.57	1.56	1.56	1.55	1.55	1.54	1.54	1.53	1.53	1.53	1.53	1.53	.25	
2.34	2.30	2.28	2.25	2.23	2.22	2.21	2.19	2.18	2.17	2.17	2.16	.10	9
3.01	2.94	2.90	2.86	2.83	2.80	2.79	2.76	2.75	2.73	2.72	2.71	.05	
4.96	4.81	4.73	4.65	4.57	4.52	4.48	4.42	4.40	4.36	4.33	4.31	.01	

Table D.5. (Continued)

Degrees of Freedom for Denominator, v_2	α	\multicolumn{12}{c}{Degrees of Freedom for Numerator, v_1}
		1	2	3	4	5	6	7	8	9	10	11	12
10	.25	1.49	1.60	1.60	1.59	1.59	1.58	1.57	1.56	1.56	1.55	1.55	1.54
	.10	3.29	2.92	2.73	2.61	2.52	2.46	2.41	2.38	2.35	2.32	2.30	2.28
	.05	4.96	4.10	3.71	3.48	3.33	3.22	3.14	3.07	3.02	2.98	2.94	2.91
	.01	10.0	7.56	6.55	5.99	5.64	5.39	5.20	5.06	4.94	4.85	4.77	4.71
11	.25	1.47	1.58	1.58	1.57	1.56	1.55	1.54	1.53	1.53	1.52	1.52	1.51
	.10	3.23	2.86	2.66	2.54	2.45	2.39	2.34	2.30	2.27	2.25	2.23	2.21
	.05	4.84	3.98	3.59	3.36	3.20	3.09	3.01	2.95	2.90	2.85	2.82	2.79
	.01	9.65	7.21	6.22	5.67	5.32	5.07	4.89	4.74	4.63	4.54	4.46	4.40
12	.25	1.46	1.56	1.56	1.55	1.54	1.53	1.52	1.51	1.51	1.50	1.50	1.49
	.10	3.18	2.81	2.61	2.48	2.39	2.33	2.28	2.24	2.21	2.19	2.17	2.15
	.05	4.75	3.89	3.49	3.26	3.11	3.00	2.91	2.85	2.80	2.75	2.72	2.69
	.01	9.33	6.93	5.95	5.41	5.06	4.82	4.64	4.50	4.39	4.30	4.22	4.16
13	.25	1.45	1.55	1.55	1.53	1.52	1.51	1.50	1.49	1.49	1.48	1.47	1.47
	.10	3.14	2.76	2.56	2.43	2.35	2.28	2.23	2.20	2.16	2.14	2.12	2.10
	.05	4.67	3.81	3.41	3.18	3.03	2.92	2.83	2.77	2.71	2.67	2.63	2.60
	.01	9.07	6.70	5.74	5.21	4.86	4.62	4.44	4.30	4.19	4.10	4.02	3.96
14	.25	1.44	1.53	1.53	1.52	1.51	1.50	1.49	1.48	1.47	1.46	1.46	1.45
	.10	3.10	2.73	2.52	2.39	2.31	2.24	2.19	2.15	2.12	2.10	2.08	2.05
	.05	4.60	3.74	3.34	3.11	2.96	2.85	2.76	2.70	2.65	2.60	2.57	2.53
	.01	8.86	6.51	5.56	5.04	4.69	4.46	4.28	4.14	4.03	3.94	3.86	3.80
15	.25	1.43	1.52	1.52	1.51	1.49	1.48	1.47	1.46	1.46	1.45	1.44	1.44
	.10	3.07	2.70	2.49	2.36	2.27	2.21	2.16	2.12	2.09	2.06	2.04	2.02
	.05	4.54	3.68	3.29	3.06	2.90	2.79	2.71	2.64	2.59	2.54	2.51	2.48
	.01	8.68	6.36	5.42	4.89	4.56	4.32	4.14	4.00	3.89	3.80	3.73	3.67
16	.25	1.42	1.51	1.51	1.50	1.48	1.47	1.46	1.45	1.44	1.44	1.44	1.43
	.10	3.05	2.67	2.46	2.33	2.24	2.18	2.13	2.09	2.06	2.03	2.01	1.99
	.05	4.49	3.63	3.24	3.01	2.85	2.74	2.66	2.59	2.54	2.49	2.46	2.42
	.01	8.53	6.23	5.29	4.77	4.44	4.20	4.03	3.89	3.78	3.69	3.62	3.55
17	.25	1.42	1.51	1.50	1.49	1.47	1.46	1.45	1.44	1.43	1.43	1.42	1.41
	.10	3.03	2.64	2.44	2.31	2.22	2.15	2.10	2.06	2.03	2.00	1.98	1.96
	.05	4.45	3.59	3.20	2.96	2.81	2.70	2.61	2.55	2.49	2.45	2.41	2.38
	.01	8.40	6.11	5.18	4.67	4.34	4.10	3.93	3.79	3.68	3.59	3.52	3.46
18	.25	1.41	1.50	1.49	1.48	1.46	1.45	1.44	1.43	1.42	1.42	1.41	1.40
	.10	3.01	2.62	2.42	2.29	2.20	2.13	2.08	2.04	2.00	1.98	1.96	1.93
	.05	4.41	3.55	3.16	2.93	2.77	2.66	2.58	2.51	2.46	2.41	2.37	2.34
	.01	8.29	6.01	5.09	4.58	4.25	4.01	3.84	3.71	3.60	3.51	3.43	3.37
19	.25	1.41	1.49	1.49	1.47	1.46	1.44	1.43	1.42	1.41	1.41	1.40	1.40
	.10	2.99	2.61	2.40	2.27	2.18	2.11	2.06	2.02	1.98	1.96	1.94	1.91
	.05	4.38	3.52	3.13	2.90	2.74	2.63	2.54	2.48	2.42	2.38	2.34	2.31
	.01	8.18	5.93	5.01	4.50	4.17	3.94	3.77	3.63	3.52	3.43	3.36	3.30
20	.25	1.40	1.49	1.48	1.46	1.45	1.44	1.43	1.42	1.41	1.40	1.39	1.39
	.10	2.97	2.59	2.38	2.25	2.16	2.09	2.04	2.00	1.96	1.94	1.92	1.89
	.05	4.35	3.49	3.10	2.87	2.71	2.60	2.51	2.45	2.39	2.35	2.31	2.28
	.01	8.10	5.85	4.94	4.43	4.10	3.87	3.70	3.56	3.46	3.37	3.29	3.23

Table D.5. (Continued)

\	Degrees of Freedom for Numerator, v_1											α	Degrees of Freedom for Denominator, v_2
15	20	24	30	40	50	60	100	120	200	500	∞		
1.53	1.52	1.52	1.51	1.51	1.50	1.50	1.49	1.49	1.49	1.48	1.48	.25	
2.24	2.20	2.18	2.16	2.13	2.12	2.11	2.09	2.08	2.07	2.06	2.06	.10	10
2.85	2.77	2.74	2.70	2.66	2.64	2.62	2.59	2.58	2.56	2.55	2.54	.05	
4.56	4.41	4.33	4.25	4.17	4.12	4.08	4.01	4.00	3.96	3.93	3.91	.01	
1.50	1.49	1.49	1.48	1.47	1.47	1.47	1.46	1.46	1.46	1.45	1.45	.25	
2.17	2.12	2.10	2.08	2.05	2.04	2.03	2.00	2.00	1.99	1.98	1.97	.10	11
2.72	2.65	2.61	2.57	2.53	2.51	2.49	2.46	2.45	2.43	2.42	2.40	.05	
4.25	4.10	4.02	3.94	3.86	3.81	3.78	3.71	3.69	3.66	3.62	3.60	.01	
1.48	1.47	1.46	1.45	1.45	1.44	1.44	1.43	1.43	1.43	1.42	1.42	.25	
2.10	2.06	2.04	2.01	1.99	1.97	1.96	1.94	1.93	1.92	1.91	1.90	.10	12
2.62	2.54	2.51	2.47	2.43	2.40	2.38	2.35	2.34	2.32	2.31	2.30	.05	
4.01	3.86	3.78	3.70	3.62	3.57	3.54	3.47	3.45	3.41	3.38	3.36	.01	
1.46	1.45	1.44	1.43	1.42	1.42	1.42	1.41	1.41	1.40	1.40	1.40	.25	
2.05	2.01	1.98	1.96	1.93	1.92	1.90	1.88	1.88	1.86	1.85	1.85	.10	13
2.53	2.46	2.42	2.38	2.34	2.31	2.30	2.26	2.25	2.23	2.22	2.21	.05	
3.82	3.66	3.59	3.51	3.43	3.38	3.34	3.27	3.25	3.22	3.19	3.17	.01	
1.44	1.43	1.42	1.41	1.41	1.40	1.40	1.39	1.39	1.39	1.38	1.38	.25	
2.01	1.96	1.94	1.91	1.89	1.87	1.86	1.83	1.83	1.82	1.80	1.80	.10	14
2.46	2.39	2.35	2.31	2.27	2.24	2.22	2.19	2.18	2.16	2.14	2.13	.05	
3.66	3.51	3.43	3.35	3.27	3.22	3.18	3.11	3.09	3.06	3.03	3.00	.01	
1.43	1.41	1.41	1.40	1.39	1.39	1.38	1.38	1.37	1.37	1.36	1.36	.25	
1.97	1.92	1.90	1.87	1.85	1.83	1.82	1.79	1.79	1.77	1.76	1.76	.10	15
2.40	2.33	2.29	2.25	2.20	2.18	2.16	2.12	2.11	2.10	2.08	2.07	.05	
3.52	3.37	3.29	3.21	3.13	3.08	3.05	2.98	2.96	2.92	2.89	2.87	.01	
1.41	1.40	1.39	1.38	1.37	1.37	1.36	1.36	1.35	1.35	1.34	1.34	.25	
1.94	1.89	1.87	1.84	1.81	1.79	1.78	1.76	1.75	1.74	1.73	1.72	.10	16
2.35	2.28	2.24	2.19	2.15	2.12	2.11	2.07	2.06	2.04	2.02	2.01	.05	
3.41	3.26	3.18	3.10	3.02	2.97	2.93	2.86	2.84	2.81	2.78	2.75	.01	
1.40	1.39	1.38	1.37	1.36	1.35	1.35	1.34	1.34	1.34	1.33	1.33	.25	
1.91	1.86	1.84	1.81	1.78	1.76	1.75	1.73	1.72	1.71	1.69	1.69	.10	17
2.31	2.23	2.19	2.15	2.10	2.08	2.06	2.02	2.01	1.99	1.97	1.96	.05	
3.31	3.16	3.08	3.00	2.92	2.87	2.83	2.76	2.75	2.71	2.68	2.65	.01	
1.39	1.38	1.37	1.36	1.35	1.34	1.34	1.33	1.33	1.32	1.32	1.32	.25	
1.89	1.84	1.81	1.78	1.75	1.74	1.72	1.70	1.69	1.68	1.67	1.66	.10	18
2.27	2.19	2.15	2.11	2.06	2.04	2.02	1.98	1.97	1.95	1.93	1.92	.05	
3.23	3.08	3.00	2.92	2.84	2.78	2.75	2.68	2.66	2.62	2.59	2.57	.01	
1.38	1.37	1.36	1.35	1.34	1.33	1.33	1.32	1.32	1.31	1.31	1.30	.25	
1.86	1.81	1.79	1.76	1.73	1.71	1.70	1.67	1.67	1.65	1.64	1.63	.10	19
2.23	2.16	2.11	2.07	2.03	2.00	1.98	1.94	1.93	1.91	1.89	1.88	.05	
3.15	3.00	2.92	2.84	2.76	2.71	2.67	2.60	2.58	2.55	2.51	2.49	.01	
1.37	1.36	1.35	1.34	1.33	1.33	1.32	1.31	1.31	1.30	1.30	1.29	.25	
1.84	1.79	1.77	1.74	1.71	1.69	1.68	1.65	1.64	1.63	1.62	1.61	.10	20
2.20	2.12	2.08	2.04	1.99	1.97	1.95	1.91	1.90	1.88	1.86	1.84	.05	
3.09	2.94	2.86	2.78	2.69	2.64	2.61	2.54	2.52	2.48	2.44	2.42	.01	

Table D.5. (Continued)

Degrees of Freedom for Denominator, v_2	α	Degrees of Freedom for Numerator, v_1											
		1	2	3	4	5	6	7	8	9	10	11	12
22	.25	1.40	1.48	1.47	1.45	1.44	1.42	1.41	1.40	1.39	1.39	1.38	1.37
	.10	2.95	2.56	2.35	2.22	2.13	2.06	2.01	1.97	1.93	1.90	1.88	1.86
	.05	4.30	3.44	3.05	2.82	2.66	2.55	2.46	2.40	2.34	2.30	2.26	2.23
	.01	7.95	5.72	4.82	4.31	3.99	3.76	3.59	3.45	3.35	3.26	3.18	3.12
24	.25	1.39	1.47	1.46	1.44	1.43	1.41	1.40	1.39	1.38	1.38	1.37	1.36
	.10	2.93	2.54	2.33	2.19	2.10	2.04	1.98	1.94	1.91	1.88	1.85	1.83
	.05	4.26	3.40	3.01	2.78	2.62	2.51	2.42	2.36	2.30	2.25	2.21	2.18
	.01	7.82	5.61	4.72	4.22	3.90	3.67	3.50	3.36	3.26	3.17	3.09	3.03
26	.25	1.38	1.46	1.45	1.44	1.42	1.41	1.39	1.38	1.37	1.37	1.36	1.35
	.10	2.91	2.52	2.31	2.17	2.08	2.01	1.96	1.92	1.88	1.86	1.84	1.81
	.05	4.23	3.37	2.98	2.74	2.59	2.47	2.39	2.32	2.27	2.22	2.18	2.15
	.01	7.72	5.53	4.64	4.14	3.82	3.59	3.42	3.29	3.18	3.09	3.02	2.96
28	.25	1.38	1.46	1.45	1.43	1.41	1.40	1.39	1.38	1.37	1.36	1.35	1.34
	.10	2.89	2.50	2.29	2.16	2.06	2.00	1.94	1.90	1.87	1.84	1.81	1.79
	.05	4.20	3.34	2.95	2.71	2.56	2.45	2.36	2.29	2.24	2.19	2.15	2.12
	.01	7.64	5.45	4.57	4.07	3.75	3.53	3.36	3.23	3.12	3.03	2.96	2.90
30	.25	1.38	1.45	1.44	1.42	1.41	1.39	1.38	1.37	1.36	1.35	1.35	1.34
	.10	2.88	2.49	2.28	2.14	2.05	1.98	1.93	1.88	1.85	1.82	1.79	1.77
	.05	4.17	3.32	2.92	2.69	2.53	2.42	2.33	2.27	2.21	2.16	2.13	2.09
	.01	7.56	5.39	4.51	4.02	3.70	3.47	3.30	3.17	3.07	2.98	2.91	2.84
40	.25	1.36	1.44	1.42	1.40	1.39	1.37	1.36	1.35	1.34	1.33	1.32	1.31
	.10	2.84	2.44	2.23	2.09	2.00	1.93	1.87	1.83	1.79	1.76	1.73	1.71
	.05	4.08	3.23	2.84	2.61	2.45	2.34	2.25	2.18	2.12	2.08	2.04	2.00
	.01	7.31	5.18	4.31	3.83	3.51	3.29	3.12	2.99	2.89	2.80	2.73	2.66
60	.25	1.35	1.42	1.41	1.38	1.37	1.35	1.33	1.32	1.31	1.30	1.29	1.29
	.10	2.79	2.39	2.18	2.04	1.95	1.87	1.82	1.77	1.74	1.71	1.68	1.66
	.05	4.00	3.15	2.76	2.53	2.37	2.25	2.17	2.10	2.04	1.99	1.95	1.92
	.01	7.08	4.98	4.13	3.65	3.34	3.12	2.95	2.82	2.72	2.63	2.56	2.50
120	.25	1.34	1.40	1.39	1.37	1.35	1.33	1.31	1.30	1.29	1.28	1.27	1.26
	.10	2.75	2.35	2.13	1.99	1.90	1.82	1.77	1.72	1.68	1.65	1.62	1.60
	.05	3.92	3.07	2.68	2.45	2.29	2.17	2.09	2.02	1.96	1.91	1.87	1.83
	.01	6.85	4.79	3.95	3.48	3.17	2.96	2.79	2.66	2.56	2.47	2.40	2.34
200	.25	1.33	1.39	1.38	1.36	1.34	1.32	1.31	1.29	1.28	1.27	1.26	1.25
	.10	2.73	2.33	2.11	1.97	1.88	1.80	1.75	1.70	1.66	1.63	1.60	1.57
	.05	3.89	3.04	2.65	2.42	2.26	2.14	2.06	1.98	1.93	1.88	1.84	1.80
	.01	6.76	4.71	3.88	3.41	3.11	2.89	2.73	2.60	2.50	2.41	2.34	2.27
∞	.25	1.32	1.39	1.37	1.35	1.33	1.31	1.29	1.28	1.27	1.25	1.24	1.24
	.10	2.71	2.30	2.08	1.94	1.85	1.77	1.72	1.67	1.63	1.60	1.57	1.55
	.05	3.84	3.00	2.60	2.37	2.21	2.10	2.01	1.94	1.88	1.83	1.79	1.75
	.01	6.63	4.61	3.78	3.32	3.02	2.80	2.64	2.51	2.41	2.32	2.25	2.18

Table D.5. (Continued)

	Degrees of Freedom for Numerator, v_1													Degrees of Freedom for Denominator, v_2
15	20	24	30	40	50	60	100	120	200	500	∞	α		
1.36	1.34	1.33	1.32	1.31	1.31	1.30	1.30	1.30	1.29	1.29	1.28	.25		
1.81	1.76	1.73	1.70	1.67	1.65	1.64	1.61	1.60	1.59	1.58	1.57	.10		22
2.15	2.07	2.03	1.98	1.94	1.91	1.89	1.85	1.84	1.82	1.80	1.78	.05		
2.98	2.83	2.75	2.67	2.58	2.53	2.50	2.42	2.40	2.36	2.33	2.31	.01		
1.35	1.33	1.32	1.31	1.30	1.29	1.29	1.28	1.28	1.27	1.27	1.26	.25		
1.78	1.73	1.70	1.67	1.64	1.62	1.61	1.58	1.57	1.56	1.54	1.53	.10		24
2.11	2.03	1.98	1.94	1.89	1.86	1.84	1.80	1.79	1.77	1.75	1.73	.05		
2.89	2.74	2.66	2.58	2.49	2.44	2.40	2.33	2.31	2.27	2.24	2.21	.01		
1.34	1.32	1.31	1.30	1.29	1.28	1.28	1.26	1.26	1.26	1.25	1.25	.25		
1.76	1.71	1.68	1.65	1.61	1.59	1.58	1.55	1.54	1.53	1.51	1.50	.10		26
2.07	1.99	1.95	1.90	1.85	1.82	1.80	1.76	1.75	1.73	1.71	1.69	.05		
2.81	2.66	2.58	2.50	2.42	2.36	2.33	2.25	2.23	2.19	2.16	2.13	.01		
1.33	1.31	1.30	1.29	1.28	1.27	1.27	1.26	1.25	1.25	1.24	1.24	.25		
1.74	1.69	1.66	1.63	1.59	1.57	1.56	1.53	1.52	1.50	1.49	1.48	.10		28
2.04	1.96	1.91	1.87	1.82	1.79	1.77	1.73	1.71	1.69	1.67	1.65	.05		
2.75	2.60	2.52	2.44	2.35	2.30	2.26	2.19	2.17	2.13	2.09	2.06	.01		
1.32	1.30	1.29	1.28	1.27	1.26	1.26	1.25	1.24	1.24	1.23	1.23	.25		
1.72	1.67	1.64	1.61	1.57	1.55	1.54	1.51	1.50	1.48	1.47	1.46	.10		30
2.01	1.93	1.89	1.84	1.79	1.76	1.74	1.70	1.68	1.66	1.64	1.62	.05		
2.70	2.55	2.47	2.39	2.30	2.25	2.21	2.13	2.11	2.07	2.03	2.01	.01		
1.30	1.28	1.26	1.25	1.24	1.23	1.22	1.21	1.21	1.20	1.19	1.19	.25		
1.66	1.61	1.57	1.54	1.51	1.48	1.47	1.43	1.42	1.41	1.39	1.38	.10		40
1.92	1.84	1.79	1.74	1.69	1.66	1.64	1.59	1.58	1.55	1.53	1.51	.05		
2.52	2.37	2.29	2.20	2.11	2.06	2.02	1.94	1.92	1.87	1.83	1.80	.01		
1.27	1.25	1.24	1.22	1.21	1.20	1.19	1.17	1.17	1.16	1.15	1.15	.25		
1.60	1.54	1.51	1.48	1.44	1.41	1.40	1.36	1.35	1.33	1.31	1.29	.10		60
1.84	1.75	1.70	1.65	1.59	1.56	1.53	1.48	1.47	1.44	1.41	1.39	.05		
2.35	2.20	2.12	2.03	1.94	1.88	1.84	1.75	1.73	1.68	1.63	1.60	.01		
1.24	1.22	1.21	1.19	1.18	1.17	1.16	1.14	1.13	1.12	1.11	1.10	.25		
1.55	1.48	1.45	1.41	1.37	1.34	1.32	1.27	1.26	1.24	1.21	1.19	.10		120
1.75	1.66	1.61	1.55	1.50	1.46	1.43	1.37	1.35	1.32	1.28	1.25	.05		
2.19	2.03	1.95	1.86	1.76	1.70	1.66	1.56	1.53	1.48	1.42	1.38	.01		
1.23	1.21	1.20	1.18	1.16	1.14	1.12	1.11	1.10	1.09	1.08	1.06	.25		
1.52	1.46	1.42	1.38	1.34	1.31	1.28	1.24	1.22	1.20	1.17	1.14	.10		200
1.72	1.62	1.57	1.52	1.46	1.41	1.39	1.32	1.29	1.26	1.22	1.19	.05		
2.13	1.97	1.89	1.79	1.69	1.63	1.58	1.48	1.44	1.39	1.33	1.28	.01		
1.22	1.19	1.18	1.16	1.14	1.13	1.12	1.09	1.08	1.07	1.04	1.00	.25		
1.49	1.42	1.38	1.34	1.30	1.26	1.24	1.18	1.17	1.13	1.08	1.00	.10		∞
1.67	1.57	1.52	1.46	1.39	1.35	1.32	1.24	1.22	1.17	1.11	1.00	.05		
2.04	1.88	1.79	1.70	1.59	1.52	1.47	1.36	1.32	1.25	1.15	1.00	.01		

Table D.6. Critical Values of the Pearson r [a]

Degrees of Freedom, $v = n - 2$ [b]	Level of Significance for a One-Tailed Test			
	0.05	0.025	0.01	0.005
	Level of Significance for a Two-Tailed Test			
	0.10	0.05	0.02	0.01
1	0.988	0.997	0.9995	0.9999
2	0.900	0.950	0.980	0.990
3	0.805	0.378	0.934	0.959
4	0.729	0.811	0.882	0.917
5	0.669	0.754	0.833	0.874
6	0.622	0.707	0.789	0.834
7	0.582	0.666	0.750	0.798
8	0.549	0.632	0.716	0.765
9	0.521	0.602	0.685	0.735
10	0.497	0.576	0.658	0.708
11	0.476	0.553	0.634	0.684
12	0.458	0.532	0.612	0.661
13	0.441	0.514	0.592	0.641
14	0.426	0.497	0.574	0.623
15	0.412	0.482	0.558	0.606
16	0.400	0.468	0.542	0.590
17	0.389	0.456	0.528	0.575
18	0.378	0.444	0.516	0.561
19	0.369	0.433	0.503	0.549
20	0.360	0.423	0.492	0.537
21	0.352	0.413	0.482	0.526
22	0.344	0.404	0.472	0.515
23	0.337	0.396	0.462	0.505
24	0.330	0.388	0.453	0.496
25	0.323	0.381	0.445	0.487
26	0.317	0.374	0.437	0.479
27	0.311	0.367	0.430	0.471
28	0.306	0.361	0.423	0.463
29	0.301	0.355	0.416	0.456
30	0.296	0.349	0.409	0.449
35	0.275	0.325	0.381	0.418
40	0.257	0.304	0.358	0.393
45	0.243	0.288	0.338	0.372
50	0.231	0.273	0.322	0.354
60	0.211	0.250	0.295	0.325
70	0.195	0.232	0.274	0.302
80	0.183	0.217	0.256	0.283
90	0.173	0.205	0.242	0.267
100	0.164	0.195	0.230	0.254
120	0.150	0.178	0.210	0.232
150	0.134	0.159	0.189	0.208
200	0.116	0.138	0.164	0.181
300	0.095	0.113	0.134	0.148
400	0.082	0.098	0.116	0.128
500	0.073	0.088	0.104	0.115

Table D.6 is taken from Table VII of Fisher and Yates: *Statistical Tables for Biological, Agricultural and Medical Research*, published by Longman Group Ltd., London (previously published by Oliver & Boyd, Edinburgh), and reprinted by permission of the authors and publishers.

[a] Discussed in Section 12.5. [b] n is the number of pairs.

Table D.7. Critical Values of r_s (Spearman Rank Correlation Coefficient)[a]

Number of Pairs, n	Level of Significance for a One-Tailed Test			
	0.05	0.025	0.01	0.005
	Level of Significance for a Two-Tailed Test			
	0.10	0.05	0.02	0.01
5	0.900	1.000	1.000	—
6	0.829	0.886	0.943	1.000
7	0.714	0.786	0.893	0.929
8	0.643	0.738	0.833	0.881
9	0.600	0.683	0.783	0.833
10	0.564	0.648	0.746	0.794
12	0.506	0.591	0.712	0.777
14	0.456	0.544	0.645	0.715
16	0.425	0.506	0.601	0.665
18	0.399	0.475	0.564	0.625
20	0.377	0.450	0.534	0.591
22	0.359	0.428	0.508	0.562
24	0.343	0.409	0.485	0.537
26	0.329	0.392	0.465	0.515
28	0.317	0.377	0.448	0.496
30	0.306	0.364	0.432	0.478

From "Distributions of Sums of Squares of Rank Differences for Small Numbers of Individuals," by E. G. Olds, *Annals of Mathematical Statistics*, 1938, *9*, 133–148, and from "The 5 Per Cent Significance Levels for Sums of Squares of Rank Differences and a Correction," by E. G. Olds, *Annals of Mathematical Statistics*, 1949, *20*, 117–118. Reprinted by permission.

[a] Discussed in Section 12.5.

Table D.8. Transformation of r to Z' [a]

r	Z'	r	Z'	r	Z'	r	Z'	r	Z'
0.000	0.000	0.200	0.203	0.400	0.424	0.600	0.693	0.800	1.099
0.005	0.005	0.205	0.208	0.405	0.430	0.605	0.701	0.805	1.113
0.010	0.010	0.210	0.213	0.410	0.436	0.610	0.709	0.810	1.127
0.015	0.015	0.215	0.218	0.415	0.442	0.615	0.717	0.815	1.142
0.020	0.020	0.220	0.224	0.420	0.448	0.620	0.725	0.820	1.157
0.025	0.025	0.225	0.229	0.425	0.454	0.625	0.733	0.825	1.172
0.030	0.030	0.230	0.234	0.430	0.460	0.630	0.741	0.830	1.188
0.035	0.035	0.235	0.239	0.435	0.466	0.635	0.750	0.835	1.204
0.040	0.040	0.240	0.245	0.440	0.472	0.640	0.758	0.840	1.221
0.045	0.045	0.245	0.250	0.445	0.478	0.645	0.767	0.845	1.238
0.050	0.050	0.250	0.255	0.450	0.485	0.650	0.775	0.850	1.256
0.055	0.055	0.255	0.261	0.455	0.491	0.655	0.784	0.855	1.274
0.060	0.060	0.260	0.266	0.460	0.497	0.660	0.793	0.860	1.293
0.065	0.065	0.265	0.271	0.465	0.504	0.665	0.802	0.865	1.313
0.070	0.070	0.270	0.277	0.470	0.510	0.670	0.811	0.870	1.333
0.075	0.075	0.275	0.282	0.475	0.517	0.675	0.820	0.875	1.354
0.080	0.080	0.280	0.288	0.480	0.523	0.680	0.829	0.880	1.376
0.085	0.085	0.285	0.293	0.485	0.530	0.685	0.838	0.885	1.398
0.090	0.090	0.290	0.299	0.490	0.536	0.690	0.848	0.890	1.422
0.095	0.095	0.295	0.304	0.495	0.543	0.695	0.858	0.895	1.447
0.100	0.100	0.300	0.310	0.500	0.549	0.700	0.867	0.900	1.472
0.105	0.105	0.305	0.315	0.505	0.556	0.705	0.877	0.905	1.499
0.110	0.110	0.310	0.321	0.510	0.563	0.710	0.887	0.910	1.528
0.115	0.116	0.315	0.326	0.515	0.570	0.715	0.897	0.915	1.557
0.120	0.121	0.320	0.332	0.520	0.576	0.720	0.908	0.920	1.589
0.125	0.126	0.325	0.337	0.525	0.583	0.725	0.918	0.925	1.623
0.130	0.131	0.330	0.343	0.530	0.590	0.730	0.929	0.930	1.658
0.135	0.136	0.335	0.348	0.535	0.597	0.735	0.940	0.935	1.697
0.140	0.141	0.340	0.354	0.540	0.604	0.740	0.950	0.940	1.738
0.145	0.146	0.345	0.360	0.545	0.611	0.745	0.962	0.945	1.783
0.150	0.151	0.350	0.365	0.550	0.618	0.750	0.973	0.950	1.832
0.155	0.156	0.355	0.371	0.555	0.626	0.755	0.984	0.955	1.886
0.160	0.161	0.360	0.377	0.560	0.633	0.760	0.996	0.960	1.946
0.165	0.167	0.365	0.383	0.565	0.640	0.765	1.008	0.965	2.014
0.170	0.172	0.370	0.388	0.570	0.648	0.770	1.020	0.970	2.092
0.175	0.177	0.375	0.394	0.575	0.655	0.775	1.033	0.975	2.185
0.180	0.182	0.380	0.400	0.580	0.662	0.780	1.045	0.980	2.298
0.185	0.187	0.385	0.406	0.585	0.670	0.785	1.058	0.985	2.443
0.190	0.192	0.390	0.412	0.590	0.678	0.790	1.071	0.990	2.647
0.195	0.198	0.395	0.418	0.595	0.685	0.795	1.085	0.995	2.994

Table D.8 is taken from Table VIIi of Fisher and Yates: *Statistical Tables for Biological, Agricultural and Medical Research*, published by Longman Group Ltd., London (previously published by Oliver & Boyd, Edinburgh), and reprinted by permission of the authors and publishers.

[a] Discussed in Section 12.5.

Table D.9. Approximate *n* Required for Testing Hypotheses about Means[a]

One-Sample Test

Effect Size, d	α	One-Tailed Hypothesis, $1-\beta$			Two-Tailed Hypothesis, $1-\beta$		
		0.80	0.90	0.95	0.80	0.90	0.95
0.2	0.05	156	215	272	198	264	326
	0.01	253	328	396	294	374	447
0.5	0.05	27	36	45	34	44	54
	0.01	43	55	66	51	63	75
0.8	0.05	12	15	19	15	19	22
	0.01	19	24	28	22	27	32

Two-Sample Test (Independent Samples)

Effect Size, d	α	0.80	0.90	0.95	0.80	0.90	0.95
0.2	0.05	310	429	542	393	526	651
	0.01	503	652	790	586	746	892
0.5	0.05	50	69	87	64	85	105
	0.01	82	105	128	95	120	144
0.8	0.05	21	27	35	26	34	42
	0.01	33	42	51	38	48	57

Two-Sample Test (Dependent Samples)

Effect Size, d	α	ρ	0.80	0.90	0.95	0.80	0.90	0.95
0.2	0.05	0.4	187	258	326	237	317	391
		0.5	156	215	272	198	264	326
		0.6	125	172	218	159	212	261
		0.7	94	130	164	119	159	197
		0.8	63	87	109	80	107	131
		0.9	32	44	55	41	54	66
	0.01	0.4	303	393	475	353	449	537
		0.5	253	328	396	294	374	447
		0.6	203	262	317	236	300	358
		0.7	153	197	239	177	225	269
		0.8	102	132	160	119	151	180
		0.9	52	67	81	60	76	91
0.5	0.05	0.4	31	42	53	39	52	63
		0.5	26	35	44	33	44	53
		0.6	21	29	36	27	35	43
		0.7	16	22	27	20	27	33
		0.8	11	15	19	14	18	22
		0.9	6	8	10	8	10	12

[a] Discussed in Sections 12.2 and 13.5. For the two-sample test (independent samples), it is assumed that $\sigma_1^2 = \sigma_2^2$ and $n_1 = n_2$; the values in the table are for each of the samples. If dependent samples are used, the values in the table are for the number of pairs of dependent elements.

Table D.9. (Continued)

			Two-Sample Test (Dependent Samples)						
		ρ							
0.8	0.01	0.4	50	65	78	58	73	88
		0.5	42	54	65	49	62	73
		0.6	34	44	52	39	50	59
		0.7	26	33	40	30	38	45
		0.8	18	23	27	21	26	31
		0.9	10	12	15	11	14	16
	0.05	0.4	13	17	21	16	21	26
		0.5	11	15	18	13	18	22
		0.6	9	12	15	11	15	18
		0.7	7	9	11	9	11	14
		0.8	5	7	8	6	8	10
		0.9	3	4	5	4	5	6
	0.01	0.4	21	26	32	24	30	35
		0.5	18	22	27	20	25	30
		0.6	15	18	22	17	21	24
		0.7	11	14	17	13	16	19
		0.8	8	10	12	9	11	13
		0.9	5	6	7	6	7	8

Table D.10. Critical Values of the Mann–Whitney U.[a] For a one-tailed test at $\alpha = 0.01$ (roman type) and $\alpha = 0.005$ (boldface type) and for a two-tailed test at $\alpha = 0.02$ (roman type) and $\alpha = 0.01$ (boldface type).

n_2 \ n_1	1	2	3	4	5	6	7	8	9	10	11	12	13	14	15	16	17	18	19	20
1	—[b]	—	—	—	—	—	—	—	—	—	—	—	—	—	—	—	—	—	—	—
2	—	—	—	—	—	—	—	—	—	—	—	—	0	0	0	0	0	0	1	1
	—	—	—	—	—	—	—	—	—	—	—	—	—	—	—	—	—	—	**0**	**0**
3	—	—	—	—	—	—	0	0	1	1	1	2	2	2	3	3	4	4	4	5
	—	—	—	—	—	—	—	—	**0**	**0**	**0**	**1**	**1**	**1**	**2**	**2**	**2**	**2**	**3**	**3**
4	—	—	—	—	0	1	1	2	3	3	4	5	5	6	7	7	8	9	9	10
	—	—	—	—	—	**0**	**0**	**1**	**1**	**2**	**2**	**3**	**3**	**4**	**5**	**5**	**6**	**6**	**7**	**8**
5	—	—	—	0	1	2	3	4	5	6	7	8	9	10	11	12	13	14	15	16
	—	—	—	—	**0**	**1**	**1**	**2**	**3**	**4**	**5**	**6**	**7**	**7**	**8**	**9**	**10**	**11**	**12**	**13**
6	—	—	—	1	2	3	4	6	7	8	9	11	12	13	15	16	18	19	20	22
	—	—	—	**0**	**1**	**2**	**3**	**4**	**5**	**6**	**7**	**9**	**10**	**11**	**12**	**13**	**15**	**16**	**17**	**18**
7	—	—	0	1	3	4	6	7	9	11	12	14	16	17	19	21	23	24	26	28
	—	—	—	**0**	**1**	**3**	**4**	**6**	**7**	**9**	**10**	**12**	**13**	**15**	**16**	**18**	**19**	**21**	**22**	**24**
8	—	—	0	2	4	6	7	9	11	13	15	17	20	22	24	26	28	30	32	34
	—	—	—	**1**	**2**	**4**	**6**	**7**	**9**	**11**	**13**	**15**	**17**	**18**	**20**	**22**	**24**	**26**	**28**	**30**
9	—	—	1	3	5	7	9	11	14	16	18	21	23	26	28	31	33	36	38	40
	—	—	**0**	**1**	**3**	**5**	**7**	**9**	**11**	**13**	**16**	**18**	**20**	**22**	**24**	**27**	**29**	**31**	**33**	**36**
10	—	—	1	3	6	8	11	13	16	19	22	24	27	30	33	36	38	41	44	47
	—	—	**0**	**2**	**4**	**6**	**9**	**11**	**13**	**16**	**18**	**21**	**24**	**26**	**29**	**31**	**34**	**37**	**39**	**42**
11	—	—	1	4	7	9	12	15	18	22	25	28	31	34	37	41	44	47	50	53
	—	—	**0**	**2**	**5**	**7**	**10**	**13**	**16**	**18**	**21**	**24**	**27**	**30**	**33**	**36**	**39**	**42**	**45**	**48**
12	—	—	2	5	8	11	14	17	21	24	28	31	35	38	42	46	49	53	56	60
	—	—	**1**	**3**	**6**	**9**	**12**	**15**	**18**	**21**	**24**	**27**	**31**	**34**	**37**	**41**	**44**	**47**	**51**	**54**
13	—	0	2	5	9	12	16	20	23	27	31	35	39	43	47	51	55	59	63	67
	—	—	**1**	**3**	**7**	**10**	**13**	**17**	**20**	**24**	**27**	**31**	**34**	**38**	**42**	**45**	**49**	**53**	**56**	**60**
14	—	0	2	6	10	13	17	22	26	30	34	38	43	47	51	56	60	65	69	73
	—	—	**1**	**4**	**7**	**11**	**15**	**18**	**22**	**26**	**30**	**34**	**38**	**42**	**46**	**50**	**54**	**58**	**63**	**67**
15	—	0	3	7	11	15	19	24	28	33	37	42	47	51	56	61	66	70	75	80
	—	—	**2**	**5**	**8**	**12**	**16**	**20**	**24**	**29**	**33**	**37**	**42**	**46**	**51**	**55**	**60**	**64**	**69**	**73**
16	—	0	3	7	12	16	21	26	31	36	41	46	51	56	61	66	71	76	82	87
	—	—	**2**	**5**	**9**	**13**	**18**	**22**	**27**	**31**	**36**	**41**	**45**	**50**	**55**	**60**	**65**	**70**	**74**	**79**
17	—	0	4	8	13	18	23	28	33	38	44	49	55	60	66	71	77	82	88	93
	—	—	**2**	**6**	**10**	**15**	**19**	**24**	**29**	**34**	**39**	**44**	**49**	**54**	**60**	**65**	**70**	**75**	**81**	**86**
18	—	0	4	9	14	19	24	30	36	41	47	53	59	65	70	76	82	88	94	100
	—	—	**2**	**6**	**11**	**16**	**21**	**26**	**31**	**37**	**42**	**47**	**53**	**58**	**64**	**70**	**75**	**81**	**87**	**92**
19	—	1	4	9	15	20	26	32	38	44	50	56	63	69	75	82	88	94	101	107
	—	**0**	**3**	**7**	**12**	**17**	**22**	**28**	**33**	**39**	**45**	**51**	**56**	**63**	**69**	**74**	**81**	**87**	**93**	**99**
20	—	1	5	10	16	22	28	34	40	47	53	60	67	73	80	87	93	100	107	114
	—	**0**	**3**	**8**	**13**	**18**	**24**	**30**	**36**	**42**	**48**	**54**	**60**	**67**	**73**	**79**	**86**	**92**	**99**	**105**

[a] Discussed in Section 18.2. To be significant for any given n_1 and n_2, obtained U must be *equal to* or *less than* the value shown in the table.

[b] Dashes in the body of the table indicate that no decision is possible at the stated level of significance.

Table D.10 (continued). Critical values for a one-tailed test at $\alpha = 0.05$ (roman type) and $\alpha = 0.025$ (boldface type) and for a two-tailed test at $\alpha = 0.10$ (roman type) and $\alpha = 0.05$ (boldface type).

n_2 \ n_1	1	2	3	4	5	6	7	8	9	10	11	12	13	14	15	16	17	18	19	20
1	—	—	—	—	—	—	—	—	—	—	—	—	—	—	—	—	—	—	0	0
	—	—	—	—	—	—	—	—	—	—	—	—	—	—	—	—	—	—	—	—
2	—	—	—	—	0	0	0	1	1	1	1	2	2	2	3	3	3	4	4	4
	—	—	—	—	—	—	—	0	0	0	0	1	1	1	1	1	2	2	2	2
3	—	—	0	0	1	2	2	3	3	4	5	5	6	7	7	8	9	9	10	11
	—	—	—	—	0	1	1	2	2	3	3	4	4	5	5	6	6	7	7	8
4	—	—	0	1	2	3	4	5	6	7	8	9	10	11	12	14	15	16	17	18
	—	—	—	0	1	2	3	4	4	5	6	7	8	9	10	11	11	12	13	13
5	—	0	1	2	4	5	6	8	9	11	12	13	15	16	18	19	20	22	23	25
	—	—	0	1	2	3	5	6	7	8	9	11	12	13	14	15	17	18	19	20
6	—	0	2	3	5	7	8	10	12	14	16	17	19	21	23	25	26	28	30	32
	—	—	1	2	3	5	6	8	10	11	13	14	16	17	19	21	22	24	25	27
7	—	0	2	4	6	8	11	13	15	17	19	21	24	26	28	30	33	35	37	39
	—	—	1	3	5	6	8	10	12	14	16	18	20	22	24	26	28	30	32	34
8	—	1	3	5	8	10	13	15	18	20	23	26	28	31	33	36	39	41	44	47
	—	0	2	4	6	8	10	13	15	17	19	22	24	26	29	31	34	36	38	41
9	—	1	3	6	9	12	15	18	21	24	27	30	33	36	39	42	45	48	51	54
	—	0	2	4	7	10	12	15	17	20	23	26	28	31	34	37	39	42	45	48
10	—	1	4	7	11	14	17	20	24	27	31	34	37	41	44	48	51	55	58	62
	—	0	3	5	8	11	14	17	20	23	26	29	33	36	39	42	45	48	52	55
11	—	1	5	8	12	16	19	23	27	31	34	38	42	46	50	54	57	61	65	69
	—	0	3	6	9	13	16	19	23	26	30	33	37	40	44	47	51	55	58	62
12	—	2	5	9	13	17	21	26	30	34	38	42	47	51	55	60	64	68	72	77
	—	1	4	7	11	14	18	22	26	29	33	37	41	45	49	53	57	61	65	69
13	—	2	6	10	15	19	24	28	33	37	42	47	51	56	61	65	70	75	80	84
	—	1	4	8	12	16	20	24	28	33	37	41	45	50	54	59	63	67	72	76
14	—	2	7	11	16	21	26	31	36	41	46	51	56	61	66	71	77	82	87	92
	—	1	5	9	13	17	22	26	31	36	40	45	50	55	59	64	67	74	78	83
15	—	3	7	12	18	23	28	33	39	44	50	55	61	66	72	77	83	88	94	100
	—	1	5	10	14	19	24	29	34	39	44	49	54	59	64	70	75	80	85	90
16	—	3	8	14	19	25	30	36	42	48	54	60	65	71	77	83	89	95	101	107
	—	1	6	11	15	21	26	31	37	42	47	53	59	64	70	75	81	86	92	98
17	—	3	9	15	20	26	33	39	45	51	57	64	70	77	83	89	96	102	109	115
	—	2	6	11	17	22	28	34	39	45	51	57	63	67	75	81	87	93	99	105
18	—	4	9	16	22	28	35	41	48	55	61	68	75	82	88	95	102	109	116	123
	—	2	7	12	18	24	30	36	42	48	55	61	67	74	80	86	93	99	106	112
19	0	4	10	17	23	30	37	44	51	58	65	72	80	87	94	101	109	116	123	130
	—	2	7	13	19	25	32	38	45	52	58	65	72	78	85	92	99	106	113	119
20	0	4	11	18	25	32	39	47	54	62	69	77	84	92	100	107	115	123	130	138
	—	2	8	13	20	27	34	41	48	55	62	69	76	83	90	98	105	112	119	127

Table D.11. Critical Values of the Wilcoxon T [a]

n	Level of Significance for a One-Tailed Test				n	Level of Significance for a One-Tailed Test			
	0.05	0.025	0.01	0.005		0.05	0.025	0.01	0.005
	Level of Significance for a Two-Tailed Test					Level of Significance for a Two-Tailed Test			
	0.10	0.05	0.02	0.01		0.10	0.05	0.02	0.01
5	0	—	—	—	28	130	116	101	91
6	2	0	—	—	29	140	126	110	100
7	3	2	0	—	30	151	137	120	109
8	5	3	1	0	31	163	147	130	118
9	8	5	3	1	32	175	159	140	128
10	10	8	5	3	33	187	170	151	138
11	13	10	7	5	34	200	182	162	148
12	17	13	9	7	35	213	195	173	159
13	21	17	12	9	36	227	208	185	171
14	25	21	15	12	37	241	221	198	182
15	30	25	19	15	38	256	235	211	194
16	35	29	23	19	39	271	249	224	207
17	41	34	27	23	40	286	264	238	220
18	47	40	32	27	41	302	279	252	233
19	53	46	37	32	42	319	294	266	247
20	60	52	43	37	43	336	310	281	261
21	67	58	49	42	44	353	327	296	276
22	75	65	55	48	45	371	343	312	291
23	83	73	62	54	46	389	361	328	307
24	91	81	69	61	47	407	378	345	322
25	100	89	76	68	48	426	396	362	339
26	110	98	84	75	49	446	415	379	355
27	119	107	92	83	50	466	434	397	373

[a] Discussed in Section 18.3. The symbol T denotes the smaller sum of ranks associated with differences that are all of the same sign. For any given n (number of ranked differences), the obtained T is significant at a given level if it is *equal to* or *less than* the value shown in the table.

Appendix E
Recommended Supplemental Readings

The following list of articles was prepared for students who want to enrich their understanding of selected statistical topics. The articles are taken from the author's *Statistical Issues*, which is a book of readings for introductory- and intermediate-level statistics courses.

Chapter in *Introductory Statistics*	Recommended Articles in *Statistical Issues*[1]
1	Webb, W. B. The choice of the problem. Pp. 35–40.
	Weitz, J. Criteria for criteria. Pp. 41–45.
	Lord, F. M. On the statistical treatment of football numbers. Pp. 52–54.
	Dudycha, A. L. & Dudycha, L. W. Behavioral statistics: An historical perspective. Pp. 2–25.
5	Binder, A. Considerations of the place of assumptions in correlational analysis. Pp. 164–171.
7	Flynn, J. C. Some basic concepts of mathematical statistics. Pp. 83–86.
9	Flynn, J. C. Some basic concepts of mathematical statistics. Pp. 87–94 (omit starred sections).
10	Flynn, J. C. Some basic concepts of mathematical statistics. Pp. 94–103 (omit starred section).
11	O'Brien, T. C. & Shapiro, B. J. Statistical significance—What? Pp. 109–112.

[1] Listed in approximate sequence of topics in *Introductory Statistics*.

Chapter in *Introductory Statistics*	Recommended Articles in *Statistical Issues*
	Jones, L. V. Tests of hypotheses: One-sided vs. two-sided alternatives. Pp. 276–278.
	Kimmel, H. D. Three criteria for the use of one-tailed tests. Pp. 285–287.
	Skipper, J. K., Jr., Guenther, A. L., & Nass, G. The sacredness of 0.05: A note concerning the uses of statistical levels of significance in social science. Pp. 141–145.
	Lykken, D. T. Statistical significance in psychological research. Pp. 150–159.
12	McMullen, L. "Student" as a man. Pp. 26–30.
	Binder, A. Further considerations on testing the null hypothesis and the strategy and tactics of investigating theoretical models. Pp. 118–126.
	Edwards, W. Tactical note on the relation between scientific and statistical hypotheses. Pp. 127–130.
	Wilson, W., Miller, H. L., & Lower, J. S. Much ado about the null hypothesis. Pp. 131–140.
13	Edgington, E. S. Statistical inference and nonrandom samples. Pp. 146–149.
	Campbell, D. T. Factors relevant to the validity of experiments in social settings. Pp. 186–199.
	Kish, L. Some statistical problems in research design. Pp. 200–211.
	Boneau, C. A. & Pennypacker, H. S. Group matching as research strategy: How not to get significant results. Pp. 212–216.
15	Natrella, M. G. The relation between confidence intervals and tests of significance. Pp. 113–117.
	Chandler, R. E. The statistical concepts of confidence and significance. Pp. 106–108.
16	Cochran, W. G. Footnote to R. A. Fisher (1890–1962): An appreciation. Pp. 31–34.
	Kendall, M. G. Hiawatha designs an experiment. Pp. 175–176.
	Kirk, R. E. Classification of ANOVA designs. Pp. 241–260.
	Stanley, J. C. Elementary experimental design—An expository treatment. Pp. 177–185.
18	Bradley, J. V. Nonparametric statistics. Pp. 329–338.
	Gaito, J. Scale classification and statistics. Pp. 48–49.
	Anderson, N. H. Scales and statistics: Parametric and nonparametric. Pp. 55–65.

Recommended Supplemental Readings

References

Anderson, N. H. Scales and statistics: Parametric and nonparametric. *Psychological Bulletin*, 1961, *58*, 305–316. [16]*

Arken, A., & Colton, R. *Graphs: How to make and use them* (2nd ed.). New York: Harper, 1938. [33]

Baggaley, A. R. *Mathematics for introductory statistics*. New York: Wiley, 1969. [368]

Bakan, D. The test of significance in psychological research. *Psychological Bulletin*, 1966, *66*, 423–437. [232]

Binder, A. Further considerations on testing the null hypothesis and the strategy and tactics of investigating theoretical models. *Psychological Review*, 1963, *70*, 107–115. [214]

Blommers, P., & Lindquist, E. F. *Elementary statistical methods in psychology and education*. Boston: Houghton Mifflin, 1960. [49]

Boik, R. J., & Kirk, R. E. A general method for partitioning sums of squares of treatments and interactions in the analysis of variance. *Educational and Psychological Measurement*, 1977, *37*, 1–9. [327]

Boneau, C. A. The effects of violations of assumptions underlying the *t* test. *Psychological Bulletin*, 1960, *57*, 49–64. [243]

Boneau, C. A. A note on measurement scales and statistical tests. *American Psychologist*, 1961, *16*, 160–161. [16]

Boneau, C. A. & Pennypacker, H. S. Group matching as research strategy: How not to get significant results. *Psychological Reports*, 1961, *8*, 143–147. [270]

Box, G. E. P., & Anderson, S. L. Permutation theory in the derivation of robust criteria and the study of departures from assumptions. *Journal of the Royal Statistical Society, Series B*, 1955, *17*, 1–34. [321]

Bradley, J. V. *Distribution-free statistical tests*. Englewood Cliffs, N.J.: Prentice-Hall, 1968. [354, 358]

Bradley, J. V. Non-parametric statistics. In R. E. Kirk (Ed.), *Statistical issues*. Monterey, Calif.: Brooks/Cole, 1972. [353]

Bradshaw, W. L. *Mathematics for statistics*. New York: Wiley, 1969. [368]

---

* *Page on which reference is cited.*

Bresnahan, J. L., & Shapiro, M. M. A general equation and technique for the exact partitioning of chi-square contingency tables. *Psychological Bulletin*, 1966, *66*, 252–262. [339]

Campbell, D. T. Factors relevant to the validity of experiments in social settings. *Psychological Bulletin*, 1957, *54*, 297–312. [242, 320]

Campbell, D. T., & Stanley, J. C. *Experimental and quasi-experimental designs for research.* Chicago: Rand McNally, 1963. [320]

Careers in statistics. Washington, D.C.: American Statistical Association. [3]

Castellan, N. J., Jr. On the partitioning of contingency tables. *Psychological Bulletin*, 1965, *64*, 330–338. [339]

Chissom, B. S. Interpretation of the kurtosis statistic. *American Statistician*, 1970, *24*(4), 19–22. [86]

Clark, V. A., & Tarter, M. E. *Preparation for basic statistics.* New York: McGraw-Hill, 1968. [368]

Cochran, W. G. Some consequences when the assumptions for the analysis of variance are not satisfied. *Biometrics*, 1947, *3*, 22–38. [321]

Cochran, W. G. Some methods for strengthening the common chi-square tests. *Biometrics*, 1954, *10*, 417–451. [334]

Cohen, Jacob. *Statistical power analysis for the behavioral sciences.* New York: Academic Press, 1969. [231, 243, 248, 251, 314]

Conover, W. J. Rejoinder. *Journal of the American Statistical Association*, 1974, *69*, 382. (a) [337]

Conover, W. J. Some reasons for not using the Yates continuity correction on 2 × 2 contingency tables. *Journal of the American Statistical Association*, 1974, *69*, 374–376. (b) [337]

Darlington, R. B. Is kurtosis really "peakedness?" *American Statistician*, 1970, *24*(2), 19–22. [86]

Dixon, W. J., & Massey, F. J. *Introduction to statistical analysis.* New York: McGraw-Hill, 1957. [243]

Dudycha, A. L., & Dudycha, L. W. Behavioral statistics: An historical perspective. In R. E. Kirk (Ed.), *Statistical issues.* Monterey, Calif.: Brooks/Cole, 1972. [18, 213]

Edgington, E. S. Statistical inference and nonrandom samples. *Psychological Bulletin*, 1966, *66*, 485–487. [260]

Edgington, E. S. A new tabulation of statistical procedures used in APA journals. *American Psychologist*, 1974, *29*, 25–26. [302]

Edwards, A. L. On "the use and misuse of the chi-square test"—the case of the 2 × 2 contingency table. *Psychological Bulletin*, 1950, *47*, 347–355. [331]

Edwards, A. L. *Statistical methods* (2nd ed.). New York: Holt, Rinehart and Winston, 1967. [120]

Edwards, A. L. *Experimental design in psychological research* (4th ed.). New York: Holt, Rinehart and Winston, 1972. [303, 307, 320, 325]

Edwards, W. Tactical note on the relation between scientific and statistical hypotheses. *Psychological Bulletin*, 1965, *63*, 400–402. [214]

Edwards, W., Lindman, W. H., & Savage, L. J. Bayesian statistical inference for psychological research. *Psychological Review*, 1963, *70*, 193–242. [167]

Fisher, R. A., & Yates, F. *Statistical tables for biological, agricultural and medical research.* London: Longman, 1974. Previously, Edinburgh: Oliver & Boyd, 1963. [404, 407, 410, 411, 418, 420]

Gaito, J. Scale classification and statistics. *Psychological Review*, 1960, *67*, 277–278. [16]

Galton, F. *Natural inheritance.* London and New York: Macmillan, 1889. [96]

Games, P. A. Multiple comparisons of means. *American Educational Research Journal*, 1971, *8*, 531–565. [322, 324]

Gardner, P. L. Scales and statistics. *Review of Educational Research*, 1975, *45*, 43–57. [16]

Glass, G. V., & Stanley, J. C. *Statistical methods in education and sychology.* Englewood Cliffs, N.J.: Prentice-Hall, 1970. [96, 110, 120]

Godard, R. H., & Lindquist, E. F. An empirical study of the effect of heterogeneous within-groups variance upon certain F-tests of significance in analysis of variance. *Psychometrika*, 1940, *5*, 263–274. [321]

Grant, D. A. Testing the null hypothesis and the strategy and tactics of investigating theoretical models. *Psychological Review*, 1962, *69*, 54–61. [214]

Grizzle, J. E. Continuity correction in the χ^2-test for 2 × 2 tables. *American Statistician*, 1967, *21*(4), 28–32. [337]

Halperin, M., Hartley, H. O., & Hoel, P. G. Recommended standards for statistical symbols and notation. *American Statistician*, 1965, *19*(3), 12–14. [47]

Hammond, K. R., Householder, J. E., & Castellan, N. J., Jr. *Introduction to the statistical method* (2nd ed.). New York: Knopf, 1970. [73]

Hays, W. L. *Statistics for the social sciences* (2nd ed.). New York: Holt, Rinehart and Winston, 1973. [96, 110, 120, 147, 180, 184, 185, 321, 340, 341]

Horsnell, G. The effect of unequal group variances on the F-test for the homogeneity of group means. *Biometrika*, 1953, *40*, 128–136. [321]

Huff, D. *How to lie with statistics*. New York: Norton, 1954. [43]

Kearney, P. A. *Programmed review of fundamental mathematics for elementary statistics*. Englewood Cliffs, N.J.: Prentice-Hall, 1970. [368]

Keppel, G. *Design and analysis: A researcher's handbook*. Englewood Cliffs, N.J.: Prentice-Hall, 1973. [307, 325]

Kirk, R. E. *Experimental design: Procedures for the behavioral sciences*. Monterey, Calif.: Brooks/Cole, 1968. [303, 312, 314, 320, 321, 322. 324. 325]

Kirk, R. E. (Ed.). *Statistical issues*, Monterey, Calif.: Brooks/Cole, 1972. (a) [16, 213, 214, 227, 242, 260, 270, 285, 320, 353]

Kirk, R. E. Classification of ANOVA designs. In R. E. Kirk (Ed.), *Statistical issues*. Monterey, Calif.: Brooks/Cole, 1972. (b) [307, 313]

Korin, B. P. *Statistical concepts for the social sciences*. Cambridge, Mass.: Winthrop, 1975. [334]

Lee, W. *Experimental design and analysis*. San Francisco: W. H. Freeman, 1975. [307, 325]

Lewis, D., & Burke, C. J. The use and misuse of the chi-square test. *Psychological Bulletin*, 1949, *46*, 433–489. [331]

Lewis, D., & Burke, C. J. Further discussion of the use and misuse of the chi-square test. *Psychological Bulletin*, 1950, *47*, 347–355. [331]

Lindman, H. R. *Analysis of variance in complex experimental designs*. San Francisco: W. H. Freeman, 1974. [307, 325]

Lindquist, E. F. *Design and analysis of experiments in psychology and education*. Boston: Houghton Mifflin, 1953. [321]

Loether, H. J., & McTavish, D. G. *Descriptive statistics for sociologists*. Boston: Allyn & Bacon, 1974. [340]

Mantel, N. Comment and suggestion. *Journal of the American Statistical Association*, 1974, *69*, 378–380. [337]

Marascuilo, L. A. Large sample multiple comparisons. *Psychological Bulletin*, 1966, *65*, 280–290. [339]

Marascuilo, L. A. *Statistical methods for behavioral science research*. New York: McGraw-Hill, 1971. [333, 343]

Marascuilo, L. A., & McSweeney, M. *Nonparametric and distribution-free methods for the social sciences*. Monterey, Calif.: Brooks/Cole, 1977. [333, 339, 354]

McGee, V. E. *Principles of statistics*. New York: Appleton-Century-Crofts, 1971. [147]

McNemar, Q. Note on the sampling error of the difference between correlated proportions or percentages. *Psychometrika*, 1947, *12*, 153–157. [279]

McNemar, Q. *Psychological statistics* (4th ed.). New York: Wiley, 1969. [129]

Miettinen, O. S. Comment. *Journal of the American Statistical Association*, 1974, *69*, 380–382. [337]

Minium, E. W. *Statistical reasoning in psychology and education*. New York: Wiley, 1970. [243]

Mueller, J. H., & Schuessler, K. F. *Statistical reasoning in sociology*. Boston: Houghton Mifflin, 1961. [73]

Myers, J. L. *Fundamentals of experimental design* (2nd ed.). Boston: Allyn & Bacon, 1972. [303, 307, 320, 325]

Natrella, M. G. The relation between confidence intervals and tests of significance. *American Statistician*, 1960, *14*(1), 20–22, 33. [285]

Novick, M. R., & Jackson, P. H. *Statistical methods for educational and psychological research*. New York: McGraw-Hill, 1974. [147]

Nunnally, J. The place of statistics in psychology. *Educational and Psychological Measurement*, 1960, *20*, 641–650. [232]

Olds, E. G. Distributions of sums of squares of rank differences for small numbers of individuals. *Annals of Mathematical Statistics*, 1938, *9*, 133–148. [419]

Olds, E. G. The 5 per cent significance levels for sums of squares of rank differences and a correction. *Annals of Mathematical Statistics*, 1949, *20*, 117–118. [419]

Pastore, N. Some comments on "The use and misuse of the chi-square test." *Psychological Bulletin*, 1950, *47*, 338–340. [331]

Pearson, E. S. The analysis of variance in cases of non-normal variation. *Biometrika*, 1931, *23*, 114–133. [321]

Pearson, E. S., & Hartley, H. O. *Biometrika tables for statisticians* (Vol. 1, 3rd ed.). New York: Cambridge, 1966. [412]

Peters, C. C. The misuse of chi-square—a reply to Lewis and Burke. *Psychological Bulletin*, 1950, *47*, 331–337. [331]

Plackett, R. L. The continuity correction in 2×2 tables. *Biometrika*, 1964, *51*, 327–337. [337]

Rozeboom, W. W. The fallacy of the null-hypothesis significance test. *Psychological Bulletin*, 1960, *57*, 416–428. [232]

Senders, V. L. *Measurement and statistics*. New York: Oxford, 1958. [15]

Siegel, S. *Nonparametric statistics for the behavioral sciences*. New York: McGraw-Hill, 1956. [15]

Sockloff, A. Behavior of the product-moment correlation coefficient when two heterogeneous subgroups are pooled. *Educational and Psychological Measurement*, 1975, *35*, 267–276. [111]

Starmer, C. J., Grizzle, J. E., & Sen, P. K. Comment. *Journal of the American Statistical Association*, 1974, *69*, 376–378. [337]

Stevens, S. S. On the theory of scales of measurement. *Science*, 1946, *103*, 667–680. [11]

Stevens, S. S. Mathematics, measurement, and psychophysics. In S. S. Stevens (Ed.), *Handbook of experimental psychology*. New York: Wiley, 1951. [15]

Stevens, S. S. Measurement, statistics, and the schemapiric view. *Science*, 1968, *161*, 849–856. [16]

Stilson, D. W. *Probability and statistics in psychological research*. San Francisco: Holden-Day, 1966. [184]

Taylor, P. A. *An introduction to statistical methods*. Itasca, Ill.: J. E. Peacock, 1972. [334, 341]

Tippett, L. H. C. On the extreme individuals and the range of samples from a population. *Biometrika*, 1925, *17*, 386. [82]

Ury, H. In response to Noether's letter, "Needed—a new name." *American Statistician*, 1967, *21*(4), 53. [353]

Wainer, H. Estimating coefficients in linear models: It don't make no nevermind. *Psychological Bulletin*, 1976, *83*, 213–217. [16]

Walker, H. *Mathematics essential for elementary statistics*. New York: Holt, Rinehart and Winston, 1951. [368]

Walpole, R. E. *Introduction to statistics*. New York: Macmillan, 1968. [184]

Wilson, W. R., & Miller, H. A note on the inconclusiveness of accepting the null hypothesis. *Psychological Review*, 1964, *71*, 238–242. [214]

Wilson, W., Miller, H. L., & Lower, J. S. Much ado about the null hypothesis. *Psychological Bulletin*, 1967, *67*, 188–196. [214]

Winer, B. J. *Statistical principles in experimental design* (2nd ed.). New York: McGraw-Hill, 1971. [303, 307, 320, 325]

Index

Abscissa, 33
Absolute value, 371
Addition rule, 152–154
Alienation, coefficient of, 138
Alpha, specifying, 220–221, 227–233
Alternative hypothesis (*see* Hypothesis)
Analysis of variance:
   assumptions, 318–321
   basic concepts, 303–312
   completely randomized design, 313–316
   degrees of freedom, 308–309
   expectations of means square, 312
   factorial design, 325
   fixed effects model, 319–320
   interaction in, 326
   partition of total sum of squares, 307–308
   purposes, 302–303
   random effects model, 319–320
ANOVA, 302 (*see also* Analysis of variance)
Arithmetic mean (*see* Mean)
Association, measures (*see* Correlation)
Associative law, 375
Assumption-freer tests:
   comparison with parametric tests, 365–366
   introduction to, 353–354
Asymmetrical distribution, 41
Asymptotic efficiency, 358
Average deviation, 70

Bar graph, 33
Bayesian inference, 147
Bayes' theorem, 168
Bernoulli, J., 19, 182
Bernoulli trial, 182
Bimodal distribution, 41
Binomial distribution, 183
   expected value, 185
   standard deviation, 185
Bivariate normal population, 251

Causal relationship, 259
Centile, 68
Central limit theorem, 206
Central tendency, measures of (*see* Geometric mean; Mean; Median; Mode)
Chi-square distribution, 245
   table, 411
Chi-square test for frequency data (*see* Pearson's chi-square statistic)
Chi-square test for σ^2, 244–247
   assumptions, 245, 247
   degrees of freedom, 245
   sampling distribution, 245
Class interval, 23
   midpoint, 36
   nominal limits, 24
   open, 25
   preferred number, 25

Class interval (continued)
  preferred size, 26
  real limits, 24
Coefficient of alienation, 138
  assumptions, 140
Coefficient of determination, 105
Coefficient of nondetermination, 105
Coefficient of predictive efficiency, 138
Combination ($_nC_r$), 166
Commutative law, 375
Comparison between means, 322
Completely randomized design, 313–316
Concomitant relationship, 108, 259
Conditional probability, 155
Confidence coefficient, 287
Confidence interval, 285
  comparison with hypothesis testing, 288–289
  interpretation, 288–289
  one-sided, 289
  for p, 295
  for $p_1 - p_2$, 295
  for σ^2, 292–293
  for σ_1^2/σ_2^2, 293–294
  for ρ, 296–297
  for $\rho_1 - \rho_2$, 297
  for μ, 287–288
  for $\mu_1 - \mu_2$, 289–291
Confidence limits, 287
Constant, 9
Contingency table, 336
Continuity, correction for, 249
  Pearson's chi-square test, 334–335, 337, 351
  T test, 362
  U test, 355
  z test for p, 249
  z test for $p_1 = p_2$, 278, 280
Continuous variable, 10
Contrast, 322
Control group, 242
Correlation:
  biserial, 120
  and causality, 108
  coefficient, 98
  Cramer's measure of association, 340–341
  cross product, information in, 102–103
  distinguished from regression, 95
  eta squared, 110
  fourfold, 120
  Kendall's tau, 120
  Pearson product-moment correlation coefficient, 100
    interpretation, 105–107
    errors in interpreting, 107–108

Correlation (continued)
    factors that affect size, 109–116
    and truncated range, 110
  phi coefficient, 120
  point biserial, 120
  ratio, 110
  Spearman rank correlation coefficient, 116–118
    and tied ranks, 118
  tetrachoric, 120
Covariance, 103
Cramer's measure of association, 340–341
Critical region, 221
Cross product, 102
Cumulative polygon, 38

Data snooping test, 322
Decision rule, 221
Degrees of freedom, 238–239 (*see also* specific tests, such as Chi-square, t, and F)
Denominator, 372
Dependent samples, 265–266
Descriptive statistics, nature of, 5–6
Discrete variable, 10
Dispersion, measures of (*see* Index of dispersion; Range; Semi-interquartile range; Standard deviation)
Distribution:
  asymmetrical, 41
  bimodal, 41
  binomial, 183
  cumulative frequency, 29–30
  discontinuous, 113
  frequency (*see* Frequency distribution)
  hypergeometric, 186
  J, 41
  leptokurtic, 41, 86
  mesokurtic, 41, 86
  multimodal, 41
  multinomial, 185
  negatively skewed, 41, 84
  normal, 40, 190–192
    history of, 19, 190–191
    inflection points, 82
  open-ended, 56
  platykurtic, 41, 86
  positively skewed, 41, 84
  probability, 176
  rectangular, 41
  relative frequency, 27
  sampling, 182
  skewed, 41
  symmetrical, 40

Distribution (continued)
   U, 41
   uniform, 41
Distribution-free tests, 353
Distributive law, 375

Effect size, 231
Equation:
   algebraic, 376
   arithmetic, 376
   conditional, 376
   identity, 376
   permissible operations, 376–377
   roots of, 376
Equivalence class, 11, 23
Error effect, 305
Estimate:
   interval, 202, 285
   point, 202, 285
Estimation, 202
Estimator, 202
   minimum variance, 206
   properties of good estimators, 206–207
   unbiased, 206
Eta squared, 110
Euler diagram, 150
Euler, Leonhard, 150
Events:
   compound, 149
   exhaustive, 154
   intersection of, 152
   mutually exclusive, 153
   simple, 149
      graph of, 150
   statistically independent, 156
   union of, 152
Expected value:
   binomial distribution, 185
   discrete random variable, 178–179
   continuous random variable, 180
Experiment, 149
Experimental design considerations, 242, 257–260, 265–270
Exponents, operations with, 373–374
Ex post facto experiment, 269

Factorial design, 325–328
   advantages and disadvantages, 328
   concept of interaction, 326
Factorials, 375
Factoring, 374–375
F distribution, 275–276
   table, 412–417
Fermat, P. de, 19
Fisher, R. A., 20, 252, 275, 302

Fisher's r to Z' transformation, 252
   table, 420
Fractions, operations with, 372–373
Frequency distribution, 23–30
   bivariate, 96
   cumulative, 29
   grouped, 24–27
      advantages and disadvantages, 27
   for qualitative variables, 29–30
   for quantitative variables, 23–29
   relative, 27
   rules for constructing, 25–26
   ungrouped, 23–24
Frequency polygon, 36–38
F test for $\sigma_1^2 = \sigma_2^2$, 275–277
   assumptions, 275
   degrees of freedom, 275
Fundamental counting rule, 163

Galton, F., 19, 95–96
Gauss, C. F., 19
Geometric mean, 133
Goodness of fit, test for, 332–335
Gosset, W. S., 19, 238
Graph:
   bar, 33
   cumulative polygon, 38
   frequency polygon, 36–38
   histogram, 36
   misleading, 42–43
   pie chart, 33–35
   for qualitative variables, 33–35
   for quantitative variables, 36–38
Graunt, J., 18
Grouping error, 49, 73
Group matching, 269–270

Helmert, F. R., 245
Heterogeneity, 114
Heteroscedasticity, 114
Histogram, 36
History of statistics, 18–20
Homogeneity of variance, 321
Homoscedasticity, 140
Huygens, C., 19
Hypergeometric distribution, 186
Hypothesis:
   alternative, 214
   directional, 224
   exact, 227
   inexact, 227
   nondirectional, 225
   null, 214
   one-tailed, 224
   scientific, 213

Hypothesis (continued)
   statistical, 214, 227
   two-tailed, 224
Hypothesis testing, 213–221
   comparison with confidence interval, 288–289
   errors in, 227–230, 232–233
   and method of indirect proof, 215
   role of logic in, 216
   steps, 218

Identity, 376
Independence, statistical, 156
   Pearson's chi-square test for, 336–340
Index of dispersion, 73–75
   derivation of formula, 91–93
   properties, 80
   relative merits, 79
Index of forecasting efficiency, 138
   assumptions, 140
Index of summation, 48
Inequality:
   defined, 377
   permissible operations, 377–378
Inferential statistics, nature of, 6
Interaction, 326
Intersection of events, 152
Interval measurement, 13

J distribution, 41
Joint probability, 156

Kurtosis, 40–41, 86

Laplace, P. S. de, 19
Law of large numbers, 204, 206
Least squares, principle of, 129
Leptokurtic distribution, 41, 86
Level of significance, 221
Levels of measurement (see Measurement)
Linear function, 96, 129
Linear model, 304–305

Mann-Whitney U test, 354–357
   assumptions, 354
   efficiency, 358
   normal distribution approximation, 355–357
Marginal probability, 156
Mathematics, review of, 370–378
Mean, 47–49
   of combined subgroups, 60–61
   formula for grouped distribution, 48–49
   formula for ungrouped distribution, 48
   merits, 55–56

Mean (continued)
   properties, 58, 63–64
   variance error, 207
Mean deviation, 70
Mean square, 309
Measurement, 11–15
   interval, 13
   levels, 11
   nominal, 11–12
   ordinal, 12–13
   ratio, 14–15
Median, 51–54
   merits, 56–57
   properties, 58
   variance error, 207
Mesokurtic distribution, 41, 86
Midpoint of class interval, 36
Mid-range, 67
Mode, 46–47
   merits, 57
   properties, 58
Moivre, A. de, 19
Monotonic relationship, 117–118
Multimodal distribution, 41
Multinomial distribution, 185, 349–351
Multiple comparisons among means, 322–324
Multiplication principle, 163
Multiplication rule, 155–158

Neyman, J., 20
n-factorial, 163
Nominal measurement, 11–12
Nonparametric test, 353
Nonrandom sampling, 173
Normal distribution, 190–192 (see also Distribution, normal)
   approximation to binomial distribution, 195–196
   characteristics, 191–192
   finding area under, 192–194
   history of, 19
   table, 407–409
Null hypothesis (see Hypothesis)
Numerator, 372

Ogive, 38
One-tailed hypothesis, 224
Ordinal measurement, 12–13
Ordinate, 33

Parameter, 48
Parametric tests, comparison with assumption-freer tests, 365–366
Pascal, B., 19

Pearson, K., 19, 100, 245, 331
Pearson, E., 20
Pearson's chi-square statistic:
   applications, 331
   characteristics of test statistic, 333–334
   for equality of proportions, 342–345
      comparison with test for independence, 343–344
      degrees of freedom, 343
      with more than two response categories, 344–345
   for goodness to fit, 332–335
      assumptions, 334
      comparison with exact test, 349–351
      comparison with z test for $p = p_0$, 348–349
      degrees of freedom, 332, 334
      Yates' correction, 334–335
   for independence, 336–340
      assumption, 341
      Cramer's measure of association, 340–341
      degrees of freedom, 337–338
Pearson's product-moment correlation coefficient, 100 (*see also* Correlation)
Percentage frequency, 27
Percentile point, 68
Percentile rank, 68, 194–195
   interpreting scores in terms of, 198
Permutation:
   $_nP_n$, 163
   $_nP_r$, 164
   $_nP_{r_1, r_2, \ldots, r_k}$, 165
Pie chart, 33–35
Placebo, 6
Platykurtic distribution, 41, 86
Point estimate, 202, 285
Population, 5, 173
   conceptual, 5
   concrete, 5
   defining the, 173
   element, 5, 173
   finite, 5
   infinite, 5
Posterior probability, 168
Power, 226, 228
   calculation, 229
Power efficiency, 357
Prediction error, 129
   and standard error of estimate, 136
Prior probability, 167
Probability:
   addition rule, 152–154
   classical or logical view, 147–148
   of combined events, 152–158

Probability (continued)
   common errors in applying, 158
   conditional, 155
   empirical relative-frequency view, 148
   formal properties, 151
   history, 18–19
   joint, 156
   marginal, 156
   multiplication rule, 155–158
   posterior, 168
   prior, 167
   subjective-personalistic view, 147
Probability distribution, 176
Product of two numbers, 371
Proportionate frequency, 27
Proportions, test for equality, 342–345 (*see also* Pearson's chi-square statistic)

Qualitative variable, 9–10
Quantitative variable, 10
Quartile point:
   first, 67–68
   third, 67–68
Quetelet, L. A. J., 19
Quotient, 371

r to Z' transformation (*see* Fisher's r to Z' transformation)
Radicals, operations with, 374
Random assignment, 173, 258–260
Random numbers, 174
   table, 404–405
Random sample, 173 (*see also* Sample)
Random sampling, 6–7, 173
   procedures, 174
   versus random assignment, 259–260
Random variable, 175–176
   continuous, 176
   discrete, 176
   expected value, 176–180
   probability distribution, 176
Range, 67
   properties, 80
   relationship to standard deviation, 82–83, 91
   relative merits, 79
Rank correlation, 116–118
Ratio measurement, 14–15
Rectangular distribution, 41
Regression coefficients and r, 133–134
Regression distinguished from correlation, 95
Regression line, 96, 129, 132
   assumptions, 140
   of best fit, 129
   predicting X from Y, 132

Regression (continued)
  predicting Y from X, 129–130
  slope of, 130
Regression toward the mean, 96
Relationship:
  causal, 108
  concomitant, 108
  linear, 96, 129
  monotonic, 117–118
  nonlinear, 109–110
  strictly monotonic, 118
Reliability, test-retest, 107–108
Reversion, 96, 98
Robustness:
  of chi-square test, quantitative data, 247
  of t test, 243
Roots of equation, 376
Rounding numbers, 370

Sample, 5
  nonrandom, 6–7, 173
  point, 150
  random, 6–7, 173
  space, 150
Sample size, determining, 219, 230–232
  derivation of formula, 235–236
  for one-sample z statistic, 230–232
  table, 421–422
  for two sample t statistic, 262, 269
Sampling:
  distribution, 182, 202–203
  fluctuation, 7
  random, 6–7, 173
    procedures, 174
  stability, 55
  systematic, 175
  without replacement, 157, 174
  with replacement, 157, 174
Sampling distribution, 182
  of the mean, 203–206
  of z test statistic, 219–220
  under H_0 and under H_1, 228–229
Scatter diagram (scattergram), 96
  and prediction, 128–129
Scheffé's S statistic, 322–324
  assumptions, 322
  degrees of freedom, 323
Semi-interquartile range, 67–68
  properties, 79–80
  relationship to standard deviation, 82
  relative merits, 78
Sheppard's correction, 73
Significance level, 221
Skewness, 41, 84
Slope of regression line, 130

Snedecor, G. W., 275
Spearman's rank correlation coefficient, 116–118
Standard deviation, 70–73
  of binomial distribution, 185
  deviation formula, 71
  of discrete random variable, 180–181
  minimum and maximum values, 83, 91
  properties, 79
  raw score formula, 72
  relationship to mean, 82
  relationship to range, 82
  relationship to semi-interquartile range, 82
  relative merits, 78
  Sheppard's correction, 73
Standard error:
  defined, 206
  of the difference between two means, 256
  of mean, 207
  of one-sample t statistic, 238
Standard error of estimate, 136–137
  assumptions in, 140
  and standard deviation of prediction errors, 142–143
Standard score, 192, 197–198
  advantages over percentile rank, 198–199
  interpreting scores in terms of, 197–198, 200–201
  kinds, 199–200
Statistic, 48
Statistical independence (see Independence, statistical)
Statistical significance versus practical significance, 232, 242
Statistical test, 216
Statisticians, types, 3
Statistics, 2
  descriptive, 5
  experimental, 19–20
  history, 18–20
  how to study, 3–5
  inferential, 6, 201–202
  national, 18
Student, 19, 238
Summation, 48
  index of, 48
  rules, 61–62
Sum of squares, 307–308
Symmetrical distribution, 40

t distribution, 238–239
  table, 410
t test for μ, 238–243
  assumptions, 240–243
  comparison with z, 239–240

t test for μ (continued)
  degrees of freedom, 238–239
  estimating sample size, 243
  sampling distribution, 238–239
t test for ρ, 250–252
  assumptions, 251
  degrees of freedom, 251
t test for $\mu_1 - \mu_2$:
  dependent samples, 266–269
    degrees of freedom, 267
    estimating sample sizes, 269
  independent samples, 261–263
    degrees of freedom, 261
    estimating sample sizes, 262
t test for $\sigma_1^2 = \sigma_2^2$, 277–278
Tchebycheff's theorem, 83
Test statistic, 207
  specifying the, 219
Transformation:
  Fisher's r to Z', 252
  monotonic, 12
  one-to-one, 12
  percentage frequency, 27
  positive linear, 13
  proportionate frequency, 27
  z score, 192
Treatment combination, 325
Treatment level, 304
Truncated range, 110
Two-tailed hypothesis, 224
Type I and II errors, 227–230, 232–233

U distribution, 41
Unbiased estimator, 206
Uniform distribution, 41
Union of events, 152

Validity, 108
Variable, 9
  qualitative, 9–10
    ordered, 10
    unordered, 10

Variable (continued)
  quantitative, 10
    continuous, 10
    discrete, 10
  random, 175–176 (*see also* Random variable)
  range of, 9
Variance, 70
  of chi-square distribution, 245
  of t distribution, 239

Wilcoxon rank-sum test, 354
Wilcoxon T test, 360–363
  assumptions, 360
  efficiency, 360
  normal distribution approximation, 362–363

X axis, 33

Yates' correction for continuity, 334–335 (*see also* Correction for continuity)
Y axis, 33

Zero, operations with, 372
z score (*see* Standard score)
z test for p, 247–250
  assumptions, 248
  and binomial distribution 247–248
  comparison with Pearson's chi-square test, 348–349
  for finite population, 249–250
z test for $p_1 = p_2$:
  dependent samples, 279–280
  independent samples, 278
z test for $p_1 - p_2 = \delta_0$, 279
z test for ρ, 252–253
z test for μ, 222–223
z test for $\mu_1 - \mu_2$:
  dependent samples, 266
  independent samples, 256–257